ANALYTICAL GEOMETRY

and

VECTOR ALGEBRA

[For Degree Course (General) of Calcutta, Burdwan, Vidyasagar, North Bengal and Other Universities]

Dr Ratan Mohan Khan

Formerly Reader, Department of Mathematics
City College and Rammohan College, Kolkata
Formerly Lecturer in Mathematics
B N Mahavidyalaya, Itachuna, Hooghly

New Central Book Agency (P) Ltd

LONDON

HYDERABAD ERNAKULAM BHUBANESWAR

NEW DELHI KOLKATA PUNE GUWAHATI

NCBA

REGD OFFICE

8/1 Chintamoni Das Lane, Kolkata 700 009, India

email: ncbapvtltd@eth.net

OVERSEAS

NCBA (UK) Ltd, 149 Park Avenue North, Northampton, NN3 2HY, UK

email: ncbauk@yahoo.com

EXPORT

NCBA Exports Pvt. Ltd, 212 Shahpur Jat, New Delhi 110 049

email: ncbaexports@ncbapvtltd.com

BRANCHES

212 Shahpur Jat
New Delhi 110 049
email: ncbadel@ncbapvtltd.com

House No. 3-1-315, 1st Floor
Nimboliadda, Kachiguda, **Hyderabad** 500 027
email: ncbahydb@ncbapvtltd.com

Shop Nos. 3 & 4, Vinayak Towers
681/B Budhwar Peth, Appa Balwant Chowk
Pune 411 002
email: ncbapune@ncbapvtltd.com

GSS Shopping Complex, 1st Floor
Opposite College Post Office, Convent Road
Ernakulam 682 035
0484-4021054/53

Shop No. 15, Bharati Towers, Ground Floor
Forest Park
Bhubaneswar 751 001
email: ncbabub@ncbapvtltd.com

Radhamayee Complex (Second Floor)
Opp. SBI, Panbazar, Jaswanta Road
Guwahati 781 001
email: ncbaguwahati@ncbapvtltd.com

ANALYTICAL GEOMETRY AND VECTOR ALGEBRA • Dr Ratan Mohan Khan

First Published: 1994
Second Edition: 2001
Reprinted with Additions: 2004
Reprinted: 2006, 2009, 2010, 2013, 2015

PUBLISHER AND TYPESETTER
New Central Book Agency (P) Ltd
8/1 Chintamoni Das Lane, Kolkata 700 009

PRINTER
New Central Book Agency (P) Ltd
Web-Offset Division, Dhulagarh, Sankrail, Howrah

COVER DESIGNER
Dulal Halder

COVER PRINTER
Biswanath Studio, Kolkata

PROJECT TEAM
Abhijit Sen, Subir Pal, and Prabhat Jas

ISBN 978 81 7381 016 9

Price ₹ 235.00

ANALYTICAL GEOMETRY
and
VECTOR ALGEBRA

Preface to the Second Edition

In this edition the book has been revised according to the new syllabi of some Indian universities. I hope that this edition will be more useful to all concerned.

Calcutta
15th August, 2001 **R. M. KHAN**

Preface to the First Edition

This book has been written as a standard text-book of Analytical Geometry and Vector Algebra for the B.A., B.Sc. and Technical Examinations of some reputed Universities of India and abroad. The book is self-contained and knowledge of school mathematics is sufficient to understand the basic ideas and methods of the matter contained in this book. My endeavour is confined to the discussion of the subject-matter, as comprehensive, rigorous and lucid as possible and permissible within the scope of a college text-book. It gives the ideas of straight lines, planes, different conics with their properties in Cartesian and polar co-ordinates, different quadrics with usual properties, general equations, generating lines and vector algebra needful for various branches of science. A good number of typical examples, mostly taken from University questions, are either worked out or given in exercises for the students to grasp the subject easily and clearly.

My attempt will be successful, if this treatise is helpful to my students and it invites a good criticism from all corners. Any suggestion for the improvement of this book will be gratefully acknowledged.

Calcutta
1 Baisakh 1401 **R. M. KHAN**

Preface to the Second Edition

In this edition the book has been revised according to the new syllabi of some Indian universities. I hope that this edition will be more useful to all concerned.

Calcutta
15th August 2001 **S. M. KHAN**

Preface to the First Edition

This book has been written as a standard text book of Analytical Geometry and Vector Algebra for the B.A., B.Sc. and Technical Examinations of some reputed Universities of India and abroad. The book is self contained and knowledge of school mathematics is sufficient to understand the basic ideas and methods of the matter contained in this book. My endeavour is confined to the discussion of the subject-matter, as comprehensive, rigorous and lucid as possible and permissible within the scope of a college text book. It gives the ideas of straight lines, planes, different conics with their properties in Cartesian and polar co-ordinates, different quadrics with usual properties, general equations, generating lines and vector algebra needful for various branches of sciences. A good number of typical examples, mostly taken from University questions, are either worked out or given in exercises for the students to grasp the subject easily and clearly.

My attempt will be successful, if this treatise is helpful to my students and it invites a good criticism from all corners. Any suggestion for the improvement of this book will be gratefully acknowledged.

Calcutta
1 August 1997 **S. M. KHAN**

Contents

Part I

Geometry of Two Dimensions

Chapter 1

Orthogonal Transformation and Invariants

1.1 Transformation of Axes

The co-ordinates of a point depend on the position of axes. Thus the co-ordinates of a point and consequently the equation of a locus will be changed with the alteration of origin without the alteration of direction of axes, or by altering the direction of axes and keeping the origin fixed, or by altering the origin and also the direction of axes. Either of these processes is known as transformation of axes or transformation of co-ordinates.

1.2 Change of Origin without change of Direction of Axes

Let (x, y) be the co-ordinates of P with respect to rectangular axes OX and OY and (x', y') be the co-ordinates of it with respect to a new set of axes $O'X'$ and $O'Y'$ which are parallel to the original axes OX and OY respectively.

Let (α, β) be the co-ordinates of the new origin O' with respect to axes OX and OY. PN is perpendicular to OX and it meets $O'X'$ at N'. $O'T$ is perpendicular to OX.

$$\therefore \; ON = x,\; NP = y;\; O'N' = x',$$
$$N'P = y';\; OT = \alpha,\; TO' = \beta.$$

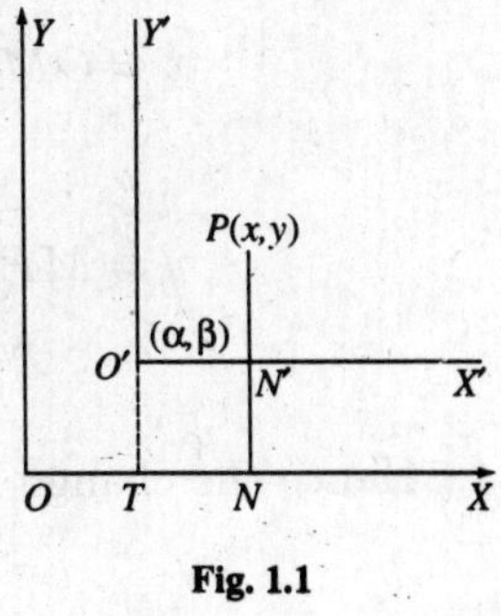

Fig. 1.1

Now

$$x = ON = OT + TN = OT + O'N' = \alpha + x',$$
$$y = NP = NN' + N'P = TO' + N'P = \beta + y'.$$

Hence the required transformation formulae are given by

$$x = x' + \alpha, \quad y = y' + \beta.$$

This transformation is also known as *translation* or *parallel displacement.*

NOTE. In the equation of a locus referred to original system of axes (x, y) will be replaced by $(x' + \alpha, y' + \beta)$ when the equation is referred to new pair of axes. Inversely (x', y') will be replaced by $(x - \alpha, y - \beta)$.

▶ **Example 1** *Find the form of the equation* $3x + 4y = 5$ *due to change of origin to the point* $(3, -2)$ *only.*

The transformed equation is

$$3(x' + 3) + 4(y' - 2) = 5 \quad \text{or} \quad 3x' + 4y' = 4.$$

▶ **Example 2** *Find the transformed form of* $\frac{x}{a} + \frac{y}{b} = 2$ *when the origin is shifted to* (a, b).

The transformed equation is $\frac{x'+a}{a} + \frac{y'+b}{b} = 2$, or $\frac{x'}{a} + \frac{y'}{b} = 0$.

1.3 Rotation of Rectangular Axes in their Own Plane without Changing the Origin

Let the original axes OX and OY be rotated through an angle θ in the anticlockwise direction. In the adjoining figure OX' and OY' are the new set of axes. Let (x, y) and (x', y') be the co-ordinates of the same point P referred to OX, OY and OX', OY' respectively.

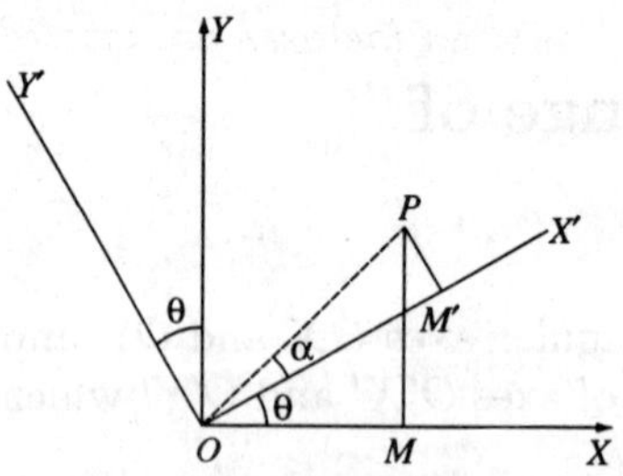

Fig. 1.2

PM and PM' are perpendiculars to OX and OX' respectively. PO is joined. $\angle XOX' = \theta$. Let $\angle X'OP = \alpha$,

From the fig. 1.2, $OM = x$, $MP = y$; $OM' = x'$, $M'P = y'$.

Now

$$\begin{aligned} x = OM &= OP\cos(\theta + \alpha) = OP\cos\alpha\cos\theta - OP\sin\alpha\sin\theta \\ &= OM'\cos\theta - M'P\sin\theta = x'\cos\theta - y'\sin\theta, \\ y = MP &= OP\sin(\theta + \alpha) = OP\cos\alpha\sin\theta + OP\sin\alpha\cos\theta \\ &= OM'\sin\theta + M'P\cos\theta = x'\sin\theta + y'\cos\theta. \end{aligned}$$

Hence the change from (x, y) to (x', y') is given by

$$\left.\begin{aligned} x &= x'\cos\theta - y'\sin\theta \\ \text{and} \quad y &= x'\sin\theta + y'\cos\theta \end{aligned}\right\}. \tag{1.3.1}$$

From (1.3.1) we can easily deduce that

$$\left.\begin{aligned} x' &= x\cos\theta + y\sin\theta \\ \text{and}\quad y' &= y\cos\theta - x\sin\theta \end{aligned}\right\}. \tag{1.3.2}$$

Both the transformations (1.3.1) and (1.3.2) can be remembered by the scheme;

	x'	y'
x	$\cos\theta$	$-\sin\theta$
y	$\sin\theta$	$\cos\theta$

► **Example 3** *Find the equation of the line $y = \sqrt{3}x$ when the axes are rotated through an angle $\frac{\pi}{3}$.*

The transformed equation is

$$x'\sin\frac{\pi}{3} + y'\cos\frac{\pi}{3} = \sqrt{3}\left(x'\cos\frac{\pi}{3} - y'\sin\frac{\pi}{3}\right),$$

$$\text{or}\quad \frac{\sqrt{3}}{2}x' + \frac{1}{2}y' = \sqrt{3}\left(\frac{1}{2}x' - \frac{\sqrt{3}}{2}y'\right),$$

$$\text{or}\quad \frac{1}{2}y' + \frac{3}{2}y' = 0, \quad \text{or}\quad y' = 0.$$

► **Example 4** *What will be the form of $x\cos\alpha + y\sin\alpha = p$, when the axes are rotated through an angle α?*

The transformed form is

$$(x'\cos\alpha - y'\sin\alpha)\cos\alpha + (x'\sin\alpha + y'\cos\alpha)\sin\alpha = p,$$

$$\text{or}\quad x'(\cos^2\alpha + \sin^2\alpha) = p, \quad \text{or}\quad x' = p.$$

1.4 Combination of Translation and Rotation of Axes

If the origin O of a set of rectangular axes (OX, OY) is shifted to $O'(\alpha, \beta)$ [referred to OX and OY] without changing the direction of axes and then the axes are rotated through an angle θ in the anticlockwise direction, the total effective changes in the co-ordinates (x, y) of a point are given by

$$\left.\begin{aligned} x &= \alpha + x''\cos\theta - y''\sin\theta \\ \text{and}\quad y &= \beta + x''\sin\theta + y''\cos\theta. \end{aligned}\right\}$$

(x'', y'') are the co-odinates of the point referred to the final set of axes.

► **Example 5** *What does the equation $x^2-3xy+y^2+7x-18y+32=0$ become when the origin is moved to the point* (4, 5) *and the axes are turned through an angle of* 45°?

[B.U. '96]

Replacing x and y by $4+(x\cos 45°-y\sin 45°)$ and $5+(x\sin 45°+y\cos 45°)$, i.e., $4+\frac{1}{\sqrt{2}}(x-y)$ and $5+\frac{1}{\sqrt{2}}(x+y)$ respectively, the transformed equation is

$$\left\{4+\frac{1}{\sqrt{2}}(x-y)\right\}^2-3\left\{4+\frac{1}{\sqrt{2}}(x-y)\right\}\left\{5+\frac{1}{\sqrt{2}}(x+y)\right\}+\left\{5+\frac{1}{\sqrt{2}}(x+y)\right\}^2$$

$$+7\left\{4+\frac{1}{\sqrt{2}}(x-y)\right\}-18\left\{5+\frac{1}{\sqrt{2}}(x+y)\right\}+32=0,$$

or $$16+\frac{1}{2}(x-y)^2+4\sqrt{2}(x-y)-3\left\{20+2\sqrt{2}(x+y)+\frac{5\sqrt{2}}{2}(x-y)+\frac{1}{2}(x^2-y^2)\right\}$$

$$+25+\frac{1}{2}(x+y)^2+5\sqrt{2}(x+y)+28+\frac{7}{2}\sqrt{2}(x-y)-90-9\sqrt{2}(x+y)+32=0,$$

or $\frac{5}{2}y^2-\frac{1}{2}x^2-10\sqrt{2}(x+y)+49=0,$

or $5y^2-x^2-20\sqrt{2}(x+y)+98=0.$

1.5 Transformation of co-ordinates when the equations of New Axes are given

Let (x, y) be the co-ordinates of a point P referred to rectangular axes OX and OY and (x', y') be the co-ordinates of the same point referred to a new set of rectangular axes $O'X'$ and $O'Y'$ whose equations are $lx+my+n=0$ and $mx-ly+k=0$ with respect to OX and OY.

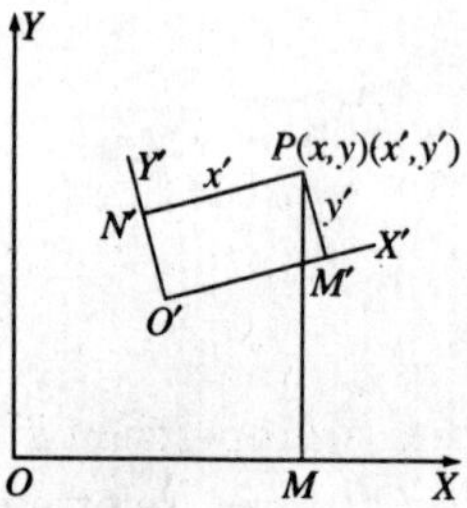

Fig. 1.3

Perpendicular distance from P to $mx-ly+k=0$ is

$$N'P=x'=\pm\frac{mx-ly+k}{\sqrt{l^2+m^2}}. \tag{1.5.1}$$

Perpendicular distance from P to $lx+my+n=0$ is

$$M'P=y'=\pm\frac{lx+my+n}{\sqrt{l^2+m^2}}. \tag{1.5.2}$$

The same sign is taken in both cases according to convenience. By (1.5.1) and (1.5.2) values of x and y are found out in terms of x' and y'.

NOTE. The transformation is helpful when the equation can be expressed in terms of

$$\frac{mx-ly+k}{\sqrt{l^2+m^2}} \quad\text{and}\quad \frac{lx+my+n}{\sqrt{l^2+m^2}}.$$

▶ **Example 6** *Find the transformed equation of the curve* $(3x + 4y + 7)(4x - 3y + 5) = 50$ *when the new axes are* $3x + 4y + 7 = 0$ *and* $4x - 3y + 5 = 0$.

If (x', y') be the co-ordinates of a point $P(x, y)$ referred to the new set of axes then

$$y' = \frac{3x + 4y + 7}{\sqrt{3^2 + 4^2}} = \frac{3x + 4y + 7}{5} \quad \text{and} \quad x' = \frac{4x - 3y + 5}{\sqrt{4^2 + 3^2}} = \frac{4x - 3y + 5}{5}.$$

The given equation can be written as

$$\frac{3x + 4y + 7}{5} \cdot \frac{4x - 3y + 5}{5} = 2. \quad \therefore \quad x'y' = 2.$$

Hence the transformed equation is $xy = 2$.

1.6 Orthogonal Transformation

We have discussed three types of transformation of axes or transformation of co-ordinates, namely—(i) translation, (ii) rotation, (iii) translation and rotation.

These are called orthogonal transformations when both the system of axes are rectangular. The combination of translation and rotation is called a rigid body motion.

Let (x, y) be the co-ordinates of a point in the old system, (x', y') the co-ordinates of the same point in the new system, (α, β) the co-ordinates of the new origin with respect to the old system and θ the angle of rotation.

Writing

$$X = \begin{bmatrix} x \\ y \end{bmatrix}, \quad X' = \begin{bmatrix} x' \\ y' \end{bmatrix}, \quad A = \begin{bmatrix} \alpha \\ \beta \end{bmatrix}, \quad I = \begin{bmatrix} 1 & 0 \\ 0 & 1 \end{bmatrix} \quad \text{and} \quad S = \begin{bmatrix} \cos\theta & -\sin\theta \\ \sin\theta & \cos\theta \end{bmatrix}$$

the formulae for co-ordinate transformation due to (i) translation, (ii) rotation and (iii) translation and rotation can be written in matrix notation as

(i) $X = IX' + A$, (ii) $X = SX'$ and (iii) $X = SX' + A$ respectively.

From these formulae, the inverse relations are

$$X' = I^{-1}X - I^{-1}A, X' = S^{-1}X \text{ and } X' = S^{-1}X - S^{-1}A.$$

I^{-1} and S^{-1} are inverses of the matrices I and S.

Here $S^{-1} = \begin{bmatrix} \cos\theta & \sin\theta \\ -\sin\theta & \cos\theta \end{bmatrix}$ and $I^{-1} = I$.

NOTE 1. I and S are orthogonal matrices and $|I| = 1 = |S|$. For this property of I and S each transformation is called an orthogonal transformation.

NOTE 2. The transformation by the formulae

$$\left.\begin{aligned} x &= \lambda x' - \mu y' + v \\ y &= \mu x' + \lambda y' + \zeta \end{aligned}\right\}, \quad \text{where } \lambda^2 + \mu^2 = 1$$

is an orthogonal transformation.

If $\lambda = 1, \mu = 0$ then it is *translation.*

If $v = 0 = \zeta$ then it is *rotation.*

1.7 Invariants

Some expressions remain unchanged under an orthogonal transformation. These are known as *invariants of orthogonal transformation.*

(i) *The degree of an equation is an invariant under orthogonal transformation.*

Let a polynomial equation be $f(x, y) = 0$. By an orthogonal transformation of the co-ordinate axes the equation transforms to $f(\lambda x' - \mu y' + v, \mu x' + \lambda y' + \zeta) = 0$, where $\lambda^2 + \mu^2 = 1$. If $\boldsymbol{ax^p y^q}$ is the highest degree term in $f(x, y) = 0$, then in the transformed equation it becomes $\boldsymbol{a(\lambda x' - \mu y' + v)^p(\mu x' + \lambda y' + \zeta)^q}$. The highest degree of the term containing x' or y' or $x'y'$ is $p + q$. It is equal to the degree of $\boldsymbol{ax^p y^q}$. Thus the degree of an equation is an invariant under orthogonal transformation.

(ii) *The distance between two points is an invariant under an orthogonal transformation.*

Let (x_1, y_1) and (x_2, y_2) be the co-ordinates of two points in the old system. In the new system the co-ordinates of these points are (x_1', y_1') and (x_2', y_2').

The formulae for transformation are

$$\left.\begin{aligned} x &= \lambda x' - \mu y' + v \\ y &= \mu x' + \lambda y' + \zeta \end{aligned}\right\}, \quad \text{where } \lambda^2 + \mu^2 = 1.$$

If $\boldsymbol{d}$ is the distance between the points then $d^2 = (x_2 - x_1)^2 + (y_2 - y_1)^2$.

By transformation

$$\begin{aligned} (x_2 - x_1)^2 + (y_2 - y_1)^2 &= \{(\lambda x_2' - \mu y_2' + v) - (\lambda x_1' - \mu y_1' + v)\}^2 \\ &\quad + \{(\mu x_2' + \lambda y_2' + \zeta) - (\mu x_1' + \lambda y_1' + \zeta)\}^2 \\ &= \{\lambda(x_2' - x_1') - \mu(y_2' - y_1')\}^2 + \{\mu(x_2' - x_1') + \lambda(y_2' - y_1')\}^2 \\ &= (\lambda^2 + \mu^2)\{(x_2' - x_1')^2 + (y_2' - y_1')^2\} \\ &= (x_2' - x_1')^2 + (y_2' - y_1')^2. \quad [\because \quad \lambda^2 + \mu^2 = 1] \end{aligned}$$

$$\therefore \quad d^2 = (x_2' - x_1')^2 + (y_2' - y_1')^2.$$

Hence the result follows.

(iii) *The coefficients of* $\boldsymbol{x^2}$, $\boldsymbol{xy}$ *and* $\boldsymbol{y^2}$ *and* Δ *obtained from*

$$\boldsymbol{ax^2 + 2hxy + by^2 + 2gx + 2fy + c}$$

are invariants under translation.

Let the the origin be shifted to (α, β). The expression transforms to

$$\begin{aligned} & a(x' + \alpha)^2 + 2h(x' + \alpha)(y' + \beta) + b(y' + \beta)^2 + 2g(x' + \alpha) + 2f(y' + \beta) + c \\ &= ax'^2 + 2hx'y' + by'^2 + 2(a\alpha + h\beta + g)x' + 2(h\alpha + b\beta + f)y' \\ &\quad + a\alpha^2 + 2h\alpha\beta + b\beta^2 + 2g\alpha + 2f\beta + c \\ &= a'x'^2 + 2h'x'y' + b'y'^2 + 2g'x' + 2f'y' + c' \text{(say)}. \end{aligned}$$

Here

$$a' = a, \quad b' = b, \quad h' = h, \quad g' = a\alpha + h\beta + g, \quad f' = h\alpha + b\beta + f,$$
$$c' = a\alpha^2 + 2h\alpha\beta + b\beta^2 + 2g\alpha + 2f\beta + c.$$

We see that the coefficients of x^2, y^2 and xy, i.e., **a, b** and **h** remain invariant due to translation.

Let us consider the invariance of $\Delta = \begin{vmatrix} a & h & g \\ h & b & f \\ g & f & c \end{vmatrix}$.

After translation Δ changes to Δ' (say).

Now

$$\Delta' = \begin{vmatrix} a' & h' & g' \\ h' & b' & f' \\ g' & f' & c' \end{vmatrix} = \begin{vmatrix} a & h & a\alpha + h\beta + g \\ h & b & h\alpha + b\beta + f \\ a\alpha + h\beta + g & h\alpha + b\beta + f & a\alpha^2 + 2h\alpha\beta + \ldots + c \end{vmatrix}$$

$$= \begin{vmatrix} a & h & g \\ h & b & f \\ a\alpha + h\beta + g & h\alpha + b\beta + f & g\alpha + f\beta + c \end{vmatrix}$$

(on subtracting α times the elements of the first column and β times the elements of the second column from those of the third column)

$$= \begin{vmatrix} a & h & g \\ h & b & f \\ g & f & c \end{vmatrix}$$

(on subtracting α times the elements of the first row and β times the elements of the second row from those of the third row).

$\therefore \quad \Delta$ is also an invariant.

(iv) $\boldsymbol{a + b}$, $\boldsymbol{ab - h^2}$, $\boldsymbol{f^2 + g^2}$ *and* Δ *obtained from* $\boldsymbol{ax^2 + 2hxy + by^2 + 2gx + 2fy + c = 0}$ *remain invariants under transformation of rotation.*

Let the axes be rotated through an angle θ. The expression $ax^2 + 2hxy + by^2 + 2gx + 2fy + c = 0$ transforms to

$$a(x' \cos\theta - y' \sin\theta)^2 + 2h(x' \cos\theta - y' \sin\theta)(x' \sin\theta + y' \cos\theta)$$
$$+ b(x' \sin\theta + y' \cos\theta)^2 + 2g(x' \cos\theta - y' \sin\theta) + 2f(x' \sin\theta + y' \cos\theta) + c$$
$$= a'x'^2 + 2h'x'y' + b'y'^2 + 2g'x' + 2f'y' + c' \text{ (say)}.$$

Here

$$\begin{aligned}
a' &= a\cos^2\theta + 2h\sin\theta\cos\theta + b\sin^2\theta,\\
b' &= a\sin^2\theta - 2h\sin\theta\cos\theta + b\cos^2\theta,\\
h' &= (b-a)\sin\theta\cos\theta + h(\cos^2\theta - \sin^2\theta),\\
g' &= g\cos\theta + f\sin\theta,\\
f' &= -g\sin\theta + f\cos\theta, c' = c.
\end{aligned}$$

If we put

$$a_1 = a\cos\theta + h\sin\theta,\quad b_1 = h\cos\theta + b\sin\theta,$$
$$a_2 = -a\sin\theta + h\cos\theta,\; b_2 = -h\sin\theta + b\cos\theta,$$

then

$$a' = a_1\cos\theta + b_1\sin\theta,\quad b' = -a_2\sin\theta + b_2\cos\theta,$$
$$h' = a_2\cos\theta + b_2\sin\theta = -a_1\sin\theta + b_1\cos\theta.$$

[a_2 and b_2 are the derivatives of a_1 and b_1 w.r.t. θ.]

$$\begin{aligned}
a' + b' &= a(\cos^2\theta + \sin^2\theta) + 2h\sin\theta\cos\theta - 2h\sin\theta\cos\theta + b(\sin^2\theta + \cos^2\theta)\\
&= a + b. \qquad (1.7.1)
\end{aligned}$$

$$\begin{aligned}
a'b' - h'^2 &= \begin{vmatrix} a' & h' \\ h' & b' \end{vmatrix} = \begin{vmatrix} a_1\cos\theta + b_1\sin\theta & -a_1\sin\theta + b_1\cos\theta \\ a_2\cos\theta + b_2\sin\theta & -a_2\sin\theta + b_2\cos\theta \end{vmatrix}\\
&= \begin{vmatrix} a_1 & b_1 \\ a_2 & b_2 \end{vmatrix} \times \begin{vmatrix} \cos\theta & \sin\theta \\ -\sin\theta & \cos\theta \end{vmatrix} = \begin{vmatrix} a_1 & b_1 \\ a_2 & b_2 \end{vmatrix}\\
&= \begin{vmatrix} a\cos\theta + h\cos\theta & h\cos\theta + b\sin\theta \\ -a\sin\theta + h\cos\theta & -h\sin\theta + b\cos\theta \end{vmatrix}\\
&= \begin{vmatrix} a & h \\ h & b \end{vmatrix} \times \begin{vmatrix} \cos\theta & \sin\theta \\ -\sin\theta & \cos\theta \end{vmatrix} = \begin{vmatrix} a & h \\ h & b \end{vmatrix} = ab - h^2. \qquad (1.7.2)
\end{aligned}$$

$$g'^2 + f'^2 = (g\cos\theta + f\sin\theta)^2 + (f\cos\theta - g\sin\theta)^2 = g^2 + f^2. \qquad (1.7.3)$$

$$\begin{aligned}
\Delta' &= \begin{vmatrix} a' & h' & g' \\ h' & b' & f' \\ g' & f' & c' \end{vmatrix} = g'\begin{vmatrix} h' & g' \\ b' & f' \end{vmatrix} - f'\begin{vmatrix} a' & g' \\ h' & f' \end{vmatrix} + c'\begin{vmatrix} a' & h' \\ h' & b' \end{vmatrix}\\
&= g'\begin{vmatrix} a_2\cos\theta + b_2\sin\theta & g\cos\theta + f\sin\theta \\ -a_2\sin\theta + b_2\cos\theta & -g\sin\theta + f\cos\theta \end{vmatrix}\\
&\quad - f'\begin{vmatrix} a_1\cos\theta + b_1\sin\theta & g\cos\theta + f\sin\theta \\ -a_1\sin\theta + b_1\cos\theta & -g\sin\theta + f\cos\theta \end{vmatrix} + c(ab - h^2) \quad \text{[by (1.7.2)]}
\end{aligned}$$

$$= g'\begin{vmatrix} a_2 & b_2 \\ g & f \end{vmatrix} \times \begin{vmatrix} \cos\theta & \sin\theta \\ -\sin\theta & \cos\theta \end{vmatrix} - f'\begin{vmatrix} a_1 & b_1 \\ g & f \end{vmatrix}$$

$$\times \begin{vmatrix} \cos\theta & \sin\theta \\ -\sin\theta & \cos\theta \end{vmatrix} + c(ab - h^2)$$

$$= g'\begin{vmatrix} a_2 & b_2 \\ a & f \end{vmatrix} - f'\begin{vmatrix} a_1 & b_1 \\ g & f \end{vmatrix} + c(ab - h^2)$$

$$= g'(a_2 f - b_2 g) - f'(a_1 f - b_1 g) + c(ab - h^2)$$

$$= g(b_1 f' - b_2 g') - f(a_1 f' - a_2 g') + c(ab - h^2)$$

$$= g\begin{vmatrix} b_1 & b_2 \\ g' & f' \end{vmatrix} - f\begin{vmatrix} a_1 & a_2 \\ g' & f' \end{vmatrix} + c(ab - h^2)$$

$$= g\begin{vmatrix} h\cos\theta + b\sin\theta & -h\sin\theta + b\cos\theta \\ g\cos\theta + f\sin\theta & -g\sin\theta + f\cos\theta \end{vmatrix}$$

$$- f\begin{vmatrix} a\cos\theta + h\sin\theta & -a\sin\theta + h\cos\theta \\ g\cos\theta + f\sin\theta & -g\sin\theta + f\cos\theta \end{vmatrix} + c(ab - h^2)$$

$$= g\begin{vmatrix} h & b \\ g & f \end{vmatrix} \times \begin{vmatrix} \cos\theta & \sin\theta \\ -\sin\theta & \cos\theta \end{vmatrix} - f\begin{vmatrix} a & h \\ g & f \end{vmatrix}$$

$$\times \begin{vmatrix} \cos\theta & \sin\theta \\ -\sin\theta & \cos\theta \end{vmatrix} + c(ab - h^2)$$

$$= g\begin{vmatrix} h & b \\ g & f \end{vmatrix} - f\begin{vmatrix} a & h \\ g & f \end{vmatrix} + c\begin{vmatrix} a & h \\ h & b \end{vmatrix} = \begin{vmatrix} a & h & g \\ h & b & f \\ g & f & c \end{vmatrix} = \Delta. \quad (1.7.4)$$

$\therefore \quad a + b,\ ab - h^2,\ g^2 + f^2$ and Δ are invariants.

Worked Out Examples

1. The equation $3x^2 + 2xy + 3y^2 - 18x - 22y + 50 = 0$ is reduced to $4x^2 + 2y^2 = 1$ when referred to rectangular axes through the point $(2, 3)$. Find the inclination of the latter axes to the former.

   Due to translation the equation transforms to

$$3(x' + 2)^2 + 2(x' + 2)(y' + 3) + 3(y' + 3)^2 - 18(x' + 2) - 22(y' + 3) + 50 = 0$$

or $\quad 3x'^2 + 2x'y' + 3y'^2 = 1.$

Let the axes be rotated through an angle θ. The equation changes to

$$3(x''\cos\theta - y''\sin\theta)^2 + 2(x''\cos\theta - y''\sin\theta)(x''\sin\theta + y''\cos\theta)$$
$$+3(x''\sin\theta + y''\cos\theta)^2 = 1$$
$$\text{or}\quad (3+\sin 2\theta)x''^2 + 2\cos 2\theta x''y'' + (3-\sin 2\theta)y''^2 = 1.$$

To remove the term of $x''y''$, we take $\cos 2\theta = 0$ or $2\theta = 90°$ or $\theta = 45°$.
For this value of θ the equation reduces to $4x''^2 + 2y''^2 = 1$.
Thus the required inclination is $45°$.

2. The co-ordinates of new origin are $(2, 1)$ and the axes are rotated through an angle $60°$. If the co-ordinates of a point in the new system are

$$\left(\frac{3-4\sqrt{3}}{2}, -\frac{4+3\sqrt{3}}{2}\right),$$

find the co-ordinates of it in the old system.

Let (x, y) be the co-ordinates of the point in the old system. By the formulae for translation and rotation

$$x = \frac{3-4\sqrt{3}}{2}\cos 60° + \frac{4+3\sqrt{3}}{2}\sin 60° + 2$$

$$\text{and}\quad y = \frac{3-4\sqrt{3}}{2}\sin 60° - \frac{4+3\sqrt{3}}{2}\cos 60° + 1 \quad\text{or}\quad x = 5 \text{ and } y = -3.$$

Thus the required co-ordinates are $(5, -3)$.

3. Find the angle through which the axes must be turned so that the equation $lx - my + n = 0 (m \neq 0)$ may be reduced to the form $ay + b = 0$. **[B.U. '90]**

Let the axes be rotated through an angle θ. The equation transforms to

$$l(x'\cos\theta - y'\sin\theta) - m(x'\sin\theta + y'\cos\theta) + n = 0,$$
$$\text{or}\quad (l\cos\theta - m\sin\theta)x' - (l\sin\theta + m\cos\theta)y' + n = 0.$$

Here the coefficient of x' is zero, i.e., $l\cos\theta - m\sin\theta = 0$, or $\tan\theta = \frac{l}{m}$, or $\theta = \tan^{-1}\frac{l}{m}$.
For this value of θ,

$$\sin\theta = \frac{l}{\sqrt{l^2+m^2}} \quad\text{and}\quad \cos\theta = \frac{m}{\sqrt{l^2+m^2}}$$

The equation reduces to

$$-\left(\frac{l^2}{\sqrt{l^2+m^2}} + \frac{m^2}{\sqrt{l^2+m^2}}\right)y + n = 0, \text{ or } -\sqrt{l^2+m^2}y + n = 0$$

$$\text{or}\quad \sqrt{l^2+m^2}y - n = 0. \quad \text{Thus } a = \sqrt{l^2+m^2}, b = -n.$$

4. The co-ordinates of two points are $(3, -2)$ and $(3+3\sqrt{3}, 1)$. The origin is shifted to the point $(3, -2)$ and the new x-axis is the line joining the given points. Find the formulae for this orthogonal transformation.

For translation and rotation

$$x = x' \cos\theta - y' \sin\theta + \alpha,$$

$$y = x' \sin\theta + y' \cos\theta + \beta.$$

By the given condition $\alpha = 3$, $\beta = -2$.

In the new system the other point lies on the x-axis. Therefore, the co-ordinates of this point are of the form $(c, 0)$. Thus putting $x' = c$ and $y' = 0$,

$$3 + 3\sqrt{3} = c\cos\theta + 3 \text{ and } 1 = c\sin\theta - 2$$

$$\text{or} \quad c\cos\theta = 3\sqrt{3} \text{ and } c\sin\theta = 3$$

$$\text{or} \quad \tan\theta = \frac{3}{3\sqrt{3}} = \frac{1}{\sqrt{3}}, \quad \text{i.e.,} \quad \theta = \frac{\pi}{6}.$$

Thus the required formulae are

$$x = x'\frac{\sqrt{3}}{2} - y'\frac{1}{2} + 3, \quad y = x' \cdot \frac{1}{2} + y' \cdot \frac{\sqrt{3}}{2} - 2.$$

5. If $ax+by$ transforms to $a'x'+b'y'$ under rotation of axes then show that $a^2+b^2 = a'^2+b'^2$.

[B.U. '91, '96, '99; C.U. '92, 2000]

If $A = [a\ b], X = \begin{bmatrix} x \\ y \end{bmatrix}$ and $X' = \begin{bmatrix} x' \\ y' \end{bmatrix}$ then $ax + by = AX$.

Let us consider the rotation given by $X = SX'$, where

$$S = \begin{bmatrix} \lambda & -\mu \\ \mu & \lambda \end{bmatrix}, \lambda^2 + \mu^2 = 1.$$

Now

$$AX = ASX' = [a\ \ b]\begin{bmatrix} \lambda & -\mu \\ \mu & \lambda \end{bmatrix}\begin{bmatrix} x' \\ y' \end{bmatrix}$$

$$= (a\lambda + b\mu)x' + (-a\mu + b\lambda)y' = a'x' + b'y'.$$

$\therefore\ \ a' = a\lambda + b\mu$ and $b' = -a\mu + b\lambda$.

$$\therefore\ \ a'^2 + b'^2 = (a\lambda + b\mu)^2 + (-a\mu + b\lambda)^2 = (\lambda^2 + \mu^2)(a^2 + b^2) = a^2 + b^2.$$

6. If $ax^2 + 2hxy + by^2$ and $Ax^2 + 2Hxy + By^2$ transform to $a'x'^2 + 2h'x'y' + b'y'^2$ and $A'x'^2 + 2H'x'y' + B'y'^2$ under rotation of axes show that $aA + bB + 2hH$ is an invariant.

Let the axes be rotated through an angle θ.

By section 1.7 (iv),

$$\begin{aligned} a'A' + b'B' + 2h'H' &= (a_1\cos\theta + b_1\sin\theta)(A_1\cos\theta + B_1\sin\theta) \\ &\quad +(-a_2\sin\theta + b_2\cos\theta)(-A_2\sin\theta + B_2\cos\theta) \\ &\quad +(-a_1\sin\theta + b_1\cos\theta)(-A_1\sin\theta + B_1\cos\theta) \\ &\quad +(a_2\cos\theta + b_2\sin\theta)(A_2\cos\theta + B_2\sin\theta) \\ &= a_1A_1 + b_1B_1 + a_2A_2 + b_2B_2 \\ &= (a\cos\theta + h\sin\theta)(A\cos\theta + H\sin\theta) \\ &\quad +(h\cos\theta + b\sin\theta)(H\cos\theta + B\sin\theta) \\ &\quad +(-a\sin\theta + h\cos\theta)(-A\sin\theta + H\cos\theta) \\ &\quad +(-h\sin\theta + b\cos\theta)(-H\sin\theta + b\cos\theta) \\ &= aA + bB + 2hH. \end{aligned}$$

Exercise

1. Reduce the following in the form $lx + my = 0$ by choice of new origin (i) on the x-axis, (ii) on the y-axis without rotation of axes.

   (a) $2x + 3y - 6 = 0$;
   (b) $y = mx + c$;
   (c) $x\cos\alpha + y\sin\alpha = p$.

2. Reduce the following equation in the form $ax^2 + by^2 = 1$ by proper translation of axes without rotation.

   (a) $5x^2 - 2y^2 - 30x + 8y = 0$;
   (b) $4x^2 + 3y^2 - 2x - 3y - 7 = 0$. [B.U. '84]

3. Choose a new origin in such a way that the following equations referred to this new set of axes will turn out to be homogeneous in (x, y) of second degree.

   (a) $2x^2 + 4xy + 5y^2 - 4x - 22y + 29 = 0$;
   (b) $12x^2 - 10xy + 2y^2 + 11x - 5y + 2 = 0$; [B.U. '89]
   (c) $x^2 - y^2 + (h + k)x - (h - k)y + hk = 0$.

4. Find a new origin without rotating the axes in order that the equation $y^2 + 2y - 8x + 25 = 0$ will turn out to be $y^2 = 4ax$. Determine a.

5. The origin is shifted to the point $(3, -3)$ without changing the directions of axes. If the co-ordinates of P, Q, R are $(5, 5), (-2, 4)$ and $(7, -7)$ respectively in the new system, find the co-ordinates of these points in the old system.

6. The axes are rotated through an angle of 60° without change of origin. The co-ordinates of a point are $(4, \sqrt{3})$ in the new system. Find the co-ordinates of it in the old system.

7. The origin is shifted to the point $(3, -1)$ and the axes are rotated through an angle $\tan^{-1}\frac{3}{4}$. If the co-ordinates of a point are $(5, 10)$ in the new system, find its co-ordinates in the old system.

8. The co-ordinates of A and B are $(5, -1)$ and $(3, 1)$. The origin is shifted to A and the axes are rotated in such a way that the new x-axis coincides with AB. If the rotation is made in the positive direction, find the formulae for transformation.

9. (a) What does the equation $4x^2 + 2\sqrt{3}xy + 2y^2 = 1$ become when the axes are rotated through an angle $30°$? **[B.U. '85]**

   (b) Transform the equation $x^2 + 2cxy + y^2 = a^2$ by turning the rectangular axes through an angle $\frac{\pi}{4}$ without changing the origin. **[B.U., '84]**

10. (a) Find the angle of rotation of the axes for which the equation $x^2 - y^2 = a^2$ will reduce to $xy = c^2$. Determine c^2. **[C.U. '93]**

    (b) What is the angle of rotation of the axes for which the equation $3x^2 - 5xy + 3y^2 = 1$ will reduce to one being wanted of the xy term. **[B.U. '93]**

    (c) Find the angle through which the axes are rotated, so that the equation $17x^2 + 18xy - 7y^2 = 1$ may reduce to the form $Ax^2 + By^2 = 1, A > 0$. Determine the nature of the curve. **[C.U. '87; N.B. '90; B.U. '92]**

11. Transform the equation $x^2 - 2xy + y^2 + x - 3y = 0$ to axes through the point $(-1, 0)$ parallel to the lines bisecting the angles between the axes.

12. (a) Find the equations of the following when $ax + by + c = 0$ and $bx - ay + d = 0$ are considered as axes of x and y respectively.

    (i) $(ax + by + c)^2 = a^2 + b^2$,

    (ii) $(ax + by + c)(bx - ay + d) = a^2 + b^2$.

    (b) Transform the equation $3(12x - 5y + 39)^2 + 2(5x + 12y - 26)^2 = 169$, taking the lines $12x - 5y + 39 = 0$ and $5x + 12y - 26 = 0$ as the new axes of x and y respectively. **[C.U. '96]**

13. Show that if the origin is transferred to $(0, 1)$ and the axes are rotated through $45°$, the equation $5x^2 - 2xy + 5y^2 + 2x - 10y - 7 = 0$ referred to new axes becomes $x'^2/3 + y'^2/2 = 1$.

14. If $ax + by$ and $cx + dy$ are changed into $a'x' + b'y'$ and $c'x' + d'y'$ respectively for rotation of axes, show that $ad - bc = a'd' - b'c'$.

15. If by a transformation of rotation of axes the expression $ax^2 + 2hxy + by^2$ is changed into $Ax^2 + 2Hxy + By^2$, then prove that

    (a) $A + B = a + b$

    (b) $AB - H^2 = ab - h^2$. **[B.U. '97]**

16. Show that the radius of a circle remains unchanged due to any rigid body motion.

17. Show that there is only one point whose co-ordinates do not alter due to a rigid motion.

[HINTS. Let $\left.\begin{array}{l} x = \lambda x' - \mu y' + \nu \\ y = \mu x' + \lambda y' + \zeta \end{array}\right\}, \lambda^2 + \mu^2 = 1$ be the rigid motion.

For unalteration of co-ordinates, $\left.\begin{array}{l} x = \lambda x - \mu y + \nu \\ y = \mu x + \lambda y + \zeta \end{array}\right\}$.

Solving for x and y, the point is

$$\left\{\frac{\nu(1-\lambda) - \mu\zeta}{2(1-\lambda)}, \frac{\zeta(1-\lambda) + \mu\nu}{2(1-\lambda)}\right\}]$$

18. If (x_1, y_1), (x_2, y_2), (x_3, y_3) be three non-collinear points in the plane, show that the expression $\begin{vmatrix} x_1 & y_1 & 1 \\ x_2 & y_2 & 1 \\ x_3 & y_3 & 1 \end{vmatrix}$ remains invariant under an orthogonal transformation.

[B.U.'81,'83]

[Under any rigid motion the area of a triangle is an invariant.]

[HINTS. Let the orthogonal transformation be $\left.\begin{matrix} x = \lambda x' - \mu y' + \nu \\ y = \mu x' + \lambda y' + \zeta \end{matrix}\right\}$, $\lambda^2 + \mu^2 = 1$.

Now

$$\begin{vmatrix} x_1 & y_1 & 1 \\ x_2 & y_2 & 1 \\ x_3 & y_3 & 1 \end{vmatrix} = \begin{vmatrix} \lambda x_1' - \mu y_1' + \nu & \mu x_1' + \lambda y_1' + \zeta & 1 \\ \lambda x_2' - \mu y_2' + \nu & \mu x_2' + \lambda y_2' + \zeta & 1 \\ \lambda x_3' - \mu y_3' + \nu & \mu x_3' + \lambda y_3' + \zeta & 1 \end{vmatrix} = \begin{vmatrix} \lambda & -\mu & \nu \\ \mu & \lambda & \zeta \\ 0 & 0 & 1 \end{vmatrix} \times \begin{vmatrix} x_1' & y_1' & 1 \\ x_2' & y_2' & 1 \\ x_3' & y_3' & 1 \end{vmatrix}$$

$$= (\lambda^2 + \mu^2) \begin{vmatrix} x_1' & y_1' & 1 \\ x_2' & y_2' & 1 \\ x_3' & y_3' & 1 \end{vmatrix} = \begin{vmatrix} x_1' & y_1' & 1 \\ x_2' & y_2' & 1 \\ x_3' & y_3' & 1 \end{vmatrix}].$$

Answers

1. (a) (3, 0), (0, 2); (b) $\left(-\frac{c}{m}, 0\right)$, $(0, c)$; (c) $\left(\frac{p}{\cos\alpha}, 0\right)$, $\left(0, \frac{p}{\sin\alpha}\right)$. **2.** (a) $\frac{5}{37}x^2 - \frac{2}{37}y^2 = 1$, $(3, 2)$; (b) $\frac{x^2}{2} + \frac{3}{8}y^2 = 1$, $\left(\frac{1}{4}, \frac{1}{2}\right)$. **3.** (a) $(-2, 3)$; (b) $\left(-\frac{3}{2}, -\frac{5}{2}\right)$; (c) $\left(-\frac{h+k}{2}, -\frac{h-k}{2}\right)$. **4.** $(3, -1)$, $a = 2$. **5.** $(8, 2)$, $(1, 1)$, $(10, -10)$. **6.** $\left(\frac{1}{2}, \frac{5\sqrt{3}}{2}\right)$. **7.** (1, 10). **8.** $x = -\frac{1}{\sqrt{2}}(x' + y') + 5$, $y = \frac{1}{\sqrt{2}}(x' - y') - 1$. **9.** (a) $5x^2 + y^2 = 1$; (b) $(1 + c)x'^2 + (1 - c)y'^2 = a^2$. **10.** (a) $-\frac{\pi}{2}$, $\frac{a^2}{2}$; (b) $\pm\frac{\pi}{4}$; (c) $\left(5 + \frac{369}{\sqrt{613}}\right)x'^2 - \left(\frac{369}{\sqrt{613}} - 5\right)y'^2 = 1$, hyperbola. **11.** $\sqrt{2}y^2 = x$. **12.** (a) (i) $y^2 = 1$; (ii) $xy = 1$; (b) $3y^2 + 2x^2 = 1$.

Chapter 2

Pair of Straight Lines

2.1 Second Degree Homogeneous Equation

(i) *A second degree homogeneous equation of the form* $ax^2 + 2hxy + by^2 = 0$ *represents a pair of straight lines through the origin.*

Considering $ax^2 + 2hxy + by^2 = 0$ as a quadratic equation in x, we have

$$x = \frac{-h \pm \sqrt{h^2 - ab}}{a} y, \quad \text{i.e.,} \quad ax + \left(h \pm \sqrt{h^2 - ab}\right) y = 0.$$

It shows that the equation will represent

(A) two real and distinct lines through the origin if $h^2 - ab > 0$;

(B) two coincident lines through the origin if $h^2 - ab = 0$;

(C) two imaginary lines if $h^2 - ab < 0$.

The equations of the pair of lines are

$$ax + \left(h + \sqrt{h^2 - ab}\right) y = 0 \quad \text{and} \quad ax + \left(h - \sqrt{h^2 - ab}\right) y = 0.$$

These are of the form $l_1x + m_1y = 0$ and $l_2x + m_2y = 0$.

(ii) **Angle between the lines** $ax^2 + 2hxy + by^2 = 0$.

Let $ax^2 + 2hxy + by^2 = (l_1x + m_1y)(l_2x + m_2y)$.

Equating the coefficients of x^2, xy and y^2, we have

$$l_1l_2 = a, \quad l_1m_2 + l_2m_1 = 2h, \quad m_1m_2 = b.$$

The lines are $l_1x + m_1y = 0$ and $l_2x + m_2y = 0$.

If θ be the angle between the lines,

$$\tan^2\theta = \frac{(l_1m_2 - l_2m_1)^2}{(l_1l_2 + m_1m_2)^2} = \frac{(l_1m_2 + l_2m_1)^2 - 4l_1l_2m_1m_2}{(l_1l_2 + m_1m_2)^2} = \frac{4(h^2 - ab)}{(a + b)^2}.$$

$$\therefore \quad \tan\theta = \pm\frac{2\sqrt{h^2-ab}}{a+b} \quad \text{or} \quad \theta = \tan^{-1}\left(\pm\frac{2\sqrt{h^2-ab}}{a+b}\right).$$

+ or − sign is taken according as θ is the acute or obtuse angle between the lines.

COROLLARY 1. **Condition of coincidence.** If $\theta = 0$, the lines are coincident and in this case $h^2 = ab$.

COROLLARY 2. **Condition of perpendicularity.** If $\theta = \pi/2$, the lines are at right angle and in this case

$$\cot\theta = 0 = \frac{a+b}{2\sqrt{h^2-ab}}, \quad \text{i.e.,} \quad a+b=0.$$

[Coefficient of x^2+ coefficient of $y^2 = 0$.]

▶ **Example 1** *Find the angle between the lines* $4x^2 - 3xy - y^2 = 0$.

Here $a = 4, h = -\frac{3}{2}, b = -1$.

$$\therefore \quad \theta = \tan^{-1}\frac{2\sqrt{h^2-ab}}{a+b} = \tan^{-1}\frac{2\sqrt{\frac{9}{4}+4}}{4-1} = \tan^{-1}\frac{5}{3}.$$

(iii) **Bisectors of the angles between the lines**

$$ax^2 + 2hxy + by^2 = 0.$$

Let $ax^2 + 2hxy + by^2 = (l_1x + m_1y)(l_2x + m_2y)$.

Equating the coefficients of x^2, xy and y^2 we have

$$l_1l_2 = a, \quad l_1m_2 + l_2m_1 = 2h, \quad m_1m_2 = b.$$

The lines are $l_1x + m_1y = 0$ and $l_2x + m_2y = 0$. The equations of the bisectors of the angles between the lines are

$$\frac{l_1x+m_1y}{\sqrt{l_1^2+m_1^2}} = \pm\frac{l_2x+m_2y}{\sqrt{l_2^2+m_2^2}}.$$

In pair form

$$\left(\frac{l_1x+m_1y}{\sqrt{l_1^2+m_1^2}} + \frac{l_2x+m_2y}{\sqrt{l_2^2+m_2^2}}\right)\left(\frac{l_1x+m_1y}{\sqrt{l_1^2+m_1^2}} - \frac{l_2x+m_2y}{\sqrt{l_2^2+m_2^2}}\right) = 0,$$

or $$\frac{(l_1x+m_1y)^2}{l_1^2+m_1^2} - \frac{(l_2x+m_2y)^2}{l_2^2+m_2^2} = 0,$$

or $$(l_2^2+m_2^2)(l_1x+m_1y)^2 - (l_1^2+m_1^2)(l_2x+m_2y)^2 = 0,$$

or $$(l_1^2m_2^2 - l_2^2m_1^2)(x^2-y^2) = 2(l_1m_2 - l_2m_1)(l_1l_2 - m_1m_2)xy,$$

or $$(l_1m_2 + l_2m_1)(x^2-y^2) = 2(l_1l_2 - m_1m_2)xy,$$

or $$2h(x^2-y^2) = 2(a-b)xy, \quad \text{or} \quad \frac{x^2-y^2}{a-b} = \frac{xy}{h}.$$

► **Example 2** *Find the equation of the bisector of the angles between the lines* $x^2 - 4xy - y^2 = 0$.

The equation is

$$\frac{x^2 - y^2}{1 - (-1)} = \frac{xy}{-2} \quad \text{or} \quad \frac{x^2 - y^2}{2} = \frac{xy}{-2} \quad \text{or} \quad x^2 + xy - y^2 = 0.$$

2.2 General Second Degree Equation (Non-Homogeneous)

The necessary and sufficient conditions for the general equation of the second degree $ax^2 + 2hxy + by^2 + 2gx + 2fy + c = 0$ *to represent a pair of real straight lines are* $abc + 2fgh - af^2 - bg^2 - ch^2 = 0$ *and* $h^2 - ab \geq 0$.

Proof. (i) *Conditions are necessary.*

If the equation $ax^2 + 2hxy + by^2 + 2gx + 2fy + c = 0$ represent a pair of real straight lines, the expression on the L.H.S. must be factorised into two linear factors.

Let $ax^2 + 2hxy + by^2 + 2gx + 2fy + c = (l_1x + m_1y + n_1)(l_2x + m_2y + n_2)$.

Equating the coefficients, $l_1l_2 = a$, $l_1m_2 + l_2m_1 = 2h$, $m_1m_2 = b, l_1n_2 + l_2n_1 = 2g, m_1n_2 + m_2n_1 = 2f$, $n_1n_2 = c$.

Conditions are obtained by eliminating l_1, m_1, n_1, l_2, m_2, n_2 from the above relations. For elimination we consider the following product.

$$\begin{vmatrix} l_1 & l_2 & 0 \\ m_1 & m_2 & 0 \\ n_1 & n_2 & 0 \end{vmatrix} \times \begin{vmatrix} l_2 & l_1 & 1 \\ m_2 & m_1 & 1 \\ n_2 & n_1 & 1 \end{vmatrix} = \begin{vmatrix} 2l_1l_2 & l_1m_2 + l_2m_1 & l_1n_2 + l_2n_1 \\ l_1m_2 + l_2m_1 & 2m_1m_2 & m_1n_2 + m_2n_1 \\ n_1l_2 + n_2l_1 & m_1n_2 + m_2n_1 & 2n_1n_2 \end{vmatrix}$$

$$= \begin{vmatrix} 2a & 2h & 2g \\ 2h & 2b & 2f \\ 2g & 2f & 2c \end{vmatrix} = 8 \begin{vmatrix} a & h & g \\ h & b & f \\ g & f & c \end{vmatrix}.$$

$$\because \begin{vmatrix} l_1 & l_2 & 0 \\ m_1 & m_2 & 0 \\ n_1 & n_2 & 0 \end{vmatrix} = 0, \text{ the product is zero.}$$

$$\therefore \begin{vmatrix} a & h & g \\ h & b & f \\ g & f & c \end{vmatrix} = 0 \quad \text{or} \quad abc + 2fgh - af^2 - bg^2 - ch^2 = 0.$$

Again $4(h^2 - ab) = (l_1m_2 + l_2m_1)^2 - 4l_1l_2m_1m_2 = (l_1m_2 - l_2m_1)^2$.

$$\therefore \; h^2 - ab \geq 0.$$

(ii) *Conditions are sufficient.*

If $\begin{vmatrix} a & h & g \\ h & b & f \\ g & f & c \end{vmatrix} = 0$, the equations $\left.\begin{aligned} ax + hy + g = 0 \\ hx + by + f = 0 \\ gx + fy + c = 0 \end{aligned}\right\}$ have a solution. Let the solution be (x', y').

Under translation, $x = X + x', y = Y + y'$ the general equation transforms to

$$a(X + x')^2 + 2h(X + x')(Y + y') + b(Y + y')^2 + 2g(X + x') + 2f(Y + y') + c = 0,$$

$$\text{or}\quad aX^2 + 2hXY + bY^2 + 2(ax' + hy' + g)X + 2(hx' + by' + f)Y$$
$$+ax'^2 + 2hx'y' + by'^2 + 2gx' + 2fy' + c = 0,$$

$$\text{or}\quad aX^2 + 2hXY + bY^2 + 2(ax' + hy' + g)X + 2(hx' + by' + f)Y$$
$$+(ax' + hy' + g)x' + (hx' + by' + f)y' + (gx' + fy' + c) = 0,$$

$$\text{or}\quad aX^2 + 2hXY + bY^2 = 0. \quad \text{[by the help of linear equations]}$$

Since $h^2 - ab \geq 0$, the above homogeneous equation represents two real lines. Consequently the given equation represents two real lines.

NOTE. $\begin{vmatrix} a & h & g \\ h & b & f \\ g & f & c \end{vmatrix}$ is known as the discriminant of the general equation and it will be denoted by Δ. If $\Delta = 0$, the general equation represents a pair of lines (real or imaginary).

ALTERNATIVE METHOD. The equation can be written as

$$ax^2 + 2(hy + g)x + by^2 + 2fy + c = 0.$$

If $a \neq 0$ then solving for x we get

$$x = \frac{-(hy + g) \pm \sqrt{\{(hy + g)^2 - a(by^2 + 2fy + c)\}}}{a},$$

$$\text{or}\quad ax + hy + g = \pm\sqrt{\{(hy + g)^2 - a(by^2 + 2fy + c)\}}$$
$$= \pm\sqrt{\{(h^2 - ab)y^2 + 2(gh - af)y + g^2 - ca\}}.$$

To represent a pair of real straight lines, the expression under radical sign must be a perfect square. This gives

$$4(gh - af)^2 - 4(h^2 - ab)(g^2 - ca) = 0,$$
$$\text{or}\quad a(abc + 2fgh - af^2 - bg^2 - ch^2) = 0 \text{ and } h^2 - ab > 0,$$

$\therefore$ the required conditions are $abc + 2fgh - af^2 - bg^2 - ch^2 = 0$ and $h^2 - ab > 0$. If $h^2 - ab = 0$ then the lines are parallel.

NOTE. The conditions obtained are necessary. These are sufficient. Since $(hy + g)^2 - a(by^2 + 2fy + c)$ is a perfect square when the conditions are satisfied.

▶ **Example 3** *Show that* $14x^2 + 29xy + 12y^2 - 31x - 14y - 10 = 0$ *represents a pair of straight lines.*

Since

$$14 \cdot 12 \cdot (-10) + 2 \cdot (-7) \cdot \left(-\frac{31}{2}\right)\left(\frac{29}{2}\right) - 14 \cdot (-7)^2 - 12 \cdot \left(-\frac{31}{2}\right)^2 - (-10) \cdot \left(\frac{29}{2}\right)^2$$

$$= -1680 + \frac{6293}{2} - 686 - 2883 + \frac{4205}{2}$$

$$= 5249 - 5249 = 0 \text{ and } \left(\frac{29}{2}\right)^2 - 14 \times 12 > 0,$$

the equation represents a pair of straight lines.

▶ **Example 4** *For what value of* λ *does* $\lambda xy - 8x + 9y - 12 = 0$ *represent a pair of lines?*

To represent a pair of lines

$$\begin{vmatrix} 0 & \lambda/2 & -4 \\ \lambda/2 & 0 & 9/2 \\ -4 & 9/2 & -12 \end{vmatrix} = 0, \quad \text{or} \quad \frac{\lambda}{2}(-18 + 6\lambda) - 4 \cdot \frac{\lambda}{2} \cdot \frac{9}{2} = 0,$$

$$\text{or} \quad 3\lambda^2 - 18\lambda = 0, \quad \text{or} \quad 3\lambda(\lambda - 6) = 0, \quad \text{or} \quad \lambda = 0, 6.$$

If $\lambda = 0$, the equation will not be of second degree.

Thus $\lambda = 6$.

(ii) Angle between the lines.

Let $ax^2 + 2hxy + by^2 + 2gx + 2fy + c = (l_1x + m_1y + n_1)(l_2x + m_2y + n_2)$.

Equating the coefficients, $l_1l_2 = a, l_1m_2 + l_1m_1 = 2h, m_1m_2 = b$,

$$l_1n_2 + l_2n_1 = 2g, \quad m_1n_2 + m_2n_1 = 2f, \quad n_1n_2 = c.$$

The lines are $l_1x + m_1y + n_1 = 0$ and $l_2x + m_2y + n_2 = 0$.

If θ be the angle between the lines,

$$\tan^2\theta = \frac{(l_1m_2 - l_2m_1)^2}{(l_1l_2 + m_1m_2)^2} = \frac{(l_1m_2 + l_2m_1)^2 - 4l_1l_2m_1m_2}{(l_1l_2 + m_1m_2)^2} = \frac{4(h^2 - ab)}{(a+b)^2}.$$

$$\therefore \quad \tan\theta = \pm\frac{2\sqrt{h^2 - ab}}{a+b}, \quad \text{or} \quad \theta = \tan^{-1}\left(\pm\frac{2\sqrt{h^2 - ab}}{a+b}\right),$$

$+$ or $-$ sign is taken according as θ is acute or obtuse.

▶ **Example 5** *Find the angle between the lines* $y^2 + xy - 2x^2 - 5x - y - 2 = 0$.

The equation is $2x^2 - xy - y^2 + 5x + y + 2 = 0$.

$$\therefore \text{ the required angle} = \tan^{-1}\frac{2\sqrt{\left(-\frac{1}{2}\right)^2 - 2.(-1)}}{2-1} = \tan^{-1}3.$$

(iii) (A) **Condition of perpendicularity of the pair of lines.**

The lines will be perpendicular, if $a + b = 0$, i.e., coefficient of x^2+ coefficient of $y^2 = 0$.

(B) **Condition of parallelism of the pair of lines.**

Let the parallel lines be $lx + my + n_1 = 0$ and $lx + my + n_2 = 0$.

In this case

$$ax^2 + 2hxy + by^2 + 2gx + 2fy + c = \lambda(lx + my + n_1)(lx + my + n_2),$$

where λ is a constant.

Equating the coefficients, $\lambda l^2 = a$, $\lambda lm = h$, $\lambda m^2 = b$, $\lambda l(n_1 + n_2) = 2g$, $\lambda m(n_1 + n_2) = 2f$, $\lambda n_1 n_2 = c$.

Now

$$\frac{a}{h} = \frac{l^2}{lm} = \frac{l}{m} = \frac{lm}{m^2} = \frac{h}{b} \quad \text{and} \quad \frac{a}{h} = \frac{l}{m} = \frac{l(n_1+n_2)}{m(n_1+n_2)} = \frac{g}{f}.$$

$\therefore$ **the required condition is** $\frac{a}{h} = \frac{h}{b} = \frac{g}{f}$.

With $\Delta = 0$, the conditions for parallelism can be written as

$$h^2 = ab, \quad hf = bg, \quad bg^2 = af^2.$$

(C) **Condition of coincidence of the pair of lines.**

If the lines are coincident then

$$ax^2 + 2hxy + by^2 + 2gx + 2fy + c = \lambda(lx + my + n)^2.$$

Equating the coefficients, $\lambda l^2 = a$, $\lambda lm = h$, $\lambda m^2 = b$, $\lambda ln = g$, $\lambda mn = f$, $\lambda n^2 = c$.

Now

$$\frac{a}{h} = \frac{l^2}{lm} = \frac{l}{m} = \frac{lm}{m^2} = \frac{h}{b}, \quad \frac{a}{h} = \frac{l}{m} = \frac{ln}{mn} = \frac{g}{f}. \therefore \quad \frac{a}{h} = \frac{h}{b} = \frac{g}{f}. \quad (2.2.1)$$

$$\frac{h}{g} = \frac{lm}{ln} = \frac{m}{n} = \frac{m^2}{mn} = \frac{b}{f} = \frac{mn}{n^2} = \frac{f}{c}. \quad \therefore \quad \frac{h}{g} = \frac{b}{f} = \frac{f}{c}. \quad (2.2.2)$$

$$\frac{g}{a} = \frac{ln}{l^2} = \frac{n}{l} = \frac{nm}{lm} = \frac{f}{h} = \frac{n^2}{nl} = \frac{c}{g}. \quad \therefore \quad \frac{g}{a} = \frac{f}{h} = \frac{c}{g}. \quad (2.2.3)$$

With $\Delta = 0$, the condition for coincidence can be taken as

$$h^2 = ab, \; f^2 = bc, \; g^2 = ca \text{ or } gh - af = 0, \; hf - bg = 0, \; fg - ch = 0.$$

► **Example 6** *Show that* $6x^2 - 5xy - 6y^2 + 14x + 5y + 4 = 0$ *represents a pair of perpendicular lines.*

Since

$$6.(-6).4 + 2.\frac{5}{2}.7(-5/2) - 6\left(\frac{5}{2}\right)^2 - (-6).7^2 - 4.(-5/2)^2$$

$$= -144 - \frac{175}{2} - \frac{75}{2} + 294 - 25 = 294 - 294 = 0,$$

the equation represents a pair of lines.

Again coefficient of $x^2 +$ coefficient of $y^2 = 6 - 6 = 0$.

Therefore, the lines are perpendicular to each other.

► **Example 7** *Show that* $x^2 + 6xy + 9y^2 - 5x - 15y + 6 = 0$ *represents a pair of parallel lines and find the distance between them.* **[B.U. '76; C.U. 2000]**

$$x^2 + 6xy + 9y^2 - 5x - 15y + 6 = 0,$$

$$\text{or} \quad x^2 + (6y - 5)x + 9y^2 - 15y + 6 = 0,$$

$$\text{or} \quad x = \frac{1}{2}\left[-(6y-5) \pm \sqrt{\{(6y-5)^2 - 4(9y^2 - 15y + 6)\}}\right] = \frac{1}{2}[-(6y-5) \pm 1].$$

Thus the lines are $x + 3y - 3 = 0$ and $x + 3y - 2 = 0$. Obviously these are parallel.

Distance between the lines $= \frac{3-2}{\sqrt{10}} = \frac{1}{\sqrt{10}}$ units.

► **Example 8** *Show that* $4x^2 + 24xy + 36y^2 + 4x + 12y + 1 = 0$ *represents a pair of coincident lines.*

Since $4x^2 + 24xy + 36y^2 + 4x + 12y + 1 = (2x + 6y + 1)^2$, the equation represents a pair of coincident lines.

(iv) **Point of intersection of the lines.**

Let (α, β) be the point of intersection of the lines represented by the general equation and the origin be transferred to (α, β) without rotation of axes. Replacing x and y by $x' + \alpha$ and $y' + \beta$ the equation takes the form

$$a(x' + \alpha)^2 + 2h(x' + \alpha)(y' + \beta) + b(y' + \beta)^2 + 2g(x' + \alpha) + 2f(y' + \beta) + c = 0,$$

$$\text{or} \quad ax'^2 + 2hx'y' + by'^2 + 2(a\alpha + h\beta + g)x' + 2(h\alpha + b\beta + f)y' + a\alpha^2 + 2h\alpha\beta$$
$$+ b\beta^2 + 2g\alpha + 2f\beta + c = 0.$$

Since this equation represents a pair of lines through the origin, this must be a second degree homogeneous equation in x' and y'. For this

$$a\alpha + h\beta + g = 0, \tag{2.2.4}$$

$$h\alpha + b\beta + f = 0 \tag{2.2.5}$$

$$\text{and} \quad a\alpha^2 + 2h\alpha\beta + b\beta^2 + 2g\alpha + 2f\beta + c = 0. \tag{2.2.6}$$

From (2.2.4) and (2.2.5),

$$\alpha = \frac{hf - bg}{ab - h^2}, \quad \beta = \frac{gh - af}{ab - h^2}.$$

Here $ab - h^2 \neq 0$, since the lines are not parallel.

COROLLARY 1. Condition for the pair of straight lines:

Subtracting α times (2.2.4) and β times (2.2.5) from (2.2.6) we have

$$g\alpha + f\beta + c = 0. \tag{2.2.7}$$

Eliminating α and β from (2.2.4), (2.2.5) and (2.2.7) we have

$$\Delta = \begin{vmatrix} a & h & g \\ h & b & f \\ g & f & c \end{vmatrix} = 0.$$ It is the condition for pair of lines.

COROLLARY 2. If A, H, G, ... are the cofactors of a, h, g, ... of Δ, then

$$\alpha = \frac{G}{C}, \quad \beta = \frac{F}{C}.$$

COROLLARY 3. Distance of the point of intersection from the origin.

If we take the lines as $l_1x + m_1y + n_1 = 0$ and $l_2x + m_2y + n_2 = 0$, then the point of intersection is

$$\left(\frac{m_1n_2 - m_2n_1}{l_1m_2 - l_2m_1}, \frac{n_1l_2 - n_2l_1}{l_1m_2 - l_2m_1}\right), \text{ or } \left(\sqrt{\frac{f^2 - bc}{h^2 - ac}}, \sqrt{\frac{g^2 - ca}{h^2 - ab}}\right), \text{ or } \left(\sqrt{\frac{A}{C}}, \sqrt{\frac{B}{C}}\right).$$

$$\therefore \text{ the distance } = \sqrt{\left(\frac{A}{C} + \frac{B}{C}\right)} = \sqrt{\frac{A+B}{C}} = \sqrt{\frac{c(a+b) - f^2 - g^2}{ab - h^2}}.$$

COROLLARY 4. Condition for the point of intersection to lie on the x-axis or the y-axis.

On the x-axis the y-co-ordinate of a point is zero. Thus if the lines represented by the general equation intersect on the x-axis then

$$gh - af = 0 \text{ and } g^2 - ca = 0, \text{ i.e., } \frac{g}{a} = \frac{f}{h} = \frac{c}{g}.$$

Similarly if the lines intersect on the y-axis then

$$hf - bg = 0 \text{ and } f^2 - bc = 0, \text{ i.e., } \frac{f}{b} = \frac{g}{h} = \frac{c}{f}.$$

(v) **Bisectors of the angles between the lines.**

Let (α, β) be the point of intersection and the origin be shifted to this point. The equation will reduce to

$$ax'^2 + 2hx'y' + by'^2 = 0, \quad \text{when } x = x' + \alpha, \ y = y' + \beta.$$

The equation of the bisectors of the angles between the lines is

$$\frac{x'^2 - y'^2}{a-b} = \frac{x'y'}{h}. \quad \text{[See section 2.1(iii)]}$$

Thus the equation in the original set of axes is

$$\frac{(x-\alpha)^2 - (y-\beta)^2}{a-b} = \frac{(x-\alpha)(y-\beta)}{h}.$$

► **Example 9** *Find the equations of the bisectors of the angles between the lines* $x^2 - 5xy + 4y^2 + x + 2y - 2 = 0$.

Let (α, β) be the point of intersection. Changing the origin to (α, β), the equation transforms to

$$(x'+\alpha)^2 - 5(x'+\alpha)(y'+\beta) + 4(y'+\beta)^2 + (x'+\alpha) + 2(y'+\beta) - 2 = 0,$$

i.e., $x'^2 - 5x'y' + 4y'^2 = 0$ with $2\alpha - 5\beta + 1 = 0$ and $-5\alpha + 8\beta + 2 = 0$.

Solving for α and β, $\dfrac{\alpha}{-10-8} = \dfrac{\beta}{-5-4} = \dfrac{1}{16-25} \Rightarrow \alpha = 2, \beta = 1.$

Thus the pair of bisectors is

$$\frac{(x-2)^2 - (y-1)^2}{1-4} = \frac{(x-2)(y-1)}{-5/2} \quad \text{or} \quad (x-2)^2 - (y-1)^2 = \frac{6}{5}(x-2)(y-1).$$

2.3 Equation of a Pair of Straight Lines passing through the Origin and the Points of Intersection of a Locus represented by a Second Degree Equation with a Straight Line

Let $ax^2 + 2hxy + by^2 + 2gx + 2fy + c = 0$ be the equation of the locus and $lx + my + n = 0$ be the equation of the straight line.

The equation of the line can be written as

$$\frac{lx+my}{-n} = 1. \tag{2.3.1}$$

Let us make the equation of the locus homogeneous in x and y of second degree by the help of (2.3.1) in the following way.

$$ax^2 + 2hxy + by^2 + 2(gx+fy)\left(\frac{lx+my}{-n}\right) + c\left(\frac{lx+my}{-n}\right)^2 = 0. \tag{2.3.2}$$

It is of the form $Ax^2 + 2Hxy + By^2 = 0$.

The equation (2.3.2) obviously represents a pair of straight lines through the origin. Moreover it passes through the points of intersection of the locus and the line. Hence the equation (2.3.2) is the required equation.

NOTE 1. If the lines are coincident, $H^2 - AB = 0$.

NOTE 2. If the lines are perpendicular, $A + B = 0$.

► **Example 10** *Prove that the angle between the lines joining the origin to the intersection of the line* $y = 3x + 2$ *with the curve*

$x^2 + 2xy + 3y^2 + 4x + 8y - 11 = 0$ *is* $\tan^{-1}\frac{2\sqrt{2}}{3}$. **[B.U. '77; C.U. '93]**

The equation of the pair of lines through the origin and the points of intersection of the given line and curve is the homogeneous equation

$$x^2 + 2xy + 3y^2 + (4x + 8y)\frac{y - 3x}{2} - 11\left(\frac{y - 3x}{2}\right)^2 = 0,$$

$$\text{or}\quad 7x^2 - 2xy - y^2 = 0.$$

$$\text{Angle between these lines} = \tan^{-1}\frac{2\sqrt{1+7}}{7-1} = \tan^{-1}\frac{2\sqrt{2}}{3}.$$

Worked Out Examples

1. Show that the two straight lines $x^2(\tan^2\theta + \cos^2\theta) - 2xy\tan\theta + y^2\sin^2\theta = 0$ make with the x-axis angles such that the difference of their tangents is 2.

$$x^2(\tan^2\theta + \cos^2\theta) - 2xy\tan\theta + y^2\sin^2\theta = 0,$$

$$\text{or}\quad y^2 - 2xy\sec\theta\,\text{cosec}\,\theta + x^2(\sec^2\theta + \cot^2\theta) = 0.$$

Let $y^2 - 2xy\sec\theta\,\text{cosec}\,\theta + x^2(\sec^2\theta + \cot^2\theta) = (y - m_1x)(y - m_2x)$.
Comparing the coefficients $m_1 + m_2 = 2\sec\theta\,\text{cosec}\,\theta, m_1m_2 = \sec^2\theta + \cot^2\theta$.
m_1 and m_2 are the tangents of the angles made by the lines with the x-axis.
Now

$$\begin{aligned}(m_1 - m_2)^2 &= (m_1 + m_2)^2 - 4m_1m_2 = 4\sec^2\theta\,\text{cosec}^2\theta - 4(\sec^2\theta + \cot^2\theta)\\ &= 4\{\sec^2\theta(\text{cosec}^2\theta - 1) - \cot^2\theta\} = 4(\sec^2\theta\cot^2\theta - \cot^2\theta)\\ &= 4\cot^2\theta(\sec^2\theta - 1) = 4\cot^2\theta\tan^2\theta = 4.\end{aligned}$$

$$\therefore\quad m_1 - m_2 = 2.$$

2. Show that the equation of the lines through the origin, each of which makes an angle α with the line $y = x$ is $x^2 - 2xy\sec 2\alpha + y^2 = 0$.

The line $y = x$ makes an angle 45° with the x-axis. The line which makes angle α with $y = x$ makes an angle $45° + \alpha$ or $45° - \alpha$ with the x-axis. The equation to the pair is

$$\{y - x\tan(45° + \alpha)\}\{y - x\tan(45° - \alpha)\} = 0$$

$$\text{or}\quad \left\{y - \frac{1+\tan\alpha}{1-\tan\alpha}x\right\}\left\{y - \frac{1-\tan\alpha}{1+\tan\alpha}x\right\} = 0$$

$$\text{or}\quad y^2 - \left(\frac{1+\tan\alpha}{1-\tan\alpha} + \frac{1-\tan\alpha}{1+\tan\alpha}\right)xy + x^2 = 0$$

$$\text{or}\quad y^2 - 2xy\sec 2\alpha + x^2 = 0.$$

3. **Find the equation to the pair of straight lines through the origin, perpendicular to the pair of straight lines given by $2x^2 + 5xy + 2y^2 + 10x + 5y = 0$.**

$$2x^2 + 5xy + 2y^2 + 10x + 5y = 0 \quad \text{or} \quad (2x+y)(x+2y) + 10x + 5y = 0.$$

The lines are parallel to $2x + y = 0$ and $x + 2y = 0$.

Therefore, the lines passing through the origin and perpendicular to the given lines are $x - 2y = 0$ and $2x - y = 0$.

In pair form the required equation is

$$(x - 2y)(2x - y) = 0 \quad \text{or} \quad 2x^2 - 5xy + 2y^2 = 0.$$

4. **Find the values of a and f for which the equation $ax^2 - 12xy + 9y^2 + 20x - 2fy - 11 = 0$ may represent a pair of parallel lines.**

If the equation represents a pair of parallel lines then

$$\begin{vmatrix} a & -6 & 10 \\ -6 & 9 & -f \\ 10 & -f & -11 \end{vmatrix} = 0 \qquad (1)$$

and

$$\begin{vmatrix} a & -6 \\ -7 & 9 \end{vmatrix} = 0. \qquad (2)$$

From (2), $9a - 36 = 0$ or $a = 4$.

From (1),

$$\begin{vmatrix} 4 & -6 & 10 \\ -6 & 9 & -f \\ 10 & -f & -11 \end{vmatrix} = 0, \quad \text{or} \quad \begin{vmatrix} 2 & -3 & 5 \\ -6 & 9 & -f \\ 10 & -f & -11 \end{vmatrix} = 0, \quad \text{or} \quad \begin{vmatrix} 2 & -3 & 5 \\ 0 & 0 & 15-f \\ 10 & -f & 11 \end{vmatrix} = 0,$$

$$\text{or}\quad (15 - f)(-2f + 30) = 0, \quad \text{or} \quad f = 15.$$

$$\therefore\ a = 4, \quad f = 15.$$

5. **Prove that $bx^2 - 2hxy + ay^2 = 0$ represents two straight lines at right angles to the lines represented by $ax^2 + 2hxy + by^2 = 0$. [B.U. '87; N.B. '70; C.U. '97]**

Let $ax^2 + 2hxy + by^2 = (l_1x + m_1y)(l_2x + m_2y)$.

Equating the coefficients, $l_1l_2 = a, l_1m_2 + l_2m_1 = 2h, m_1m_2 = b$.

The lines are $l_1x + m_1y = 0$ and $l_2x + m_2y = 0$.

The lines passing through the origin and perpendicular to the above lines are $m_1x - l_1y = 0$ and $m_2x - l_2y = 0$.

In pair form the equation is

$$(m_1x - l_1y)(m_2x - l_2y) = 0, \text{ or } m_1m_2x^2 - (l_1m_2 + l_2m_1)xy + l_1l_2y^2 = 0,$$

$$\text{or } bx^2 - 2hxy + ay^2 = 0.$$

Hence the result follows.

6. **If pairs of lines $x^2 - 2pxy - y^2 = 0$ and $x^2 - 2qxy - y^2 = 0$ be such that each pair bisects the angles between the other pair, prove that $pq + 1 = 0$.**

[C.U. '65; B.U. '66, '69]

$$x^2 - 2pxy - y^2 = 0 \quad (1)$$

$$x^2 - 2qxy - y^2 = 0 \quad (2)$$

Bisectors of the angles between the lines of (1) are

$$\frac{x^2 - y^2}{1-(-1)} = \frac{xy}{-p} \quad \text{or} \quad x^2 + \frac{2}{p}xy - y^2 = 0.$$

It is identical with (2). $\therefore$ $\frac{2}{p} = -2q$, or $pq + 1 = 0$.

7. **Find the product of the perpendiculars from (x_1, y_1) to the lines represented by $ax^2 + 2hxy + by^2 = 0$.** **[B.H. '66]**

Let $ax^2 + 2hxy + by^2 = (l_1x + m_1y)(l_2x + m_2y)$.

Equating the coefficients $l_1l_2 = a, l_1m_2 + l_2m_1 = 2h, m_1m_2 = b$.

The lines are $l_1x + m_1y = 0$ and $l_2x + m_2y = 0$.

The perpendicular distances from (x_1, y_1) to the lines are

$$\frac{l_1x_1 + m_1y_1}{\sqrt{l_1^2 + m_1^2}} \quad \text{and} \quad \frac{l_2x_1 + m_2y_1}{\sqrt{l_2^2 + m_2^2}}.$$

$$\text{Product of them} = \frac{(l_1x_1 + m_1y_1)(l_2x_1 + m_2y_1)}{\sqrt{l_1^2 + m_1^2} \cdot \sqrt{l_2^2 + m_2^2}}$$

$$= \frac{l_1l_2x_1^2 + (l_1m_2 + l_2m_1)x_1y_1 + m_1m_2y_1^2}{\sqrt{(l_1^2l_2^2 + m_1^2m_2^2 + l_1^2m_2^2 + l_2^2m_1^2)}}$$

$$= \frac{ax_1^2 + 2hx_1y_1 + by_1^2}{\sqrt{\{a^2 + b^2 + (l_1m_2 + l_2m_1)^2 - 2l_1l_2m_1m_2\}}}$$

$$= \frac{ax_1^2 + 2hx_1y_1 + by_1^2}{\sqrt{(a^2 + b^2 + 4h^2 - 2ab)}} = \frac{ax_1^2 + 2hx_1y_1 + by_1^2}{\sqrt{\{(a-b)^2 + 4h^2\}}}.$$

8. **Show that the area of the triangle formed by the lines $ax^2 + 2hxy + by^2 = 0$ and $lx + my + n = 0$ is $\frac{n^2\sqrt{h^2-ab}}{am^2-2hlm+bl^2}$.** **[N.B. '80]**

Let $ax^2 + 2hxy + by^2 = (l_1x + m_1y)(l_2x + m_2y)$.

Equating the coefficients $l_1l_2 = a, l_1m_2 + l_2m_1 = 2h, m_1m_2 = b$.

The triangle is bounded by the lines

$$l_1x + m_1y = 0, \tag{2.3.1}$$

$$l_2x + m_2y = 0 \tag{2.3.2}$$

$$\text{and } lx + my + n = 0. \tag{2.3.3}$$

The lines (1) and (2) meet at $(0, 0)$ and the line (3) meets (1) and (2) at

$$\left(\frac{-nm_1}{lm_1 - ml_1}, \frac{nl_1}{lm_1 - ml_1}\right) \text{ and } \left(\frac{-nm_2}{lm_2 - ml_2}, \frac{nl_2}{lm_2 - ml_2}\right).$$

The area of the triangle whose vertices are the the above points

$$= \frac{1}{2}\left[\frac{-n^2l_2m_1 + n^2l_1m_2}{(lm_1 - ml_1)(lm_2 - ml_2)}\right] = \frac{1}{2}\frac{n^2(l_1m_2 - l_2m_1)}{l^2m_1m_2 - lm(l_1m_2 + l_2m_1) + m^2l_1l_2}$$

$$= \frac{1}{2}\frac{n^2\sqrt{\{(l_1m_2 + l_2m_1)^2 - 4l_1l_2m_1m_2\}}}{bl^2 - 2hlm + am^2} = \frac{n^2\sqrt{h^2 - ab}}{bl^2 - 2hlm + am^2}.$$

9. **Show that the lines $(A^2 - 3B^2)x^2 + 8ABxy + (B^2 - 3A^2)y^2 = 0$ form with the line $Ax + By + C = 0$ an equilateral triangle of area $\frac{C^2}{\sqrt{3(A^2+B^2)}}$.** **[B.H. '63, C.U. '99]**

Let OLM be the triangle formed by the given lines and the equation of LM be $Ax + By + C = 0$. O is the origin. Let the equation of OL or OM be $y = mx$. The triangle OLM will be equilateral, if each of $\angle L$ and $\angle M$ is 60°, i.e., the lines OL and OM make angles 60° and 60° with LM.

$$\therefore \quad \tan(\pm 60°) = \frac{m + \frac{A}{B}}{1 - \frac{mA}{B}}.$$

$$\left(-\frac{A}{B} \text{ is the slop of } LM\right) \quad \text{or} \quad \pm\sqrt{3} = \frac{A + Bm}{B - Am}.$$

$$\therefore \quad 3(B - Am)^2 = (A + Bm)^2.$$

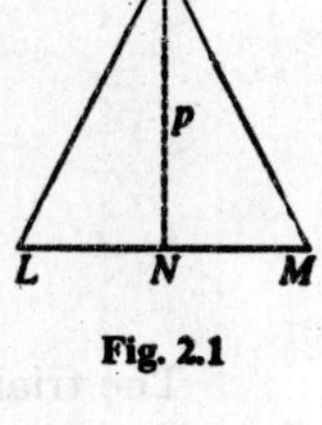

Fig. 2.1

The equation of OL and OM in pair form is

$$3\left(B - A\frac{y}{x}\right)^2 = \left(A + B\frac{y}{x}\right)^2$$

$$\text{or} \quad (A^2 - 3B^2)x^2 + 8ABxy + (B^2 - 3A^2)y^2 = 0.$$

Hence the triangle formed by the given lines is an equilateral.

If p is perpendicular from O on LM, then

$$p = \frac{C}{\sqrt{A^2 + B^2}}.$$

$$\text{The area of the triangle} = \frac{1}{2}LM \cdot p = \frac{1}{2} \cdot 2LN \cdot p$$

$$= p \cdot \tan 30^\circ \cdot p = p^2 \cdot \frac{1}{\sqrt{3}} = \frac{C^2}{\sqrt{3}(A^2 + B^2)}.$$

10. Show that if one of the lines given by equation $ax^2 + 2hxy + by^2 = 0$ be perpendicular to one of the lines given by $a'x^2 + 2h'xy + b'y^2 = 0$, then $(aa' - bb')^2 + 4\,(ah' + hb')(ha' + bh') = 0$. **[B.U. '84; C.U. '97; C.H. '97]**

$$ax^2 + 2hxy + by^2 = 0 \tag{2.3.1}$$

$$a'x^2 + 2h'xy + b'y^2 = 0. \tag{2.3.2}$$

Let $y = mx$ and $y = -\frac{1}{m}x$ be one of the lines of (1) and (2) respectively.

Putting $y = mx$ in (1) and $y = -\frac{1}{m}x$ in (2) we have

$$bm^2 + 2hm + a = 0 \text{ and } a'm^2 - 2h'm + b' = 0.$$

By cross-multiplication

$$\frac{m^2}{2(hb' + h'a)} = \frac{m}{aa' - bb'} = \frac{1}{-2(h'b + ha')}$$

$$\text{or} \quad m = \frac{2(hb' + h'a)}{aa' - bb'} = \frac{aa' - bb'}{-2(h'b + ha')}.$$

$$\therefore \quad (aa' - bb')^2 + 4(ha' + h'b)(h'a + hb') = 0.$$

11. Show that the triangle formed by the lines $ax^2 + 2hxy + by^2 = 0$ and $lx + my = 1$ is right-angled, if $(a + b)(al^2 + 2hlm + bm^2) = 0$. **[N. B. '69, '84, '89]**

Let $ax^2 + 2hxy + by^2 = (l_1x + m_1y)(l_2x + m_2y)$.

Equating the coefficients $l_1l_2 = a, l_1m_2 + l_2m_1 = 2h, m_1m_2 = b$.

Lines are

$$l_1x + m_1y = 0, \tag{2.3.1}$$

$$l_2x + m_2y = 0 \tag{2.3.2}$$

$$\text{and} \quad lx + my = 1. \tag{2.3.3}$$

The triangle will be right-angled, if any two of the above lines are at right-angle.

The lines (1) and (2) will be at right-angle if $l_1l_2 + m_1m_2 = 0$.

The lines (2) and (3) will be at right-angle if $ll_2 + mm_2 = 0$.

The lines (3) and (1) will be at right-angle if $ll_1 + mm_1 = 0$.

Combining these we have

$$(l_1l_2 + m_1m_2)(ll_1 + mm_1)(ll_2 + mm_2) = 0,$$

$$\text{or} \quad (a + b)\{l^2l_1l_2 + lm(l_1m_2 + l_2m_1) + m^2m_1m_2\} = 0,$$

$$\text{or} \quad (a + b)(al^2 + 2hlm + bm^2) = 0.$$

12. If $ax^2 + 2hxy + by^2 = 0$ be the equation of two adjacent sides of a parallelogram and $lx+my = 1$ be the equation of one diagonal, show that the equation of its other diagonal is $y(bl - hm) = x(am - hl)$. [C.U. '98]

Let $ax^2 + 2hxy + by^2 = (l_1x + m_1y)(l_2x + m_2y)$.

Equating the coefficients, $l_1l_2 = a, l_1m_2 + l_2m_1 = 2h, m_1m_2 = b$.

Lines are

$$l_1x + m_1y = 0 \tag{1}$$

$$\text{and } l_2x + m_2y = 0. \tag{2}$$

One diagonal is

$$lx + my - 1 = 0. \tag{3}$$

If (x_1, y_1) is the point of intersection between (1) and (3) then

$$x_1 = \frac{m_1}{lm_1 - ml_1}, \quad y_1 = \frac{-l_1}{lm_1 - ml_1}.$$

If (x_2, y_2) is the point of intersection between (2) and (3) then

$$x_2 = \frac{m_2}{lm_2 - ml_2}, \quad y_2 = \frac{-l_2}{lm_2 - ml_2}.$$

If (x_3, y_3) is the mid-point of the line segment joining the points (x_1, y_1) and (x_2, y_2) then

$$x_3 = \frac{x_1 + x_2}{2} = \frac{1}{2}\left(\frac{m_1}{lm_1 - ml_1} + \frac{m_2}{lm_2 - ml_2}\right)$$

$$= \frac{1}{2}\frac{m_1(lm_2 - ml_2) + m_2(lm_1 - ml_1)}{(lm_1 - ml_1)(lm_2 - ml_2)} = \frac{bl - hm}{lb^2 - 2lmh + ma^2}$$

$$\text{and } y_3 = \frac{1}{2}\left(\frac{-l_1}{lm_1 - ml_1} + \frac{-l_2}{lm_2 - ml_2}\right) = \frac{am - hl}{lb^2 - 2lmh + ma^2}.$$

Since the given adjacent sides meet at (0, 0), the other diagonal passes through the origin and the point (x_3, y_3). Hence the equation of the other diagonal is $yx_3 = xy_3$ i.e. $y(bl - hm) = x(am - hl)$.

13. Show that the orthocentre of the triangle formed by the lines $ax^2 + 2hxy + by^2 = 0$ and $lx + my = 1$ is given by $\frac{x}{l} = \frac{y}{m} = \frac{a+b}{am^2 - 2hlm + bl^2}$. [B.H. '69; C.H. 2000]

Let $ax^2 + 2hxy + by^2 = (l_1x + m_1y)(l_2x + m_2y)$.

Equating the coefficient $l_1l_2 = a, l_1m_2 + l_2m_1 = 2h, m_1m_2 = b$.

The lines are

$$l_1x + m_1y = 0, \tag{1}$$

$$l_2x + m_2y = 0 \tag{2}$$

$$\text{and } lx + my = 1. \tag{3}$$

The lines (1) and (2) meet at $(0, 0)$ and the line (3) meets (1) at

$$\left(\frac{m_1}{lm_1 - ml_1}, \frac{-l_1}{lm_1 - ml_1}\right).$$

The equation of the line perpendicular to (3) and passing through the point of intersection between (1) and (2) [i.e., (0,0)] is $mx - ly = 0$

$$\text{or} \quad \frac{x}{l} = \frac{y}{m} = k \text{ (say)}. \tag{2.3.4}$$

The orthocentre lines on (4). Let the orthocentre be (kl, km).

The line passing through (kl, km) and $\left(\frac{m_1}{lm_1 - ml_1}, \frac{-l_1}{lm_1 - ml_1}\right)$ is perpendicular to (2). The slope of this line is

$$\frac{km + \frac{l_1}{lm_1 - ml_1}}{kl - \frac{m_1}{lm_1 - ml_1}} \text{ and that of (2) is } -\frac{l_2}{m_2}.$$

As these two lines are at right angle

$$\frac{l_2}{m_2} \cdot \frac{km + \frac{l_1}{lm_1 - ml_1}}{kl - \frac{m_1}{lm_1 - ml_1}} = 1,$$

$$\text{or} \quad \{km(lm_1 - ml_1) + l_1\}l_2 - \{kl(lm_1 - ml_1) - m_1\}m_2 = 0,$$

$$\text{or} \quad k\{lm(l_1m_2 + l_2m_1) - m^2l_1l_2 - l^2m_1m_2\} + l_1l_2 + m_1m_2 = 0,$$

$$\text{or} \quad k\{lm.2h - m^2.a - l^2.b\} + a + b = 0,$$

$$\text{or} \quad k = \frac{a+b}{am^2 - 2hlm + bl^2}. \quad \therefore \quad \frac{x}{l} = \frac{y}{m} = \frac{a+b}{am^2 - 2hlm + bl^2}.$$

14. Prove that $ax^2 + 2hxy + by^2 + 2gx + 2fy + c = 0$ represents two parallel straight lines if $h^2 = ab$ and $bg^2 = af^2$. Also show that the distance between the lines is $2\sqrt{\frac{g^2-ac}{a(a+b)}}$.

[C.U. '2000; C.H. '95; N.B. '84, '89]

Let $lx + my + n_1 = 0$ and $lx + my + n_2 = 0$ be the lines represented by the given equation and $ax^2 + 2hxy + by^2 + 2gx + 2fy + c = \lambda(lx + my + n_1)(lx + my + n_2)$, where λ is a suitable constant.

Equating the coefficients, $\lambda l^2 = a, \lambda lm = h, \lambda m^2 = b,$

$$\lambda l(n_1 + n_2) = 2g, \lambda m(n_1 + n_2) = 2f, \lambda n_1n_2 = c.$$

Now

$$\frac{a}{h} = \frac{\lambda l^2}{\lambda lm} = \frac{l}{m} = \frac{\lambda lm}{\lambda m^2} = \frac{h}{b} \quad \text{or} \quad h^2 = ab.$$

Again

$$\frac{a}{b} = \frac{\lambda l^2}{\lambda m^2} = \frac{\lambda^2l^2(n_1 + n_2)^2}{\lambda^2m^2(n_1 + n_2)^2} = \frac{g^2}{f^2} \quad \text{or} \quad af^2 = bg^2.$$

The distances of the lines from (0,0) are $\frac{n_1}{\sqrt{l^2+m^2}}$ and $\frac{n_2}{\sqrt{l^2+m^2}}$.

$\therefore$ Distance between the lines

$$= \frac{n_1 \sim n_2}{\sqrt{l^2+m^2}} = \sqrt{\frac{(n_1+n_2)^2 - 4n_1n_2}{l^2+m^2}} = \sqrt{\left\{\frac{\lambda^2l^2(n_1+n_2)^2 - 4\lambda^2l^2n_1n_2}{\lambda^2l^2(l^2+m^2)}\right\}}$$

$$= \sqrt{\frac{4g^2 - 4ac}{a(a+b)}} = 2\sqrt{\frac{g^2 - ac}{a(a+b)}}.$$

15. If $ax^2+2hxy+by^2+2gx+2fy+c=0$ represents two lines equidistant from the origin, show that $f^4-g^4=c(bf^2-ag^2)$. [C.U. '60, 62; C.H. '92; B.H. '62]

Let $ax^2+2hxy+by^2+2gx+2fy+c=(l_1x+m_1y+n_1)(l_2x+m_2y+n_2)$.

Comparing the coefficients

$$l_1l_2=a, m_1m_2=b, l_1m_2+l_2m_1=2h, l_1n_2+l_2n_1=2g,$$

$$m_1n_2+m_2n_1=2f, n_1n_2=c.$$

The lines are $l_1x+m_1y+n_1=0$ and $l_2x+m_2y+n_2=0$.

Since the lines are equidistant from the origin,

$$\frac{n_1}{\sqrt{l_1^2+m_1^2}}=\frac{n_2}{\sqrt{l_2^2+m_2^2}},$$

or $n_1^2(l_2^2+m_2^2)=n_2^2(l_1^2+m_1^2)$,

or $n_1^2l_2^2-n_2^2l_1^2=n_2^2m_1^2-n_1^2m_2^2$,

or $(n_1l_2+n_2l_1)(n_1l_2-n_2l_1)=(n_2m_1+n_1m_2)(n_2m_1-n_1m_2)$,

or $2g(n_1l_2-n_2l_1)=2f(n_2m_1-n_1m_2)$,

or $g^2(n_1l_2-n_2l_1)^2=f^2(n_2m_1-n_1m_2)^2$,

or $g^2\{(n_1l_2+n_2l_1)^2-4n_1n_2l_1l_2\}=f^2\{(n_2m_1+n_1m_2)^2-4n_1n_2m_1m_2\}$,

or $g^2(4g^2-4ca)=f^2(4f^2-4bc)$,

or $f^4-g^4=c(bf^2-ag^2)$.

16. If the equation $ax^2+2hxy+by^2+2gx+2fy+c=0$ represents a pair of straight lines, prove that the equation to the third pair of straight lines passing through the points, where these meet the axes is

$$ax^2-2hxy+by^2+2gx+2fy+c+\frac{4fg}{c}xy=0.$$

[C.U. '62; C.H. '62; B.H. '64]

$$ax^2+2hxy+by^2+2gx+2fy+c=0. \quad (1)$$

Combined equation of the axes is

$$xy=0. \quad (2)$$

The equation of a curve through the points of intersection between (1) and (2) is $ax^2+2hxy+by^2+2gx+2fy+c+2\lambda xy=0$, where λ is a constant. If it represents a pair of straight lines, then

$$abc+2fg(h+\lambda)-af^2-bg^2-c(h+\lambda)^2=0. \quad (3)$$

As the equation (1) represents a pair of straight lines,

$$abc+2fgh-af^2-bg^2-ch^2=0. \quad (4)$$

By (3) and (4) $\lambda=\frac{2(fg-ch)}{c}$.

Hence the required equation is

$$ax^2-2hxy+by^2+2gx+2fy+c+\frac{4fg}{c}xy=0.$$

17. Show that the lines joining the origin to the points of intersection of the curves $ax^2 + 2hxy + by^2 + 2gx = 0$ and $a'x^2 + 2h'xy + b'y^2 + 2g'x = 0$ will be at right angle if $g(a' + b') = g'(a + b)$. **[B.H. '66]**

$$ax^2 + 2hxy + by^2 + 2gx = 0, \tag{1}$$

$$a'x^2 + 2h'xy + b'y^2 + 2g'x = 0. \tag{2}$$

Multiplying (1) and (2) by g' and g respectively and then subtracting we have

$$(ag' - a'g)x^2 + 2(hg' - h'g)xy + (bg' - b'g)y^2 = 0.$$

This homogeneous equation in x and y is the required equation.

The straight lines will be at right angle if coefficient of x^2+ coefficient of $y^2 = 0$, i.e., $(ag' - a'g) + (bg' - b'g) = 0$ or $g'(a + b) = g(a' + b')$.

18. Find the equations of the lines $x^2 + 8\sqrt{2}xy + 5y^2 = 0$ referred to the bisectors of the angles between them as axes of co-ordinates.

$$x^2 + 8\sqrt{2}xy + 5y^2 = 0. \tag{1}$$

The equation of the bisectors is

$$y^2 - x^2 - \frac{1}{\sqrt{2}}xy = 0. \tag{2}$$

Let the lines make angles α and $90° + \alpha$ with the x-axis. The equation of the lines in pair form is

$$(y - \tan\alpha \cdot x)(y + \cot\alpha \cdot x) = 0 \quad \text{or} \quad y^2 - x^2 - (\tan\alpha - \cot\alpha)xy = 0.$$

Comparing it with (2), $\tan\alpha - \cot\alpha = \frac{1}{\sqrt{2}}$ or $\sqrt{2}\tan^2\alpha - \tan\alpha - \sqrt{2} = 0$ or $\tan\alpha = \sqrt{2}, -\frac{1}{\sqrt{2}}$.

Taking

$$\tan\alpha = \sqrt{2}, \quad \sin\alpha = \frac{\sqrt{2}}{\sqrt{3}}, \quad \cos\alpha = \frac{1}{\sqrt{3}}.$$

For rotation of axes through angle α, the equation (1) transforms to

$$(x\cos\alpha - y\sin\alpha)^2 + 8\sqrt{2}(x\cos\alpha - y\sin\alpha)(x\sin\alpha + y\cos\alpha)$$
$$+5(x\sin\alpha + y\cos\alpha)^2 = 0,$$

$$\text{or} \quad (x - \sqrt{2}y)^2 + 8\sqrt{2}(x - \sqrt{2}y)(\sqrt{2}x + y) + 5(\sqrt{2}x + y)^2 = 0,$$

$$\text{or} \quad 27x^2 - 9y^2 = 0, \quad \text{or} \quad 3x^2 - y^2 = 0.$$

Taking

$$\tan\alpha = -\frac{1}{\sqrt{2}}, \quad \cos\alpha = -\frac{\sqrt{2}}{\sqrt{3}}, \quad \sin\alpha = \frac{1}{\sqrt{3}}.$$

For rotation of axes through this angle, the equation (1) will reduce to

$$x^2 - 3y^2 = 0.$$

Exercise

1. Find the angle between the pair of lines

   (a) $x^2 + 2hxy - y^2 = 0$;
   (b) $4x^2 - 24xy + 11y^2 = 0$;
   (c) $x^2 + 2xy \operatorname{cosec} \alpha + y^2 = 0$;
   (d) $(x^2 + y^2) \sin^2 \alpha = (x \cos \theta - y \sin \theta)^2$;
   (e) $(x^2 + y^2)(\cos^2 \theta \sin^2 \alpha + \sin^2 \theta) = (x \tan \alpha - y \sin \theta)^2$;
   (f) $y^2(\cos \alpha + \sqrt{3} \sin \alpha) \cos \alpha - xy(\sin 2\alpha - \sqrt{3} \cos 2\alpha) + x^2(\sin \alpha - \sqrt{3} \cos \alpha) \sin \alpha = 0$.

2. Show that the equation of the lines passing through the origin and perpendicular to $5x^2 - 7xy - 3y^2 = 0$ is $3x^2 - 7xy - 5y^2 = 0$.

3. Show that the lines $a^2x^2 + 2h(a + b)xy + b^2y^2 = 0$ are equally inclined to the lines $ax^2 + 2hxy + by^2 = 0$.

4. Prove that the straight lines $ax^2 + 2hxy + by^2 + \lambda(x^2 + y^2) = 0$ have the same pair of bisectors. Interpret the case when $\lambda = -(a + b)$.

5. Show that if one of the lines given by the equation $ax^2 + 2hxy + by^2 = 0$ coincides with one of those given by $a'x^2 + 2h'xy + b'y^2 = 0$, then $(ab' - a'b)^2 = 4(ha' - h'a)(bh' - b'h)$. **[N.B. '97]**

6. Prove that the equation $y^3 - x^3 + 3xy(y - x) = 0$ represents three straight lines equally inclined to one another.
   [HINTS. $y^3 - x^3 + 3xy(y - x) = (y - x)(y^2 + x^2 + 4xy) = (y - x)\{(y + 2x)^2 - 3x^2\}$
   $= (y - x)\{y + (2 - \sqrt{3})x\}\{y + (2 + \sqrt{3})x\}$.]

7. Show that the following equations represent pair of straight lines and find the points of intersections and angles between them.

   (a) $8x^2 + 10xy + 3y^2 + 26x + 16y + 21 = 0$; **[B.U. '67, '77]**
   (b) $12x^2 + 7xy - 10y^2 + 13x + 45y - 35 = 0$;
   (c) $2x^2 - 7xy + 3y^2 + 26x - 33y + 72 = 0$.

8. Find the value of k so that the following equations may represent pair of straight lines.

   (a) $kx^2 - 3xy - 2y^2 + x + 13y - 15 = 0$;
   (b) $x^2 + kxy - 2y^2 + 3y - 1 = 0$; **[B.U. '68]**
   (c) $xy + 5x + ky + 15 = 0$; **[C.U. '99]**
   (d) $x^2 + \frac{10}{3}xy + y^2 - 5x - 7y + k = 0$;
   (e) $x^2 + 6xy + 9y^2 + kx + 12y - 5 = 0$. **[N.B. '82]**

9. Show that the following equations represent pair of parallel lines and find the distance between them.

   (a) $x^2 + 6xy + 9y^2 + 4x + 12y - 5 = 0$; **[N.B. '98]**
   (b) $4x^2 + 12xy + 9y^2 + 2x + 3y - 42 = 0$.

10. Show that $2x^2 + 3xy - 2y^2 + 7x - y + 3 = 0$ represents a pair of perpendicular lines.

11. Show that

 (a) $x^2 + 6xy + 9y^2 + 4x + 12y + 4 = 0$ and

 (b) $4x^2 + 4xy + y^2 + 4x + 2y + 1 = 0$ represent coincident lines.

12. Find the equation of the lines passing through $(2, 3)$ and perpendicular to the lines $2x^2 + xy - 3y^2 + 3x + 2y + 1 = 0$.

13. The equations to the pair of opposite sides of a rectangle are $x^2 - 7x + 6 = 0$ and $y^2 - 14y + 40 = 0$. Find the equation to its diagonals.

14. Prove that the equations to the straight lines passing through the origin which make an angle θ with the straight line $y + x = 0$ are given by the equation $x^2 + 2xy \sec 2\theta + y^2 = 0$. **[N.B. '85]**

15. Show that the lines $x + y + 1 = 0$ and $(x + y)^2 - 3(x - y)^2 = 0$ form an equilateral triangle.

16. If the pair of lines $ax^2 + 2hxy - ay^2 = 0$ and $bx^2 + 2gxy - by^2 = 0$ be such that each bisects the angles between the other then $hg + ab = 0$.

17. Show that the lines joining the origin to the points common to $x^2 + hxy - y^2 + gx + fy = 0$ and $fx - gy = \lambda$ are at right angle whatever λ may be.

18. Find the angle between the chords of the circle $x^2 + y^2 = ax + by$, obtained by joining the origin to the points of intersection of the straight line $\frac{x}{a} + \frac{x}{b} = 1$ and the above circle. **[C.U. '96]**

19. Find the angle between the lines joining the origin to the points common to $5x^2 + 12xy - 8y^2 + 8x - 4y + 12 = 0$ and $x - y = 2$. **[C.U. '91]**

20. If the angle between the lines joining the origin to the points of the intersection of $lx + my = 1$ and $x^2 + y^2 = a^2$ be $45°$, show that

$$4[a^2(l^2 + m^2) - 1] = [a^2(l^2 + m^2) - 2]^2.$$

21. Show that the distance from the origin to the orthocentre of the triangle formed by the lines $\frac{x}{\alpha} + \frac{y}{\beta} = 1$ and $ax^2 + 2hxy + by^2 = 0$ is

$$\frac{(a + b)\alpha\beta(\alpha^2 + \beta^2)^{1/2}}{a\alpha^2 - 2h\alpha\beta + b\beta^2}.$$

22. If one of the lines given by the equation $ax^2 + 2hxy + by^2 = 0$ coincides with one of the lines given by $a'x^2 + 2h'xy + b'y^2 = 0$ and the other lines are perpendicular, then

$$\frac{ha'b'}{b' - a'} = \frac{h'ab}{b - a} = \frac{1}{2}\sqrt{(-aa'bb')}.$$

Answers

1 (a) $\frac{\pi}{2}$, (b) $\tan^{-1}\frac{4}{3}$, (c) $\frac{\pi}{2} - \alpha$, (d) 2α (e) 2α (f) $\frac{\pi}{3}$. **7.** (a) $(-1, -1)$, $\tan^{-1}\frac{2}{11}$, (b) $(-\frac{25}{23}, \frac{43}{23})$, $\tan^{-1}\frac{23}{2}$, (c) $(-3, 2)$, $\frac{\pi}{4}$. **8.** (a) 2 (b) ± 1, (c) 3, (d) 6, (e) 4. **9.** (a) $\frac{6}{\sqrt{10}}$, (b) $\sqrt{13}$. **12.** $3x^2 + xy - 2y^2 - 15x + 10y = 0$. **13.** $6x - 5y + 14 = 0, 6x + 5y - 56 = 0$. **18.** $\frac{\pi}{2}$. **19.** $\tan^{-1}\frac{4}{3}$.

Chapter 3

General Equation of Second Degree

3.1 Introduction

The general equation of second degree in x and y is usually written in the form

$$ax^2 + 2hxy + by^2 + 2gx + 2fy + c = 0. \tag{3.1.1}$$

The curve represented by this equation is a *conic section* or simply a *conic*. The curve is also called a *second order curve*. The nature of the conic is determined by the quantities

$$\Delta = \begin{vmatrix} a & h & g \\ h & b & f \\ g & f & c \end{vmatrix}, \quad D = ab - h^2,\ P = a + b.$$

In case of rectangular co-ordinate axes the quantities Δ, D and P are invariants under any orthogonal transformation.

(i) If $\Delta = 0$, the equation (1) represents a pair of lines, which has been discussed earlier.

(ii) If $a = b$ and $h = 0$, the equation represents a circle.

(iii) If $\Delta \neq 0$, the equation represents a *proper conic*. It will be seen that the values of the quantity $D = ab - h^2$ determines the nature of the conic.

(a) When $D = 0$, i.e., $ab = h^2$, the conic is a parabola. In this case the second degree terms form a perfect square;

(b) When $D > 0$, i.e., $ab > h^2$, the conic is an ellipse.

(c) When $D < 0$, i.e., $ab < h^2$, the conic is a hyperbola. If $a + b = 0$, the conic is a rectangular hyperbola.

Conditions for proper conic

Let LM be the directrix, $S(\alpha, \beta)$ be the focus and $P(x, y)$ be any point on the conic whose eccentricity is e. The equation of LM is $lx+my+n=0$. If PM is perpendicular to the directrix then by the definition of a conic,

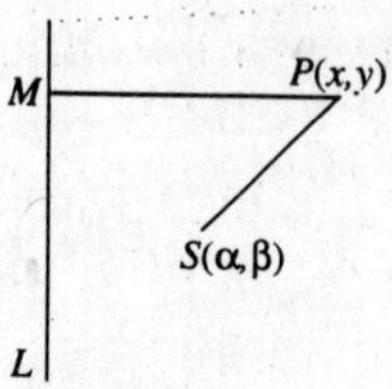

Fig. 3.1

$$SP^2 = e^2 PM^2,$$

$$\text{or } (x-\alpha)^2 + (y-\beta)^2 = e^2 \frac{(lx+my+n)^2}{l^2+m^2},$$

$$\text{or } \{l^2(1-e^2)+m^2\}x^2 - 2lme^2xy + \{l^2+m^2(1-e^2)\}y^2$$

$$-2\{(l^2+m^2)\alpha + lne^2\}x - 2\{(l^2+m^2)\beta + mne^2\}y$$

$$+(l^2+m^2)(\alpha^2+\beta^2) - n^2e^2 = 0.$$

If this equation represents the equation (1) then we can write that

$$a = l^2(1-e^2)+m^2, \qquad b = l^2+m^2(1-e^2),$$
$$h = -lme^2, \qquad g = -\{(l^2+m^2)\alpha + lne^2\},$$
$$f = -\{(l^2+m^2)\beta + mne^2\}, \; c = (l^2+m^2)(\alpha^2+\beta^2) - n^2e^2.$$

Now

$$D = ab - h^2 = \{l^2(1-e^2)+m^2\}\{l^2+m^2(1-e^2)\} - l^2m^2e^4$$
$$= (l^2+m^2)^2(1-e^2).$$

For the parabola, $e = 1$, i.e., $D = 0$.

For the ellipse, $e < 1$, i.e., $D > 0$.

For the hyperbola, $e > 1$, i.e., $D < 0$.

3.2 Canonical Form

The general equation of a second degree curve can be reduced to the standard equation of a conic by suitable transformation of co-ordinates. This standard equation is also called the *canonical equation* or *normal/canonical form* of the equation.

To find the canonical form from the general equation the following transformations are made successively.

(i) The term in xy is removed by suitable rotation of axes;

(ii) One or both (when possible) the terms in x and y are removed by translation.

(iii) The constant term is removed if possible.

3.3 Reduction to Canonical Form (Tracing of Conics)

Let the axes be rotated through an angle θ. The new co-ordinates are related as

$$x = x'\cos\theta - y'\sin\theta \text{ and } y = x'\sin\theta + y'\cos\theta.$$

Substituting these values of x and y in the equation (3.1.1) of section 3.1 we have

$$\begin{aligned}
&a(x'\cos\theta - y'\sin\theta)^2 + 2h(x'\cos\theta - y'\sin\theta)(x'\sin\theta + +y'\cos\theta)\\
&+b(x'\sin\theta + y'\cos\theta)^2 + 2g(x'\cos\theta - y'\sin\theta)\\
&+2f(x'\sin\theta + y'\cos\theta) + c = 0,\\
\text{or}\quad &(a\cos^2\theta + 2h\sin\theta\cos\theta + b\sin^2\theta)x'^2 + 2\{h(\cos^2\theta - \sin^2\theta)\\
&-(a-b)\sin\theta\cos\theta\}x'y' + (a\sin^2\theta - 2h\sin\theta\cos\theta + b\cos^2\theta)y'^2\\
&+2(g\cos\theta + f\sin\theta)x' + 2(f\cos\theta - g\sin\theta)y' + c = 0. \qquad (3.3.1)
\end{aligned}$$

Let us choose θ in such a way that the coefficient of $x'y'$ in (3.3.1) will be zero. To satisfy this condition, we have

$$h(\cos^2\theta - \sin^2\theta) = (a-b)\sin\theta\cos\theta,$$

$$\text{or}\quad \tan 2\theta = \frac{2h}{a-b}, \quad \text{i.e.,}\quad \theta = \frac{1}{2}\tan^{-1}\frac{2h}{a-b}.$$

For this value of θ the equation (3.3.1) will be of the form

$$Ax'^2 + By'^2 + 2Gx' + 2Fy' + c = 0. \qquad (3.3.2)$$

By the property of invariants,

$$\Delta = \begin{vmatrix} A & 0 & G \\ 0 & B & F \\ G & F & C \end{vmatrix}, \quad D = AB \text{ and } a + b = A + B.$$

(I) If $\Delta \neq 0$ but $D = 0$ then the equation (3.1.1) represents a parabola.

$$D = 0 \Rightarrow AB = 0.$$

There are three possibilities.

(i) $A = 0$, $B = 0$; (ii) $A = 0$, $B \neq 0$; (iii) $A \neq 0$, $B = 0$.

If $A = 0$ and $B = 0$, then $\Delta = 0$. Possibility (i) is thus ruled out.

For the possibility (ii) $\Delta \neq 0$, if $G \neq 0$.

In this case the equation (3.3.2) reduces to

$$By'^2 + 2Gx' + 2Fy' + c = 0. \qquad (3.3.3)$$

It is a parabola having its axis parallel to the new x-axis.

From (3.3.3),

$$B\left(y'^2 + \frac{2F}{B}y'\right) = -2Gx' - c, \quad \text{or} \quad \left(y' + \frac{F}{B}\right)^2 = -\frac{2G}{B}\left(x' + \frac{Bc - F^2}{2GB}\right).$$

Changing the origin to $\left(-\dfrac{Bc - F^2}{2GB}, -\dfrac{F}{B}\right)$, the equation reduces to

$$y''^2 = -\frac{2G}{B}x''. \tag{3.3.4}$$

It is the *canonical form* of the equation (1) when $\Delta \neq 0$ but $D = 0$.

For the possibility (iii) $\Delta \neq 0$, if $F \neq 0$.

In this case, the equation (3.3.2) reduces to

$$Ax'^2 + 2Gx' + 2Fy' + c = 0. \tag{3.3.5}$$

It is a parabola having its axis parallel to the y' axis, i.e., new y-axis.

From (6),

$$x'^2 + \frac{2G}{A}x' = -\frac{2F}{A}y' - \frac{c}{A}, \quad \text{or} \quad \left(x' + \frac{G}{A}\right)^2 = -\frac{2F}{A}\left(y' + \frac{cA - G^2}{2FA}\right).$$

Changing the origin to $\left(-\dfrac{G}{A}, -\dfrac{cA - G^2}{2FA}\right)$, the equation raduces to

$$x''^2 = -\frac{2F}{A}y''. \tag{3.3.6}$$

It is the *canonical equation* of the equation (3.1.1).

NOTE 1. Since $h^2 \doteq ab$, the terms of second degree in the equation (3.1.1) form a perfect square.

NOTE 2. If $A = 0 = B$ then $a = 0 = b = h$ and the equation represents a line.

NOTE 3. If $A = 0 = G$ but $B \neq 0$ then the equation (3.3.2) reduces to

$$By'^2 + 2Fy' + c = 0, \quad \text{or} \quad \left(y' + \frac{F}{B}\right)^2 = \frac{F^2 - Bc}{B^2}.$$

Here the equation (3.1.1) represents a pair of parallel lines, a pair of coincident lines or no geometric locus according as $F^2 - Bc >=< 0$.

NOTE 4. If $B = 0 = F$ but $A \neq 0$ then the equation (3.3.2) reduces to

$$Ax'^2 + 2Gx' + c = 0, \quad \text{or} \quad \left(x' + \frac{G}{A}\right)^2 = \frac{G^2 - cA}{A^2}.$$

Here the equation (3.1.1) represents a pair of parallel lines, a pair of coincident lines or no geometric locus according as $G^2 - cA >=< 0$.

▶ **Example 1** *Discuss the nature of the conic represented by*

$$9x^2 - 24xy + 16y^2 - 18x - 101y + 19 = 0$$

and reduce it to the canonical form.

Here $a = 9$, $h = -12$, $b = 16$, $g = -9$, $f = -\frac{101}{2}$, $c = 19$,

$$\Delta = \begin{vmatrix} 9 & -12 & -9 \\ -12 & 16 & -(101/2) \\ -9 & -(101/2) & 19 \end{vmatrix} \neq 0, \quad D = 9.16 - (-12)^2 = 0.$$

Therefore, the given equation represents a parabola.

Let the axes be rotated through an acute angle θ. The equation transforms to

$$9(x'\cos\theta - y'\sin\theta)^2 - 24(x'\cos\theta - y'\sin\theta)(x'\sin\theta + y'\cos\theta)$$
$$+16(x'\sin\theta + y'\cos\theta)^2 - 18(x'\cos\theta - y'\sin\theta)$$
$$-101(x'\sin\theta + y'\cos\theta) + 19 = 0,$$
or $$(9\cos^2\theta - 24\sin\theta\cos\theta + 16\sin^2\theta)x'^2$$
$$2\{12(\cos^2\theta - \sin^2\theta) - 7\sin\theta\cos\theta\}x'y'$$
$$+(16\cos^2\theta + 24\sin\theta\cos\theta + 9\sin^2\theta)y'^2$$
$$-(18\cos\theta + 101\sin\theta)x' + (18\sin\theta - 101\cos\theta)y' + 19 = 0,$$
or $$(3\cos\theta - 4\sin\theta)^2x'^2 - 2\{12(\cos^2\theta - \sin^2\theta) - 7\sin\theta\cos\theta\}x'y'$$
$$+(4\cos\theta + 3\sin\theta)^2y'^2 - (18\cos\theta + 101\sin\theta)x'$$
$$+(18\sin\theta - 101\cos\theta)y' + 19 = 0.$$

Let us choose θ in such a way that

$$12(\cos^2\theta - \sin^2\theta) - 7\sin\theta\cos\theta = 0.$$

From this $12\tan^2\theta + 7\tan\theta - 12 = 0$, or $\tan\theta = \frac{3}{4}, -\frac{4}{3}$.

Since θ is acute, $\tan\theta = \frac{3}{4}$. Hence $\sin\theta = \frac{3}{5}$ and $\cos\theta = \frac{4}{5}$.

For these values of $\cos\theta$ and $\sin\theta$ the equation reduces to

$$25y'^2 - 75x' - 70y' + 19 = 0,$$

or $$\left(y' - \frac{7}{5}\right)^2 = 3\left(x' + \frac{2}{5}\right).$$

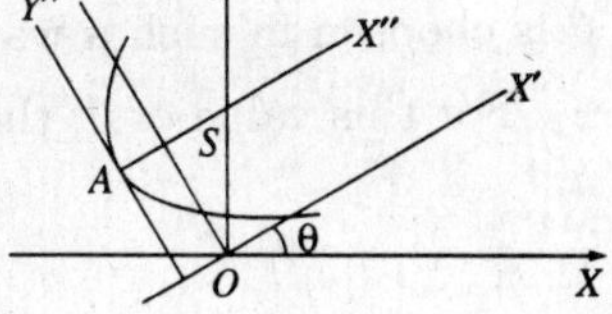

Fig. 3.2

Changing the origin to $\left(-\frac{2}{5}, \frac{7}{5}\right)$, the equation further reduces to $y''^2 = 3x''$. It is the required canonical form.

The parabola can be traced now as shown in Fig. 3.2. In the figure, $\theta = \tan^{-13}/4$, vertex A is at $\left(-\frac{2}{5}, \frac{7}{5}\right)$ w.r.t. OX', OY' and $AS = 3/4$. S being the focus of the parabola.

(II) The equation (3.1.1) represents an ellipse when $\Delta \neq 0$ but $D > 0$.

We have $D = AB$. If $D > 0$, none of A and B is zero and both of them are positive or negative. Without any loss of generality we may assume that both of A and B are positive.

From (3.3.2) $A\left(x' + \frac{G}{A}\right)^2 + B\left(y' + \frac{F}{B}\right)^2 = \frac{G^2}{A} + \frac{F^2}{B} - c = k$ (say).

By translation $x' = x'' - \frac{G}{A}, y' = y'' - \frac{F}{B}$, the equation reduces to

$$Ax''^2 + By''^2 = k, \quad \text{or} \quad \frac{x''^2}{k/A} + \frac{y''^2}{k/B} = 1. \tag{3.3.7}$$

It is the equation of an ellipse in the canonical form.

NOTE 1. The centre of the ellipse represented by the equation (3.1.1) is

$$\left(-\frac{G}{A}\cos\theta + \frac{F}{B}\sin\theta, \quad -\frac{G}{A}\sin\theta - \frac{F}{B}\cos\theta\right).$$

NOTE 2. The equation (8) represents a real ellipse, a point (null) ellipse or an imaginary (without any real trace) ellipse according as $k >=< 0$.

NOTE 3. $\Delta = -ABk$. It is $<=> 0$ according as the equation (8) represents a real ellipse, a point ellipse or an imaginary ellipse.

► **Example 2** *Reduce the equation* $3x^2 + 2xy + 3y^2 - 16x + 20 = 0$ *to normal form.*

[B.U. '82, '88; N.B. '82]

Here $a = 3,\ h = 1,\ b = 3,\ g = -8,\ f = 0,\ c = 20$,

$$\Delta = \begin{vmatrix} 3 & 1 & -8 \\ 1 & 3 & 0 \\ -8 & 0 & 20 \end{vmatrix} = -32 \text{ and } D = 3.3 - 1^2 = 8 > 0.$$

Therefore, the given equation represents an ellipse.

By rotating the axes through an acute angle θ, the equation transforms to

$$3(x'\cos\theta - y'\sin\theta)^2 + 2(x'\cos\theta - y'\sin\theta)(x'\sin\theta + y'\cos\theta)$$
$$+3(x'\sin\theta + y'\cos\theta)^2 - 16(x'\sin\theta - y'\cos\theta) + 20 = 0,$$

or $\quad (3 + \sin 2\theta)x'^2 + 2\cos 2\theta x'y' + (3 - \sin 2\theta)y'^2 - 16\sin\theta x' + 16\cos\theta y' + 20 = 0.$

θ is choosen in such a way that $\cos 2\theta = 0$, or $2\theta = 90°$, or $\theta = 45°$.

For this value of θ, the equation takes the form

$$4x'^2 + 2y'^2 - 8\sqrt{2}x' + 8\sqrt{2}y' + 20 = 0,$$

$$\text{or} \quad 4(x' - \sqrt{2})^2 + 2(y' + 2\sqrt{2})^2 = 4.$$

Changing the origin to $(\sqrt{2}, -2\sqrt{2})$, the equation reduces to

$$4x''^2 + 2y''^2 = 4,$$

$$\text{or} \quad x''^2 + y''^2/2 = 1.$$

It is the required normal form. The conic is an ellipse with semiaxes 1 and $\sqrt{2}$.

The ellipse can be traced now as shown in fig. 3.3. In the figure $\theta = 45°$, the centre C is at $(\sqrt{2}. - 2\sqrt{2})$ w.r.t. OX', OY'.

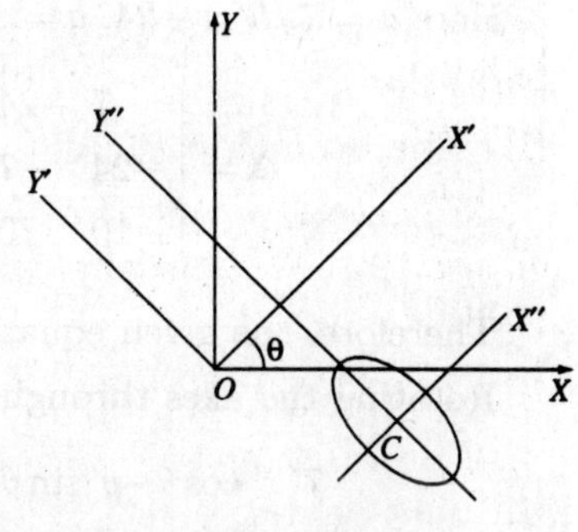

Fig. 3.3

(III) **The equation (3.1.1) represents a hayperbola when $\Delta \neq 0$ but $D < 0$.**

If $D < 0$ then $AB < 0$. Consequently none of A and B is zero.

Moreover A and B have opposite signs. Without any loss of generality we may assume that $A > 0$ and $B < 0$. Proceeding as (II) the equation (3.3.2) reduces to

$$\frac{x''^2}{k/A} + \frac{y''^2}{k/B} = 1 \quad \text{when } k \text{ is not zero.}$$

If $k > 0$ then it can be written as

$$\frac{x''^2}{\alpha^2} - \frac{y''^2}{\beta^2} = 1. \tag{3.3.8}$$

It is the equation of a hyperbola in the canonical form.

If $k < 0$ then the equation can be written as

$$\frac{x''^2}{\alpha^2} - \frac{y''^2}{\beta^2} = -1. \tag{3.3.9}$$

It is also the canonical equation of a hyperbola.

NOTE 1. Hyperbolas represented by (9) and (10) are conjugate to each other.

NOTE 2. The centre of the hyperbola represented by the equation (3.1.1) is

$$\left(-\frac{G}{A}\cos\theta + \frac{F}{B}\sin\theta, \quad -\frac{G}{A}\sin\theta - \frac{F}{B}\cos\theta\right).$$

NOTE 3. If $k = 0$, the equation (3.3.2) reduces to $Ax''^2 + By''^2 = 0$. It represents a pair of straight lines which are asymptotes to the hyperbolas represented by (3.3.8) or (3.3.9).

NOTE 4. If $\alpha^2 = \beta^2$, the hyperbola is rectangular.

NOTE 5. Here $\Delta = -ABk$. It may be $<$ or $>$ 0 for the real hyperbola.

► **Example 3** *Show that the equation*

$$7x^2 - 48xy - 7y^2 - 20x + 140y + 300 = 0$$

represents a hyperbola and find its canonical equation.

Here $a = 7$, $h = -24$, $b = -7$, $g = -10$, $f = 70$, $c = 300$,

$$\Delta = \begin{vmatrix} 7 & -24 & -10 \\ -24 & -7 & 70 \\ -10 & 70 & 300 \end{vmatrix} \neq 0, \quad D = 7\cdot(-7) - (-24)^2 < 0.$$

Therefore, the given equation represents a hyperbola.

Rotating the axes through an acute angle θ, the equation transforms to

$$\begin{aligned} 7(x'\cos\theta - y'\sin\theta)^2 &- 48(x'\cos\theta - y'\sin\theta)(x'\sin\theta + y'\cos\theta) \\ &-7(x'\sin\theta + y'\cos\theta)^2 - 20(x'\cos\theta - y'\sin\theta) \\ &+140(x'\sin\theta + y'\cos\theta) + 300 = 0, \\ \text{or}\quad (7\cos^2\theta &- 7\sin^2\theta - 48\sin\theta\cos\theta)x'^2 \\ &-\{48(\cos^2\theta - \sin^2\theta) + 28\sin\theta\cos\theta\}x'y' \\ &-(7\cos^2\theta - 7\sin^2\theta - 48\sin\theta\cos\theta)y'^2 - 20(\cos\theta - 7\sin\theta)x' \\ &+20(\sin\theta + 7\cos\theta)y' + 300 = 0. \end{aligned}$$

To make the coefficient of $x'y'$ zero,

$$48(\cos^2\theta - \sin^2\theta) + 28\sin\theta\cos\theta = 0,$$

$$\text{or}\quad 12\tan^2\theta - 7\tan\theta - 12 = 0, \quad \text{or}\quad \tan\theta = \frac{4}{3}, -\frac{3}{4}.$$

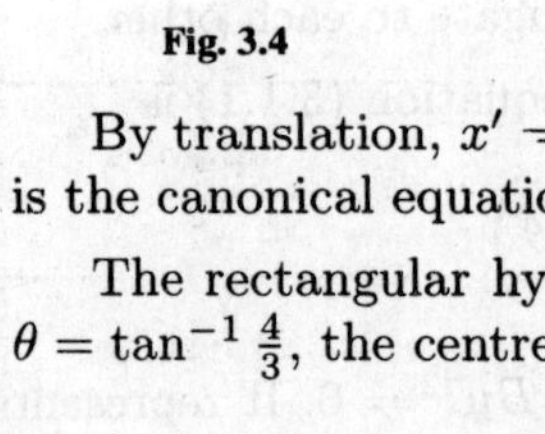

Fig. 3.4

Taking $\tan\theta = \frac{4}{3}$, $\sin\theta = \frac{4}{5}$ and $\cos\theta = \frac{3}{5}$.

For those values of $\sin\theta$ and $\cos\theta$, the above equation takes the form

$$y'^2 - x'^2 + 4x' + 4y' + 12 = 0,$$

$$\text{or}\quad (x'-2)^2 - (y'+2)^2 = 12.$$

By translation, $x' = x'' + 2$, $y' = y'' - 2$, the equation reduces to $x''^2 - y''^2 = 12$. It is the canonical equation of the hyperbola which is rectangular one in this case.

The rectangular hyperbola can be traced now as shown in fig. 3.4. In the figure $\theta = \tan^{-1}\frac{4}{3}$, the centre C is at $(2, -2)$ w.r.t. OX', OY'.

3.4 Rank and Classification of a Second Order Curve

The rank of a second order curve $ax^2 + 2hxy + by^2 + 2gx + 2fy + c = 0$ is the rank of the matrix

$$\begin{bmatrix} a & h & g \\ h & b & f \\ g & f & c \end{bmatrix}.$$

A second order curve is classified as *non-singular* or *non-degenerate* and *singular* or *degenerate* according as its rank is 3 and 2 or 1 respectively. A circle, a parabola, an ellipse and a hyperbola are non-singular. A point ellipse, a pair of intersecting lines, a pair of parallel lines and a pair of coincident lines are singular or degenerate. The rank of a pair of coincident lines is one.

NOTE. The general second degree equation represents (i) a degenerate conic if $\Delta = 0$ and (ii) a non-degenerate conic if $\Delta \neq 0$. The non-degenerate conics are mainly divided into three classes:

(a) *elliptic* if $D > 0$, (b) *parabolic* if $D = 0$ and (c) *hyperbolic* if $D < 0$.

Circle is a special case of the ellipse when the major and minor axes are equal.

3.5 Table for Metric Classification

Δ	D	Canonical form	Name	Rank	Class
$\Delta < 0$	$D > 0$	$\frac{x^2}{\alpha^2} + \frac{y^2}{\beta^2} = 1$	ellipse	3	non-singular
$\Delta < 0$	$D > 0$	$x^2 + y^2 = \alpha^2$	circle	3	non-singular
$\Delta > 0$	$D > 0$	$\frac{x^2}{\alpha^2} + \frac{y^2}{\beta^2} = -1$	imaginaryellipse	3	non-singular
$\Delta > 0$	$D < 0$	$\frac{x^2}{\alpha^2} - \frac{y^2}{\beta^2} = 1$	hyperbola	3	non-singular
$\Delta < 0$	$D < 0$	$\frac{x^2}{\alpha^2} - \frac{y^2}{\beta^2} = -1$	hyperbola	3	non-singular
$\Delta \neq 0$	$D = 0$	$y^2 = 4\alpha x$ $x^2 = 4\beta y$	parabola	3	non-singular
$\Delta = 0$	$D > 0$	$Ax^2 + By^2 = 0$	pair of imaginary lines or point ellipse	2	singular
$\Delta = 0$	$D < 0$	$y^2 - k^2x^2 = 0$	pair of inter-secting lines	2	singular
$\Delta = 0$	$D = 0$	$y^2 = k^2$ $x^2 = l^2$	pair of parallel lines	2	singular
$\Delta = 0$	$D = 0$	$y^2 = 0$ $x^2 = 0$	pair of coinci-dent lines	1	singular

3.6 Centre of the Central Conic

DEFINITION. *If any chord of a conic through a particular point is bisected by that point then the conic is said to be central and that particular point is called the centre of this conic. If the origin is the centre, the extremities of the chord through the origin will* *be of the form (x, y) and $(-x, -y)$. Thus if the point (x, y) lies on the conic, the point $(-x, -y)$ will also lie on the conic. Consequently there will be no term of odd degree in the equation of the conic.*

To determine the centre: Let (α, β) be the centre of the central conic.

$$ax^2 + 2hxy + by^2 + 2gx + 2fy + c = 0. \tag{3.6.1}$$

Changing the origin to (α, β) with the axes remaining parallel, the equation transforms to

$$a(x'+\alpha)^2 + 2h(x'+\alpha)(y'+\beta) + b(y'+\beta)^2$$
$$+2g(x'+\alpha) + 2f(y'+\beta) + c = 0,$$

or
$$ax'^2 + 2hx'y' + by'^2 + 2(a\alpha + h\beta + g)x' + 2(h\alpha + b\beta + f)y'$$
$$+a\alpha^2 + 2h\alpha\beta + b\beta^2 + 2g\alpha + 2f\beta + c = 0. \tag{3.6.2}$$

Since (α, β) is the centre, the coefficients of x' and y' must vanish.

$$\therefore\ a\alpha + h\beta + g = 0 \tag{3.6.3}$$
$$h\alpha + b\beta + f = 0. \tag{3.6.4}$$

From (3.6.3) and (3.6.4),

$$\alpha = \frac{fh - bg}{ab - h^2}, \quad \beta = \frac{gh - af}{ab - h^2}.$$

Thus (α, β) are determined.

The constant term in (3.6.2)

$$= a\alpha^2 + 2h\alpha\beta + b\beta^2 + 2g\alpha + 2f\beta + c$$
$$= \alpha(a\alpha + h\beta + g) + \beta(h\alpha + b\beta + f) + g\alpha + f\beta + c$$
$$= g\alpha + f\beta + c \quad \text{[by (3) and (4)]}$$
$$= \frac{g(fh - bg) + f(gh - af) + c(ab - h^2)}{ab - h^2}$$
$$= \frac{abc + 2fgh - af^2 - bg^2 - ch^2}{ab - h^2} = \frac{\Delta}{D}.$$

Thus the equation (3.6.2) can be written as

$$ax'^2 + 2hx'y' + by'^2 + \frac{\Delta}{D} = 0.$$

NOTE 1. **If $S = 0$ is the equation of the conic then the centre is the point of intersection of the lines $\frac{\partial S}{\partial x} = 0$ and $\frac{\partial S}{\partial y} = 0$.**

NOTE 2. **Non-central conic.** If $ab - h^2 = 0$, $fh - bg \neq 0$ and $gh - af \neq 0$ then the conic represented by the equation (3.1.1) is non-central, i.e., it has no centre.

NOTE 3. **Infinitely many-central conic.** If $ab - h^2 = 0$, $fh - bg = 0$ and $gh - af = 0$ then the equation (3.1.1) represents an infinitely many central conic, i.e., it has infinitely many centres.

NOTE 4. For the reduction of a central conic the rotation after translation to the centre is advantageous. In case of a parabola the technique adopted in the worked out example 1 is worthwhile. The form will be $Y^2 = \lambda x$ or $X^2 = \mu Y$ according as the coefficient of $xy <$ or > 0.

3.7 Common Points

(i) **Conic through the common points of two conics.**

If $S = 0$ and $S' = 0$ are two conics then $S + \lambda S' = 0$ is the equation of a conic through the common points of $S = 0$ and $S' = 0$. λ is an arbitrary constant. It is found out by the given condition.

(ii) **Conic through the common points of a conic and two lines.**

If $S = 0$ is the conic and the lines are $u = 0$ and $v = 0$ then the equation of the required conic is $S + \lambda uv = 0$. λ is an arbitrary constant. It is determined by the given condition.

► **Example 4** *Find the equation of the conic which passes through the point* $(-1, -1)$ *and also through the intersections of the conic*

$$25x^2 - 14xy + 25y^2 + 64x - 64y - 228 = 0$$

with the lines $x + 3y - 2 = 0$ *and* $3x + y - 4 = 0$. *Find also the parabolas passing through the same points of intersection.*

The equation of the conic is of the form

$$25x^2 - 14xy + 25y^2 + 64x - 64y - 228 + \lambda(x + 3y - 2)(3x + y - 4) = 0. \qquad (3.7.1)$$

If it passes through the point $(-1, -1)$ then

$$25 - 14 + 25 - 64 + 64 - 228 + \lambda(-1 - 3 - 2)(-3 - 1 - 4) = 0, \quad \text{i.e.,} \quad \lambda = 4.$$

$\therefore$ the required equation is

$$37x^2 + 26xy + 37y^2 + 24x - 120y - 196 = 0.$$

The equation of the parabola will be also of the form (i), i.e.,

$$(25 + 3\lambda)x^2 + (10\lambda - 14)xy + (25 + 3\lambda)y^2 + (64 - 10\lambda)x$$
$$- (64 + 14\lambda)y + 8\lambda - 228 = 0.$$

It represents a parabola, if

$$(5\lambda - 7)^2 - (25 + 3\lambda)^2 = 0,$$

$$\text{or} \quad (5\lambda - 7 + 25 + 3\lambda)(5\lambda - 7 - 25 - 3\lambda) = \text{υ}, \quad \text{or} \quad \lambda = -\frac{9}{4}, 16.$$

For $\lambda = -\frac{9}{4}$, the equation (1) takes the form

$$73x^2 - 146xy + 73y^2 + 346x - 130y - 984 = 0.$$

For $\lambda = 16$, the equation (1) takes the form

$$73x^2 + 146xy + 73y^2 - 96x - 228y - 100 = 0.$$

These are the required equations of parabolas.

Worked Out Examples

1. Discuss the nature of the conic $16x^2 - 24xy + 9y^2 - 104x - 172y + 44 = 0.$ **[C.U. '89]**

Here

$$\Delta = \begin{vmatrix} 16 & -12 & -52 \\ -12 & 9 & -86 \\ -52 & -86 & 44 \end{vmatrix} \neq 0.$$

Therefore, the equation does not represent a pair of straight lines.

The given equation can be written as

$$(4x - 3y)^2 = 104x + 172y - 44. \tag{1}$$

Since the terms of second degree are a perfect square, the equation represents a parabola. Let us write the equation in the following way:

$$(4x - 3y + \lambda)^2 = 104x + 172y - 44 + 2\lambda(4x - 3y) + \lambda^2,$$

$$\text{or} \quad (4x - 3y + \lambda)^2 = (104 + 8\lambda)x + (172 - 6\lambda)y + \lambda^2 - 44. \tag{2}$$

The constant λ is chosen in such a way that the lines $4x - 3y + \lambda = 0$ and $(104 + 8\lambda)x + (172 - 6\lambda)y + \lambda^2 - 44 = 0$ will be at right angle. For this

$$4 \cdot (104 + 8\lambda) - 3 \cdot (172 - 6\lambda) = 0, \quad \text{or} \quad \lambda = 2.$$

Putting $\lambda = 2$ in (2), $(4x - 3y + 2)^2 = 40(3x + 4y - 1)$,

$$\text{or} \quad \left(\frac{4x - 3y + 2}{\sqrt{4^2 + 3^2}}\right)^2 = \frac{40}{\sqrt{4^2 + 3^2}} \cdot \frac{3x + 4y - 1}{\sqrt{4^2 + 3^2}}. \tag{3}$$

Therefore, the axis of the parabola is $4x - 3y + 2 = 0$ and the tangent at the vertex is $3x + 4y - 1 = 0$.

The length of the latus rectum $= \dfrac{40}{\sqrt{4^2 + 3^3}} = \dfrac{40}{5} = 8.$

Vertex is the point of intersection of the lines $4x - 3y + 2 = 0$ and $3x + 4y - 1 = 0$. Thus the vertex is $(-\frac{1}{5}, \frac{2}{5})$.

Writing the equation (3) in the form $Y^2 = 4AX$, where

$$A = 2, \quad Y = \frac{4x - 3y + 2}{5}, \quad X = \frac{3x + 4y - 1}{5},$$

the focus is $(A, 0)$, i.e., $X = A$ and $Y = 0$,

$$\text{or} \quad \frac{3x + 4y - 1}{5} = 2, \quad \frac{4x - 3y + 2}{5} = 0,$$

$$\text{or} \quad 3x + 4y - 11 = 0, \quad 4x - 3y + 2 = 0.$$

Solving these two we get the focus as $(1, 2)$.

Equation of the latus rectum is $X = A$,

$$\text{or} \quad \frac{3x + 4y - 1}{5} = 2, \quad \text{or} \quad 3x + 4y - 11 = 0.$$

Equation of the directrix is $X + A = 0$, or $3x + 4y + 9 = 0$.

The canonical form of the equation is $Y^2 = 8X$.

2. Discuss the nature of the conic $11x^2 - 4xy + 14y^2 - 58x - 44y + 71 = 0$.

[C.U. '64; B.U. '87, '97]

Here

$$\Delta = \begin{vmatrix} 11 & -2 & -29 \\ -2 & 14 & -22 \\ -29 & -22 & 71 \end{vmatrix} \neq 0.$$

$\therefore$ the equation does not represent a pair of straight lines.

Since $D = 11.14 - 2^2 > 0$, the equation represents an ellipse.

The centre of the ellipse is obtained from the equations

$$11\alpha - 2\beta - 29 = 0, \quad 2\alpha - 14\beta + 22 = 0.$$

Solving these two equations, the centre is $(3, 2)$.

Shifting the origin to $(3, 2)$, the equation reduces to

$$11x^2 - 4xy + 14y^2 - 60 = 0. \tag{1}$$

Rotating the axes through an angle θ, where

$$\tan 2\theta = \frac{2 \cdot (-2)}{11 - 14} = \frac{4}{3}, \quad \text{the equation (1) reduces to}$$

$$11(x\cos\theta - y\sin\theta)^2 - 4(x\cos\theta - y\sin\theta)(x\sin\theta + y\cos\theta)$$
$$+14(x\sin\theta + y\cos\theta)^2 - 60 = 0,$$

$$\text{or} \quad (11\cos^2\theta - 4\sin\theta\cos\theta + 14\sin^2\theta)x^2$$
$$+(11\sin^2\theta + 4\sin\theta\cos\theta + 14\cos^2\theta)y^2 - 60 = 0,$$

$$\text{or} \quad \left\{11 + \frac{3}{2}(1 - \cos 2\theta) - 2\sin 2\theta\right\}x^2 + \left\{11 + \frac{3}{2}(1 + \cos 2\theta) + 2\sin 2\theta\right\}y^2 - 60 = 0,$$

$$\text{or} \quad \left(11 + \frac{3}{2} - \frac{3}{2} \cdot \frac{3}{5} - 2 \cdot \frac{4}{5}\right)x^2 + \left(11 + \frac{3}{2} + \frac{3}{2} \cdot \frac{3}{5} + 2 \cdot \frac{4}{5}\right)y^2 - 60 = 0,$$

$$\text{or} \quad 10x^2 + 15y^2 = 60, \quad \text{or} \quad \frac{x^2}{6} + \frac{y^2}{4} = 1. \quad \text{[It is the canonical form.]}$$

Thus the semi-axes are $\sqrt{6}$ and 2.

[The equation of the axes are $(x-3)\cos\theta + (y-2)\sin\theta = 0$ and $(y-2)\cos\theta - (x-3)\sin\theta = 0$.]

3. Discuss the nature of the conic $3x^2 - 8xy - 3y^2 + 10x - 13y + 8 = 0$.

Here

$$\Delta = 3.(-3).8 + 2.\left(-\frac{13}{2}\right).5.(-4) - 3.\left(-\frac{13}{2}\right)^2 - (-3).5^2 - 8.(-4)^2 = \frac{33}{4} \neq 0.$$

$$D = 3.(-3) - (-4)^2 = -9 - 16 = -25 < 0.$$

$\therefore$ the equation represents a hyperbola.

Again coefficient of x^2+ coefficient of $y^2 = 3 - 3 = 0$.

$\therefore$ the hyperbola is rectangular.

Solving the equations $3\alpha - 4\beta + 5 = 0$ and $8\alpha + 6\beta + 13 = 0$, we have the centre $(-\frac{41}{25}, \frac{1}{50})$.

Changing the origin to the centre the equation reduces to

$$3x^2 - 8xy - 3y^2 + \frac{\Delta}{D} = 0, \quad \text{or} \quad 3x^2 - 8xy - 3y^2 - \frac{33}{100} = 0. \tag{1}$$

The axes are rotated through an angle θ, where

$$\tan 2\theta = \frac{-2.4}{3-(-3)} = -\frac{4}{3}.$$

The equation (1) transforms to

$$3(x\cos\theta - y\sin\theta)^2 - 8(x\cos\theta - y\sin\theta)$$

$$(x\sin\theta + y\cos\theta) - 3(x\sin\theta + y\cos\theta)^2 = \frac{33}{100},$$

$$\text{or} \quad (3\cos 2\theta - 4\sin 2\theta)x^2 - (3\cos 2\theta - 4\sin 2\theta)y^2 = \frac{33}{100},$$

$$\text{or} \quad x^2 - y^2 = -\frac{33}{500}. \quad \text{[It is the canonical form.]}$$

Thus the semi-axis is $\sqrt{\frac{38}{500}}$.

4. Reduce the equation $8x^2 - 12xy + 17y^2 - 4x - 22y + 13 = 0$ into its normal form and discuss its nature. [N.B. '98]

Here

$$\Delta = \begin{vmatrix} 8 & -6 & -2 \\ -6 & 17 & -11 \\ -2 & -11 & 13 \end{vmatrix} = 4\begin{vmatrix} 2 & -3 & -1 \\ -3 & 17 & -11 \\ -1 & -11 & 13 \end{vmatrix}$$

$$= 4\{2(221 - 121) - 3(11 + 39) - (33 + 17)\}$$

$$= 4(200 - 150 - 50) = 0$$

$$D = 8.17 - 36 = 136 - 36 = 100.$$

Since $\Delta = 0$ and $D > 0$, the equation represents a pair of imaginary lines or a point ellipse.

Shifting the origin to the centre the equation reduces to

$$8x^2 - 12xy + 17y^2 = 0 \qquad (1)$$

Now rotating the axes through an acute angle θ where $\tan 2\theta = \frac{-12}{8-17} = \frac{4}{3}$, the equation (1) transforms to

$$8(x'\cos\theta - y'\sin\theta)^2 - 12(x'\cos\theta - y'\sin\theta)(x'\sin\theta + y'\cos\theta)$$
$$+17(x'\sin\theta + y'\cos\theta)^2 = 0$$

or $(8\cos^2\theta - 12\sin\theta\cos\theta + 17\sin^2\theta)x'^2 + (8\sin^2\theta + 12\sin\theta\cos\theta + 17\cos^2\theta)y'^2 = 0$

(For $\tan 2\theta = \frac{4}{3}$, the coefficient of $x'y'$ is 0.)

or $\left(8 - 6\sin 2\theta + 9\frac{1-\cos 2\theta}{2}\right)x'^2 + \left(8 + 6\sin 2\theta + 9\frac{1+\cos 2\theta}{2}\right)y'^2 = 0,$

or $\left(8 - 6\cdot\frac{4}{5} + \frac{9}{2} - \frac{9}{2}\cdot\frac{3}{5}\right)x'^2 + \left(8 + 6\cdot\frac{4}{5} + \frac{9}{2} + \frac{9}{2}\cdot\frac{3}{2}\right)y'^2 = 0,$ or $x'^2 + 4y'^2 = 0.$

It is the canonical form or normal form of the given equation.

5. Reduce the equation $3x^2 - 5xy + 6y^2 + 11x - 17y + 13 = 0$ to the canonical form and state the nature of the conic. [C.U. '99]

Here

$$\Delta = \begin{vmatrix} 3 & -\frac{5}{2} & \frac{11}{2} \\ -\frac{5}{2} & 6 & -\frac{17}{2} \\ \frac{11}{2} & -\frac{17}{2} & 13 \end{vmatrix} = -\frac{47}{4}, \quad D = 18 - \frac{25}{4} = \frac{47}{4}$$

Since $\Delta \neq 0$, $D > 0$, the equation represents an ellipse. Changing the origin to the centre of the ellipse the equation reduces to $3x^2 - 5xy + 6y^2 + \frac{\Delta}{D} = 0$, i.e.,

$$3x^2 - 5xy + 6y^2 - 1 = 0. \qquad (2)$$

Rotating the axes through an acute angle θ where $\tan 2\theta = \frac{-5}{3-6} = \frac{5}{3}$, the equation (1) transforms to

$$3(x'\cos\theta - y'\sin\theta)^2 - 5(x'\cos\theta - y'\sin\theta)(x'\sin\theta + y'\cos\theta)$$
$$+6(x'\sin\theta + y'\cos\theta)^2 = 1,$$

or $(3\cos^2\theta - 5\sin\theta\cos\theta + 6\sin^2\theta)x'^2 + (3\sin^2\theta + 5\sin\theta\cos\theta + 6\cos^2\theta)y'^2 = 1,$

(For $\tan 2\theta = \frac{5}{3}$, the coefficient of $x'y'$ is zero.)

or $\left(3 - \frac{5}{2}\sin 2\theta + 3\frac{1-\cos 2\theta}{2}\right)x'^2 + \left(3 + \frac{5}{2}\sin 2\theta + 3\frac{1+\cos 2\theta}{2}\right)y'^2 = 1,$

or $\left(3 - \frac{5}{2}\cdot\frac{5}{\sqrt{34}} + \frac{3}{2} - \frac{3}{2}\cdot\frac{3}{\sqrt{34}}\right)x'^2 + \left(3 + \frac{5}{2}\cdot\frac{5}{\sqrt{34}} + \frac{3}{2} + \frac{3}{2}\cdot\frac{3}{\sqrt{34}}\right)y'^2 = 1,$

or $\frac{9-\sqrt{34}}{2}x'^2 + \frac{9+\sqrt{34}}{2}y'^2 = 1.$

It is the canonical form.

6. Reduce the equation $4x^2+4xy+y^2-4x-2y+a=0$ to its canonical form and determine the type of the conic represented by it for different values of a. [C.H. '82; '91]

Here

$$\Delta = \begin{vmatrix} 4 & 2 & -2 \\ 2 & 1 & -1 \\ -2 & -1 & a \end{vmatrix} = 2\begin{vmatrix} 2 & 1 & -1 \\ 2 & 1 & -1 \\ -2 & -1 & a \end{vmatrix} = 0$$

($\because$ two rows are identical)

Therefore, the equation represents a pair of straight lines.

Rotating the axes through an acute angle θ, the equation transforms to

$$4(x'\cos\theta - y'\sin\theta)^2 + 4(x'\cos\theta - y'\sin\theta)(x'\sin\theta + y'\cos\theta)$$
$$+(x'\sin\theta + y'\cos\theta)^2 - 4(x'\cos\theta - y'\sin\theta) - 2(x'\sin\theta + y'\cos\theta) + a = 0,$$

or $$(4\cos^2\theta + 4\sin\theta\cos\theta + \sin^2\theta)x'^2 + (4\cos^2\theta - 4\sin^2\theta - 6\sin\theta\cos\theta)x'y'$$
$$+(4\sin^2\theta - 4\sin\theta\cos\theta + \cos^2\theta)y'^2 - (4\cos\theta + 2\sin\theta)x'$$
$$+(4\sin\theta - 2\cos\theta)y' + a = 0.$$

To make the coefficient of $x'y'$ zero, we get

$$4\cos^2\theta - 4\sin^2\theta - 6\sin\theta\cos\theta = 0, \quad \text{or} \quad 2\tan^2\theta + 3\tan\theta - 2 = 0,$$
$$\text{or} \quad (\tan\theta + 2)(2\tan\theta - 1) = 0, \text{ or } \tan\theta = 1/2, -2.$$

Since θ is acute $\tan\theta = 1/2$, or $\sin\theta = \dfrac{1}{\sqrt{5}}$ and $\cos\theta = \dfrac{2}{\sqrt{5}}$.

With these values of $\sin\theta$ and $\cos\theta$ the equation takes the form

$$5x'^2 - 2\sqrt{5}x' + a = 0.$$

It can be written as $\left(x' - \dfrac{1}{\sqrt{5}}\right)^2 + \dfrac{a-1}{5} = 0.$

By translation $x' = x'' + \dfrac{1}{\sqrt{5}}$, $y' = y''$, the equation reduces to

$$x''^2 + \frac{a-1}{5} = 0. \tag{1}$$

It is the canonical form of the given equation.

If $a < 1$, the given equation represents two parallel lines.

If $a = 1$, the equation (1) reduces to $x''^2 = 0$ and the given equation represents two coincident lines.

If $a > 1$, the given equation represents two imaginary lines.

7. Determine the values of h and g so that the equation $x^2 - 2hxy + 4y^2 + 2gx - 12y + 9 = 0$ may represent (i) a conic having no centre, (ii) a conic having infinitely may centres. In the last case find the type of the conic.

The centre of the conic lies on $x - hy + g = 0$ and $-hx + 4y - 6 = 0$.

By cross-multiplication,

$$\frac{x}{6h - 4g} = \frac{y}{-gh + 6} = \frac{1}{4 - h^2}.$$

(i) The given equation will represent a conic having no centre if $4 - h^2 = 0, 6h - 4g \neq 0$ and $6 - gh \neq 0$.

From $4 - h^2 = 0$, or $h = \pm 2$.

Again $6 - gh \neq 0$, or $6 \mp 2g \neq 0$, or $g \neq \pm 3$.

(ii) The given equation will represent a conic having infinitely many centres if $4 - h^2 = 0, 6h - gh = 0$ and $6 - gh = 0$.

From $4 - h^2 = 0$, $h = \pm 2$. For $h = \pm 2$, $g = \pm 3$.

For these values of h and g, the equation of the conic is

$$x^2 \mp 4xy + 4y^2 \pm 6x - 12y + 9 = 0, \quad \text{or} \quad (x \mp 2y \pm 3)^2 = 0.$$

Therefore, the given equation represents a pair of coincident lines.

8. If the equation $13x^2 + 10xy + by^2 - 62\sqrt{2}x - 46\sqrt{2}y + 170 = 0$ represents a point ellipse, find the value of b and the point.

Here

$$\Delta = \begin{vmatrix} 13 & 5 & -31\sqrt{2} \\ 5 & b & -23\sqrt{2} \\ -31\sqrt{2} & -23\sqrt{2} & 170 \end{vmatrix} = 0,$$

$$\text{or} \quad (2210b - 13754) + (7130 - 4250) + (7130 - 1922b) = 0,$$

$$\text{or} \quad 288b - 3744 = 0, \quad \text{or} \quad b = 13.$$

Again $D = 13b - 25 = 169 - 25 > 0$.

$\therefore$ the equation represents a point ellipse when $b = 13$.

The point is the centre. If (α, β) is the centre then

$$13\alpha + 5\beta - 31\sqrt{2} = 0$$

$$\text{and} \quad 5\alpha + 13\beta - 23\sqrt{2} = 0.$$

By cross-multiplication,

$$\frac{\alpha}{288\sqrt{2}} = \frac{\beta}{144\sqrt{2}} = \frac{1}{144}, \quad \text{or} \quad \alpha = 2\sqrt{2} \quad \text{and} \quad \beta = \sqrt{2}.$$

Thus the point is $(2\sqrt{2}, \sqrt{2})$.

Exercise

1. Find the canonical equations and describe the principal properties of the following by proper translation.

   (a) $y^2 + 8x - 6y + 1 = 0$;

   (b) $x^2 - 8x - 12y - 20 = 0$;

   (c) $9x^2 + 16y^2 - 54x + 64y + 1 = 0$;

   (d) $3x^2 - 2y^2 + 6x - 8y - 17 = 0$.

2. Determine the centre of the following conics.

(a) $3x^2 + 4y^2 - 12x + 8y + 4 = 0$;

(b) $3x^2 + 3y^2 - 2xy - 2x + 6y + 2 = 0$;

(c) $2x^2 - 5xy - 3y^2 - x - 4y + 6 = 0$. [B.H. '83]

3. Determine whether the following conic have a single centre, or infinitely may centres or no centre.

(a) $12x^2 + 4y^2 + 14xy - 2x - 3y + 7 = 0$;

(b) $4x^2 - 4xy + y^2 - 12x + 10y - 19 = 0$;

(c) $9x^2 - 6xy + y^2 + 24x - 8y + 16 = 0$;

(d) $4x^2 + 12xy + 9y^2 + 3x + 2y + 1 = 0$.

4. Reduce the following to the canonical form and state the type of the conic.

(a) $6x^2 - 5xy - 6y^2 + 14x + 5y + 4 = 0$; [B.H. '92; C.U. '82, '96]

(b) $8x^2 - 12xy + 17y^2 + 16x - 12y + 3 = 0$; [N.B. '81]

(c) $4x^2 - 3xy - 18 = 0$;

(d) $8x^2 + 12xy + 13y^2 = 884$;

(e) $4x^2 - 12xy + 9y^2 - 52x + 26y + 81 = 0$;

(f) $x^2 + y^2 - xy - 6x = 0$;

(g) $x^2 - 2xy + 2y^2 - 4x - 6y + 3 = 0$; [B.U. '90; B.H. '91; C.U. 2000]

(h) $2x^2 - 4xy - y^2 + 20x - 2y + 17 = 0$;

(i) $x^2 + 4xy + 4y^2 + 4x + y - 15 = 0$; [C.U. '80; B.U. '93, '97]

(j) $3x^2 + 10xy + 3y^2 - 2x - 14y - 5 = 0$;

(k) $12x^2 + 24xy + 19y^2 - 12x - 40y + 31 = 0$;

(l) $5x^2 - 20xy - 5y^2 - 16x + 8y - 7 = 0$.

5. Discuss the nature of the following conics.

(a) $9x^2 - 6xy + y^2 - 14x - 2y + 12 = 0$;

(b) $3x^2 - 5xy + 6y^2 + 11x - 17y + 13 = 0$; [N.B. '87; C.U. '99]

(c) $x^2 - 4xy - 2y^2 + 10x + 4y = 0$;

(d) $4x^2 - 4xy + y^2 - 8x - 6y + 5 = 0$; [B.U. '87, '96; C.H. '93]

(e) $32x^2 + 52xy - 7y^2 - 64x - 52y - 148 = 0$;

(f) $41x^2 + 24xy + 34y^2 + 30x - 40y + 25 = 0$.

6. (a) Reduce the equation $x^2 - 4xy + 4y^2 + 2x - 4y + c = 0$ to its canonical form and determine the type of the conic represented by it for different values of c.

(b) Find the values of a and g for which the curve $ax^2 + 8xy + 4y^2 + 2gx + 4y + 1 = 0$ represents (i) a conic having no centre, (ii) a conic having infinitely many centres.

(c) Find the values of a and f for which the curve $ax^2 - 6xy + 9y^2 + 4x + 2fy + 1 = 0$ may represent (i) a central conic. (ii) a parabola and (iii) a conic with infinitely many centres.

[HINTS. The centre lies on $ax - 3y + 2 = 0$ and $3x - 9y + f = 0$.

By cross-multiplication $\dfrac{x}{3(f+6)} = \dfrac{y}{6+af} = \dfrac{1}{9(1-a)}$.

(a) The given equation will represent a central conic if $1 - a \neq 0$, i.e., $a \neq 1$.

(b) In case of parabola $1 - a = 0, f + 6 \neq 0, 6 + af \neq 0$, i.e., $a = 1$ and $f \neq -6$.

(c) For infinitely many centres $1 - a = 0, f + 6 = 0, 6 + af = 0$, i.e., $a = 1$ and $f = -6$.]

(d) Find the values of h, f and c for which the equation

$$4x^2 + 2hxy + 36y^2 + 4x + 2fy + c = 0$$

will represent a pair of coincident lines.]

[HINTS. For the coincident lines $\begin{vmatrix} 4 & h \\ h & 36 \end{vmatrix} = 0$, $\begin{vmatrix} 36 & f \\ f & c \end{vmatrix} = 0$ and $\begin{vmatrix} 4 & 2 \\ 2 & c \end{vmatrix} = 0$.

From these $c = 1, h = \pm 12$ and $f = \pm 6$.

For these values the equation becomes $(2x \pm 6y + 1)^2 = 0$.]

(e) If the equation $5x^2 + 4xy + y^2 + 2gx - 2y + 2 = 0$ represents a point ellipse, find g.

[HINTS. For the point ellipse $\Delta = 0$, i.e., $g^2 + 4g + 3 = 0$, or $g = -3, -1$.

For $g = -3$, the point lies on $5x + 2y - 3 = 0$ and $2x + y - 1 = 0$ and the point is $(1, -1)$.

For $g = -1$, the point lies on $5x + 2y - 1 = 0$ and $2x + y - 1 = 0$ and the point is $(-1, 3)$]

7. (a) Show that the equation of the conic passing through the point $(1, 1)$ and the points of intersection of the conic $x^2 + 2xy + 5y^2 - 7x - 8y + 6 = 0$ with the lines $2x - y - 5 = 0$ and $3x + y - 11 = 0$ is $34x^2 + 55xy + 139y^2 - 233x - 218y + 233 = 0$.

(b) Find the equations of the parabolas passing through the common points of $x^2 + 6xy - y^2 + 2x - 3y - 5 = 0$ and $2x^2 - 8xy + 3y^2 + 2y - 1 = 0$.

8. (a) By reducing the equation $20x^2 + 15xy + 9x + 3y + 1 = 0$ to its canonical form, show that it represents a pair of intersecting straight lines which are equidistant from the origin. Find their point of intersection. **[C.H. '84]**

[HINTS. Here $\Delta = 0$ and $D < 0$. Therefore, the equation represents two intersecting lines. The point of intersection lies on $40x + 15y + 9 = 0$ and $5x + 1 = 0$. It is $\left(-\frac{1}{5}, -\frac{1}{15}\right)$. Changing the origin to this point the equation reduces to $4x'^2 + 3x'y' = 0$. Now rotating the axes through an acute angle θ, where $\tan 2\theta = \frac{3}{4}$, the required canonical form is $9x''^2 - y''^2 = 0$. Obviously it represents two intersecting lines. The bisectors of the angles between these lines are $x'' = 0$ and $y'' = 0$.

The co-ordinates of the old origin in the new system are

$$\left(\frac{\cos\theta}{5} + \frac{\sin\theta}{15}, \frac{\cos\theta}{15} - \frac{\sin\theta}{5}\right), \text{ i.e., } \left(\frac{\sqrt{10}}{5}, 0\right) \text{ since } \tan\theta = \frac{1}{3}.$$

Thus the old origin lies on one of the bisectors. Consequently the lines represented by the given equation are equidistant from the origin.]

(b) Show that the equation $(a^2 + 1)x^2 + 2(a + b)xy + (b^2 + 1)y^2 = c$ represents an ellipse whose area is $\frac{\pi c}{ab-1}$.

[HINTS. Here $\Delta \neq 0$ and $D = (ab - 1)^2 > 0$. $\therefore$ the given equation represents an ellipse. Rotating the axes through an acute angle θ, where $\tan 2\theta = \frac{2}{a-b}$, the equation transforms to

$$\left\{\frac{a^2+b^2+2}{2}+\frac{a+b}{2}\sqrt{(a-b)^2+4}\right\}x'^2+\left\{\frac{a^2+b^2+2}{2}-\frac{a+b}{2}\sqrt{(a-b)^2+4}\right\}y'^2=c$$

$\therefore$ the area of the ellipse

$$=\frac{\pi c}{\sqrt{\left[\left(\frac{a^2+b^2+2}{2}\right)^2-\left(\frac{a+b}{2}\right)^2\{(a-b)^2+4\}\right]}}=\frac{\pi c}{ab-1}]$$

(c) Show that $(a^2+b^2)(x^2+y^2)=(ax+by-ab)^2$ represents a parabola of latus rectum $\frac{2ab}{\sqrt{a^2+b^2}}$.

Answers

1. (a) Parabola, normal form $y'^2=-8x'$, length of latus rectum $=8$, vertex $(1,3)$, focus $(-1,3)$, axis $y-3=0$ directrix $x-3=0$.

(b) Parabola, normal form $x'^2=12y'$, length of latus rectum $=12$, vertex $(4,-3)$, focus $(4,0)$, axis $x=4$, directrix $y+6=0$.

(c) Ellipse, normal form $\frac{x'^2}{16}+\frac{y'^2}{9}=1$, centre $(3,-2)$, foci $(3\pm\sqrt{7},-2)$, eccentricity $=\frac{\sqrt{7}}{4}$, length of latus rectum $=9/2$.

(d) Hyperbola, normal form $\frac{x'^2}{4}-\frac{y'^2}{6}=1$, centre $(-1,-2)$, eccentricity $=\sqrt{\frac{5}{2}}$, foci $(-1\pm\sqrt{10},-2)$, length of latus rectum $=6$, length of transverse axis $=4$, length of conjugate axis $=2\sqrt{6}$.

2. (a) $(2,-1)$; (b) $(0,-1)$; (c) $\left(-\frac{2}{7},-\frac{3}{7}\right)$.

3. (a) Single centre $\left(\frac{13}{2},-11\right)$; (b) no centre; (c) infinitely many centres; (d) no centre.

4. (a) $x''^2-y''^2=0$, pair of intersecting lines (b) $x''^2+4y''^2=1$, ellipse;

(c) $9y'^2-x'^2=36$, hyperbola; (d) $\frac{x'^2}{52}+\frac{y'^2}{221}=1$, ellipse;

(e) $y''^2=\frac{8}{\sqrt{13}}x''$, parabola; (f) $\frac{x''^2}{24}+\frac{y''^2}{8}=1$, ellipse;

(g) $(3-\sqrt{5})x''^2+(3+\sqrt{5})y''^2=52$, ellipse;

(h) $\frac{x''^2}{3}-\frac{y''^2}{2}=-1$, hyperbola; (i) $x''^2=\frac{7}{5\sqrt{5}}y''$, parabola;

(j) $4x''^2-y''^2=0$, pair of intersecting lines;

(k) $28x''^2+3y''^2=0$, point ellipse, $(-3/2,2)$;

(l) $x''^2-y''^2=-\frac{351}{125\sqrt{5}}$, rectangular hyperbola.

5. (a) Parabola, vertex $(1,1)$, focus $\left(\frac{21}{20},\frac{23}{20}\right)$, axis $3x-y-2=0$, tangent at the vertex $x+3y-4=0$, directrix $x+3y=\frac{7}{2}$, latus rectum $=\frac{2}{\sqrt{10}}$, $x+3y=\frac{9}{2}$. (b) Ellipse, centre $(-1,1)$, semi-axes $\sqrt{\frac{2}{9-\sqrt{34}}}$, $\sqrt{\frac{2}{9+\sqrt{34}}}$. (c) Hyperbola, centre $(-1,2)$, semi-axes (transverse $\frac{1}{\sqrt{2}}$, conjugate $\frac{1}{\sqrt{3}}$). (d) Parabola, axis $2x-y-1=0$, tangent at the vertex $x+2y-1=0$, vertex $\left(\frac{3}{5},\frac{1}{5}\right)$, focus $\left(\frac{4}{5},\frac{3}{5}\right)$, latus rectum $=\frac{4}{\sqrt{5}}$, $x+2y-2=0$, directrix $x+2y=0$ (e) Hyperbola, centre $(1,0)$, transverse axis $x-2y-1=0$, length 4; conjugate axis $2x+y-2=0$, length 6; foci $\left(1\pm2\sqrt{\frac{13}{5}},\pm\sqrt{\frac{13}{5}}\right)$. (f) Point ellipse, $\left(-\frac{3}{5},\frac{4}{5}\right)$.

6. (a) $y''^2+\frac{c-1}{5}=0$, parallel, coincident and imaginary lines according as $c<=>1$; (b) (i) $a=4$, $g\neq2$, (ii) $a=4$, $g=2$.

7. (b) $5x^2-10xy+5y^2+2x+y-7=0$, $4x^2+4xy+y^2+4x+4y-9=0$.

Chapter 4

Tangent, Normal, Pole, Polar, Chord of Contact and Pair of Tangents

4.1 Equation of the Tangent

DEFINITION. *If a line (secant) meets a curve at coincident points then the line is called the tangent to the curve at the meeting point and the point is called the point of contact.*

Let

$$ax^2 + 2hxy + by^2 + 2gx + 2fy + c = 0 \tag{4.1.1}$$

be the equation of the conic and (x_1, y_1) be a point in the plane of the conic. The equation of a line through (x_1, y_1) can be written as

$$\frac{x - x_1}{l} = \frac{y - y_1}{m} \quad (= r \text{ say}), \tag{4.1.2}$$

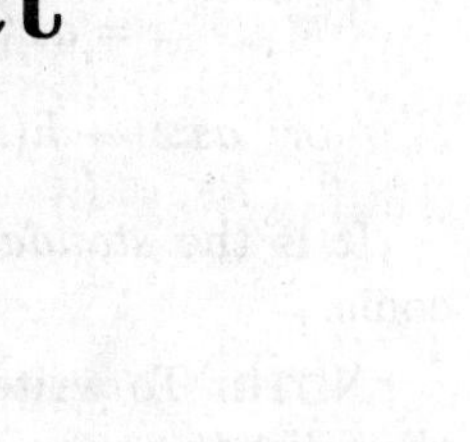

Fig. 4.1

where l and m are the cosines of the angles made by the line with x and y-axis respectively and r is the algebraical distance between (x, y) and (x_1, y_1) on the line.

From (4.1.2) $x = lr + x_1, y = mr + y_1$.

To find the point of intersection between (4.1.1) and (4.1.2), we have

$$\begin{aligned} a(lr + x_1)^2 &+ 2h(lr + x_1)(mr + y_1) + b(mr + y_1)^2 \\ &+ 2g(lr + x_1) + 2f(mr + y_1) + c = 0, \\ \text{or} \quad (al^2 &+ 2hlm + bm^2)r^2 + 2\{(ax_1 + hy_1 + g)l + (hx_1 + by_1 + f)m\}r \\ &+ ax_1^2 + 2hx_1y_1 + by_1^2 + 2gx_1 + 2fy_1 + c = 0. \end{aligned} \tag{4.1.3}$$

It is a quadratic equation in r. Let the roots be r_1 and r_2, where r_1 and r_2 are the distances of the two points of intersection between (1) and (2) from (x_1, y_1). If the point (x_1, y_1) is on the conic and the line (4.1.2) is the tangent at this point then $r_1 = 0 = r_2$.

In this case $r_1 + r_2 = 0$ and $r_1 r_2 = 0$.

By the equation (3)

$$(ax_1 + hy_1 + g)l + (hx_1 + by_1 + f)m = 0 \quad (4.1.4)$$

$$\text{and} \quad ax_1^2 + 2hx_1y_1 + by_1^2 + 2gx_1 + 2fy_1 + c = 0. \quad (4.1.5)$$

Eliminating l and m from (4.1.4) by (4.1.2)

$$(ax_1 + hy_1 + g)(x - x_1) + (hx_1 + by_1 + f)(y - y_1) = 0,$$

or $axx_1 + h(xy_1 + yx_1) + byy_1 + gx + fy$

$$= ax_1^2 + 2hx_1y_1 + by_1^2 + gx_1 + fy_1 = -gx_1 - fy_1 - c \quad \text{[by (4.1.5)]},$$

or $$axx_1 + h(xy_1 + yx_1) + byy_1 + g(x + x_1) + f(y + y_1) + c = 0. \quad (4.1.6)$$

It is the *standard form* of the equation of the tangent at a point (x_1, y_1) on the conic.

NOTE. To write down the equation of the tangent at (x_1, y_1) to the conic (4.1.1), the following rules are to be remembered.

Change x^2 into xx_1, y^2 into yy_1, $2xy$ into $xy_1 + yx_1$, $2x$ into $x + x_1$ and $2y$ into $y + y_1$.

Tangents of the Standard Equations

	Tangent at (x_1, y_1)
(1) **Circle** $x^2 + y^2 + 2gx + 2fy + c = 0$ $x^2 + y^2 = a^2$	 $xx_1 + yy_1 + g(x + x_1) + f(y + y_1) + c = 0$ $xx_1 + yy_1 = a^2$
(2) **Parabola** $y^2 = 4ax$	$yy_1 = 2a(x + x_1)$
(3) **Ellipse** $\frac{x^2}{a^2} + \frac{y^2}{b^2} = 1$	$\frac{xx_1}{a^2} + \frac{yy_1}{b^2} = 1$
(4) **Hyperbola** $\frac{x^2}{a^2} - \frac{y^2}{b^2} = 1$	$\frac{xx_1}{a^2} - \frac{yy_1}{b^2} = 1$
(5) **Rectangular hyperbola** $xy = k^2$	$xy_1 + yx_1 = 2k^2$.

4.2 To Find the Condition that a given Line may touch the Conic

Let the equations of the conic and the line be

$$ax^2 + 2hxy + by^2 + 2gx + 2fy + c = 0 \quad (4.2.1)$$

$$\text{and} \quad lx + my + n = 0 \quad (4.2.2)$$

respectively.

From (4.2.2)

$$y = -\frac{lx+n}{m}.$$

Putting in (4.2.1), we have

$$x^2(am^2 - 2hlm + bl^2) - 2x(hmn - bln - gm^2 + flm) + bn^2 - 2fmn + cm^2 = 0. \quad (4.2.3)$$

If the line (4.2.2) be a tangent, the roots of the equation (4.2.3) must be equal.

$$\therefore \quad (hmn - bln - gm^2 + flm)^2 = (am^2 - 2hlm + bl^2)(bn^2 - 2fmn + cm^2),$$

$$\text{or,} \quad l^2(bc - f^2) + m^2(ca - g^2) + n^2(ab - h^2) + 2mn(gh - af) + 2nl(hf - bg)$$
$$+2lm(fg - ch) = 0. \quad (4.2.4)$$

It is the required condition.

DEDUCTIONS. (1) Circle : $x^2 + y^2 = a^2$.

Condition : $a^2(l^2 + m^2) - n^2 = 0$, i.e., $n = \pm a\sqrt{l^2 + m^2}$.

(2) Parabola : $y^2 = 4ax$.

Condition : $am^2 - ln = 0$, i.e., $n = \dfrac{am^2}{l}$.

(3) Ellipse : $\dfrac{x^2}{a^2} + \dfrac{y^2}{b^2} = 1$.

Condition : $a^2l^2 + b^2m^2 - n^2 = 0$, i.e., $n = \pm\sqrt{a^2l^2 + b^2m^2}$.

(4) Hyperbola : $\dfrac{x^2}{a^2} - \dfrac{y^2}{b^2} = 1$.

Condition : $a^2l^2 - b^2m^2 - n^2 = 0$, i.e., $n = \pm\sqrt{a^2l^2 - b^2m^2}$.

NOTE. $y = mx + c$ will touch either (1), (2), (3) or (4), if $c = \pm a\sqrt{1+m^2}$,

$c = \dfrac{a}{m}$, $c = \pm\sqrt{a^2m^2 + b^2}$, or $c = \pm\sqrt{a^2m^2 - b^2}$ respectively.

4.3 To find the Equation of the Normal to

$$ax^2 + 2hxy + by^2 + 2gx + 2fy + c = 0 \text{ at } (x_1, y_1).$$

DEFINITION. *The normal to a curve is the straight line perpendicular to the tangent at the point of contact.*

The equation of the tangent at (x_1, y_1) is

$$axx_1 + h(xy_1 + yx_1) + byy_1 + g(x + x_1) + f(y + y_1) + c = 0,$$

$$\text{or} \quad (ax_1 + hy_1 + g)x + (hx_1 + by_1 + f)y + gx_1 + fy_1 + c = 0. \quad (4.3.1)$$

Since the normal is perpendicular to (4.3.1) and passes through the point (x_1, y_1), its equation is

$$y - y_1 = \frac{hx_1 + by_1 + f}{ax_1 + by_1 + g}(x - x_1). \quad (4.3.2)$$

4.4 Particular Cases

(1) **Parabola : $y^2 = 4ax$.**

The equation of the tangent at (x_1, y_1) is $yy_1 = 2a(x + x_1)$, or $y = \frac{2a}{y_1}(x + x_1)$.

$\therefore$ the equation of the normal at (x_1, y_1) is

$$y - y_1 = -\frac{y_1}{2a}(x - x_1).$$

COROLLARY. If we put $-\frac{y_1}{2a} = m$, then $x_1 = \frac{y_1^2}{4a} = am^2$.

Thus the equation of the normal at $(am^2, -2am)$ is

$$y = mx - 2am - am^3.$$

(Ia) Show that three normals can be drawn to a parabola from a given point and the sum of the ordinates of the feet of the normals is zero.

$(at^2, 2at)$ is a point on the parabola $y^2 = 4ax$.

The equation of the normal at this point is

$$y - 2at = -\frac{2at}{2a}(x - at^2), \quad \text{or} \quad y + tx = 2at + at^3. \tag{4.4.1}$$

If this normal passes through a fixed point (h, k), then

$$k + th = 2at + at^3, \quad \text{or} \quad at^3 + (2a - h)t - k = 0. \tag{4.4.2}$$

It is a cubic equation in t. Therefore, it has three roots. Corresponding to each of these three roots we have on substitution, the equation of a normal passing through the point (h, k).

Hence in general three normals can be drawn to a parabola through a given point.

Let t_1, t_2, t_3 be the roots of the equation (2). From the relation between roots and coefficients, $t_1 + t_2 + t_3 = 0$. If y_1, y_2, y_3 be the ordinates of the feet of the normals, then $y_1 + y_2 + y_3 = 2a(t_1 + t_2 + t_3) = 0$. Hence the result follows.

NOTE 1. Reality of the normals depends on the nature of the roots of (4.4.2).

NOTE 2. If the normal at (x_1, y_1) passes through the point (h, k), then $k - y_1 = -\frac{y_1}{2a}(h - x_1)$. Hence the locus of the feet of the normals through the point (h, k) is $k - y = -\frac{y}{2a}(h - x)$.

(ii) **Ellipse** : $\frac{x^2}{a^2} + \frac{y^2}{b^2} = 1$.

The equation of the tangent at (x_1, y_1) is

$$\frac{xx_1}{a^2} + \frac{yy_1}{b^2} = 1, \quad \text{or} \quad y = -\frac{b^2}{a^2}\frac{x_1}{y_1}x + \frac{b^2}{y_1}.$$

$\therefore$ the equation of the normal is

$$y - y_1 = \frac{a^2 y_1}{b^2 x_1}(x - x_1), \quad \text{or} \quad \frac{x - x_1}{x_1/a^2} = \frac{y - y_1}{y_1/b^2}. \tag{4.4.3}$$

(IIa) Show that four normals can be drawn to an ellipse through a given point and the feet of the normals lie on a rectangular hyperbola.

Let the normal (4.4.3) pass through a given point (h, k).

Then

$$\frac{h - x_1}{x_1/a^2} = \frac{k - y_1}{y_1/b^2} = \lambda \text{ (say)}, \quad \text{or} \quad \frac{x_1}{a} = \frac{ha}{a^2 + \lambda}, \ \frac{y_1}{b} = \frac{kb}{b^2 + \lambda}$$

Since (x_1, y_1) is on the ellipse,

$$\frac{x_1^2}{a^2} + \frac{y_1^2}{b^2} = 1, \quad \text{or} \quad \frac{a^2h^2}{(a^2 + \lambda)^2} + \frac{b^2k^2}{(b^2 + \lambda)^2} = 1.$$

It is a biquadratic equation in λ and gives four values of λ. These four values correspond to four points on the ellipe and the normals at which pass through the given point.

Again

$$a^2 + \lambda = \frac{a^2h}{x_1}, \quad b^2 + \lambda = \frac{b^2k}{y_1}.$$

Subtracting $a^2 - b^2 = \dfrac{a^2h}{x_1} - \dfrac{b^2k}{y_1}$. The point (x_1, y_1) is the foot of a normal. Hence the locus of the feet of the normals is

$$a^2 - b^2 = \frac{a^2h}{x} - \frac{b^2k}{y}, \quad \text{or} \quad (a^2 - b^2)xy = a^2hy - b^2kx.$$

which is a rectangular hyperbola.

(III) **Hyperbola** : $\dfrac{x^2}{a^2} - \dfrac{y^2}{b^2} = 1.$

The equation of the tangent at (x_1, y_1) is

$$\frac{xx_1}{a^2} - \frac{yy_1}{b^2} = 1, \quad \text{or} \quad y = \frac{b^2}{a^2}\frac{x_1}{y_1}x - \frac{b^2}{y_1}.$$

$\therefore$ the equation of the normal at (x_1, y_1) is

$$y - y_1 = -\frac{a^2y_1}{b^2x_1}(x - x_1), \quad \text{or} \quad \frac{x - x_1}{x_1/a^2} = -\frac{y - y_1}{y_1/b^2}. \tag{4.4.4}$$

(IIIa) Show that four normals can be drawn to a hyperbola through a given point and the locus of the feet of these normals is a rectangular hyperbola.

Let the normal (4.4.4) pass through a fixed point (h, k).

Then

$$\frac{h - x_1}{x_1/a^2} = -\frac{k - y_1}{y_1/b^2} = \lambda \text{ (say)}, \quad \text{or} \quad \frac{x_1}{a} = \frac{ah}{a^2 + \lambda}, \ \frac{y_1}{b} = \frac{bk}{b^2 - \lambda}.$$

We have

$$\frac{x_1^2}{a^2} - \frac{y_1^2}{b^2} = 1. \quad \therefore \frac{a^2h^2}{(a^2+\lambda)^2} - \frac{b^2k^2}{(b^2-\lambda)^2} = 1.$$

It is a biquadratic equation in λ and gives four values of it. These four values correspond to four points on the hyperbola at which normals pass through the fixed point.

Again

$$a^2 + \lambda = \frac{a^2h}{x_1}, \quad b^2 - \lambda = \frac{b^2k}{y_1}.$$

Adding

$$a^2 + b^2 = \frac{a^2h}{x_1} + \frac{b^2k}{y_1}.$$

(x_1, y_1) is a foot of one of the four normals. Hence the locus of the feet of the normals is

$$a^2 + b^2 = \frac{a^2h}{x} + \frac{b^2k}{y}, \quad \text{or} \quad (a^2+b^2)xy = a^2hy + b^2kx.$$

It is a rectangular hyperbola.

NOTE. **Co-normal points:**

The points on a curve at which normals pass through a fixed point are known as co-normal points.

4.5 Pair of Tangents: Director Circle

(a) *A pair of tangents can be drawn to a conic from a point not lying on the conic.*

Let

$$ax^2 + 2hxy + by^2 + 2gx + 2fy + c = 0 \tag{4.5.1}$$

be the equation of the conic and (x_1, y_1) be a point not lying on the conic. Let a line through (x_1, y_1) touch the conic at (x_2, y_2). The equation of the tangent at (x_2, y_2) is

$$axx_2 + h(xy_2 + yx_2) + byy_2 + g(x + x_2) + f(y + y_2) + c = 0.$$

As it passes through (x_1, y_1), we have

$$ax_1x_2 + h(x_1y_2 + y_1x_2) + by_1y_2 + g(x_1 + x_2) + f(y_1 + y_2) + c = 0. \tag{4.5.2}$$

Again (x_2, y_2) is on the conic. Therefore,

$$ax_2^2 + 2hx_2y_2 + by_2^2 + 2gx_2 + 2fy_2 + c = 0. \tag{4.5.3}$$

From (4.5.2) and (4.5.3) generally two values of x_2 and y_2 are obtained. Thus there will be two points of contact of tangents from (x_1, y_1) but they may not be real in all cases.

(b) **Equation of the pair of tangents:**

The equation of a line through (x_1, y_1) can be written as

$$\frac{x-x_1}{l} = \frac{y-y_1}{m} (= r \text{ say}). \tag{4.5.4}$$

From (4.5.4), $x = lr + x_1, y = mr + y_1$. Putting these values of x and y in the equation (4.5.1) we have

$$(al^2 + 2hlm + bm^2)r^2 + 2\{(ax_1 + hy_1 + g)l + (hx_1 + by_1 + f)m\}r$$
$$+ax_1^2 + 2hx_1y_1 + by_1^2 + 2gx_1 + 2fy_1 + c = 0. \tag{4.5.5}$$

It is a quadratic equation in r. If the line (4.5.4) touches the conic (4.5.1), the equation (4.5.5) must have equal roots. By the condition of equal roots

$$\{(ax_1 + hy_1 + g)l + (hx_1 + by_1 + f)m\}^2$$
$$= (al^2 + 2hlm + bm^2)(ax_1^2 + 2hx_1y_1 + by_1^2 + 2gx_1 + 2fy_1 + c).$$

Eliminating l and m by (4.5.4),

$$\{ax_1 + hy_1 + g)(x - x_1) + (hx_1 + by_1 + f)(y - y_1)\}^2$$
$$= \{a(x - x_1)^2 + 2h(x - x_1)(y - y_1) + b(y - y_1)^2\}$$
$$\times(ax_1^2 + 2hx_1y_1 + by_1^2 + 2gx_1 + fy_1 + c). \tag{4.5.6}$$

If we write

$$S = ax^2 + 2hxy + by^2 + 2gx + 2fy + c,$$
$$S_1 = ax_1^2 + 2hx_1y_1 + by_1^2 + 2gx_1 + 2fy_1 + c$$
$$\text{and } \quad T = axx_1 + h(xy_1 + yx_1) + byy_1 + g(x + x_1) + f(y + y_1) + c,$$

then the equation (6) can be written as

$$(T - S_1)^2 = (S + S_1 - 2T).S_1, \quad \text{or} \quad SS_1 = T^2. \tag{4.5.7}$$

It is the required equation.

(C) **Director Circle** : *(Locus of the points of intersection of pair of perpendicular tangents.)*

Director Circle : *If the locus of the points of intersection of pair of perpendicular tangents to a conic is a circle, then that circle is called the director circle of the conic.*

(i) **Circle** : $x^2 + y^2 = a^2$.

The equation of the pair of tangents from (x_1, y_1) to the circle is

$$(x^2 + y^2 - a^2)(x_1^2 + y_1^2 - a^2) = (xx_1 + yy_1 - a^2)^2.$$

If these lines are at right angle, then the coefficient of x^2+ the coefficient of $y^2 =$ 0, i.e., $(x_1^2 + y_1^2 - a^2 - x_1^2) + (x_1^2 + y_1^2 - a^2 - y_1^2) = 0$, or $x_1^2 + y_1^2 = 2a^2$.

Hence the locus of (x_1, y_1), i.e., the equation of the director circle is

$$x^2 + y^2 = 2a^2.$$

(ii) **Parabola** : $y^2 = 4ax$.

The equation of the pair of tangents from (x_1, y_1) to the parabola is

$$(y^2 - 4ax)(y_1^2 - 4ax_1) = (yy_1 - 2ax - 2ax_1)^2.$$

If these lines are at right angle, then

$$4a^2 + y_1^2 - y_1^2 + 4ax_1 = 0, \quad \text{or} \quad x_1 + a = 0.$$

Hence the locus of (x_1, y_1) is $x + a = 0$. It is a straight line representing the directrix of the parabola. Thus the points of intersection of perpendicular tangents to a parabola lie on the directrix.

(iii) **Ellipse** : $\dfrac{x^2}{a^2} + \dfrac{y^2}{b^2} = 1$.

The equation of the pair of tangents from (x_1, y_1) to the ellipse is

$$\left(\frac{x^2}{a^2} + \frac{y^2}{b^2} - 1\right)\left(\frac{x_1^2}{a^2} + \frac{y_1^2}{b^2} - 1\right) = \left(\frac{xx_1}{a^2} + \frac{yy_1}{b^2} - 1\right)^2.$$

If these lines are perpendicular to each other, then

$$\frac{1}{a^2}\left(\frac{x_1^2}{a^2} + \frac{y_1^2}{b^2} - 1\right) - \frac{x_1^2}{a^4} + \frac{1}{b^2}\left(\frac{x_1^2}{a^2} + \frac{y_1^2}{b^2} - 1\right) - \frac{y_1^2}{b^4} = 0,$$

$$\text{or} \quad x_1^2 + y_1^2 = a^2 + b^2.$$

Hence the director circle of the ellipse is $x^2 + y^2 = a^2 + b^2$.

(iv) **Hyperbola** : $\dfrac{x^2}{a^2} - \dfrac{y^2}{b^2} = 1$.

The equation of the pair of tangents from (x_1, y_1) to the hyperbola is

$$\left(\frac{x^2}{a^2} - \frac{y^2}{b^2} - 1\right)\left(\frac{x_1^2}{a^2} - \frac{y_1^2}{b^2} - 1\right) = \left(\frac{xx_1}{a^2} - \frac{yy_1}{b^2} - 1\right)^2.$$

For the pair of perpendicular tangents,

$$\frac{1}{a^2}\left(\frac{x_1^2}{a^2} - \frac{y_1^2}{b^2} - 1\right) - \frac{x_1^2}{a^4} - \frac{1}{b^2}\left(\frac{x_1^2}{a^2} - \frac{y_1^2}{b^2} - 1\right) - \frac{y_1^2}{b^4} = 0,$$

$$\text{or} \quad x_1^2 + y_1^2 = a^2 - b^2.$$

Hence the equation of the director circle is $x^2 + y^2 = a^2 - b^2$.

4.6 Chord of Contact

DEFINITION. *It is a chord joining the points of contact of tangents to a conic from a given point not lying on the conic.*

Let

$$ax^2 + 2hxy + by^2 + 2gx + 2fy + c = 0 \tag{4.6.1}$$

be the equation of the conic and (x_1, y_1) be the point from which tangents are drawn to the conic.

Let (x_2, y_2) and (x_3, y_3) be the points of contact. The equation of tangents at (x_2, y_2) and (x_3, y_3) are

$$axx_2 + h(xy_2 + yx_2) + byy_2 + g(x + x_2) + f(y + y_2) + c = 0 \tag{4.6.2}$$

$$\text{and} \quad axx_3 + h(xy_3 + yx_3) + byy_3 + g(x + x_3) + f(y + y_3) + c = 0. \tag{4.6.3}$$

Since these two tangents pass through the point (x_1, y_1),

$$ax_1x_2 + h(x_1y_2 + y_1x_2) + by_1y_2 + g(x_1 + x_2) + f(y_1 + y_2) + c = 0$$

$$\text{and} \quad ax_1x_3 + h(x_1y_3 + y_1x_3) + by_1y_3 + g(x_1 + x_3) + f(y_1 + y_3) + c = 0.$$

The above two conditions suggest that the line

$$axx_1 + h(xy_1 + yx_1) + byy_1 + g(x + y_1) + f(y + y_1) + c = 0$$

passes through (x_2, y_2) and (x_3, y_3). Hence it is the equation of the chord of contact of tangents through (x_1, y_1).

NOTE. This equation is identical with the equation of the tangent at (x_1, y_1) on the conic. But here the point (x_1, y_1) does not lie on the conic.

4.7 Pole and Polar

Method I.

DEFINITION. *The polar of a point with respect to a conic is the locus of the points of intersection of tangents at the extremities of the chords through that point while the point itself is called the pole of its polar.*

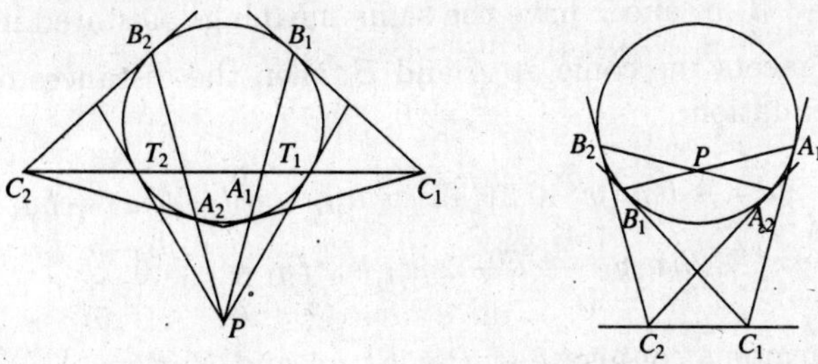

(i) (ii)

Fig. 4.2

Let the equation of the conic be

$$ax^2 + 2hxy + by^2 + 2gx + 2fy + c = 0 \tag{4.7.1}$$

and (x_1, y_1) be the pole. Let the tangents at the extremities of a chord through (x_1, y_1) meet at (x_2, y_2). According to the definition of the polar (x_2, y_2) lies on the polar of (x_1, y_1).

The chord of contact of (x_2, y_2) is

$$axx_2 + h(xy_2 + yx_2) + byy_2 + g(x + x_2) + f(y + y_2) + c = 0.$$

As (x_1, y_1) lies on it,

$$ax_1x_2 + h(x_1y_2 + y_1x_2) + by_1y_2 + g(x_1 + x_2) + f(y_1 + y_2) + c = 0.$$

It shows that the locus of (x_2, y_2) is the line

$$axx_1 + h(xy_1 + yx_1) + byy_1 + g(x + x_1) + f(y + y_1) + c = 0. \tag{4.7.2}$$

It is the polar of (x_1, y_1). (In the fig. 4.2, C_1C_2 is the polar of P).

Method II.

DEFINITION. *If a secant of a conic from a point P meets the conic at A and B and the point C on the secant is the harmonic conjugate of P with respect to A and B then the locus of C is the polar of P and P is the pole of this polar.*

$$\left[\frac{1}{PA} + \frac{1}{PB} = \frac{2}{PC}\right].$$

Let the equation of the conic be

$$ax^2 + 2hxy + by^2 + 2gx + 2fy + c = 0 \tag{4.7.3}$$

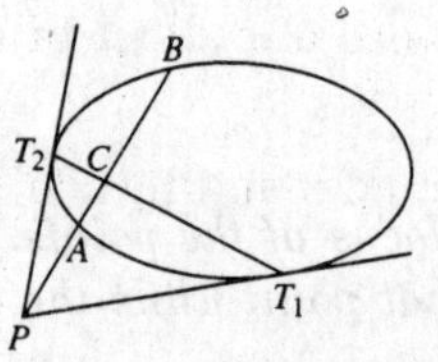

Fig. 4.3

and the co-ordinates of P be (x_1, y_1).

Let the equation of the secant through P be

$$\frac{x - x_1}{l} = \frac{y - y_1}{m} (= r \text{ say}). \tag{4.7.4}$$

[l, m and r have the same meanings as stated in section 3.1.]

If this secant meets the conic at A and B, then the distances of PA and PB are the roots of the equation

$$(al^2 + 2hlm + bm^2)r^2 + 2\{(ax_1 + hy_1 + g)l + (hx_1 + by_1 + f)m\}r$$
$$+ax_1^2 + 2hx_1y_1 + by_1^2 + 2gx_1 + 2fy_1 + c = 0. \tag{4.7.5}$$

If C be the harmonic conjugate of P w.r.t. A and B, then

$$\frac{2}{PC} = \frac{1}{PA} + \frac{1}{PB}.$$

Let $PC = \rho, PA = r_1$, and $PB = r_2$. [r_1 and r_2 are the roots of (3)].

Now

$$\frac{1}{r_1} + \frac{1}{r_2} = \frac{2}{\rho}, \quad \text{or} \quad \frac{r_1 + r_2}{r_1 r_2} = \frac{2}{\rho}.$$

From (4.7.5),

$$\frac{-2\{(ax_1 + hy_1 + g)l + (hx_1 + by_1 + f)m\}}{ax_1^2 + 2hx_1y_1 + by_1^2 + 2gx_1 + 2fy_1 + c} = \frac{2}{\rho}. \tag{4.7.6}$$

If (α, β) be the co-ordinate of C, then

$$\frac{\alpha - x_1}{l} = \frac{\beta - y_1}{m} = \rho. \quad [\because \quad C \text{ lies on (4.7.4)}] \tag{4.7.7}$$

From (4.7.7)

$$l = \frac{\alpha - x_1}{\rho}, \quad m = \frac{\beta - y_1}{\rho}.$$

Putting in (4.7.6)

$$-(ax_1 + hy_1 + g)(\alpha - x_1) - (hx_1 + by_1 + f)(\beta - y_1)$$
$$= ax_1^2 + 2hx_1y_1 + by_1^2 + 2gx_1 + 2fy_1 + c,$$
$$\text{or} \quad a\alpha x_1 + h(\alpha y_1 + \beta x_1) + b\beta y_1 + g(\alpha + x_1) + f(\beta + y_1) + c = 0.$$

Hence the locus of (α, β), i.e., the equation of the polar of (x_1, y_1) is

$$axx_1 + h(xy_1 + yx_1) + byy_1 + g(x + x_1) + f(y + y_1) + c = 0.$$

(In the fig. 4.3, T_1T_2 is the polar of P. PT_1 and PT_2 are the pair of tangents through P.)

NOTE 1. The polar of a point w.r.t. a conic coincides with the chord of contact of tangents from that point to the conic when the point is outside the conic.

NOTE 2. If P is the midpoint of the chord AB then r_1 and r_2 are equal in magnitude but opposite in sign. In this case

$$\frac{2}{\rho} = \frac{1}{r_1} + \frac{1}{r_2} = \frac{1}{r_1} - \frac{1}{r_1} = 0. \quad \therefore \quad \rho = \infty.$$

It indicates that the harmonic conjugate point of P is far away from the conic. Therefore, there will be no real polar of a point which is the midpoint of all chords through it. Hence for a central conic the centre has no real polar.

4.8 Properties of Pole and Polar

(i) *If the polar of P passes through Q then the polar of Q will pass through P. These two points are called conjugate points.*

Let the equation of the conic be

$$ax^2 + 2hxy + by^2 + 2gx + 2fy + c = 0. \tag{4.8.1}$$

The polar of $P(x_1, y_1)$ is

$$axx_1 + h(xy_1 + yx_1) + byy_1 + g(x + x_1) + f(y + y_1) + c = 0. \tag{4.8.2}$$

As $Q(x_2, y_2)$ lies on it,

$$ax_1x_2 + h(x_1y_2 + x_2y_1) + by_1y_2 + g(x_1 + x_2) + f(y_1 + y_2) + c = 0. \tag{4.8.3}$$

The polar of $Q(x_2, y_2)$ is

$$axx_2 + h(xy_2 + yx_2) + byy_2 + g(x + x_2) + f(y + y_2) + c = 0. \tag{4.8.4}$$

The condition (4.8.3) sugggest that P lies on (4.8.4). Hence the result follows.

NOTE. It is the reciprocal property of pole and polar.

(ii) *If the pole of a line L w.r.t. a conic lies on another line L', then the pole of L' w.r.t. this conic lies on L. These two lines are called conjugate lines.*

Let the pole of L w.r.t. the conic (4.8.1) be (x_1, y_1) and that of L' w.r.t. this conic be (x_2, y_2).

Therefore, the equation of L is

$$axx_1 + h(xy_1 + yx_1) + byy_1 + g(x + x_1) + f(y + y_1) + c = 0 \tag{4.8.5}$$

and the equation of L' is

$$axx_2 + h(xy_2 + yx_2) + byy_2 + g(x + x_2) + f(y + y_2) + c = 0. \tag{4.8.6}$$

As (x_1, y_1) lies on (4.8.6),

$$ax_1x_2 + h(x_1y_2 + y_1x_2) + by_1y_2 + g(x_1 + x_2) + f(y_1 + y_2) + c = 0.$$

It is also the condition for (x_2, y_2) to lie on (4.8.5). Thus the proposition is proved.

(iii) *If two lines L and L' meet at C then the polar of C w.r.t. a conic passes through the poles of L and L' w.r.t. this conic.*

Let P and Q be the poles of L and L' w.r.t. a conic S.

Therefore, the polar of P which is L passes through C and also the polar of Q which is L' passes through C. By property (i) the polar of C passes through P and Q. Hence the proposition is proved.

(iv) **Some special properties for circle:**

(a) *The distances of two points $P(x_1, y_1)$ and $Q(x_2, y_2)$ from the centre of a circle are proportional to the distance of each from the polar of the other. [Salmon's theorem]*

Let $x^2 + y^2 = a^2$ be the equation of the circle whose centre is the origin O.

$$\therefore \quad OP = \sqrt{x_1^2 + y_1^2}, \quad OQ = \sqrt{x_2^2 + y_2^2}.$$

The polars of P and Q are $xx_1 + yy_1 - a^2 = 0$ and $xx_2 + yy_2 - a^2 = 0$.

The distances of P and Q from the polars of Q and P are

$$\frac{x_1x_2 + y_1y_2 - a^2}{\sqrt{x_2^2 + y_2^2}} = d_1 \text{ (say)} \quad \text{and} \quad \frac{x_1x_2 + y_1y_2 - a^2}{\sqrt{x_1^2 + y_1^2}} = d_2 \text{ (say)}.$$

Now

$$\frac{d_1}{d_2} = \frac{\sqrt{x_1^2 + y_1^2}}{\sqrt{x_2^2 + x_2^2}} = \frac{OP}{OQ}.$$

(b) *The polar of a point w.r.t. a circle is perpendicular to the line joining the centre and the point.*

Let $x^2 + y^2 = a^2$ be the equation of the circle. Its centre is $(0, 0)$.

Let $P(x_1, y_1)$ be the point. The polar of P is $xx_1 + yy_1 = a^2$. The equation of the line through $(0, 0)$ and P is $xy_1 - yx_1 = 0$. Obviously these two lines are perpendicular to each other.

Hence the proposition is proved.

(c) *If O be the centre of a circle, P be any point and OP (produced when necessary) meets the polar of P w.r.t. the circle at Q then $OP.OQ =$ (radius)^2.*

Let $x^2 + y^2 = a^2$ be the equation of the circle. Then O is $(0, 0)$. The polar of $P(x_1, y_1)$ is

$$xx_1 + yy_1 = a^2. \tag{4.8.7}$$

Since OP is perpendicular to the polar of P, OQ is the perpendicular distance of O from (4.8.6).

$$\therefore \quad OQ = \pm\frac{a^2}{\sqrt{x_1^2 + y_1^2}} = \pm\frac{a^2}{OP}. \quad \therefore \quad OP.OQ = (\text{radius})^2.$$

NOTE. **Self-conjugate or self-polar triangle:**

If three points A, B, C are such that any two of them are conjugate points, then the triangle ABC is called a self-conjugate (or self-polar) triangle. Since the polar of A passes through B and C, BC is the polar of A. Therefore, each side is the polar of the opposite vertex and also any two sides are conjugate lines.

4.9 Pole of a given Line w.r.t. a Conic

Let the equation of the conic be

$$ax^2 + 2hxy + by^2 + 2gx + 2fy + c = 0 \tag{4.9.1}$$

and (x_1, y_1) be the pole of the given line $lx + my + n = 0 \ldots (4.9.2)$ w.r.t. the conic (4.9.1).

The polar of (x_1, y_1) w.r.t. the conic (4.9.1) is

$$axx_1 + h(xy_1 + yx_1) + byy_1 + g(x + x_1) + f(y + y_1) + c = 0,$$

$$\text{or} \quad (ax_1 + hy_1 + g)x + (hx_1 + by_1 + f)y + gx_1 + fy_1 + c = 0.$$

It is identical with (4.9.2).

$$\therefore \quad \frac{ax_1 + hy_1 + g}{l} = \frac{hx_1 + by_1 + f}{m} = \frac{gx_1 + fy_1 + c}{n}.$$

► **Example 1** *Find the pole of the line $x + 2y + 3 = 0$ w.r.t. the circle*

$$x^2 + y^2 - 2x + 5 = 0.$$

Let (x_1, y_1) be the pole. The polar of it w.r.t. the given circle is

$$x_1 + yy_1 - (x + x_1) + 5 = 0, \quad \text{or} \quad (x_1 - 1)x + y_1 y + 5 - x_1 = 0.$$

Comparing it with the given line

$$\frac{x_1 - 1}{1} = \frac{y_1}{2} = \frac{5 - x_1}{3}.$$

From these equations

$$3x_1 - 3 = 5 - x_1, \quad \text{or} \quad x_1 = 2; y_1 = 2x_1 - 2 = 4 - 2 = 2.$$

$\therefore$ (2, 2) is the required pole.

4.10 Lengths of Subtangent, Subnormal, Normal and Tangent

DEFINITION. *If the tangent and normal at any point P of a conic meet the x-axis at T and N respectively and PM is the ordinate of P, then*

length of subtangent $= MT$,

length of subnormal $= MN$,

length of tangent $= PT$

and length of normal $= PN$.

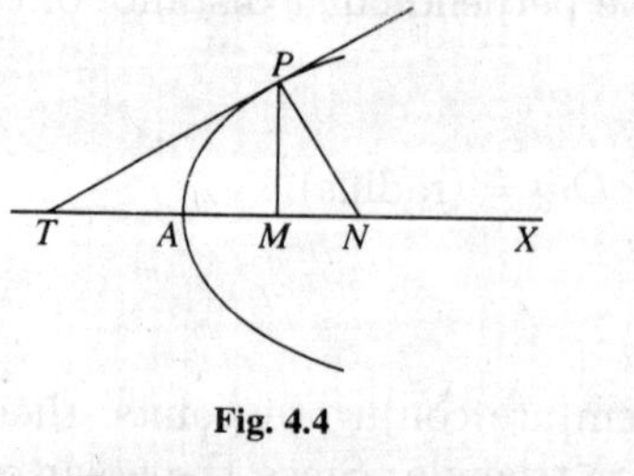

Fig. 4.4

Length of the tangent (in general sense) : *If the tangent to a conic through a point P (outside the conic) meets the conic at T, then PT is called the length of the tangent.*

Length of the tangent to a circle:

Let

$$x^2 + y^2 + 2gx + 2fy + c = 0 \tag{4.10.1}$$

be the equation of the circle, $P(x_1, y_1)$ be a point outside the circle and PT be the tangent to it. $C(-g, -f)$ is the centre of the circle and the radius CT is perpendicular to PT. From the fig. 4.5,

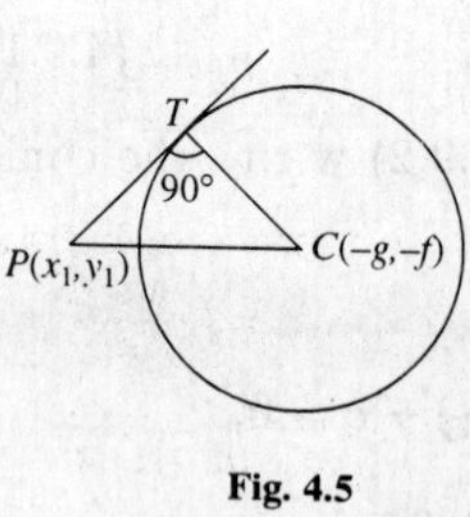

Fig. 4.5

$$PT^2 = PC^2 - CT^2 = (x_1 + g)^2 + (y_1 + f)^2 - (g^2 + f^2 - c)$$

$$= x_1^2 + y_1^2 + 2gx_1 + 2fy_1 + c.$$

$$\therefore \quad PT = \sqrt{(x_1^2 + y_1^2 + 2gx_1 + 2fy_1 + c)}$$

$$= \text{length of the tangent.}$$

Worked Out Examples

1. Find the equation of the tangent and the normal at $(1, -1)$ to the conic

$$y^2 - xy - 2x^2 - 5y + x - 6 = 0.$$

The point $(1, -1)$ is on the conic. Therefore, the equation of the tangent at $(1, -1)$ is

$$y(-1) - \frac{1}{2}\{x(-1) + y.1\} - 2x.1 - \frac{5}{2}(y - 1) + \frac{1}{2}(x + 1) - 6 = 0,$$

or $x + 4y + 3 = 0$.

Since the normal is perpendicular to the tangent at $(1, -1)$, its equation is $y + 1 = 4(x - 1)$, or $4x - y - 5 = 0$.

2. Find the equations to the tangents to the conic

$$x^2 + 4xy + 3y^2 - 5x - 6y + 3 = 0,$$

which are parallel to $x + 4y = 0$. **[B.U. '81; C.U. '67]**

Let the equation of any such tangent be $x + 4y + c = 0$, or $x = -(y + c)$.

Putting this value of x in the equation of the conic,

$$(4y + c)^2 - 4y(4y + c) + 3y^2 + 5(4y + c) - 6y + 3 = 0,$$

$$\text{or} \quad 3y^2 + 2(2c + 7)y + c^2 + 5c + 3 = 0.$$

Since the line is the tangent to the conic, the above equation in y must have equal roots.

$$\therefore \quad 4(2c + 7)^2 - 4.3(c^2 + 5c + 3) = 0,$$

$$\text{or} \quad c^2 + 13c + 40 = 0, \quad \text{or} \quad (c + 5)(c + 8) = 0, \quad \text{or} \quad c = -5, -8.$$

$\therefore$ the required tangents are $x + 4y - 5 = 0$ and $x + 4y - 8 = 0$.

3. Prove that the normal chord of a parabola at the point whose ordinate is equal to its abscissa, subtends a right angle at the focus.

Let

$$y^2 = 4ax \tag{1}$$

be the equation of the parabola. The point, whose ordinate is equal to abscissa, is obtained by putting $y = x$ in (1). $x^2 = 4ax$, or $x = 0, 4a$. Therefore, the point is $(0, 0)$, or $(4a, 4a)$. At $(0, 0)$ x-axis is the normal which passes through the focus, so it will not serve our purpose. The normal at $(4a, 4a)$ is

$$y - 4a = -\frac{4a}{2a}(x - 4a), \quad \text{or} \quad y + 2x = 12a. \tag{2}$$

Solving (1) and (2) we get the ends of this normal chord, which are $(4a, 4a)$ and $(9a, -6a)$. The gradient of the line joining the focus $(a, 0)$ and $(4a, 4a) = \dfrac{4a}{4a - a} = \dfrac{4}{3}$ and the gradient of the line joining the focus $(a, 0)$ and $(9a, -6a) = \dfrac{-6a}{9a - a} = -\dfrac{3}{4}$.

The product of these two gradients $= -1$. Hence the result follows.

4. Show that the tangents at the extremities of a focal chord of a parabola meet at right angle on the directrix.

Let $y^2 = 4ax$ be the equation of the parabola and the ends of a focal chord be $(at^2, 2at)$ and $\left(\frac{a}{t^2}, -\frac{2a}{t}\right)$.

The tangents at these points are

$$yt = x + at^2 \tag{1}$$

$$\text{and} \quad -\frac{y}{t} = x + \frac{a}{t^2}, \quad \text{or} \quad -ty = xt^2 + a. \tag{2}$$

Obviously (1) and (2) are at right angle. Again adding (1) and (2) $(x+a)\,(1+t^2) = 0$, or $x + a = 0$.

$\therefore$ the tangents meet on the directrix $x + a = 0$ at right angle.

5. Find the equation to the common tangents of the circle $x^2 + y^2 = 4ax$ and the parabola $y^2 = 4ax$.

Let $y = mx + c$ be the common tangent. If it is tangent to $y^2 = 4ax$ then $c = \frac{a}{m}$.

$$\therefore \quad y = mx + \frac{a}{m}. \tag{1}$$

To find the point of intersection between (1) and the circle,

$$x^2 + \left(mx + \frac{a}{m}\right)^2 = 4ax, \quad \text{or} \quad (1+m^2)x^2 - 2ax + \frac{a^2}{m^2} = 0.$$

If the line (1) be the tangent, the above equation must have equal roots. From the condition of equal roots

$$4a^2 - 4(1+m^2)\cdot\frac{a^2}{m^2} = 0, \quad \text{or} \quad \frac{a^2}{m^2} = 0.$$

Here $a \neq 0$. $\therefore \quad \frac{1}{m} = 0$.

Form (1)

$$\frac{y}{m} = x + \frac{a}{m^2}, \quad \text{or} \quad x = 0.$$

$\therefore$ $x = 0$ is the common tangent.

6. Show that the subnormal at any point of a parabola is equal to its semi-latus rectum. [C.U. '66]

Let $y^2 = 4ax$ be the equation of the parabola and $(at^2, 2at)$ be a point on it.

The equation of the normal at this point is

$$y - 2at = -t(x - at^2).$$

It meets the x-axis at $(2a + at^2, 0)$.

The length of the subnormal is the distance between $(at^2, 0)$ and the point, where the normal meets the x-axis, i.e., $(2a + at^2, 0)$. The distance between these points $= 2a$. Length of the latus rectum $= 4a$. Hence the length of the subnormal at any point of the parabola is equal to semi-latus rectum.

7. Find the pole of the focal chord of the parabola $y^2 = 4ax$, passing through $(9a, 6a)$.

The equation of the chord passing through the focus $(a, 0)$ and $(9a, 6a)$ is

$$y = \frac{6a}{9a-a}(x-a), \quad \text{or} \quad 3x - 4y - 3a = 0. \tag{1}$$

Let (α, β) be the pole of it. The equation of the polar of (α, β) is $y\beta = 2a(x+\alpha)$, or $2ax - y\beta + 2a\alpha = 0$.

It is identical with (1). Comparing the coefficients

$$\frac{2a}{3} = \frac{\beta}{4} = \frac{2a\alpha}{-3a}. \quad \therefore \quad \alpha = -a, \beta = \frac{8a}{3}.$$

8. Show that the locus of the poles of tangents of the parabola $y^2 = 4ax$ w.r.t. the parabola $y^2 = 4bx$ is the parabola $y^2 = \frac{4b^2}{a}x$.

Let (α, β) be the pole. The polar of it w.r.t. $y^2 = 4bx$ is

$$y\beta = 2b(x+\alpha), \quad \text{or} \quad y = \frac{2b}{\beta}x + \frac{2b\alpha}{\beta}.$$

If it is tangent to the parabola $y^2 = 4ax$, then

$$\frac{2b\alpha}{\beta} = \frac{a}{2b/\beta}, \quad \text{or} \quad \beta^2 = \frac{4b^2}{a}\alpha.$$

Hence the required locus is $y^2 = \frac{4b^2}{a}x$.

9. Prove that the two parabolas $y^2 = 4ax$ and $x^2 = 4by$ cut one another at an angle

$$\tan^{-1}\frac{3(ab)^{1/3}}{2(a^{2/3}+b^{2/3})}.$$

$$y^2 = 4ax \tag{1}$$

$$x^2 = 4by. \tag{2}$$

Let (x_1, y_1) be the common point of (1) and (2). The tangents at this point are

$$yy_1 = 2a(x+x_1) \tag{3}$$

$$\text{and} \quad xx_1 = 2b(y+y_1). \tag{4}$$

If θ be the angle between (3) and (4) then

$$\tan\theta = \frac{4ab - x_1y_1}{-2(ax_1+by_1)}. \tag{5}$$

We have $y_1^2 = 4ax_1$ and $x_1^2 = 4by_1$. From these two $y_1^4 = 64a^2by_1$.

$$\therefore \quad y_1 = 0, \quad \text{or} \quad a^{2/3}b^{1/3}.$$

Similarly $x_1 = 0$, or $4a^{1/3}b^{2/3}$.

Putting $x_1 = 4a^{1/3}b^{2/3}$ and $y_1 = 4a^{2/3}b^{1/3}$ in (5)

$$\tan\theta = \frac{4ab - 16ab}{-8(a^{4/3}b^{2/3} + a^{2/3}b^{4/3})} = \frac{3(ab)^{1/3}}{2(a^{2/3} + b^{2/3})}.$$

$$\therefore \quad \theta = \tan^{-1}\frac{3(ab)^{1/3}}{2(a^{2/3} + b^{2/3})}.$$

For $x_1 = 0, y_1 = 0, \tan\theta = \infty$.

$\therefore \quad \theta = \frac{\pi}{2}$. Thus the parabolas cut at right angle at the origin.

10. Show that the line $lx + my = n$ is a normal to the ellipse

$$\frac{x^2}{a^2} + \frac{y^2}{b^2} = 1, \quad \text{if} \quad \frac{a^2}{l^2} + \frac{b^2}{m^2} = \frac{(a^2 - b^2)^2}{n^2}.$$

[C.U. '62; B.U. '68]

Let $lx + my = n$ be the normal to the ellipse at $(a\cos\phi, b\sin\phi)$. The equation of the normal at this point is

$$\frac{x - a\cos\phi}{\frac{a\cos\phi}{a^2}} = \frac{y - b\sin\phi}{\frac{b\sin\phi}{b^2}}, \quad \text{or} \quad a\sec\phi x - b\operatorname{cosec}\phi y = a^2 - b^2.$$

It is identical with the given line. Comparing these two

$$\frac{a\sec\phi}{l} = \frac{-b\operatorname{cosec}\phi}{m} = \frac{a^2 - b^2}{n}. \quad \therefore \quad \cos\phi = \frac{an}{l(a^2 - b^2)}, \sin\phi =. -\frac{bn}{m(a^2 - b^2)}.$$

Squaring and adding,

$$1 = \left(\frac{a^2}{l^2} + \frac{b^2}{m^2}\right)\frac{n^2}{(a^2 - b^2)^2}, \quad \text{or} \quad \frac{a^2}{l^2} + \frac{b^2}{m^2} = \frac{(a^2 - b)^2}{n^2}.$$

11. Show that the foot of the perpendicular from the focus on any tangent to the hyperbola lies on the auxiliary circle.

Let $(a\sec\theta, b\tan\theta)$ be a point on hyperpola

$$\frac{x^2}{a^2} - \frac{y^2}{b^2} = 1. \tag{1}$$

The equation of the tangent at this point is

$$\frac{x}{a}\sec\theta - \frac{y}{b}\tan\theta = 1. \tag{2}$$

The equation of the line perpendicular to (2) and passing through the focus $(ae, 0)$ is

$$\frac{x}{b}\tan\theta + \frac{y}{a}\sec\theta = \frac{ae}{b}\tan\theta. \tag{3}$$

Here e is the eccentricity. To find the locus of the intersection between (2) and (3) we have to eliminate θ from these two equations. Squaring (2) and (3) and then adding we have

$$(x^2 + y^2)\left(\frac{\sec^2\theta}{a^2} + \frac{\tan^2\theta}{b^2}\right) = 1 + \frac{a^2e^2}{b^2}\tan^2\theta = 1 + \frac{a^2 + b^2}{b^2}\tan^2\theta$$

$$= (1 + \tan^2\theta) + \frac{a^2}{b^2}\tan^2\theta = a^2\left(\frac{\sec^2\theta}{a^2} + \frac{\tan^2\theta}{b^2}\right),$$

or $x^2 + y^2 = a^2$. It is the auxiliary circle.

Similarly considering the other focus $(-ae, 0)$ the same result is obtained.

12. An ellipse slides between two straight lines at right angles to each other. Show that the locus of its centre is a circle.

Let

$$\frac{x^2}{a^2} + \frac{y^2}{b^2} = 1 \qquad (1)$$

be the equation of the ellipse.

The lines on which the ellipse slides must be perpendicular tangents to (1). Let the equations of these tangents be

$$y = mx + \sqrt{a^2m^2 + b^2} \quad \text{and} \quad my = -x + \sqrt{a^2 + b^2m^2},$$

If (α, β) be the centre of the ellipse referred to these tangents as axes then α and β are equal to distances from $(0, 0)$ to these lines.

$$\therefore \quad \alpha = \sqrt{\frac{a^2m^2 + b^2}{1 + m^2}} \quad \text{and} \quad \beta = \sqrt{\frac{a^2 + b^2m^2}{1 + m^2}}.$$

Thus

$$\alpha^2 + \beta^2 = \frac{a^2m^2 + b^2 + a^2 + b^2m^2}{1 + m^2} = \frac{(a^2 + b^2)(1 + m^2)}{1 + m^2} = a^2 + b^2.$$

Hence the required locus is $x^2 + y^2 = a^2 + b^2$.

13. If the normal at any end of a latus rectum of an ellipse passes through one end of the minor axis, then prove that $e^4 + e^2 = 1$, when e is the eccentricity.

[N.B. '87, '95; C.H. '93]

Let

$$\frac{x^2}{a^2} + \frac{y^2}{b^2} = 1$$

be the equation of an ellipse with a and b as semi-major and semi-minor axes. The co-ordinates of one end of a latus rectum can be taken as $\left(ae, \frac{b^2}{a}\right)$. The equation of the normal at this point is

$$\frac{x - ae}{\frac{ae}{a^2}} = \frac{y - \frac{b^2}{a}}{\frac{b^2}{a}/b^2}, \quad \text{or} \quad \frac{x}{e} - a = y - \frac{b^2}{a}.$$

This normal passes through $(0, -b)$, $\quad \therefore \quad -a = -b - \frac{b^2}{a}$.

We have

$$b^2 = a^2(1 - e^2). \quad \therefore \quad a = b + a(1 - e^2),$$

$$\text{or} \quad ae^2 = b, \quad \text{or} \quad a^2e^4 = b^2,$$

$$\text{or} \quad a^2e^4 = a^2(1 - e^2), \quad \text{or} \quad e^4 + e^2 = 1.$$

14. Prove that the locus of the pole w.r.t. the ellipse $\frac{x^2}{a^2} + \frac{y^2}{b^2} = 1$ of any tangent to the auxiliary circle is the curve $\frac{x^2}{a^4} + \frac{y^2}{b^4} = \frac{1}{a^2}$.

Let (α, β) be the pole. Its polar w.r.t. the given ellipse is

$$\frac{x\alpha}{a^2} + \frac{y\beta}{b^2} = 1. \tag{1}$$

If it is tangent to the auxiliary circle $x^2 + y^2 = a^2$ then the perpendicular distance from $(0, 0)$ to (1) is a.

$$\therefore \quad \frac{1}{\sqrt{\dfrac{\alpha^2}{a^4} + \dfrac{\beta^2}{b^4}}} = a, \quad \text{or} \quad \frac{\alpha^2}{a^4} + \frac{\beta^2}{b^4} = \frac{1}{a^2}.$$

Hence the required locus is

$$\frac{x^2}{a^4} + \frac{y^2}{b^4} = \frac{1}{a^2}.$$

15. Find the locus of the poles of the normal chords of

(a) parabola $y^2 = 4ax$; and **[B.U. '66]**

(b) ellipse $\dfrac{x^2}{a^2} + \dfrac{y^2}{b^2} = 1$.

(a) If (α, β) be the pole, then the polar of it w.r.t. $y^2 = 4ax$ is

$$y\beta = 2a(x + \alpha). \tag{1}$$

Let it be normal to the parabola at $(at^2, 2at)$. The equation of the normal at this point is

$$y - 2at = -t(x - at^2). \tag{2}$$

Comparing (1) and (2)

$$\frac{\beta}{1} = \frac{2a}{-t} = \frac{2a\alpha}{at^3 + 2at}, \quad \text{or} \quad t = -\frac{2a}{\beta} \quad \text{and} \quad at^2 + 2a = -\alpha.$$

Eliminating t,

$$\frac{4a^3}{\beta^2} + 2a = -\alpha, \quad \text{or} \quad \beta^2(\alpha + 2a) + 4a^3 = 0.$$

Hence the required locus is $y^2(x + 2a) + 4a^3 = 0$.

(b) If (α, β) be the pole, then the polar of it w.r.t.

$$\frac{x^2}{a^2} + \frac{y^2}{b^2} = 1 \text{ is } \frac{x\alpha}{a^2} + \frac{y\beta}{b^2} = 1. \tag{1}$$

Let it be normal to the ellipse at $(a\cos\phi, b\sin\phi)$.

The equation of the normal at this point is

$$a\sec\phi\, x - b\operatorname{cosec}\phi\, y = a^2 - b^2. \tag{2}$$

Comparing (1) and (2),

$$\frac{a\sec\phi}{\alpha/a^2} = \frac{-b\operatorname{cosec}\phi}{\beta/b^2} = \frac{a^2 - b^2}{1},$$

$$\text{or} \quad \cos\phi = \frac{a^3}{\alpha(a^2 - b^2)}, \quad \sin\phi = \frac{-b^3}{\beta(a^2 - b^2)}.$$

Squaring and adding,

$$1 = \left(\frac{a^6}{\alpha^2} + \frac{b^6}{\beta^2}\right)\frac{1}{(a^2-b^2)^2}, \quad \text{or} \quad \frac{a^6}{\alpha^2} + \frac{b^6}{\beta^2} = (a^2-b^2)^2.$$

Hence the required locus is

$$\frac{a^6}{x^2} + \frac{b^6}{y^2} = (a^2-b^2)^2.$$

16. Find the equation and length of the common tangent to the hyperbolas

$$\frac{x^2}{a^2} - \frac{y^2}{b^2} = 1 \quad \text{and} \quad \frac{x^2}{b^2} - \frac{y^2}{a^2} = -1.$$

The hyperbolas are

$$\frac{x^2}{a^2} - \frac{y^2}{b^2} = 1 \tag{1}$$

$$\text{and} \quad \frac{x^2}{b^2} - \frac{y^2}{a^2} = -1. \tag{2}$$

Let the tangents at $(a\sec\theta, b\tan\theta)$ and $(b\tan\phi, a\sec\phi)$ to (1) and (2) respectively be identical.

These equations are

$$\frac{x}{a}\sec\theta - \frac{y}{b}\tan\theta = 1 \quad \text{and} \quad \frac{x}{b}\tan\phi - \frac{y}{a}\sec\phi = -1.$$

Comparing these two equations

$$\frac{\sec\theta}{a} = -\frac{\tan\phi}{b} \text{ and } \frac{\tan\theta}{b} = -\frac{\sec\phi}{a}.$$

$$\therefore \quad \sec^2\theta - \tan^2\theta = \frac{a^2}{b^2}\tan^2\phi - \frac{b^2}{a^2}\sec^2\phi,$$

$$\text{or} \quad 1 = \left(\frac{a^2}{b^2} - \frac{b^2}{a^2}\right)\tan^2\phi - \frac{b^2}{a^2},$$

$$\text{or} \quad \tan^2\phi = \frac{b^2(a^2+b^2)}{a^4-b^4} = \frac{b^2}{a^2-b^2} \text{ and } \sec^2\phi = \frac{a^2}{a^2-b^2}.$$

$$\tan^2\theta = \frac{b^2}{a^2}\sec^2\phi = \frac{b^2}{a^2-b^2} \text{ and } \sec^2\theta = \frac{a^2}{a^2-b^2}.$$

$\therefore$ the points of contact are

$$\left(\pm\frac{a^2}{\sqrt{a^2-b^2}}, \pm\frac{b^2}{\sqrt{a^2-b^2}}\right) \text{ and } \left(\mp\frac{b^2}{\sqrt{a^2-b^2}}, \mp\frac{a^2}{\sqrt{a^2-b^2}}\right).$$

Length of the common tangent is the distance between these points of contact. Hence the required length

$$= \sqrt{\left\{\frac{(a^2+b^2)^2}{a^2-b^2} + \frac{(a^2+b^2)^2}{a^2-b^2}\right\}} = \sqrt{2}\frac{a^2+b^2}{\sqrt{a^2-b^2}}.$$

The equation of the common tangents are $x \pm y = \pm\sqrt{a^2-b^2}$.

17. If the normal to the rectangular hyperbola $xy = c^2$ at $\left(ct_1, \frac{c}{t_1}\right)$ meets the curve at $\left(ct_2, \frac{c}{t_2}\right)$ then show that $t_1^3 t_2 = -1$. **[N.B. '90, '98]**

The tangent as $\left(ct_1, \frac{c}{t_1}\right)$ is

$$\frac{1}{2}\left(x\frac{c}{t_1} + yct_1\right) = c^2, \quad \text{or} \quad y = -\frac{x}{t_1^2} + \frac{2c}{t_1}.$$

Therefore, the equation of the normal at this point is $y - \frac{c}{t_1} = t_1^2(x - ct_1)$.

If it passes through $\left(ct_2, \frac{c}{t_2}\right)$, then

$$\frac{c}{t_2} - \frac{c}{t_1} = t_1^2(ct_2 - ct_1), \quad \text{or} \quad (t_1 - t_2)\left(\frac{1}{t_1 t_2} + t_1^2\right) = 0.$$

$$\because\ t_1 \neq t_2, \quad \frac{1}{t_1 t_2} + t_1^2 = 0, \quad \text{or} \quad t_1^3 t_2 = -1.$$

18. Show that the straight lines $x + y + 1 = 0$ and $2x + 3y + 4 = 0$ are conjugate w.r.t. the conic $y^2 = 4x$.

Let (α, β) be the pole of $x + y + 1 = 0$ w.r.t. $y^2 = 4x$.

The polar of (α, β) w.r.t. $y^2 = 4x$ is $y \cdot \beta = 2(x + \alpha)$, or $2x - \beta y + 2\alpha = 0$. It is identical with $x + y + 1 = 0$.

$$\therefore\ \frac{2}{1} = \frac{-\beta}{1} = \frac{2\alpha}{1}, \text{ or } \alpha = 1, \beta = -2.$$

Since $2.1 + 3 \cdot (-2) + 4 = 0$, the pole of $x + y + 1 = 0$ lies on $2x + 3y + 4 = 0$. Here the given lines are conjugate to each other.

19. The normals at the ends of the latus rectum of the parabola $y^2 = 4ax$ meet the curve again in Q and Q'. Prove that $QQ' = 12a$. **[C.H. '92]**

The co-ordinates of the ends of letus rectum are $(a, 2a)$ and $(a, -2a)$. From $y^2 = 4ax$, $\frac{dy}{dx} = \frac{2a}{y}$.

The equation of the normal at $(a, 2a)$ is

$$y - 2a = -1(x - a), \quad \text{or} \quad x = 3a - y \tag{1}$$

and the equation of the normal at $(a, -2a)$ is

$$y + 2a = (x - a), \quad \text{or} \quad x = y + 3a. \tag{2}$$

To find the point of intersection between (1) and $y^2 = 4ax$,

$$y^2 = 4a(3a - y), \quad \text{or} \quad y^2 + 4ay - 12a^2 = 0,$$

$$\text{or} \quad (y + 6a)(y - 2a) = 0, \quad \text{or} \quad y = 2a, -6a$$

For $y = -6a, x = 9a$.

If this normal meets the parabola at $y^2 = 4ax$ at Q then the co-ordinates of Q are $(9a, -6a)$.

Considering the points of intersection between (2) and $y^2 = 4ax$,

$$y^2 = 4a(y+3a), \quad \text{or} \quad y^2 - 4ay - 12a^2 = 0,$$

$$\text{or} \quad (y-6a)(y+2a) = 0, \quad \text{or} \quad y = -2a, 6a$$

For $y = 6a, x = 6a + 3a = 9a$.

$$\therefore \text{ the co-ordinates of } Q' \text{ are } (9a, 6a)$$

$$\therefore \ QQ' = \sqrt{(9a-9a)^2 + (6a+6a)^2} = 12a.$$

20. Show that the curves $\dfrac{x^2}{a^2+k_1} + \dfrac{y^2}{b^2+k_1} = 1$ and $\dfrac{x^2}{a^2+k_2} + \dfrac{y^2}{b^2+k_2} = 1$ intersect orthogonally.

Let

$$a^2 + k_1 = \frac{1}{A}, \quad b^2 + k_1 = \frac{1}{B}, \quad a^2 + k_2 = \frac{1}{A'}, \quad b^2 + k_2 = \frac{1}{B'}.$$

Then the equations are

$$Ax^2 + By^2 = 1 \tag{1}$$

$$\text{and} \quad A'x^2 + B'y^2 = 1 \tag{2}$$

From (1) $\dfrac{dy}{dx} = -\dfrac{A}{B}\dfrac{x}{y}$ and from (2) $\dfrac{dy}{dx} = -\dfrac{A'}{B'}\dfrac{x}{y}$

If the curves (1) and (2) meet at (x_1, y_1) then $Ax_1^2 + By_1^2 = 1, A'x_1^2 + B'y_1^2 = 1$ and gradients of tangents at the common point are

$$m_1 = -\frac{A}{B}\frac{x_1}{y_1}, \quad m_2 = -\frac{A'}{B'}\frac{x_1}{y_1}.$$

Now

$$Ax_1^2 + By_1^2 = A'x_1^2 + B'y_1^2, \quad \text{or} \quad (A-A')x_1^2 = (B'-B)y_1^2,$$

$$\text{or} \quad \frac{x_1^2}{y_1^2} = \frac{B'-B}{A-A'}$$

$$\text{and} \quad m_1 m_2 = \frac{AA'}{BB'}\frac{x_1^2}{y_1^2} = \frac{AA'}{BB'}\frac{B'-B}{A-A'} = \frac{B'-B}{BB'} \cdot \frac{1}{\frac{A-A'}{AA'}}$$

$$= \left(\frac{1}{B} - \frac{1}{B'}\right) \Big/ \left(\frac{1}{A'} - \frac{1}{A}\right) = (k_1 - k_2)/(k_2 - k_1) = -1.$$

$\therefore$ The curves cut orthogonally.

21. Show that the pole of any tangent to the rectangular hyperbola $xy = c^2$ w.r.t. the circle $x^2 + y^2 = a^2$ lies on a concentric and similar hyperbola.

Let $\left(ct, \dfrac{c}{t}\right)$ be a point on the hyperbola.

The equation of the tangent at this point is

$$x + t^2 y - 2ct = 0. \tag{1}$$

Let (α, β) be the pole of it w.r.t. the circle. The equation of the polar for this pole is

$$x\alpha + y\beta - a^2 = 0. \tag{2}$$

(1) and (2) are identical.

$$\therefore \quad \frac{\alpha}{1} = \frac{\beta}{t^2} = \frac{a^2}{2ct}, \quad t = \frac{a^2}{2c\alpha} = \frac{2c\beta}{a^2}, \quad \text{or} \quad \alpha\beta = \frac{a^4}{4c^2}.$$

$\therefore$ Locus of (α, β) is $xy = \dfrac{a^4}{4c^2}$ which is similar to $xy = c^2$.

22. Tangents are drawn from (h, k) to the circle $x^2 + y^2 = a^2$. Prove that the area of the triangle formed by them and the straight line joining their points of contact is $\dfrac{a(h^2 + k^2 - a^2)^{3/2}}{h^2 + k^2}$. **[N.B. '88, C.U. '63]**

The equation of the pair of tangents from (h, k) to the given circle is

$$(x^2 + y^2 - a^2)(h^2 + k^2 - a^2) = (xh + yk - a^2)^2,$$

$$\text{or} \quad (k^2 - a^2)x^2 + (h^2 - a^2)y^2 - 2hkxy + 2a^2hx + 2a^2ky - a^2(h^2 + k^2) = 0.$$

If θ be the angle between these tangents, then

$$\tan\theta = \frac{2\sqrt{\{h^2k^2 - (k^2 - a^2)(h^2 - a^2)\}}}{h^2 + k^2 - 2a^2} = \frac{2\sqrt{\{a^2(h^2 + k^2) - a^4\}}}{h^2 + k^2 - 2a^2}.$$

$$\therefore \quad \sin\theta = \frac{2\sqrt{\{a^2(h^2 + k^2) - a^4\}}}{\sqrt{\{4a^2(h^2 + k^2) - 4a^4 + (h^2 + k^2 - 2a^2)^2\}}} = \frac{2a\sqrt{(h^2 + k^2 - a^2)}}{h^2 + k^2}.$$

Length of the tangent $= \sqrt{(h^2 + k^2 - a^2)}$.

Now two sides of the triangle are equal to the length of the tangent and the angle between them is θ.

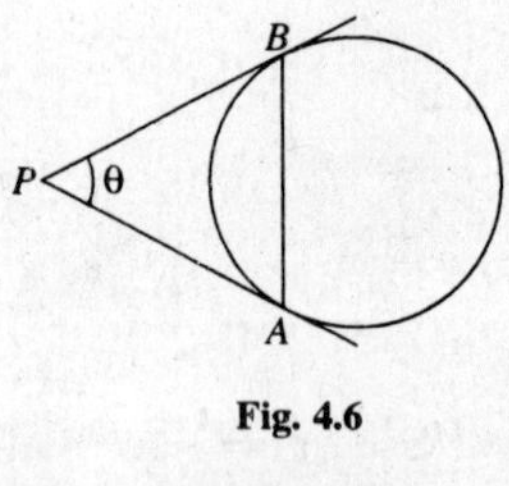

Fig. 4.6

$$\therefore \quad \text{area} = \frac{1}{2}\left(\sqrt{h^2 + k^2 - a^2}\right)^2 \cdot \sin\theta$$

$$= \frac{a(h^2 + k^2 - a^2)^{3/2}}{h^2 + k^2}.$$

23. Two tangents drawn to $y^2 = 4ax$ meet at angle $45°$. Find the locus of their point of intersection.

Let (h, k) be the point of intersection. The equation of the pair of tangents to $y^2 = 4ax$ through (h, k) is

$$(y^2 - 4ax)(k^2 - 4ah) = \{yk - 2a(x + h)\}^2,$$

$$\text{or} \quad a^2x^2 - akxy + ahy^2 + a(k^2 + 2ah)x - ahky + a^2h^2 = 0.$$

Since the angle between these lines is 45°,

$$\tan 45^\circ = \frac{2\sqrt{\left(\frac{a^2k^2}{4} - a^3h\right)}}{a^2 + ah}, \quad \text{or} \quad (a^2 + ah)^2 = 4\left(\frac{a^2k^2}{4} - a^3h\right),$$

$$\text{or} \quad (a + h)^2 = k^2 - 4ah.$$

Hence the required locus is $(x + a)^2 = y^2 - 4ax$.

Exercise

1. Find the equation of the tangents and normals to the following curves:

   (a) $x^2 + y^2 - 4x + 6y - 36 = 0$ at (2.4).

   (b) $2x^2 + 5xy + 3y^2 + 4x - 10y - 4 = 0$ at $(1, 1)$.

   (c) $13x^2 + 16xy - y^2 - 70x - 40y + 82 = 0$ at $(1, -25)$.

2. Find the points of contact and equations of tangents to

   (a) $x^2 + y^2 + 6x - 10y - 15 = 0$, which are parallel to x-axis;

   (b) $y^2 = 4ax$, which makes angle 60° with the x-axis.

3. (a) Show that $3x + 4y + 7 = 0$ touches the circle $x^2 + y^2 - 4x - 6y - 12 = 0$ and find the point of contact.

   (b) Find the condition so that $y = mx + c$ will be the tangent to $y^2 = 4(x + a)$.

   (c) Find the condition for which $lx + my + n = 0$ will be the normal to $\frac{x^2}{a^2} - \frac{y^2}{b^2} = 1$.

   (d) Prove that if the straight line $\lambda x + \mu y + v = 0$ touches the parabola $y^2 - 4px + 4pq = 0$, then $\lambda^2 q + \lambda v - p\mu^2 = 0$. **[C.U. '62]**

4. Prove that $x\cos\alpha + y\sin\alpha = a$ and $x\sin\alpha - y\cos\alpha = a$ are tangents to $x^2 + y^2 = a^2$ whatever α may be. Hence find the locus of the point of intersection from which two perpendicular tangents can be drawn to the circle $x^2 + y^2 = a^2$.

5. Find the equations to the tangents at the ends of the latera recta of $\frac{x^2}{a^2} + \frac{y^2}{b^2} = 1$ and show that they pass through the intersection of the axis and directrices.

6. (a) The normal at $(at_1^2, 2at_1)$ meets the parabola $y^2 = 4ax$ again at $(at_2^2, 2at_2)$. Show that $t_2 = -t_1 - \frac{2}{t_1}$. **[C.U. '66, '70]**

   (b) If a normal chord of a parabola subtends a right angle at the vertex, show that its inclination to the x-axis is $\tan^{-1}\sqrt{2}$.

   [HINTS. If the chord is normal at $(at^2, 2at)$ to the parabola $y^2 = 4ax$ then the other end of the chord is $(at_1^2, 2at_1)$, where $t_1 = -t - \frac{2}{t}$. Here the vertex is the origin.

   The gradient of the line joining the vertex and the point $(at^2, 2at)$ is $\frac{2}{t}$. Similarly the gradient of the line joining the vertex and the point $(at_1^2, 2at_1)$ is $\frac{2}{t_1}$.

   Here $\frac{2}{t} \cdot \frac{2}{t_1} = -1$, or $\frac{4}{t} \cdot \frac{t}{t^2+2} = 1$, or $t^2 + 2 = 4$, or $t^2 = 2$.

   The equation of the chord is $y = -tx + 2at + at^3$.

   Thus we can say that the inclination of the chord to the x-axis is $\tan^{-1}\sqrt{2}$.]

(c) If the normal to the ellipse $\frac{x^2}{14}+\frac{y^2}{5}=1$ at the point whose eccentric angle is ϕ cuts the curve again at the point whose eccentric angle is 2ϕ, show that $\cos\phi=-\frac{2}{3}$. [C.H. '78]

[HINTS. The normal at ϕ is $\frac{\sqrt{14}(x-\sqrt{14}\cos\phi)}{\cos\phi}=\frac{\sqrt{5}(y-\sqrt{5}\sin\phi)}{\sin\phi}$.

The point $(\sqrt{14}\cos 2\phi, \sqrt{5}\sin 2\phi)$ lies on it.

$$\therefore\quad \frac{14(\cos 2\phi-\cos\phi)}{\cos\phi}=\frac{5(\sin 2\phi-\sin\phi)}{\sin\phi}.$$

On simplification the result is obtained.]

(d) Show that the normal $y=mx-2am-am^3$ of $y^2=4ax$ intersects the parabola again at an angle $\tan^{-1}\left(\frac{m}{2}\right)$.

7. If the tangent and normal to a rectangular hyperbola cut off intercepts a_1 and a_2 on one axis and b_1 and b_2 on the other, show that $a_1a_2+b_1b_2=0$.

8. Prove that the locus of the foot of the perpendicular from the focus to any tangent to $y^2=4ax$ is the tangent at the vertex.

9. Find the locus of the point of intersection of two normals to the parabola which are at right angles to one another. [C.U. '63]

10. Find the chord of contact of the point

(a) $(2,3)$ w.r.t. $2x^2+6xy+4y^2-8x+7=0$;

(b) $(-1,2)$ w.r.t. $7x^2-8xy+5y^2-4x-6y+5=0$.

11. (a) Find the polar of the point

(i) $(2,3)$ w.r.t. $2x^2-3xy+y^2-2x+3=0$;

(ii) $(0,-3)$ w.r.t. $xy+2x+3y-1=0$;

(iii) $(0,0)$ w.r.t. $ax^2+2hxy+by^2+2gx+2fy=0$.

(b) Prove that the directrix is the polar of the corresponding focus of a hyperbola;

(c) Prove that the polars of the point $(1,-2)$ w.r.t. the circles whose equations are $x^2+y^2+6y+5=0$ and $x^2+y^2+2x+8y+5=0$ coincide, prove also that there is another point the polars of which w.r.t. these circles are the same and find its co-ordinates.

[HINTS. The polars of the point (x_1,y_1) w.r.t. the given circles are $x_1x+(y_1+3)\,y+3y_1+5=0$ and $(x_1+1)x+(y_1+4)y+x_1+4y_1+5=0$.

These are identical. $\therefore\quad \frac{x_1}{x_1+1}=\frac{y_1+3}{y_1+4}=\frac{3y_1+5}{x_1+4y_1+5}$.

By the first and second

$$x_1-y_1-3=0 \tag{1}$$

and by the first and third

$$x_1^2+x_1y_1-3y_1-5=0. \tag{2}$$

By (1) and (2), $x_1=1, y_1=-2$ and $x_1=2, y_1=-1$.

Therefore, the other point is $(2,-1)$.]

12. (a) Find the pole of the line

(i) $5x-4y-14=0$ w.r.t. $x^2+y^2+6x-10y-15=0$;

(ii) $x+y-1=0$ w.r.t. $3x^2+4xy-2y^2-5x+7y-10=0$.

(b) Show that the points (1, 3) and (23, 12) are conjugate points w.r.t. the conic

$$x^2 - 4xy + 4y^2 - 6x + 8y + 7 = 0.$$

13. Find the length of the tangent to $5x^2 + 5y^2 + 8x + 7y - 10 = 0$ from the point $(-1, 2)$.

14. Find the tangents to

(a) $x^2 + y^2 - 6x - 6y + 14 = 0$ from $(5, 5)$; **[C.U. '62]**

(b) $3x^2 - 2xy + 7y^2 - 4x = 0$ from $(1, -1)$.

15. The polar of the point P w.r.t. the circle $x^2 + y^2 = a^2$ touches the circle $4x^2 + 4y^2 = a^2$, show that the locus of P is the circle $x^2 + y^2 = 4a^2$.

[C.U. '66, '93]

16. (a) Find the locus of the poles of the chords of the an ellipse $\frac{x^2}{a^2} + \frac{y^2}{b^2} = 1$ which subtend right angles at the centre.

(b) Show that the locus of the poles of chords of the hyperbola $\frac{x^2}{a^2} - \frac{y^2}{b^2} = 1$ which subtends a right angle at the centre is $\frac{x^2}{a^4} + \frac{y^2}{b^4} = \frac{1}{a^2} - \frac{1}{b^2}$.

(c) If the polar of a point (α, β) with respect to the parabola $y^2 = 4ax$ touches the parabola $x^2 = 4by$ then show that the locus of the point (α, β) is $xy + 2ab = 0$.

[N.B. '97; C.U. '97, 2000]

[HINTS: The polar of (α, β) w.r.t. $y^2 = 4ax$ is

$$y\beta = 2a(x + \alpha). \quad (1)$$

To find the point of intersection between (1) and $x^2 = 4by$,

$$x^2 = 4b \cdot \frac{2a(x + \alpha)}{\beta}, \quad \text{or} \quad \beta x^2 - 8abx - 8ab\alpha = 0. \quad (2)$$

If the line (1) touches the parabola $x^2 = 4by$, then the equation (2) must have equal roots. For this $(8ab)^2 + 4.8ab\alpha \cdot \beta = 0$, or $\alpha\beta + 2ab = 0$. Hence the required locus is $xy + 2ab = 0$.]

17. Prove that the locus of the poles of normal chords of the rectangular hyperbola $xy = c^2$ is $(x^2 - y^2)^2 + 4c^2xy = 0$.

18. (a) Show that the polar of a point on $x^2 + 4ay = 0$ w.r.t. $xy = c^2$ will touch the parabola $y^2 = 4ax$.

(b) Show that the polar of any point on the circle $x^2 + y^2 - 2ax - 3a^2 = 0$ w.r.t. the circle $x^2 + y^2 + 2ax - 3a^2 = 0$ will touch the parabola $y^2 = -4ax$. **[N.B. '85]**

[HINTS. If (x_1, y_1) is a point on $x^2 + y^2 - 2ax - 3a^2 = 0$ then

$$x_1^2 + y_1^2 - 2ax_1 - 3a^2 = 0. \quad (1)$$

The polar of the point (x_1, y_1) w.r.t. the circle $x^2 + y^2 + 2ax - 3a^2 = 0$ is

$$xx_1 + yy_1 + a(x + x_1) - 3a^2 = 0, \quad \text{or} \quad y = -\frac{x_1 + a}{y_1}x + \frac{3a^2 - ax_1}{y_1}. \quad (2)$$

Now $\frac{3a^2 - ax_1}{y_1} - \frac{ay_1}{x_1 + a} = \frac{-a(x_1^2 + y_1^2 - 2ax_1 - 3a^2)}{(x_1 + a)y_1} = 0$. [by (1)]

Therefore, the polar touches the given parabola.]

19. Find the locus of the poles of tangents to the director circle of an ellipse w.r.t. this ellipse.

20. Find the locus of the poles of tangents to the auxiliary circle of the hyperbola $\frac{x^2}{a^2} - \frac{y^2}{b^2} = 1$ w.r.t. the hyperbola itself. **[N.B. '92]**

21. (a) Find the locus of the point of intersection of two tangents to a parabola such that their chord of contact subtends a right angle at the vertex. **[N.B. '70, '89]**

(b) Two lines are at right angles to one another, one of them touches $y^2 = 4a(x + a)$ and the other touches $y^2 = 4a'(x + a')$. Show that the point of intersection of the lines will lie on the line $x + a + a' = 0$. **[N.B. '84; C.U. '96]**

(c) Prove that the locus of the middle points of all tangents drawn from points on the directrix to the parabola $y^2 = 4ax$ is $y^2(2x + a) = a(3x + a)^2$. **[N.B. '90]**

22. Show that the circle described on the line joining any point on the parabola to the focus as diameter touches the tangent at the vertex.

23. Prove analytically that the difference of the squares of the tangents drawn from a point to two concentric circles is independent of the position of the point. **[C.U. '62]**

24. Tangents are drawn from (a, b) to the circle $x^2 + y^2 = a^2$. Show that the area of the triangle formed by them and the straight line joining their points of contact is $\frac{ab^3}{a^2 + b^2}$. **[C.U. '67]**

25. Prove that the area of the triangle formed by the tangents from (h, k) to the parabola $y^2 = 4ax$ and chord of contact is $\frac{(k^2 - 4ah)^{3/2}}{2a}$.

26. (a) Show that the subtangent at a point on a parabola is bisected at the vertex.

(b) Show that the subtangent and subnormal of a point (x_1, y_1) on the ellipse $\frac{x^2}{a^2} + \frac{y^2}{b^2} = 1$ are $\left(\frac{a^2}{x_1} - x_1\right)$ and $\frac{b^2}{a^2}x_1$ respectively.

27. Find the common tangents to the circles $x^2 + y^2 + 4x + 2y - 4 = 0$ and $x^2 + y^2 - 4x - 2y + 4 = 0$.

28. (a) The normal at a point P on the ellipse, $\frac{x^2}{a^2} + \frac{y^2}{b^2} = 1$ meets the axes in G and g. Find the equation of the locus of the middle point of Gg and show that the eccentricity of this locus is equal to that of the given ellipse. **[C.U. '64]**

(b) If $\alpha, \beta, \gamma, \delta$ be the eccentric angles of four points on the ellipse such that the normals at them are concurrent then show that $\alpha + \beta + \gamma + \delta$ is an odd multiple of π.

Answers

1. (a) $y - 4 = 0$, $x - 2 = 0$; (b) $13x + y - 14 = 0$, $x - 13y + 12 = 0$; (c) $222x - 13y - 547 = 0$, $13x + 222y + 5537 = 0$. **2.** (a) $y - 12 = 0$, $y + 2 = 0$, $(-3, 12)$, $(-3, -2)$; (b) $y = \sqrt{3}x + \frac{a}{\sqrt{3}}$, $\left(\frac{a}{3}, \frac{2a}{\sqrt{3}}\right)$. **3.** (a) $(-1, -1)$; (b) $c = am + \frac{1}{m}$; (c) $\frac{a^2}{l^2} - \frac{b^2}{m^2} = \frac{(a^2+b^2)^2}{n^2}$. **4.** $x^2 + y^2 = 2a^2$. **9.** $y^2 = a(x - 3a)$. **10.** (a) $9x + 18y - 1 = 0$; (b) $17x - 11y - 1 = 0$. **11.** (a) (i) $3x - 2 = 0$; (ii) $x - 3y + 11 = 0$; (iii) $gx + fy = 0$. **12.** (a) $(2, 1)$; (b) $\left(-\frac{34}{7}, \frac{19}{7}\right)$. **13.** $\sqrt{21}$. **14.** (a) $x - 5 = 0$, $y - 5 = 0$; (b) $5x^2 + 4xy - 2y^2 - 6x - 8y - 1 = 0$. **16.** $\frac{x^2}{a^4} + \frac{y^2}{b^4} = \frac{1}{a^2} + \frac{1}{b^2}$. **19.** $\frac{x^2}{a^4} + \frac{y^2}{b^4} = \frac{1}{a^2+b^2}$. **20.** $\frac{x^2}{a^4} + \frac{y^2}{b^4} = \frac{1}{a^2}$. **21.** $x + 4a = 0$. **27.** $x = 1$, $y = 2$, $4x - 3y - 10 = 0$ $3x + 4y - 5 = 0$.

Chapter 5

Equation of a Chord in terms of its Middle Point and Diameter

5.1 Introduction

If (x_1, y_1) *be the middle point of a chord of the conic*

$$S = ax^2 + 2hxy + by^2 + 2gx + 2fy + c = 0,$$

the equation of this chord is $T = S_1$, *where*

$$T \equiv axx_1 + h(xy_1 + yx_1) + byy_1 + g(x + x_1) + f(y + y_1) + c \text{ and}$$
$$S_1 \equiv ax_1^2 + 2hx_1y_1 + by_1^2 + 2gx_1 + 2fy_1 + c.$$

Let the equation of the chord be

$$\frac{x - x_1}{l} = \frac{y - y_1}{m} (= r \text{ say}). \tag{5.1.1}$$

l and m are cosines of the angles made by the line with the axes and r is the distance between (x_1, y_1) and (x, y) on the line. From (5.1.1)

$$x = lr + x_1, \quad y = mr + y_1.$$

Putting these values of x and y in S and arranging the terms,

$$(al^2 + 2hlm + bm^2)r^2 + 2\{(ax_1 + by_1 + g)l + (hx_1 + by_1 + f)m\}r$$
$$+ ax_1^2 + 2hx_1y_1 + by_1^2 + 2gx_1 + 2fy_1 + c = 0.$$

It is a quadratic equation in r. The roots are the distances of the points of intersection of (5.1.1) and S from (x_1, y_1). Since (x_1, y_1) is the mid-point of the chord, the roots are equal in magnitude but opposite in sign, i.e., the sum of the roots is zero.

$$\therefore \quad (ax_1 + hy_1 + g)l + (hx_1 + by_1 + f)m = 0. \tag{5.1.2}$$

Eliminating l and m by (5.1.1)

$$(ax_1 + hy_1 + g)(x - x_1) + (hx_1 + by_1 + f)(y - y_1) = 0.$$

This can be written as

$$axx_1 + h(xy_1 + yx_1) + byy_1 + gx + fy = ax_1^2 + 2hx_1y_1 + by_1^2 + gx_1 + fy_1.$$

Adding $gx_1 + fy_1 + c$ to both sides we have

$$axx_1 + h(xy_1 + yx_1) + byy_1 + g(x + x_1) + f(y + y_1) + c$$
$$= ax_1^2 + 2hx_1y_1 + by_1^2 + 2gx_1 + 2fy_1 + c, \quad \text{or} \quad T = S_1.$$

It is the equation of the chord.

Particular Cases

(i) **Parabola** : $y^2 = 4ax$.

The equation of the chord is

$$yy_1 - 2a(x + x_1) = y_1^2 - 4ax_1, \quad \text{or} \quad yy_1 - 2ax = y_1^2 - 2ax_1.$$

(ii) **Ellipse** : $\dfrac{x^2}{a^2} + \dfrac{y^2}{b^2} = 1$.

The equation of the chord is

$$\frac{xx_1}{a^2} + \frac{yy_1}{b^2} - 1 = \frac{x_1^2}{a^2} + \frac{y_1^2}{b^2} - 1, \quad \text{or} \quad \frac{xx_1}{a^2} + \frac{yy_1}{b^2} = \frac{x_1^2}{a^2} + \frac{y_1^2}{b^2}.$$

(iii) **Hyperbola** : $\dfrac{x^2}{a^2} - \dfrac{y^2}{b^2} = 1$.

The equation of the chord is

$$\frac{xx_1}{a^2} - \frac{yy_1}{b^2} = \frac{x_1^2}{a^2} - \frac{y_1^2}{b^2}.$$

(iv) **Rectangular hyperbola** : $xy = c^2$.

The equation of the chord is

$$xy_1 + yx_1 = 2x_1y_1.$$

5.2 Diameter and Conjugate Diameters

DEFINITION. *The locus of the middle points of a series of parallel chords of a conic is called a diameter of the conic.*

Let the chords be parallel to

$$\frac{x}{l} = \frac{y}{m} \tag{5.2.1}$$

and (x_1, y_1) be the middle point of one of these chords of the conic

$$ax^2 + 2hxy + by^2 + 2gx + 2fy + c = 0.$$

The equation of this chord is

$$\begin{aligned} &axx_1 + h(xy_1 + yx_1) + byy_1 + g(x + x_1) + f(y + y_1) + c \\ &= ax_1^2 + 2hx_1y_1 + by_1^2 + 2gx_1 + 2fy_1 + c, \\ \text{or}\quad &(ax_1 + hy_1 + g)x + (hx_1 + by_1 + f)y \\ &= ax_1^2 + 2hx_1y_1 + by_1^2 + gx_1 + fy_1. \end{aligned} \tag{5.2.2}$$

Since (5.2.2) is parallel to (5.2.1),

$$\frac{ax_1 + hy_1 + g}{m} = -\frac{hx_1 + by_1 + f}{l}. \tag{5.2.3}$$

Therefore, the locus of the middle points, or the equation of the diameter is

$$\frac{ax + hy + g}{m} + \frac{hx + by + f}{l} = 0,$$

$$\text{or}\quad (al + hm)x + (hl + bm)y + gl + fm = 0. \tag{5.2.4}$$

DEFINITION. **Conjugate diameters** : *Two diameters of a conic are said to be conjugate when each bisects the chords parallel to the other.*

Let the diameter (5.2.4) be parallel to $\dfrac{x}{l'} = \dfrac{y}{m'}$.

Then

$$l'(al + hm) + m'(hl + bm) = 0,$$

$$\text{or}\quad all' + h(lm' + l'm) + bmm' = 0.$$

It is the condition for conjugate diameters corresponding to the chords parallel to

$$\frac{x}{l} = \frac{y}{m} \quad \text{and} \quad \frac{x}{l'} = \frac{y}{m'}.$$

COROLLARY. The equation (5.2.4) suggests that the diameter passes through the point of intersection of the lines $ax + hy + g = 0$ and $hx + by + f = 0$.

This point is the centre of the central conic. Thus the diameter of a central conic passes through the centre of it.

5.3 Diameters of Standard Conics

(i) **Parabola** : $y^2 = 4ax$.

The equation of the chord whose middle point is (x_1, y_1) is $yy_1 - 2ax = y_1^2 - 2ax_1$. If it is parallel to $y = mx$, then

$$\frac{2a}{y_1} = m, \quad \text{or} \quad y_1 = \frac{2a}{m}.$$

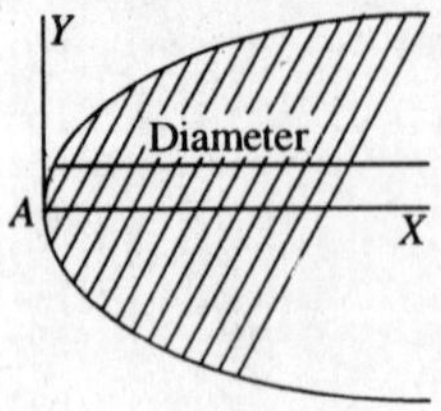

Fig 5.1

$\therefore$ the equation of the diameter corresponding to this system of chords is

$$y = \frac{2a}{m}.$$

It is a straight line parallel to the axis of the parabola.

(ii) **Ellipse** : $\dfrac{x^2}{a^2} + \dfrac{y^2}{b^2} = 1$.

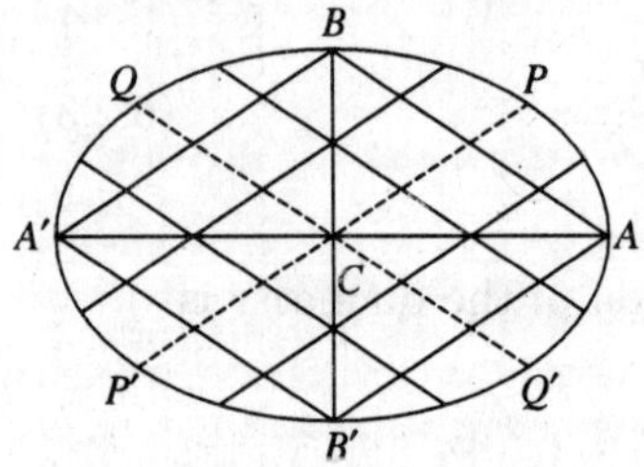

Fig 5.2

The equation of the chord whose middle point is (x_1, y_1) is

$$\frac{xx_1}{a^2} + \frac{yy_1}{b^2} = \frac{x_1^2}{a^2} + \frac{y_1^2}{b^2}.$$

If it is parallel to $y = mx$, then

$$m = -\frac{b^2}{a^2}\frac{x_1}{y_1}, \quad \text{or} \quad y_1 = -\frac{b^2}{a^2m}x_1.$$

Hence the equation of the diameter corresponding to this system of chords is

$$y = -\frac{b^2}{a^2m}x.$$

Let us consider the diameter corresponding to the chords parallel to $y = m'x$, where $m' = -\frac{b^2}{a^2m}$. The equation of this diamter is $y = mx$. It is a member of the system of chords for the diameter $y = m'x$. Therefore, $y = m'x$ and $y = mx$ are conjugate diameters. Moreover $mm' = -\frac{b^2}{a^2}$.

Hence $y = mx$ and $y = m'x$ will be two conjugate diameters of $\frac{x^2}{a^2} + \frac{y^2}{b^2} = 1$, if $mm' = -\frac{b^2}{a^2}$. In the fig. 5.2, PCP' and QCQ' are conjugate diameters.

(iii) **Hyperbola** : $\dfrac{x^2}{a^2} - \dfrac{y^2}{b^2} = 1$.

The equation of the diameter for a system of chords parallel to $y = mx$ is $y = \frac{b^2}{a^2m}x$. *The condition for which $y = mx$ and $y = m'x$ will be conjugate is that $mm' = \frac{b^2}{a^2}$.*

5.4 Properties of Diameters and Conjugate Diameters

CASE I. **Ellipse** : $\dfrac{x^2}{a^2} + \dfrac{y^2}{b^2} = 1$.

(i) *The tangent at the extremity of a diameter is parallel to the chords bisected by the diameter.*

Let the chords be parallel to $y = mx$. The equation of the diameter is $y = -\frac{b^2}{a^2m}x$. If this diameter meets the ellipse at (h, k), then

$$k = -\frac{b^2}{a^2m}h, \quad \text{or} \quad m = -\frac{b^2h}{a^2k}.$$

The equation of the tangent at (h, k) is $\frac{xh}{a^2} + \frac{yk}{b^2} = 1$.

Slope of it $= -\frac{b^2h}{a^2k}$ which is equal to m. Hence the result follows.

(ii) *The tangents at the extremities of any chord meet on the diameter which bisects this chord.*

Let the chords be parallel to $y = mx$.

The equation of the diameter is $y = -\frac{b^2}{a^2m}x$.

If the tangents at the end of a chord of this system meet at (h, k), then the equation of this chord is $\frac{xh}{a^2} + \frac{yk}{b^2} = 1$ (chord of contact).

Since the chord is parallel to $y = mx$, $m = -\frac{b^2h}{a^2k}$, or $k = -\frac{b^2}{a^2m}h$.

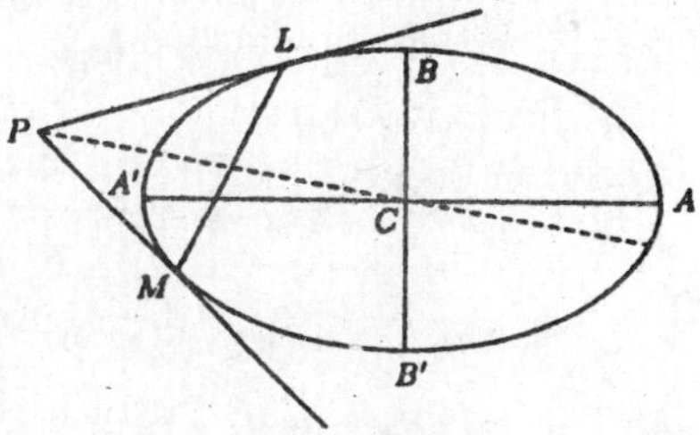

Fig. 5.3

It shows that (h, k) lies on the diameter.

(iii) *The eccentric angles of the ends of a pair of conjugate diameters differ by an odd multiple of $\frac{\pi}{2}$.*

Let PCP' and QCQ' be the two conjugate diameters and the co-ordinates of P and Q be $(a\cos\theta, b\sin\theta)$ and $(a\cos\phi, b\sin\phi)$ respectively. C is the centre of the ellipse.

Since C is the origin, the gradients of CP and CQ are

$$\frac{b\sin\theta}{a\cos\theta} \quad \text{and} \quad \frac{b\sin\phi}{a\cos\phi}.$$

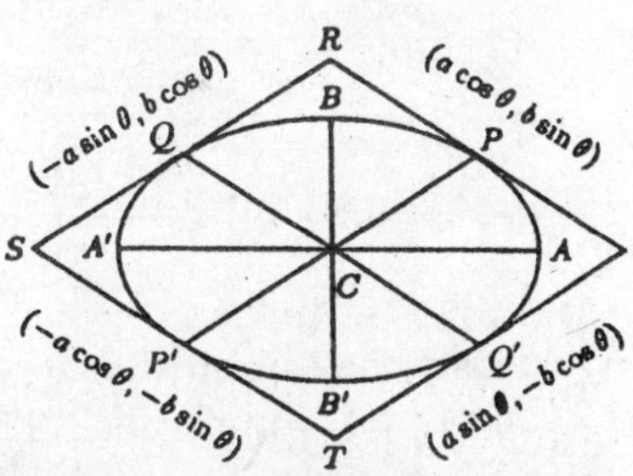

Fig. 5.4

From the condition of conjugate diameters,

$$\frac{b\sin\theta}{a\cos\theta}\cdot\frac{b\sin\phi}{a\cos\phi} = -\frac{b^2}{a^2}, \text{ or } \sin\theta\sin\phi = -\cos\theta\cos\phi,$$

$$\text{or} \quad \cos\phi\cos\theta + \sin\phi\sin\theta = 0, \text{ or } \cos(\phi - \theta) = 0. \therefore \phi - \theta = \frac{\pi}{2}.$$

In general $\phi - \theta =$ odd multiple of $\frac{\pi}{2}$.

COROLLARY. The co-ordinates of P, Q, P' and Q' are $(a\cos\theta, b\sin\theta)$,

$$\left\{a\cos\left(\frac{\pi}{2}+\theta\right), b\sin\left(\frac{\pi}{2}+\theta\right)\right\}, \quad \left\{a\cos\left(2\cdot\frac{\pi}{2}+\theta\right), b\sin\left(2\cdot\frac{\pi}{2}+\theta\right)\right\}$$

$$\text{and} \quad \left\{a\cos\left(3\cdot\frac{\pi}{2}+\theta\right), b\sin\left(3\cdot\frac{\pi}{2}+\theta\right)\right\},$$

i.e., $(a\cos\theta,\ b\sin\theta)$, $(-a\sin\theta,\ b\cos\theta)$, $(-a\cos\theta,\ -b\sin\theta)$ and $(a\sin\theta,\ -b\cos\theta)$ respectively.

(iv) *The sum of the squares of two conjugate semi-diameters is constant.* [See the fig. 5.4]

Let CP and CQ be a pair of conjugate semi-diameters. If the co-ordinates of P be $(a\cos\theta, b\sin\theta)$ then the co-ordinates of Q are $(-a\sin\theta, b\cos\theta)$.

$$\begin{aligned}\therefore\quad CP^2 + CQ^2 &= (a^2\cos^2\theta + b^2\sin^2\theta) + (a^2\sin^2\theta + b^2\cos^2\theta)\\ &= (a^2+b^2)(\cos^2\theta + \sin^2\theta) = a^2 + b^2 \text{ (constant)}.\end{aligned}$$

(v) *The tangents at the ends of a pair of conjugate diameters form a parallelogram of constant area.* [See the fig. 5.4]

Let PCP' and QCQ' be a pair of conjugate diameters and the co-ordinates of P, Q, P' and Q' be

$$(a\cos\theta, b\sin\theta),\ (-a\sin\theta, b\cos\theta),$$
$$(-a\cos\theta, -b\sin\theta) \quad\text{and}\quad (a\sin\theta, -b\cos\theta) \text{ respectively.}$$

The tangents at P, Q, P' and Q' are

$$\frac{x\cos\theta}{a} + \frac{y\sin\theta}{b} = 1, \tag{5.4.1}$$

$$\frac{x\sin\theta}{a} - \frac{y\cos\theta}{b} = -1, \tag{5.4.2}$$

$$\frac{x\cos\theta}{a} + \frac{y\sin\theta}{b} = -1, \tag{5.4.3}$$

$$\frac{x\sin\theta}{a} - \frac{y\cos\theta}{b} = 1 \quad\text{respectively.} \tag{5.4.4}$$

Obviously the tangents at P and P', i.e., (5.4.1) and (5.4.3) and at Q and Q', i.e., (5.4.2) and (5.4.4) are parallel. Therefore, these tangents form a parallelogram.

By property (i) the tangents at the extremities of a diameter are parallel to the conjugate diameter. Therefore, the tangents at P and P' are parallel to QCQ' and the tangents at Q and Q' are parallel to PCP'.

Thus by symmetry the area of the parallelogram

$RSTU$ formed by the tangents

$$= 4 \times \text{area of the parallelogram } CPRQ$$

$$= 8 \times \Delta CPQ.$$

$$\text{Area of } \Delta CPQ = \frac{1}{2}\begin{vmatrix} 0 & 0 & 1\\ a\cos\theta & b\sin\theta & 1\\ -a\sin\theta & b\cos\theta & 1\end{vmatrix} = \frac{1}{2}ab(\cos^2\theta + \sin^2\theta) = \frac{1}{2}ab.$$

$\therefore$ area of the parallelogram $RSTU = 8 \times \frac{1}{2}ab = 4ab$ (constant).

(vi) *The product of the perpendicular distance from the centre of the ellipse to the tangent at one end of a diameter and the semi-conjugate diameter is constant.*

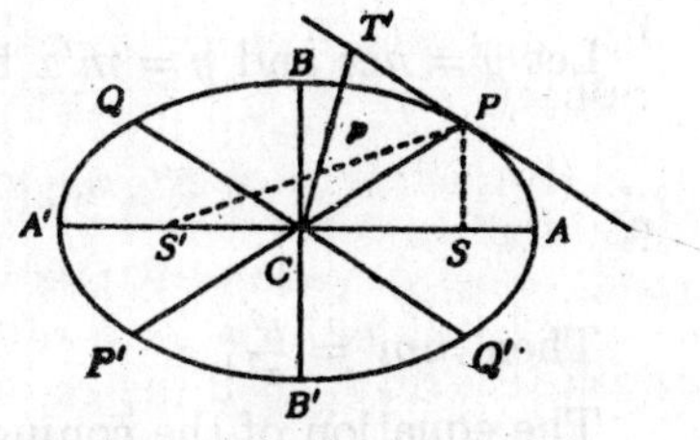

Fig. 5.5

Let PCP' and QCQ' be the conjugate diameters. If $(a\cos\theta, b\sin\theta)$ be the co-ordinates of P, then $(-a\sin\theta, b\cos\theta)$ are the co-ordinates of Q. The tangent at P is

$$\frac{x\cos\theta}{a}+\frac{y\sin\theta}{b}=1.$$

If p be the perpendicular distance from the centre $(0,0)$ to the tangent then

$$p=\frac{1}{\sqrt{\left(\frac{\cos^2\theta}{a^2}+\frac{\sin^2\theta}{b^2}\right)}}=\frac{ab}{\sqrt{(b^2\cos^2\theta+a^2\sin^2\theta)}}.$$

Semi-conjugate diameter $CQ=\sqrt{(a^2\sin^2\theta+b^2\cos^2\theta)}$.

$\therefore\quad p\cdot CQ=ab=\text{constant}.$

(vii) *The product of the focal distances of a point on an ellipse is equal to the square on the semi-diameter parallel to the tangent at this point.* [See the fig. 5.5]

Let $P(a\cos\theta, b\sin\theta)$ he a point on the ellipse.

If S and S' are the foci, then

$$SP=\sqrt{\left\{(a\cos\theta-ae)^2+b^2\sin^2\right\}}=a-ae\cos\theta,\quad [\text{using } b^2=a^2(1-e^2)]$$

$$S'P=\sqrt{\left\{(a\cos\theta+ae)^2+b^2\sin^2\theta\right\}}=a+ae\cos\theta.$$

$$\begin{aligned}\therefore\ SP.S'P &= a^2-a^2e^2\cos^2\theta=a^2+(b^2-a^2)\cos^2\theta\\ &= a^2\sin^2\theta+b^2\cos^2\theta=CQ^2.\end{aligned}$$

CQ is the semi-diameter which is conjugate to the diameter through P.

NOTE. **Equi-conjugate diameters** :

If two conjugate diameters of an ellipse are equal in length, then they are said to be equi-conjugate diameters.

In this case

$$\begin{aligned}CP^2=CQ^2,\quad \text{or}\quad & a^2\cos^2\theta+b^2\sin^2\theta=a^2\sin^2\theta+b^2\cos^2\theta,\\ \text{or}\quad & (a^2-b^2)\cos^2\theta=(a^2-b^2)\sin^2\theta,\\ \text{or}\quad & \tan^2\theta=1,\quad\text{or}\quad \tan\theta=\pm1,\quad\text{or}\quad \theta=\tfrac{\pi}{4},\tfrac{3\pi}{4},\ldots\end{aligned}$$

CASE II. **Hyperbola** : $\dfrac{x^2}{a^2}-\dfrac{y^2}{b^2}=1.$

(i) *If a pair of diameters be conjugate w.r.t. a hyperbola then they are also conjugate w.r.t. the conjugate hyperbola.*

Let $y = mx$ and $y = m'x$ be a pair of conjugate diameters of

$$\frac{x^2}{a^2} - \frac{y^2}{b^2} = 1. \tag{5.4.5}$$

Then $mm' = \frac{b^2}{a^2}$.

The equation of the conjugate hyperbola is

$$\frac{x^2}{a^2} - \frac{y^2}{b^2} = -1, \quad \text{or} \quad \frac{x^2}{-a^2} - \frac{y^2}{-b^2} = 1. \tag{5.4.6}$$

The lines $y = mx$ and $y = m'x$ are the pair of conjugate diameters of (5.4.2), since $mm' = \frac{b^2}{a^2} = \frac{-b^2}{-a^2}$.

(ii) *If a diameter meets the hyperbola at real points, then it will meet its conjugate at imaginary points.*

Let

$$y = mx \tag{5.4.7}$$

be a diameter of

$$\frac{x^2}{a^2} - \frac{y^2}{b^2} = 1. \tag{5.4.8}$$

The conjugate hyperbola is

$$\frac{x^2}{a^2} - \frac{y^2}{b^2} = -1. \tag{5.4.9}$$

The x-co-ordinates of the points of intersection between (5.4.7) and (5.4.8) and (5.4.7) and (5.4.9) are obtained from

$$x^2 = \frac{a^2b^2}{b^2 - a^2m^2} \tag{5.4.10}$$

$$\text{and} \quad x^2 = -\frac{a^2b^2}{b^2 - a^2m^2}. \tag{5.4.11}$$

If (5.4.10) gives the real values of x, then (5.4.11) must give the imaginary values of x. Hence the result follows.

(iii) *Only one of a pair of conjugate diameters of a hyperbola meets the curve in real points.*

Let

$$y = mx \tag{5.4.12}$$

$$\text{and} \quad y = m'x \tag{5.4.13}$$

be a pair of conjugate diameters of

$$\frac{x^2}{a^2} - \frac{y^2}{b^2} = 1. \tag{5.4.14}$$

Then

$$mm' = \frac{b^2}{a^2}. \tag{5.4.15}$$

The x-co-ordinates of points of intersection between (5.4.12) and (5.4.14) are obtained from

$$x^2 = \frac{a^2b^2}{b^2 - a^2m^2}. \tag{5.4.16}$$

The x-co-ordinates of points of intersection between (5.4.13) and (5.4.14) are obtained from

$$x^2 = \frac{a^2b^2}{b^2 - a^2m'^2} = \frac{a^2b^2}{b^2 - a^2.\frac{b^4}{a^4m^2}} \quad \text{[by (5.4.15)]}$$

$$= \frac{a^4m^2}{a^2m^2 - b^2} = -\frac{a^4m^2}{b^2 - a^2m^2}. \tag{5.4.17}$$

If (5.4.16) gives real values of x, then (5.4.17) gives imaginary values of x. Hence the result follows.

(iv) *If a pair of conjugate diameters meets the hyperbola and its conjugate at P and Q, then $CP^2 - CQ^2 = a^2 - b^2$, where C is the centre of the hyperbola.*

Let $(a\sec\theta, b\tan\theta)$ be the co-ordinates of P on

$$\frac{x^2}{a^2} - \frac{y^2}{b^2} = 1. \tag{5.4.18}$$

Since C is the origin, the slope of

$$CP = \frac{b\tan\theta}{a\sec\theta}. \tag{5.4.19}$$

Let $(a\tan\phi, b\sec\phi)$ be the co-ordinates of Q on the conjugate

$$\frac{x^2}{a^2} - \frac{y^2}{b^2} = -1. \tag{5.4.20}$$

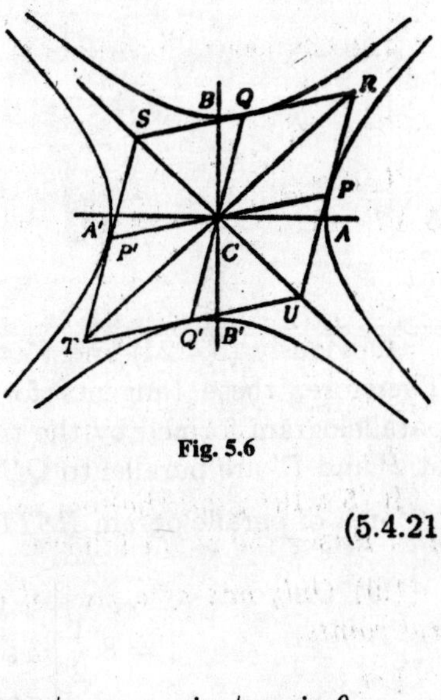

Fig. 5.6

The slope of

$$CQ = \frac{b\sec\phi}{a\tan\phi}. \tag{5.4.21}$$

Since the diameters are conjugate,

$$\frac{b\tan\theta}{a\sec\theta}\cdot\frac{b\sec\phi}{a\tan\phi} = \frac{b^2}{a^2}, \quad \text{or} \quad \tan\theta\sec\phi = \sec\theta\tan\phi, \quad \text{or} \quad \sin\phi = \sin\theta.$$

$$\therefore \quad \phi = \theta, \quad \text{or} \quad \pi - \theta.$$

θ and $\pi - \theta$ for the value of ϕ correspond to the points Q and Q'.

Now

$$CP^2 - CQ^2 = a^2\sec^2\theta + b^2\tan^2\theta - a^2\tan^2\phi - b^2\sec^2\phi$$

$$= a^2\sec^2\theta + b^2\tan^2\theta - a^2\tan^2\theta - b^2\sec^2\theta$$

$$= a^2(\sec^2\theta - \tan^2\theta) - b^2(\sec^2\theta - \tan^2\theta) = a^2 - b^2.$$

(v) *The tangents at the points, where a diameter and its conjugate meet a hyperbola and its conjugate form a parallelogram of constant area.* [Sec the fig. 5.6]

Let PCP' and QCQ' be a pair of conjugate diameters of

$$\frac{x^2}{a^2} - \frac{y^2}{b^2} = 1. \tag{5.4.22}$$

C is the centre of the hyperbola.

Let PCP' meet (5.4.22) at $P(a\sec\theta, b\tan\theta)$ and $P'(-a\sec\theta, -b\tan\theta)$ and QCQ' meet the conjugate

$$\frac{x^2}{a^2} - \frac{y^2}{b^2} = -1 \tag{5.4.23}$$

at Q and Q'.

By property (iv) the co-ordinates of Q and Q' are $(a\tan\theta, b\sec\theta)$ and $(-a\tan\theta, -b\sec\theta)$ respectively.

The tangents at P and P' to (5.4.22) and at Q and Q' to (5.4.23) are

$$\frac{x\sec\theta}{a} - \frac{y\tan\theta}{b} = 1, \tag{5.4.24}$$

$$\frac{x\sec\theta}{a} - \frac{y\tan\theta}{b} = -1, \tag{5.4.25}$$

$$\frac{x\tan\theta}{a} - \frac{y\sec\theta}{b} = 1, \tag{5.4.26}$$

$$\frac{x\tan\theta}{a} - \frac{y\sec\theta}{b} = -1. \tag{5.4.27}$$

Obviously (5.4.24) and (5.4.25) are parallel and (5.4.26) and (5.4.27) are parallel. Therefore, these tangents form a parallelogram. In the fig. 5.6 RSTU is the parallelogram formed by the tangents at ends of conjugate diameters. Since tangents at P and P' are parallel to QCQ' and the tangents at Q and Q' are parallel to PCP',

area of parallelogram $RSTU = 8 \times \Delta CPQ$

$$= 8.\frac{1}{2}\begin{vmatrix} 0 & 0 & 1 \\ a\sec\theta & b\tan\theta & 1 \\ a\tan\theta & b\sec\theta & 1 \end{vmatrix} = 4ab = \text{a constant.}$$

Worked Out Examples

1. Find the equation of the chord of the parabola $y^2 = 8x$ which is bisected at (2, –3). **[C.U. '63]**

The equation of the chord is

$$y.(-3) - 4(x+2) = (-3)^2 - 8.2,$$

$$\text{or} \quad -3y - 4x - 8 = 9 - 16, \quad \text{or} \quad 4x + 3y + 1 = 0.$$

2. Find the equation of the diameter of the ellipse $3x^2 + 4y^2 = 5$, conjugate to $y + 3x = 0$. [C.U. '63]

The equation of the ellipse in standard form is

$$\frac{x^2}{\frac{5}{3}} + \frac{y^2}{\frac{5}{4}} = 1.$$

Let $y + mx = 0$ be the required diameter.

Then $3.m = -\dfrac{5/4}{5/3} = -\dfrac{3}{4}$, or $m = -\dfrac{1}{4}$.

Thus $4y - x = 0$ is conjugate to $y + 3x = 0$.

3. If the line $\dfrac{l}{a}x + \dfrac{m}{b}y = n$ cuts the ellipse $\dfrac{x^2}{a^2} + \dfrac{y^2}{b^2} = 1$ at the ends of conjugate diameters of the ellipse, prove that $l^2 + m^2 = 2n^2$. [B.U. '66]

Let the co-ordinates of the ends of conjugate diameters be $(a\cos\theta, b\sin\theta)$ and $(-a\sin\theta, b\cos\theta)$. As the line passes through these points,

$$l\cos\theta + m\sin\theta = n \qquad (1)$$

$$\text{and} \quad -l\sin\theta + m\cos\theta = n. \qquad (2)$$

Squaring and adding $l^2 + m^2 = 2n^2$.

4. Prove that the locus of the middle points of chords of a parabola which pass through a fixed point is a parabola. [C.U. '65]

Let (x_1, y_1) be the middle point of a chord of the parabola $y^2 = 4ax$. The equation of this chord is

$$yy_1 - 2a(x + x_1) = y_1^2 - 4ax_1, \quad \text{or} \quad yy_1 - 2ax = y_1^2 - 2ax_1.$$

If this chord passes through a fixed point (h, k), then

$$ky_1 - 2ah = y_1^2 - 2ax_1, \quad \text{or} \quad y_1^2 - ky_1 = 2a(x_1 - h).$$

Hence the locus of (x_1, y_1) is $y^2 - ky = 2a(x - h)$. It is a parabola.

5. Find the locus of the middle points of the normal chords of the parabola $y^2 = 4ax$.

Let (x_1, y_1) be the middle point of a chord. The equation of this chord is

$$yy_1 - 2ax = y_1^2 - 2ax_1. \qquad (1)$$

If it is normal at $(at^2, 2at)$, then it is identical with

$$y + tx = 2at + at^3. \qquad (2)$$

Comparing (1) with (2), $\dfrac{1}{y_1} = \dfrac{t}{-2a} = \dfrac{2at + at^3}{y_1^2 - 2ax_1}$.

$$\therefore \quad t = -\frac{2a}{y_1} \qquad (3)$$

$$\text{and} \quad 2a + at^2 = \frac{y_1^2 - 2ax_1}{-2a}. \qquad (4)$$

Eliminating t from (3) and (4),

$$2a + a.\frac{4a^2}{y_1^2} = \frac{y_1^2 - 2ax_1}{-2a}, \quad \text{or} \quad y_1^4 - 2ax_1y_1^2 + 4a^2y_1^2 + 8a^4 = 0.$$

$\therefore$ the required locus is $y^4 - 2axy^2 + 4a^2y^2 + 8a^4 = 0$.

6. If the points of intersection of the ellipses $\frac{x^2}{a^2} + \frac{y^2}{b^2} = 1$ and $\frac{x^2}{\alpha^2} + \frac{y^2}{\beta^2} = 1$ be the ends of the conjugate diameters of the former, prove that $\frac{a^2}{\alpha^2} + \frac{b^2}{\beta^2} = 2$.

Let the ends of the conjugate diameters of $\frac{x^2}{a^2} + \frac{y^2}{b^2} = 1$ be
$(a\cos\theta, b\sin\theta), (-a\sin\theta, b\cos\theta), (-a\cos\theta, -b\sin\theta)$ and $(a\sin\theta, -b\cos\theta)$.
If these points lie on the second ellipse, then

$$\frac{a^2\cos^2\theta}{\alpha^2} + \frac{b^2\sin^2\theta}{\beta^2} = 1 \tag{1}$$

$$\text{and} \quad \frac{a^2\sin^2\theta}{\alpha^2} + \frac{b^2\cos^2\theta}{\beta^2} = 1. \tag{2}$$

Adding (1) and (2) we have $\frac{a^2}{\alpha^2} + \frac{b^2}{\beta^2} = 2$.

7. Show that the locus of the poles of the line joining the extrimities of two conjugate diameters of the ellipse

$$\frac{x^2}{a^2} + \frac{y^2}{b^2} = 1 \text{ is } \frac{x^2}{a^2} + \frac{y^2}{b^2} = 2.$$ **[C.H. '92; B.U. '80]**

Let $(a\cos\theta, b\sin\theta)$ and $(-a\sin\theta, b\cos\theta)$ be the ends of conjugate diameters of

$$\frac{x^2}{a^2} + \frac{y^2}{b^2} = 1.$$

The equations of the tangents at these points are

$$\frac{x}{a}\cos\theta + \frac{y}{b}\sin\theta = 1 \tag{1}$$

$$\text{and} \quad \frac{x}{a}\sin\theta - \frac{y}{b}\cos\theta = -1. \tag{2}$$

The point of intersection of these tangents is the pole of the line joining the ends of conjugate diameters. Thus eliminating θ from (1) and (2) the required locus is obtained. To eliminate θ we square and add (1) and (2). By this way we have

$$\frac{x^2}{a^2} + \frac{y^2}{b^2} = 2.$$

8. Lines are drawn through the foci of an ellipse perpendicular respectively to a pair of conjugate diameters and intersect in Q, show that the locus of Q is a concentric ellipse.

[C.U. '67]

Let $(a\cos\theta, b\sin\theta)$ and $(-a\sin\theta, b\cos\theta)$ be the ends of a pair of conjugate diameters of the ellipse $\frac{x^2}{a^2} + \frac{y^2}{b^2} = 1$. The foci are $(\pm ae, 0)$, where $b^2 = a^2(1 - e^2)$. The slopes of these diameters are $\frac{b\sin\theta}{a\cos\theta}$ and $\frac{b\cos\theta}{-a\sin\theta}$.

The equation of the line passing through $(ae, 0)$ and perpendicular to the diameter corresponding to the slope $\dfrac{b\sin\theta}{a\cos\theta}$ is

$$y = -\frac{a\cos\theta}{b\sin\theta}(x - ae). \qquad (1)$$

Similarly the equation of the other line is

$$y = \frac{a\sin\theta}{b\cos\theta}(x + ae). \qquad (2)$$

(1) and (2) meet at Q. Let (α, β) be the co-ordinates of Q.

Then from (1) and (2)

$$\tan\theta = -\frac{a}{b}\frac{\alpha - ae}{\beta} = \frac{b}{a}\frac{\beta}{\alpha + ae}, \quad \text{or} \quad -a^2(\alpha^2 - a^2e^2) = b^2\beta^2.$$

Hence the required locus is

$$a^2x^2 + b^2y^2 = a^4e^2, \quad \text{or} \quad a^2x^2 + b^2y^2 = a^2(a^2 - b^2).$$

It is concentric with $\dfrac{x^2}{a^2} + \dfrac{y^2}{b^2} = 1.$

9. Show that the locus of the middle points of the normal chords of the rectangular hyperbola

$$x^2 - y^2 = a^2 \text{ is } (y^2 - x^2)^3 = 4a^2x^2y^2.$$

Let (x_1, y_1) be the middle point of a chord of $x^2 - y^2 = a^2$.

The equation of this chord is

$$xx_1 - yy_1 = x_1^2 - y_1^2. \qquad (1)$$

Let it be normal at $(a\sec\theta, a\tan\theta)$ to the hyperbola.

The equation of the normal at this point is

$$x\cos\theta + y\cot\theta = 2a. \qquad (2)$$

Comparing (1) with (2) we have

$$\frac{\cos\theta}{x_1} = \frac{\cot\theta}{-y_1} = \frac{2a}{x_1^2 - y_1^2}. \quad \therefore \quad \cos\theta = \frac{2ax_1}{x_1^2 - y_1^2} \qquad (3)$$

$$\text{and} \quad \sin\theta = -\frac{x_1}{y_1} \qquad (4)$$

From (3) and (4)

$$\frac{4a^2x_1^2}{(x_1^2 - y_1^2)^2} + \frac{x_1^2}{y_1^2} = 1, \text{ or } \frac{4a^2x_1^2}{(x_1^2 - y_1^2)^2} + \frac{x_1^2}{y_1^2} - 1 = 0, \text{ or } \frac{4a^2x_1^2}{(x_1^2 - y_1^2)^2} + \frac{x_1^2 - y_1^2}{y_1^2} = 0,$$

$$\text{or} \quad 4a^2x_1^2y_1^2 + (x_1^2 - y_1^2)^3 = 0, \text{ or } (y_1^2 - x_1^2)^3 = 4a^2x_1^2y_1^2.$$

Hence the locus of (x_1, y_1) is $(y^2 - x^2)^3 = 4a^2x^2y^2$.

10. Prove that the locus of the middle points of chords of contact of tangents to the hyperbola $x^2 - y^2 = a^2$ from points on the auxiliary circle, is the curve $(x^2 - y^2)^2 = a^2(x^2 + y^2)$.

The auxiliary circle is $x^2 + y^2 = a^2$. Let $(a\cos\theta, a\sin\theta)$ be a point on it. The chord of contact of this point w.r.t.

$$x^2 - y^2 = a^2 \text{ is } x\cos\theta - y\sin\theta = a. \tag{1}$$

If (x_1, y_1) be the middle point of this chord, then the equation of it is identical with

$$xx_1 - yy_1 = x_1^2 - y_1^2. \tag{2}$$

Comparing (1) and (2)

$$\frac{\cos\theta}{x_1} = \frac{\sin\theta}{y_1} = \frac{a}{x_1^2 - y_1^2}, \quad \text{or} \quad \cos\theta = \frac{ax_1}{x_1^2 - y_1^2}, \quad \sin\theta = \frac{ay_1}{x_1^2 - y_1^2}.$$

Squaring and adding

$$1 = \frac{a^2(x_1^2 + y_1^2)}{(x_1^2 - y_1^2)^2}, \quad \text{or} \quad (x_1^2 - y_1^2)^2 = a^2(x_1^2 + y_1^2).$$

Hence the required locus is $(x^2 - y^2)^2 = (a^2x^2 + y^2)$.

Exercise

1. Find the equation of the chord
   (a) of $x^2 + y^2 = 81$, which is bisected at $(-2, 3)$; **[C.U. '66]**
   (b) of $y^2 = 4x + 5$, which is bisected at $(1, 4)$;
   (c) of $x^2 - y^2 = 9$, which is bisected at $(5, -3)$.

2. Find the locus of the middle points of chords
   (a) of $y^2 = 16x$, which are parallel to $x - 3y = 5$;
   (b) of $y^2 = 4ax$, passing through the vertex;
   (c) of $y^2 = 4ax$, passing through the focus; **[B.U. '67]**
   (d) of $xy = c^2$, which are parallel to $2x + 3y = 7$.

3. Find the diameter of
   (a) $\dfrac{x^2}{25} + \dfrac{y^2}{16} = 1$, which is conjugate to $5y - 4x = 0$;
   (b) $16x^2 - 9y^2 = 144$, which is conjugate to $x = 2y$. **[B.U. '70]**
   (c) For the ellipse $8x^2 + 12y^2 = 96$, find a pair of conjugate semi-diameters inclined at an angle $\tan^{-1}(7)$; **[C.U. '92]**

4. Show that $25x^2 + kxy - 36y^2 = 0$ represents a pair of conjugate diameters of $25x^2 + 36y^2 = 900$ for any value of k.

5. Show that the conjugate diameters of $\frac{x^2}{a^2}+\frac{y^2}{b^2}=1$ which are equally inclined to each of the axes are $bx+ay=0$ and $bx-ay=0$.

[HINTS. Let $y=mx$ and $y=m'x$ be the conjugate diameters of the ellipse which are equally inclined to each of the axes.

Then $m=-m'$ and $mm'=-\frac{b^2}{a^2}$. $\therefore\ m^2=\frac{b^2}{a^2}$, or $m=\pm\frac{b}{a}$.

Thus the diameters are $y=\pm\frac{b}{a}x$, or $bx+ay=0$ and $bx-ay=0$.]

6. If the pair of lines $Ax^2+2Hxy+By^2=0$ be a pair of conjugate diameters of $\frac{x^2}{a^2}+\frac{y^2}{b^2}=1$, show that $a^2A+b^2B=0$.

7. If P and Q be the extremities of conjugate diameters of the ellipse $2x^2+3y^2=24$, find the locus of the point of intersection of the tangents at P and Q. [C.U. '68]

8. If P and Q be the extremities of two conjugate diameters of the ellipse $\frac{x^2}{a^2}+\frac{y^2}{b^2}=1$, show that the locus of the middle point of PQ is $\frac{x^2}{a^2}+\frac{y^2}{b^2}=\frac{1}{2}$ and also PQ touches this ellipse.

9. If the angle between two equal conjugate diameters of an ellipse is $60°$, find the eccentricity of it.

[HINTS. Let PCP' and QCQ' be two equal conjugate diameters of $\frac{x^2}{a^2}+\frac{y^2}{b^2}=1$. We have $CP^2+CQ^2=a^2+b^2$.

$$\therefore\quad 2CP^2=a^2+b^2,\quad \because CP=CQ.$$

If ϕ is the eccentric angle of P then

$$2(a^2\cos^2\phi+b^2\sin^2\phi)=a^2+b^2,\quad\text{or}\quad a^2\cos 2\phi=b^2\cos 2\phi,$$

$$\text{or}\quad (a^2-b^2)\cos 2\phi=0,\quad\text{or}\quad \cos 2\phi=0\ \ (\because a\neq b).\quad \therefore\quad \phi=45°, 135°.$$

Now the equations of CP and CQ are $y=\frac{b}{a}x$ and $y=-\frac{b}{a}x$.

By the condition

$$\tan 60°=\frac{\frac{b}{a}+\frac{b}{a}}{1-\frac{b^2}{a^2}}=\frac{2ab}{a^2-b^2},\quad\text{or}\quad \sqrt{3}(a^2-b^2)=2ab,$$

$$\text{or}\quad 3(a^2-b^2)^2=4a^2b^2,\quad\text{or}\quad 3a^4e^4=4a^4(1-e^2),\quad\text{or}\quad 3e^4+4e^2-4=0,$$

$$\text{or}\quad e^2=\frac{2}{3},\quad\text{or}\quad e=\sqrt{\frac{2}{3}}.$$

10. If (x_1,y_1) and (x_2,y_2) be the extremities of a pair of conjugate diameters of $\frac{x^2}{a^2}+\frac{y^2}{b^2}=1$, prove that

(a) $\frac{x_1x_2}{a^2}+\frac{y_1y_2}{b^2}=0$; (b) $x_2=\pm\frac{a}{b}y_1, y_2=\mp\frac{b}{a}x_1$; (c) $x_1y_2-x_2y_1=\pm ab$.

11. Prove that the straight lines joining the centre to the intersections of the straight line $y=mx+\sqrt{\frac{a^2m^2+b^2}{2}}$ with the ellipse $\frac{x^2}{a^2}+\frac{y^2}{b^2}=1$ are conjugate diameters.

[Hints. Let $(a\cos\theta, b\sin\theta)$ and $(-a\sin\theta, b\cos\theta)$ be he ends of a pair of conjugate diameters of $\frac{x^2}{a^2}+\frac{y^2}{b^2}=1$.

The equation of the line joining these point is

$$y - b\sin\theta = \frac{b(\sin\theta-\cos\theta)}{a(\cos\theta+\sin\theta)}(x-a\cos\theta),$$

$$\text{or}\quad y = \frac{b}{a}\frac{\sin\theta-\cos\theta}{\cos\theta+\sin\theta}x + b\sin\theta - b\frac{\cos\theta(\sin\theta-\cos\theta)}{\cos\theta+\sin\theta},$$

$$\text{or}\quad y = \frac{b}{a}\frac{\sin\theta-\cos\theta}{\cos\theta+\sin\theta}x + \frac{b}{\cos\theta+\sin\theta}. \tag{1}$$

$$\text{Putting}\quad m = \frac{b}{a}\frac{\sin\theta-\cos\theta}{\cos\theta+\sin\theta},\quad \frac{a^2m^2}{b^2}+1 = \left(\frac{\sin\theta-\cos\theta}{\cos\theta+\sin\theta}\right)^2+1 = \frac{2}{(\cos\theta+\sin\theta)},$$

$$\text{or}\quad \frac{1}{\cos\theta+\sin\theta} = \pm\frac{\sqrt{a^2m^2+b^2}}{\sqrt{2}b}.$$

Taking the positive value, the equation (1) becomes $y = mx + \sqrt{\frac{a^2m^2+b}{2}}$. Hence the result follows.]

12. Find the locus of the middle points of the chords of an ellipse, the tangents at the extremities of which intersect at right angles.

[Hints. Let the tangents from (α, β) to $\frac{x^2}{a^2}+\frac{y^2}{b^2}=1$ be at right angles. The equation of this pair of tangents is

$$\left(\frac{x^2}{a^2}+\frac{y^2}{b^2}-1\right)\left(\frac{\alpha^2}{a^2}+\frac{\beta^2}{b^2}-1\right) = \left(\frac{x\alpha}{a^2}+\frac{y\beta}{b^2}-1\right)^2.$$

Since this pair of lines are at right angles, the coefficient of x^2 + the coefficient of $y^2=0$,

$$\text{i.e.,}\ \frac{1}{a^2}\left(\frac{\alpha^2}{a^2}+\frac{\beta^2}{b^2}-1\right)-\frac{\alpha^2}{a^4}+\frac{1}{b^2}\left(\frac{\alpha^2}{a^2}+\frac{\beta^2}{b^2}-1\right)-\frac{\beta^2}{b^4}=0,\ \text{or}\ \alpha^2+\beta^2=a^2+b^2. \tag{1}$$

If (x_1, y_1) is the mid-point of the chord joining the points of contact, the equation of it is

$$\frac{xx_1}{a^2}+\frac{yy_1}{b^2} = \frac{x_1^2}{a^2}+\frac{y_1^2}{b^2}. \tag{2}$$

The equation of the chord of contact corresponding to the point (α, β) is

$$\frac{x\alpha}{a^2}+\frac{y\beta}{b^2}=1. \tag{3}$$

(2) and (3) are identical.

$$\therefore \frac{\alpha}{x_1} = \frac{\beta}{y_1} = 1\Big/\left(\frac{x_1^2}{a^2}+\frac{y_1^2}{b^2}\right),\quad \text{or}\quad \alpha = \frac{x_1}{\frac{x_1^2}{a^2}+\frac{y_1^2}{b^2}}\quad\text{and}\quad \beta = \frac{y_1}{\frac{x_1^2}{a^2}+\frac{y_1^2}{b^2}}.$$

From (1), $x_1^2+y_1^2 = (a^2+b^2)\left(\frac{x_1^2}{a^2}+\frac{y_1^2}{b^2}\right)^2$.

Hence the locus is $x^2+y^2 = (a^2+b^2)\left(\frac{x^2}{a^2}+\frac{y^2}{b^2}\right)^2$.

Answers

1. (a) $2x-3y+13=0$; (b) $x-2y+7=0$; (c) $5x+3y=16$. **2.** (a) $y=24$; (b) $y^2=2ax$; (c) $y^2=2a(x-a)$; (d) $2x=3y$. **3.** (a) $4x+5y=0$; (b) $9y=32x$; (c) $y=2x, y=-x/3$. **7.** $\frac{x^2}{12}+\frac{y^2}{8}=2$. **12.** $x^2+y^2=(a^2+b^2)(\frac{x^2}{a^2}+\frac{y^2}{b^2})^2$.

Chapter 6

Asymptotes

6.1 Introduction

Definition 1 *A straight line which meets a curve at two points at infinity but which does not be wholly at infinity is called an asymptote to this curve.*

Equations of asymptotes to

$$\frac{x^2}{a^2} - \frac{y^2}{b^2} = 1. \tag{6.1.1}$$

Let

$$y = mx + c \tag{6.1.2}$$

be an asymptote.

To find the x-co-ordinates of points of intersection between (6.1.1) and (6.1.2)

$$\frac{x^2}{a^2} - \frac{(mx+c)^2}{b^2} = 1, \quad \text{or} \quad \left(\frac{1}{a^2} - \frac{m^2}{b^2}\right)x^2 - \frac{2mc}{b^2}x - \left(1 + \frac{c^2}{b^2}\right) = 0. \tag{6.1.3}$$

Since the line meets the curve at two points at infinity, both of the roots of (6.1.3) must be infinite.

$\therefore$ coefficient of x^2 and x are zero, i.e., $\frac{1}{a^2} - \frac{m^2}{b^2} = 0$ and $mc = 0$.

$\therefore$ $m = \pm\frac{b}{a}$ and $c = 0$.

Hence the asymptotes are $y = \pm\frac{b}{a}x$.

In pair from $\frac{x^2}{a^2} - \frac{y^2}{b^2} = 0$.

NOTE 1. The equation of the pair of asymptotes differs from the equation of the hyperbola by a constant only.

NOTE 2. The asymptotes pass through the centre of the hyperbola.

NOTE 3. The bisectors of the angles between the asymptotes are the axes of the hyperbola.

NOTE 4. $\frac{x^2}{a^2} - \frac{y^2}{b^2} = 0$ is the equation of the pair of asymptotes of the conjugate hyperbola $\frac{x^2}{a^2} - \frac{y^2}{b^2} = -1$.

NOTE 5. The asymptotes of $x^2 - y^2 = a^2$ are $y = \pm x$ which are at right angles.

6.2 Some Rules about Asymptotes and Hyperbola

Let $S = 0, S' = 0$ and $A = 0$ be the equations of the hyperbola, conjugate hyperbola and the pair of asymptotes.

1. *$S + \lambda = 0$ is the pair of asymptotes. λ is a constant determined by the condition of a pair of straight lines.*
2. $A + \lambda' = 0$ is the equation of the hyperbola. λ' is a constant determined by the help of a given point on the hyperbola.
3. $S + S' = 2A$. [The equation of the conjugate hyperbola is obtained by this relation.]
4. *The point of intersection of the asymptotes is the centre of the hyperbola.*
5. *The bisectors of the angles between the asymptotes are the principal axes of the hyperbola.*
6. *If the asymptotes are at right angles, the hyperbola is rectangular or equilateral.*

6.3 The Equation of a Hyperbola Referred to its Asymptotes as Co-ordinates Axes

Let

$$\frac{x^2}{a^2} - \frac{y^2}{b^2} = 1 \tag{6.3.1}$$

be the equation of the hyperbola with its axes as co-ordinate axes.

The asymptotes are $y = \pm\frac{b}{a}x$. If 2α be the angle between the asymptotes, then

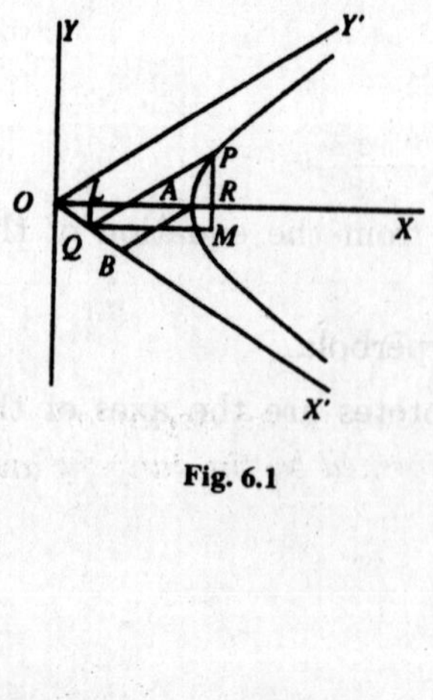

Fig. 6.1

$$\tan\alpha = \frac{b}{a}, \sin\alpha = \frac{b}{\sqrt{a^2+b^2}} \text{ and } \cos\alpha = \frac{a}{\sqrt{a^2+b^2}}. \tag{6.3.2}$$

In the fig. 6.1 OX and OY are the axes of the hyperbola, OX' and OY' are the asymptotes, A is the vertex and $\angle X'OA = \angle Y'OA = \alpha$.

METHOD 1. If the asymptotes are considered as co-ordinate-axes, then their equations must be $x = 0, y = 0$. In pair form $xy = 0$. Therefore the equation of the hyperbola must be of the form

$$xy = c^2 \quad [c^2 \text{ is a constant}]. \tag{6.3.3}$$

AB is drawn parallel to OY'. Since A is a point on $xy = c^2, OB.BA = c^2$.

$\because \angle BAO = \angle Y'OA = \angle X'OA, OB = BA. \quad \therefore \; OB^2 = c^2$, i.e., $OB = c$.

Again from ΔAOB,

$$OA^2 = OB^2 + AB^2 - 2OB.AB\cos(\pi - 2\alpha)$$

$$= 2OB^2 + 2OB^2\cos 2\alpha = 4OB^2\cos^2\alpha.$$

$$\therefore OA = 2.OB\cos\alpha, \quad \text{or } a = 2c\cos\alpha, \quad \text{or } c = \frac{\sqrt{a^2+b^2}}{2} \quad \text{[by (6.3.2)]}.$$

$\therefore\; xy = \frac{a^2+b^2}{4}$ is the equation of the hyperbola referred to asymptotes as axes.

METHOD 2. Let P be a point on the hyperbola and its co-ordinates be (x, y) w.r.t. OX and OY and (x', y') w.r.t. OX' and OY'. In the fig. 6.1, PR is perpendicular to OX. QP is parallel to OY' and QM is parallel to OX. PR (produced) meets QM at M. QL is perpendicular to OX.

$$\therefore\; OR = x,\; RP = y,\; OQ = x' \text{ and } QP = y'$$

$$\text{Now} \quad x = OR = OL + LR = OL + QM$$

$$= OQ\cos\alpha + QP\cos\alpha = (x' + y')\cos\alpha$$

$$\text{and} \quad y = RP = MP - MR = MP - QL$$

$$= QP\sin\alpha - OQ\sin\alpha = (y' - x')\sin\alpha.$$

Putting these values of x and y in $\frac{x^2}{a^2} - \frac{y^2}{b^2} = 1$,

$$\frac{(x'+y')^2\cos^2\alpha}{a^2} - \frac{(y'-x')^2\sin^2\alpha}{b^2} = 1,$$

$$\text{or} \quad \frac{(x'+y')^2}{a^2+b^2} - \frac{(y'-x')^2}{a^2+b^2} = 1, \quad \text{[by (6.3.2)]}$$

$$\text{or} \quad (x'+y')^2 - (y'-x')^2 = a^2 + b^2,$$

$$\text{or} \quad 4x'y' = a^2 + b^2, \quad \text{or} \quad x'y' = \frac{a^2+b^2}{4}$$

$\therefore$ the required equation is

$$xy = \frac{a^2+b^2}{4}, \text{ or } xy = c^2, \text{ where } c^2 = \frac{a^2+b^2}{4}.$$

6.4 Some Properties

(i) *The portion of the tangent to a hyperbola intercepted between the asymptotes is bisected at the point of contact and the area of the triangle formed by the tangent and the asymptotes is constant.*

Let (x_1, y_1) be a point on the hyperbola $xy = c^2$.

The equation of the tangent at this point is

$$\frac{1}{2}(xy_1 + yx_1) = c^2, \quad \text{or} \quad \frac{1}{2}(xy_1 + yx_1) = x_1y_1, \quad \text{or} \quad \frac{x}{2x_1} + \frac{y}{2y_1} = 1.$$

The asymptotes are $x = 0$ and $y = 0$. Therefore, the tangent meets the asymptotes at $(0, 2y_1)$ and $(2x_1, 0)$. Obviously, the mid-point of the line joining the above two points is (x_1, y_1) which is the point of contact.

If the angle between the asymptotes be 2α, the area of the triangle formed by the tangent and the asymptotes is

$$\frac{1}{2}2x_1 \cdot 2y_1 \sin 2\alpha = 2x_1y_1 \sin 2\alpha = 2c^2 \sin 2\alpha \text{ (constant)}.$$

NOTE. Area $= 2c^2 \sin 2\alpha = 4c^2 \sin\alpha \cos\alpha = 4c^2 \frac{ab}{a^2+b^2} = ab.$

(ii) *The vertices of the parallelogram formed by the tangents at the extremities of two conjugate diameters of a hyperbola lie on the asymptotes.* [See the fig. 5.3]

Let PCP' and QCQ' be a pair of conjugate diameters of

$$\frac{x^2}{a^2} - \frac{y^2}{b^2} = 1. \tag{6.4.1}$$

The co-ordinates of P, P', Q and Q' can be taken as $(a\sec\theta, b\tan\theta)$, $(-a\sec\theta, -b\tan\theta)$, $(a\tan\theta, b\sec\theta)$ and $(-a\tan\theta, -b\sec\theta)$ respectively.

The tangent at P to (6.4.1) is

$$\frac{x\sec\theta}{a} - \frac{y\tan\theta}{b} = 1. \tag{6.4.2}$$

The tangent at Q to the conjugate of (6.4.1),

$$\text{i.e.,} \quad \frac{x^2}{a^2} - \frac{y^2}{b^2} = -1 \text{ is } \frac{x\tan\theta}{a} - \frac{y\sec\theta}{b} = -1. \tag{6.4.3}$$

These two tangents meet on a line given by

$$\frac{x}{a}(\sec\theta + \tan\theta) - \frac{y}{b}(\sec\theta + \tan\theta) = 0, \quad \text{or} \quad \frac{x}{a} - \frac{y}{b} = 0.$$

It is an asymptote of the curve. Therefore the tangents at P and Q meet on one asymptote. In this way it can be shown that the tangents at P' and Q' meet on this asymptote, but the tangents at P and Q' meet on the other asymptote on which the tangents at P' and Q also meet.

Worked Out Examples

1. If the chord PP' of a hyperbola meets asymptotes at Q and Q', then show that $PQ = P'Q'$. **[C.U. '61, R.U. '60]**

Let the equation of the hyperbola be $xy = c^2$. If (x_1, y_1) be the middle point of PP', then the equation of it is

$$xy_1 + yx_1 = 2x_1y_1, \quad \text{or} \quad \frac{x}{2x_1} + \frac{y}{2y_1} = 1.$$

The intercepts on the axes which are asymptotes are $2x_1$ and $2y_1$. Therefore, the co-ordinates of Q and Q' are $(2x_1, 0)$ and $(0, 2y_1)$. Consequently the mid-point of QQ' is (x_1, y_1). Hence $QP = P'Q'$.

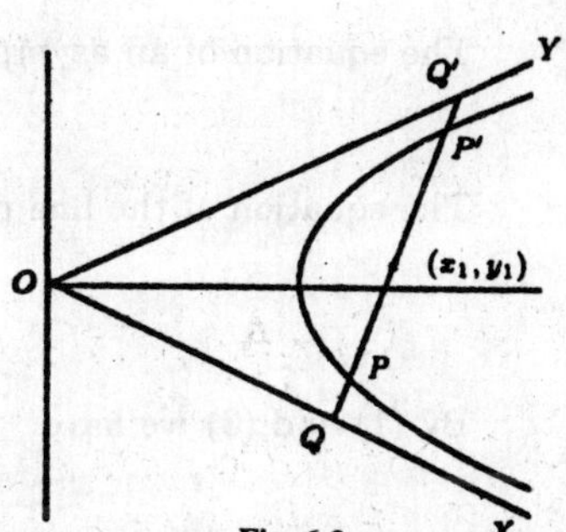

Fig. 6.2

2. If e be the eccentricity of the hyperbola $\frac{x^2}{a^2} - \frac{y^2}{b^2} = 1$, and θ be the angle made by an asymptote with the transverse axis, show that $e = \sec\theta$. [C.U. '66]

The asymptotes are $y = \pm\frac{b}{a}x$.
Since the transverse axis is the x-axis, $\tan\theta = \frac{b}{a}$.
Now

$$\sec^2\theta = 1 + \tan^2\theta = 1 + \frac{b^2}{a^2} = 1 + \frac{a^2(e^2-1)}{a^2} = e^2. \quad \therefore e = \sec\theta.$$

3. Show that the area of a triangle formed by the two asymptotes of the rectangular hyperbola $xy = c^2$ and the normal at the point (x', y') on the hyperbola is

$$\frac{1}{2}\left(\frac{x'^2 - y'^2}{c}\right)^2.$$

The equation of the normal at (x', y') is

$$y - y' = \frac{x'}{y'}(x - x'). \tag{1}$$

Here the asymptotes are co-ordinate-axes and they are at right angles. If α and β are the intercept made by the normal on the x-axis and y-axis, then

$$-y' = \frac{x'}{y'}(\alpha - x'), \quad \text{or} \quad \alpha = \frac{x'^2 - y'^2}{x'}$$

$$\text{and} \quad \beta - y' = -\frac{x'^2}{y'}, \quad \text{or} \quad \beta = \frac{y'^2 - x'^2}{y'}.$$

Area of the required triangle

$$= \frac{1}{2}\alpha\beta = -\frac{1}{2}\frac{(x'^2 - y'^2)^2}{x'y'}.$$

$$\therefore \text{area} = \frac{1}{2}\frac{(x'^2 - y'^2)^2}{c^2} = \frac{1}{2}\left(\frac{x'^2 - y'^2}{c}\right)^2 \qquad [\because x'y' = c^2]$$

4. Show that the length of the focal chord of a hyperbola drawn at right angles to an asymptote is $\frac{2ab^2}{a^2 - b^2}$, where a and b are the semiaxes of the hyperbola. [C.U. '66]

Let the equation of the hyperbola be

$$\frac{x^2}{a^2} - \frac{y^2}{b^2} = 1 \tag{1}$$

The equation of an asymptote is

$$\frac{x}{a} - \frac{y}{b} = 0. \tag{2}$$

The equation of the line perpendicular to (2) and passing through the focus $(ae, 0)$ is

$$y = -\frac{a}{b}(x - ae). \tag{3}$$

By (1) and (3) we have

$$(a^4 - b^4)x^2 - 2a^5ex + a^6e^2 + a^2b^4 = 0 \tag{4}$$

$$\text{and} \quad (a^4 - b^4)y^2 + 2a^2b^3ey + a^4b^2 - a^4b^2e^2 = 0. \tag{5}$$

If the line (3) meets (1) at (x_1, y_1) and (x_2, y_2), then (x_1, x_2) and (y_1, y_2) are the roots of (4) and (5). If d be the distance between (x_1, y_1) and (x_2, y_2), then

$$\begin{aligned} d^2 &= (x_1 - x_2)^2 + (y_1 - y_2)^2 = (x_1 + x_2)^2 - 4x_1x_2 + (y_1 + y_2)^2 - 4y_1y_2 \\ &= 4\frac{a^{10}e^2}{(a^4 - b^4)^2} - 4\frac{a^6e^2 + a^2b^4}{a^4 - b^4} + 4\frac{a^4b^6e^2}{(a^4 - b^4)^2} - 4\frac{a^4b^2 - a^4b^2e^2}{a^4 - b^4} \end{aligned}$$

Putting $a^2e^2 = a^2 + b^2$ and simplifying we get

$$d^2 = \frac{4a^2b^4}{(a^2 - b^2)^2}, \quad \therefore\ d = \frac{2ab^2}{a^2 - b^2}.$$

5. Find the equation of the asymptotes of the hyperbola

$$2x^2 - 5xy - 3y^2 - 5x - 3y - 21 = 0.$$

Hence find the centre and the principal axes of the hyperbola. **[B.U. '67]**

Let the equation of the pair of asymptotes be

$$2x^2 - 5xy - 3y^2 - 5x - 3y - 21 + \lambda = 0, \text{ where } \lambda \text{ is a constant.}$$

This equation represents a pair of straight lines. By the condition of the pair of straight lines we get $\lambda = \frac{1011}{49}$.

$\therefore$ the asymptotes are

$$2x^2 - 5xy - 3y^2 - 5x - 3y + 21 + \frac{1011}{49} = 0$$

$$\text{or} \quad 2x^2 - 5xy - 3y^2 - 5x - 3y - \frac{18}{49} = 0.$$

From this we get two lines which are

$$14x + 7y + 1 = 0 \tag{1}$$

$$\text{and} \quad 7x - 21y - 18 = 0. \tag{2}$$

The point of intersection between (1) and (2) is $\left(\frac{15}{49}, -\frac{37}{49}\right)$. It is the centre of the hyperbola.

The principal axes are

$$\frac{14x + 7y + 1}{\sqrt{14^2 + 7^2}} = \pm\frac{7x - 21y - 18}{\sqrt{7^2 + 21^2}},$$

$$\text{or}\quad 14x + 7y + 1 = \pm\frac{1}{\sqrt{2}}(7x - 21y - 18).$$

6. The centre of the hyperbola is the point $(1, 2)$ and its asymptotes are parallel to $2x + 3y = 0$ and $3x + 2y = 0$. If the hyperbola passes through the point $(5, 3)$, show that its equation is

$$(2x + 3y - 8)(3x + 2y - 7) = 154.$$

[B.U. '66]

Let the asymptotes be $2x + 3y + c_1 = 0$ and $3x + 2y + c_2 = 0$.
Since the asymptotes pass through the centre $(1, 2)$, $c_1 = -8$ and $c_2 = -7$.
$\therefore$ the asymptotes are $2x + 3y - 8 = 0$ and $3x + 2y - 7 = 0$.
Let the equation of the hyperbola be

$$(2x + 3y - 8)(3x + 2y - 7) + \lambda = 0.$$

As the hyperbola passes through $(5, 3)$,

$$(10 + 9 - 8)(15 + 6 - 7) + \lambda = 0, \quad \text{or} \quad \lambda = -154.$$

Hence the equation of the hyperbola is $(2x + 3y - 8)(3x + 2y - 7) = 154$.

Exercise

1. Show that the foot of the perpendicular from the focus on an asymptote of a hyperbola lies on the auxiliary circle.
2. Prove that the rectangle contained by the intercepts made by any tangent to a hyperbola on its asymptotes is constant.
3. The ordinate through a point P of a hyperbola meets the asymptote in Q and Q'. Show that $PQ.PQ' = (\text{semi-conjugate axis})^2$.
4. If P and Q are the extremities of a pair of conjugate diameters of $\frac{x^2}{a^2} - \frac{y^2}{b^2} = 1$, show that PQ is parallel to one of the asymptotes and bisected by the other.
5. Show that the pair of tangents of a hyperbola from the centre of it are its asymptotes.
6. Prove that the polar of any point on any asymptote of a hyperbola w.r.t. the hyperbola is parallel to that asymptote.
7. The tangent to a hyperbola $\frac{x^2}{a^2} - \frac{y^2}{b^2} = 1$ at the point $P(l, m)$ on it intersects the asymptotes of the hyperbola at R and Q respectively. Obtain the co-ordinates of Q and R and show that P is the mid-point of QR.
8. Find the asymptotes of

   (a) $2x^2 + 5xy + 2y^2 + 4x + 5y = 0$;

   (b) $3x^2 - 5xy - 2y^2 + 5x + 11y - 8 = 0$;

   (c) $xy - hx - ky = 0$.

9. Find the centre and principal axes of

(a) $6x^2 - 7xy - 3y^2 - 2x - 8y - 6 = 0$,
(b) $(2x + 3y - 1)(3x - 5y + 2) + 10 = 0$.

10. Find the equation of the hyperbola whose asymptotes are $x + 2y + 3 = 0$ and $3x + 4y + 5 = 0$ and which passess through the point $(1, -1)$. [N.B. '87; C.U. '70]

11. Find the equation of the hyperbola which has the same asymptotes as those of $3x^2 - 2xy - 5y^2 + 7x - 9y = 0$ and which passes through $(2, 2)$.

12. Find the equation of the hyperbola conjugate to

(a) $6x^2 - 7xy - 3y^2 - 2x - 8y - 6 = 0$,
(b) $3x^2 - 5xy - 2y^2 + 5x + 11y - 8 = 0$.

13. Find the equation of the hyperbola whose centre is $(1, -2)$, whose asymptotes are parallel to $x + 2y = 0$ and $3x + 4y = 0$ and which passes through $(1, -1)$.

14. Show that the centres of all hyperbolas whose asymptotes are parallel to the co-ordinate axes and which pass through the points $(2, 5)$ and $(3, 2)$ lie on the straight line $3x - y - 4 = 0$. [C.U. '60, '69]

[HINTS. Let (h, k) be the centre. The asymptotes are $x - h = 0$ and $y - k = 0$. The hyperbola is $(x - h)(y - k) = \lambda$. Since it passes through $(2, 5)$ and $(3, 2)$, $(2 - h)\ (5 - k) = \lambda$ and $(3 - h)(2 - k) = \lambda$.

$$\therefore \quad (2 - h)(5 - k) = (3 - h)(2 - k), \quad \text{or} \quad 3h - k - 4 = 0.$$

Hence the centre lies on $3x - y - 4 = 0$.]

15. (a) If an asymptote of $\frac{x^2}{a^2} - \frac{y^2}{b^2} = 1$ makes an angle θ with the transverse axis then show that the eccentricity is $\sec\theta$.

(b) If the normal at one extremity of a latus rectum to the hyperbola $\frac{x^2}{a^2} - \frac{y^2}{b^2} = 1$ is parallel to an asymptote, show that $e^2 = \frac{1+\sqrt{5}}{2}$, where e is the eccentricity of the curve.
[HINTS. $\left(ae, \frac{b^2}{a}\right)$ are the co-ordinates of an extrimity of a latus rectum. The equation of the normal at this point is

$$y - \frac{b^2}{a} = -\frac{1}{e}(x - ae).$$

The asymptotes are $y = \pm\frac{b}{a}x$. If the normal is parallel to one of the asymptotes then

$$\frac{1}{e} = \frac{b}{a} \quad \text{or} \quad -\frac{b}{a}. \quad \therefore \quad \frac{1}{e^2} = \frac{b^2}{a^2} = e^2 - 1,$$

$$\text{or} \quad e^4 - e^2 - 1 = 0, \quad \text{or} \quad e^2 = \frac{1 \pm \sqrt{5}}{2}$$

$$\therefore \ e > 0, e^2 = \frac{1 + \sqrt{5}}{2}.]$$

Answers

7. $\left(\frac{lb+ma}{b}, \frac{lb+ma}{a}\right) \cdot \left(\frac{lb-ma}{b}, \frac{ma-lb}{a}\right)$. **8.** (a) $x + 2y + 1 = 0$, $2x + y + 2 = 0$; (b) $x - 2y + 3 = 0$, $3x + y - 4 = 0$; (c) $x = k$, $y = h$. **9.** (a) $\left(-\frac{4}{11}, -\frac{10}{11}\right)$, $\frac{2x-3y-2}{\sqrt{13}} = \pm\frac{3x+y+2}{\sqrt{10}}$; (b) $\left(-\frac{1}{19}, \frac{7}{19}\right)$, $\frac{2x+3y-1}{\sqrt{13}} = \pm\frac{3x-5y+2}{\sqrt{34}}$. **10.** $(x + 2y + 3)(3x + 4y + 5) - 8 = 0$. **11.** $3x^2 - 2xy - 5y^2 + 7x - 9y + 20 = 0$. **12.** (a) $6x^2 - 7xy - 3y^2 - 2x - 8y - 2 = 0$. (b) $3x^2 - 5xy - 2y^2 + 5x + 11y - 16 = 0$.

Chapter 7

System of Circles

7.1 Orthogonal Circles

Definition 1 *The circles are said to be orthogonal when the tangents at their common paints are at right angles.*

Let us find out the condition for orthogonal circles.

Let the two circles be

$$S_1 \equiv x^2 + y^2 + 2g_1x + 2f_1y + c_1 = 0 \quad (7.1.1)$$

and $$S_2 \equiv x^2 + y^2 + 2g_2x + 2f_2y + c_2 = 0. \quad (7.1.2)$$

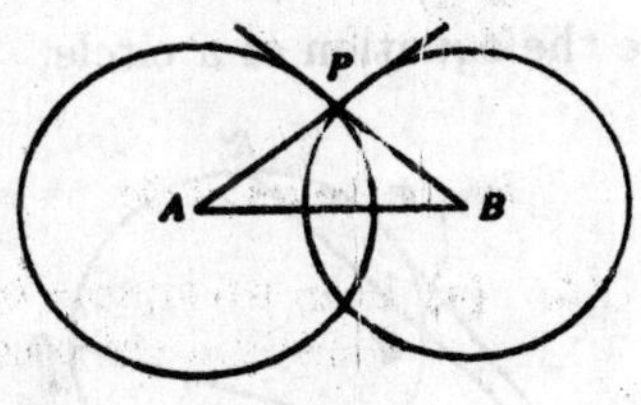

Fig. 7.1

The co-ordinates of the centres A and B are $(-g_1, -f_1)$ and $(-g_2, -f_2)$.

A line perpendicular to the tangent at the point of contact always passes through the centre ($\because$ it is the radius of the circle). For this reason the tangents at the common point P being perpendicular to each other must be PA and PB which are the radii of the circles.

From ΔPAB,

$$AB^2 = AP^2 + BP^2 \quad (\because \angle APB = 90°),$$

or $$(g_1 - g_2)^2 + (f_1 - f_2)^2 = (g_1^2 + f_1^2 - c_1) + (g_2^2 + f_2^2 - c_2),$$

or $$2g_1g_2 + 2f_1f_2 = c_1 + c_2.$$

It is the required condition for orthogonality of the two circles.

► **Example 1** *Find the equation of the circle which cuts orthogonally each of the three circles $x^2 + y^2 = 16$, $x^2 + y^2 - 14x + 40 = 0$ and $x^2 + y^2 - 12y + 32 = 0$.*

$$x^2 + y^2 = 16 \quad (1)$$

$$x^2 + y^2 - 14x + 40 = 0 \quad (2)$$

$$x^2 + y^2 - 12y + 32 = 0. \quad (3)$$

Let $x^2+y^2+2gx+2fy+c=0$ cut each of the above circles orthogonally.

Then

$$0=c-16, \quad \text{or} \quad c=16 \quad \text{[considering orthogonality with (1)]}$$

$$-2g.7=c+40, \quad \text{[considering orthogonality with (2)]}$$

$$\text{or} \quad -14g=56, \quad \text{or} \quad g=-4.$$

$$-2f.6=c+32 \quad \text{[considering orthogonality with (3)]}$$

$$\text{or} \quad -12f=48, \quad \text{or} \quad f=-4.$$

$\therefore$ the required equation is $x^2+y^2-8x-8y+16=0$.

7.2 Power of a Point w.r.t. a Circle

Let $P(x_1, y_1)$ be a point and

$$x^2+y^2+2gx+2fy+c=0 \tag{7.2.1}$$

be the equation of a circle.

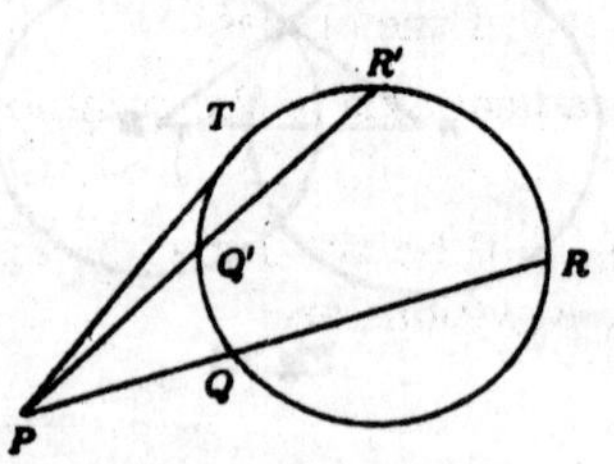

Fig. 7.2

The power of P w.r.t. (7.2.1) is defined as the product of distances of P from the points of intersection of any line through P and the circle (7.2.1).

The equation of a line through (x_1, y_1) may be written as

$$\frac{x-x_1}{\cos\theta}=\frac{y-y_1}{\sin\theta}(=r\text{say}), \tag{7.2.2}$$

where θ is the angle made by the line with x-axis and r is the distance between (x, y) and (x_1, y_1) on the line.

To find out the points of intersection between (7.2.1) and (7.2.2), we put $x=r\cos\theta+x_1$, $y=r\sin\theta+y_1$ in (7.2.1).

$$(r\cos\theta+x_1)^2+(r\sin\theta+y_1)^2+2g(r\cos\theta+x_1)+2f(r\sin\theta+y_1)+c=0,$$

$$\text{or} \quad r^2+2\{(x_1+g)\cos\theta+(y_1+f)\sin\theta\}r+x_1^2+y_1^2+2gx_1+2fy_1+c=0. \tag{7.2.3}$$

The equation (7.2.3) is a quadratic equation in r. Let the roots be r_1 and r_2. Then r_1 and r_2 are the distances of the points of intersection between (7.2.1) and (7.2.2) from $P(x_1, y_1)$.

Now

$$r_1r_2=x_1^2+y_1^2+2gx_1+2fy_1+c. \tag{7.2.4}$$

The expression (7.2.4) is constant when $P(x_1, y_1)$ and the circle are assigned and it is known as the power of P w.r.t. the circle (7.2.1).

From the fig. 7.2, $PQ.PR=PQ'.PR'=PT.PT=x_1^2+y_1^2+2gx_1+2fy_1+c$.

PT is the length of the tangent.

Thus the power of a point w.r.t. a circle is equal to the square of the length of the tangent from P.

7.3 Radical Axis

Definition 2 *The radical axis of two circles is the locus of a point whose powers w.r.t. the circles are equal.*

Let

$$S \equiv x^2 + y^2 + 2gx + 2fy + c = 0 \quad (7.3.1)$$

$$S' \equiv x^2 + y^2 + 2g'x + 2f'y + c' = 0 \quad (7.3.2)$$

be the equations of two circles and $P(\alpha, \beta)$ be a point whose powers w.r.t. (7.3.1) and (7.3.2) are equal.

$\therefore \ \alpha^2 + \beta^2 + 2g\alpha + 2f\beta + c = \alpha^2 + \beta^2 + 2g'\alpha + 2f'\beta + c'$,

or $\ 2(g - g')\alpha + 2(f - f')\beta + c - c' = 0$.

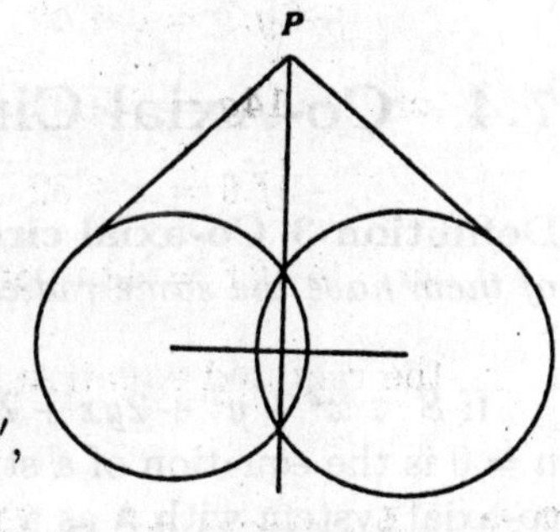

Fig. 7.3

Thus the locus of (α, β) is $2(g - g')x + 2(f - f')y + c - c' = 0$.

It is a straight line and is also the radical axis of the circles.

Note 1. The equation of the radical axis can be written as $S - S' = 0$. It suggests that the radical axis passes through the common points of (7.3.1) and (7.3.2) when the circles intersect. Further the radical axis is the common chord of the circles.

Note 2. If the circles touch with each other then the radical axis is the common tangent to these circles.

Note 3. Radical centre: Let $S_1 = 0, S_2 = 0$ and $S_3 = 0$ be the three circles. From these three circles we get three radical axes and their equations are

$$S_1 - S_2 = 0, \quad (7.3.3)$$

$$S_2 - S_3 = 0 \quad (7.3.4)$$

$$\text{and} \quad S_3 - S_1 = 0. \quad (7.3.5)$$

Since $(S_1 - S_2) + (S_2 - S_3) + (S_3 - S_1) = 0$, the lines are concurrent. Their common point is known as the radical centre w.r.t. these three circles. The radical centre is obtained by finding the point of intersection of any two radical axes.

Note 4. The slope of the radical axis of $S = 0$ and $S' = 0$ is $-\frac{g-g'}{f-f'}$. The slope of the line joining the centres $(-g, -f)$ and $(-g', f')$ is $\frac{f-f'}{g-g'}$. Product of these two slopes is -1. Therefore, the radical axis is perpendicular to the line joining the centres of the circles.

Note 5. If two circles cut a third circle orthogonally, then the centre of the third circle lies on the radical axis of the first two circles.

Note 6. The radical axis of two circles bisects each of their common tangents.

▶ **Example 2** *Find the radical centre of*

$$x^2 + y^2 - 8x + 6y + 15 = 0, \quad (1)$$

$$x^2 + y^2 - 3x - 11y + 15 = 0$$

$$\textit{and} \quad x^2 + y^2 - 6x - 10y + 15 = 0. \quad (2)$$

The radical axis of (1) and (2) is $5x - 17y = 0$.

The radical axis of (2) and (3) is $3x - y = 0$.

The two lines meet at $(0, 0)$. Therefore, the radical centre is $(0, 0)$.

7.4 Co-Axial Circles — Limiting Points

Definition 3 Co-axial circles: *A system of circles is said to be co-axial, if any two of them have the same radical axis.*

If $S \equiv x^2 + y^2 + 2gx + 2fy + c = 0$ is the equation of a circle and $L \equiv lx + my + n = 0$ is the equation of a straight line. Then $S + \lambda L = 0$ represents the equation of a co-axial system with λ as a parameter.

In this system $L = 0$ is the radical axis.

$$S + \lambda L = 0, \quad \text{or} \quad x^2 + y^2 + (2g + \lambda l)x + (2f + \lambda m)y + c + \lambda n = 0.$$

The co-ordinates of the centre are

$$\left(-\frac{2g + \lambda l}{2}, -\frac{2f + \lambda m}{2}\right)$$

$$\text{and radius} = \sqrt{\left\{\left(\frac{2g + \lambda l}{2}\right)^2 + \left(\frac{2f + \lambda m}{2}\right)^2 - (c + \lambda n)\right\}}.$$

The circle whose radius is zero is known as a *point circle. In a co-axial system the centres of the point circles are called limiting points.*

Note 1. To find out limiting points, we find the value of λ from

$$\left(\frac{2g + \lambda l}{2}\right)^2 + \left(\frac{2f + \lambda m}{2}\right)^2 - (c + \lambda n) = 0.$$

If λ_1 and λ_2 are the roots of this equation, then the limiting points are

$$\left(-\frac{2g + \lambda_1 l}{2}, -\frac{2f + \lambda_1 m}{2}\right) \quad \text{and} \quad \left(-\frac{2g + \lambda_2 l}{2}, -\frac{2f + \lambda_2 m}{2}\right).$$

Note 2. If $S_1 = 0$ and $S_2 = 0$ be the equations of two circles in the co-axial system, then $S_1 + \lambda S_2 = 0$ is the equation of the co-axial system. (λ is the parameter.)

Note 3. Since the line joining the centres of two circles is perpendicular to the radical axis, the centres of a system of co-axial circles lie on a straight line.

Note 4. If we consider the y-axis ($x = 0$) as the common radical axis and the centre of the circles lie on the x-axis, then the equation of the system can be written as $x^2 + y^2 + 2\lambda x + c = 0$. ($\lambda$ is the parameter.)

Here the centre is $(-\lambda, 0)$ and radius is $\sqrt{\lambda^2 - c}$.

For the point circle $\lambda^2 - c = 0$, or $\lambda = \pm\sqrt{c}$. Thus the limiting points are $(\sqrt{c}, 0)$ and $(-\sqrt{c}, 0)$. If $c \geq 0$, these points are real otherwise they are imaginary.

Note 5. A circle and any two circles of a co-axial system have the same radical centre.

Let the equation of a co-axial system be $x^2 + y^2 + 2\lambda x + k = 0$ and the equation of a circle be $x^2 + y^2 + 2gx + 2fy + c = 0$.

The radical axis is $2(g - \lambda)x + 2fy + c - k = 0$.

This equation holds for all values of λ. $\therefore x = 0$.

When $x = 0$,

$$y = \frac{k - c}{2f}.$$

Thus the radical centre is

$$\left(0, \frac{k-c}{2f}\right).$$

It is independent of λ. Hence the result follows.

▶ **Example 3** *Find the limiting points of the co-axial system whose two circles are* $x^2 + y^2 - 6x - 2y + 4 = 0$ *and* $3(x^2 + y^2) - 16x - 4y + 12 = 0$.

The radical axis is

$$(x^2 + y^2 - 6x - 2y + 4) - \left(x^2 + y^2 - \frac{16}{3}x - \frac{4}{3}y + 4\right) = 0, \quad \text{or} \quad x + y = 0.$$

The equation of the co-axial system is

$$x^2 + y^2 - 6x - 2y + 4 + \lambda(x + y) = 0,$$

$$\text{or} \quad x^2 + y^2 + (\lambda - 6)x + (\lambda - 2)y + 4 = 0$$

where λ is a parameter.

The centre is $\left(\frac{6-\lambda}{2}, \frac{2-\lambda}{2}\right)$ and the radius is

$$\sqrt{\left\{\left(\frac{6-\lambda}{2}\right)^2 + \left(\frac{2-\lambda}{2}\right)^2 - 4\right\}}.$$

For the point circle

$$\left(\frac{6-\lambda}{2}\right)^2 + \left(\frac{2-\lambda}{2}\right)^2 - 4 = 0,$$

$$\text{or} \quad 36 - 12\lambda + \lambda^2 + 4 - 4\lambda + \lambda^2 - 16 = 0,$$

$$\text{or} \quad \lambda^2 - 8\lambda + 12 = 0, \quad \text{or} \quad (\lambda - 2)(\lambda - 6) = 0, \quad \text{or} \quad \lambda = 2, 6.$$

$\therefore$ the limiting points are $(2, 0)$ and $(0, -2)$.

7.5 Properties of Limiting Points

(i) *The limiting points of a system of co-axial circles are conjugate points w.r.t. any circle of this system.*

Let the equation of the co-axial system be $x^2+y^2+2\lambda x+c=0$.

The limiting points are $(\sqrt{c},0)$ and $(-\sqrt{c},0)$.

The polar of $(\sqrt{c},0)$ w.r.t. the circle is

$$x\sqrt{c}+\lambda\left(x+\sqrt{c}\right)+c=0, \text{ or } x\left(\sqrt{c}+\lambda\right)+\sqrt{c}\left(\lambda+\sqrt{c}\right)=0$$

$$\text{or } \left(x+\sqrt{c}\right)\left(\lambda+\sqrt{c}\right)=0$$

$$\text{or } x+\sqrt{c}=0.$$

Obviously this polar passes through $(-\sqrt{c},0)$. Hence the result follows.

(ii) *Every circle through the limiting points of a co-axial system is orthogonal to all circles of the system.*

Let us consider the co-axial system in the form

$$x^2+y^2+2\lambda x+c=0. \tag{7.5.1}$$

The limiting points are $(\sqrt{c},0)$ and $(-\sqrt{c},0)$,

Let the circle $x^2+y^2+2gx+2fy+c'=0$ pass through these points.

Then $c+2g\sqrt{c}+c'=0$ and $c-2g\sqrt{c}+c'=0$.

From these two equations $g=0$ and $c'=-c$.

$\therefore$ The equation of this circle is

$$x^2+y^2+2fy-c=0. \tag{7.5.2}$$

With f as a parameter the equation (7.5.2) represents another co-axial system of circles whose centres lie on the y-axis and x-axis is the radical axis.

One can easily verify that (7.5.1) and (7.5.2) are orthogonal.

Note 1. The centres of the system of circles of (7.5.1) lie on the radical axis of the system (7.5.2) and conversely.

Note 2. The limiting points of the system (7.5.2) are $(0,\sqrt{-c})$ and $(0,-\sqrt{-c})$. These points lie on the circles of (7.5.1).

Note 3. The points of intersection of any two circles of (7.5.1) are $(0,\sqrt{-c})$ and $(0,-\sqrt{-c})$ and those of any two circles of (7.5.2) are $(\sqrt{c},0)$ and $(-\sqrt{c},0)$. Hence the limiting points of one system are the points of intersection of the other system.

Note 4. If $c\neq 0$, then c is either positive or negative. Hence if the circles of one system intersect in real points then the circles of the other system will intersect in imaginary points.

Note 5. The centre of the system (7.5.2) is $(0,-f)$. The length of the tangent from this point to any circle of (7.5.1) is $\sqrt{f^2+c}$. It is equal to the radius of the circle of (7.5.2). Conversely the length of the tangent from the centre of (7.5.1) to (7.5.2) is equal to the radius of the circle of (7.5.1).

Worked Out Examples

1. Find the equation of the circles cutting

$$x^2+y^2+2x-9=0, \quad x^2+y^2-8x-9=0$$

orthogonally and touching $y=x+4$.

Let $x^2+y^2+2gx+2fy+c=0$ be the equation of the circle. As it cuts the given circles orthogonally

$$2g = c-9 \qquad (1)$$

$$\text{and} \quad -8g = c-9. \qquad (2)$$

From (1) and (2) $g=0$ and $c=9$.

The equation of the circle reduces to $x^2+y^2+2fy+9=0$.

To find out the points of intersection of this circle with the given line,

$$x^2+(x+4)^2+2f(x+4)+9=0,$$

$$\text{or} \quad 2x^2+2(f+4)x+8f+25=0.$$

Since the line touches the circle, the above equation in x must have equal roots.

$$\therefore \quad 4(f+4)^2-4.2.(8f+25)=0,$$

$$\text{or} \quad f^2-8f-34=0, \quad \text{or} \quad f=4\pm5\sqrt{2}.$$

$\therefore$ the required equations are $x^2+y^2+2(4\pm5\sqrt{2})y+9=0$.

2. Prove that the length of the common chord of circles

$$(x-a)^2+(y-b)^2 = c^2$$

$$\text{and} \quad (x-b)^2+(y-a)^2 = c^2 \text{ is } \sqrt{\{4c^2-2(a-b)^2\}}.$$

$$(x-a)^2+(y-b)^2 = c^2 \qquad (1)$$

$$(x-b)^2+(y-a)^2 = c^2. \qquad (2)$$

Subtracting (2) from (1) the equation of the common chord is

$$(b-a)x+(a-b)y=0, \quad \text{or} \quad x-y=0.$$

Let AB be the common chord, O the centre of the circle (1) and ON the perpendicular to AB.

From the fig. 7.4

$$AB = 2AN = 2\sqrt{OA^2-ON^2}.$$

$$OA = \text{radius of the circle (1)} = c.$$

$$ON = \frac{a-b}{\sqrt{2}}.$$

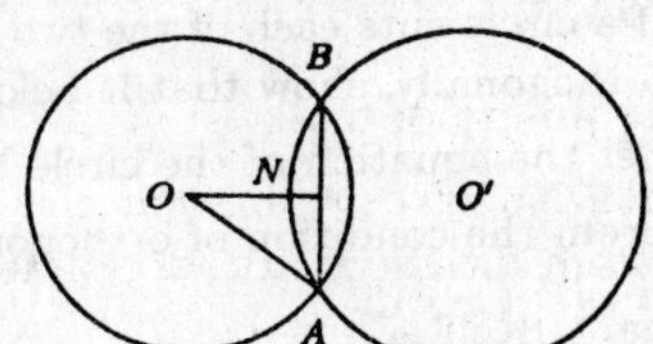

Fig. 7.4

$$\therefore \quad AB = 2\sqrt{\left\{c^2-\frac{(a-b)^2}{2}\right\}} = \sqrt{2}\sqrt{2c^2-(a-b)^2} = \sqrt{4c^2-2(a-b)^2}.$$

3. Prove that the two circles $x^2+y^2+2ax+c^2=0$ and $x^2+y^2+2by+c^2=0$, will touch, if $\frac{1}{a^2}+\frac{1}{b^2}=\frac{1}{c^2}$. **[B.U. '66]**

$$x^2+y^2+2ax+c^2=0 \quad (1)$$

$$x^2+y^2+2by+c^2=0. \quad (2)$$

The equation of the radical axis is

$$ax-by=0. \quad (3)$$

If the circles touch with each other, then the radical axis must be the common tangent to (1) and (2). If the line (3) be the tangents to (1), the distance of (3) from the centre $(-a,0)$ must be equal to the radius $\sqrt{c^2-a^2}$.

$$\therefore \quad \frac{-a^2}{\sqrt{a^2+b^2}}=\sqrt{a^2-c^2}, \quad \text{or} \quad \frac{a^4}{a^2+b^2}=a^2-c^2$$

$$\text{or} \quad a^4=(a^2+b^2)(a^2-c^2), \quad \text{or} \quad c^2(a^2+b^2)=a^2b^2, \quad \text{or} \quad \frac{1}{a^2}+\frac{1}{b^2}=\frac{1}{c^2}.$$

4. Prove that the two circles each of which passes through the points. $(0,k)$ and $(0,-k)$ and touch the line $y=mx+b$ will cut orthogonally, if $b^2=k^2(2+m^2)$.
Let the equation of one of the circles be

$$x^2+y^2+2gx+2fy+c=0. \quad (1)$$

As it passes through $(0,k)$ and $(0,-k)$, $k^2+2fk+c=0$ and $k^2-2fk+c=0$.
From these two equations in k, $f=0, c=-k^2$.
$\therefore$ the equation (1) takes the form $x^2+y^2+2gx-k^2=0$.
If the circle touches the line $y=mx+b$, the distance of it from the centre $(-g,0)$ will be equal to the radius $\sqrt{g^2+k^2}$.

$$\therefore \quad \frac{-mg+b}{\sqrt{m^2+1}}=\sqrt{g^2+k^2},$$

$$\text{or} \quad (b-mg)^2=(m^2+1)(g^2+k^2), \quad \text{or} \quad g^2+2bmg+k^2(m^2+1)-b^2=0.$$

It is a quadratic equation in g. Let the roots be g_1 and g_2. Then $g_1g_2=k^2(1+m^2)-b^2$ and the circles are

$$x^2+y^2+2g_1x-k^2=0 \quad (2)$$

$$\text{and} \quad x^2+y^2+2g_2x-k^2=0. \quad (3)$$

If (2) and (3) cut orthogonally

$$2g_1g_2=-2k^2, \text{ or } k^2(1+m^2)-b^2=-k^2 \text{ or } b^2=k^2(2+m^2).$$

5. If a circle cuts each of the two circles $x^2+y^2+2g_1x+c=0$ and $x^2+y^2+2g_2x+c=0$ orthogonally, show that it belongs to a co-axial system.
Let the equation of the circle be $x^2+y^2+2gx+2fy+c'=0$.
From the condition of orthogonality

$$2gg_1=c+c' \quad (1)$$

$$\text{and} \quad 2gg_2=c+c'. \quad (2)$$

From (1) and (2), $2g(g_1-g_2)=0$. $\because g_1\neq g_2, g=0$.
In this case $c+c'=0$, or $c'=-c$.

$\therefore$ the equation of the circle is $x^2 + y^2 + 2fy - c = 0$. Here f is a variable parameter. Therefore, the circle belongs to a co-axial system whose radical axis is $y = 0$.

6. Find the equation of the circle described upon the radical axis of a co-axial system as diameter, the limiting points of the system being $(1, 2)$ and $(3, 4)$,

The point circles of the co-axial system are

$$(x-1)^2 + (y-2)^2 = 0, \quad \text{or} \quad x^2 + y^2 - 2x - 4y + 5 = 0 \tag{1}$$

$$\text{and} \quad (x-3)^2 + (y-4)^2 = 0, \quad \text{or} \quad x^2 + y^2 - 6x - 8y + 25 = 0. \tag{2}$$

The radical axis is

$$4x + 4y - 20 = 0, \quad \text{or} \quad x + y - 5 = 0. \tag{3}$$

The equation of a circle belonging to the co-axial system is

$$x^2 + y^2 - 2x - 4y + 5 + \lambda(x + y - 5) = 0,$$

$$\text{or} \quad x^2 + y^2 - (2-\lambda)x - (4-\lambda)y + 5(1-\lambda) = 0.$$

The centre is

$$\left(\frac{2-\lambda}{2}, \frac{4-\lambda}{2}\right).$$

The centre lies on the radical axis when it is the diameter of the circle.

Thus

$$\frac{2-\lambda}{2} + \frac{4-\lambda}{2} - 5 = 0, \text{ or } \lambda = -2.$$

Hence the required circle is $x^2 + y^2 - 4x - 6y + 15 = 0$.

Exercise

1. Prove that the following circles cut orthogonally:
   (a) $x^2 + y^2 + 2x + 4y - 20 = 0, x^2 + y^2 + 6x - 8y + 10 = 0$.
   (b) $x^2 + y^2 - 2ax + 2by + c = 0, x^2 + y^2 + 2bx + 2ay - c = 0$.

2. Show that the three circles $x^2 + y^2 + 3x + 6y + 12 = 0$, $x^2 + y^2 + 2x + 8y + 16 = 0$ and $x^2 + y^2 + 12y + 24 = 0$ have a common radical axis.

3. Find the radical centre of the three circles

$$x^2 + y^2 - 4x - 6y - 12 = 0,$$

$$x^2 + y^2 + 2x + 4y - 4 = 0$$

$$\text{and} \quad x^2 + y^2 - 12x + 4y + 4 = 0.$$

4. Find the equation of the circle which cuts orthogonally the circles
   (a) $x^2 + y^2 - 2x - 4y - 20 = 0, x^2 + y^2 = 25$ and $x^2 + y^2 - 6x - 4y - 23 = 0$;
   (b) $x^2 + y^2 = a^2, (x-c)^2 + y^2 = a^2$ and $x^2 + (y-b)^2 = a^2$.

5. Find the value of k for which the circle $x^2 + y^2 - 2x + 2y + 7 = 0$ will bisect the circumference of $x^2 + y^2 - 6x + 8y - k = 0$.

6. If $x + iy = a\tan(u + iv)$, where $i = \sqrt{-1}$, prove that u = constant and v = constant represent two conjugate systems of co-axial circles.

7. (a) Find the equation of the circle whose diameter is the common chord of the circles

$$x^2 + y^2 + 2x + 3y + 1 = 0$$
$$\text{and} \quad x^2 + y^2 + 4x + 3y + 2 = 0.$$

[B.U. '66]

(b) Find the equation of the chord of the circles $(x - a)^2 + y^2 = a^2$ and $x^2 + (y - b)^2 = b^2$ and prove that the equation to the circle whose diameter is this common chord is

$$(a^2 + b^2)(x^2 + y^2) = 2ab(bx + ay).$$

[N.B. '87]

8. Find the equation of the common chord of the circles

$$x^2 + y^2 + 2gx + 2fy + c = 0$$
$$\text{and} \quad x^2 + y^2 + 2fx + 2gy + c = 0$$

and hence find the condition if they touch with each other.

9. Find the co-ordinates of the limiting points of the co-axial system determined by the pair of circles

(a) $x^2 + y^2 + 2x + 5 = 0, x^2 + y^2 + 2y + 5 = 0$;

(b) $x^2 + y^2 - 6x - 6y + 4 = 0, x^2 + y^2 - 2x - 4y + 3 = 0$.

10. It two circles cut a third circle orthogonally, show that the centre of the third circle lies on the radical axis of the two circles.

[HINTS. Let the two circles be $x^2 + y^2 + 2g_1x + \mu = 0$ and $x^2 + y^2 + 2g_2x + \mu = 0$.

Let the third circle which cuts the above circles orthogonally be

$$x^2 + y^2 + 2gx + 2fy + c = 0.$$

For orthogonality

$$2gg_1 = \mu + c \quad \text{and} \quad 2gg_2 = \mu + c.$$

Subtracting, $2g(g_1 - g_2) = 0 \Rightarrow g = 0$ since $g_1 \neq g_2$.

Hence the centre of the third circle is $(0, -f)$.

It lies on the radical axis whose equation is $x = 0$.]

11. Show that the limiting points of the system $x^2 + y^2 + 2x - 4y + 5 + \lambda(x^2 + y^2 - 4x + 6y + 13) = 0$ are $(-1, 2)$ and $(2, -3)$.

12. Find the equation of the circle which passes through the origin, has its centre on $x + y - 4 = 0$ and cuts orthogonally $x^2 + y^2 - 4x + 2y + 4 = 0$.

Answers

3. $(\frac{4}{7}, -\frac{8}{7})$. **4.** (a) $4(x^2 + y^2) + 6x - 13y + 100 = 0$, (b) $x^2 + y^2 - cx - by + a^2 = 0$. **5.** $k = -43$. **7.** (a) $2(x^2 + y^2) + 2x + 6y + 1 = 0$. **8.** $(g + f)^2 = 2c$. **9.** (a) $(1, -2), (-2, 1)$; (b) $(-1, 1), (\frac{1}{5}, \frac{8}{5})$. **12.** $x^2 + y^2 - 4x - 4y = 0$.

Chapter 8

Polar Co-ordinates

8.1 Polar Co-ordinates

Let O and OX be a fixed point and a fixed line on a given plane. A point P on this plane can be defined with reference to the fixed point and the fixed line. If r be the distance of P from O and θ be the angular distance of P from OX, then the polar co-ordinates of P are denoted by (r, θ). O and OX are called the *pole* and the *initial line* respectively. If (x, y) are the cartesian co-ordinates of P w.r.t. OX and OY, then $x = r\cos\theta$ and $y = r\sin\theta$. r is known as the *radius vector* and θ is the *vectorial angle* of P. Vectorial angle is generally measured in the anticlockwise direction.

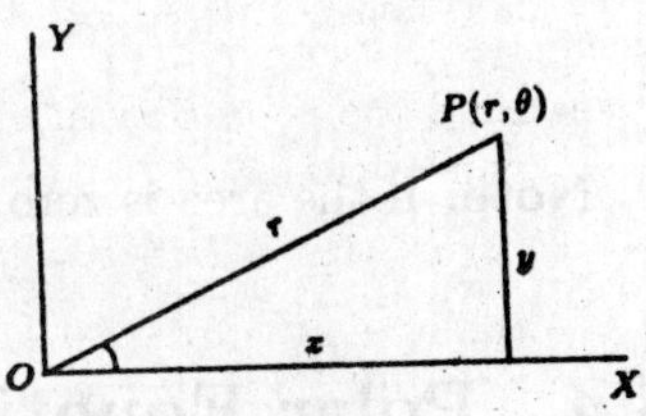

Fig. 8.1

8.2 Distance between Two Points

Let the polar co-ordinates of A and B be (r_1, θ_1) and (r_2, θ_2) w.r.t. the pole O and the initial line OX.

In $\Delta OAB, OA = r_1, OB = r_2$ and $\angle AOB = \theta_2 - \theta_1$.

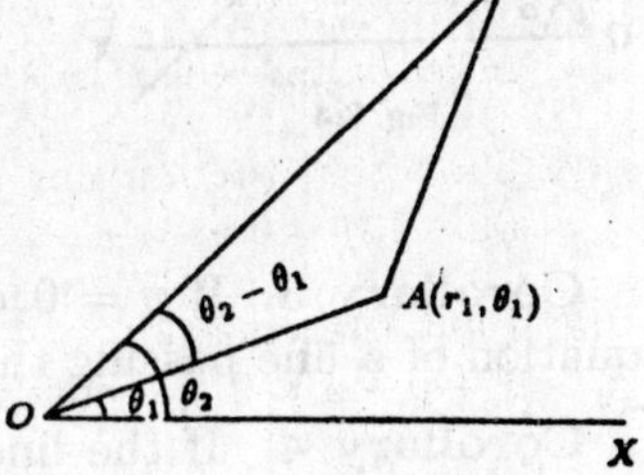

Fig. 8.2

Now

$$AB^2 = OA^2 + OB^2 - 2OA.OB\cos AOB$$
$$= r_1^2 + r_2^2 - 2r_1r_2\cos(\theta_2 - \theta_1).$$

$$\therefore \quad AB = \sqrt{\{r_1^2 + r_2^2 - 2r_1r_2\cos(\theta_2 - \theta_1)\}}.$$

8.3 Area of a Triangle

Let the polar co-ordinates of A, B and C be $(r_1, \theta_1), (r_2, \theta_2)$ and (r_3, θ_3) respectively w.r.t. the pole O and the initial line OX.

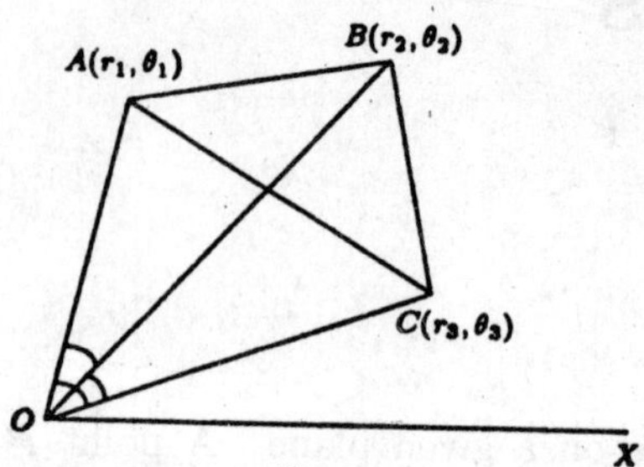

Fig. 8.3

From the fig. 8.3,

$$\begin{aligned}\Delta ABC &= \Delta OAB + \Delta OBC - \Delta OCA \\ &= \frac{1}{2} OA \cdot OB \sin A\hat{O}B \\ &\quad + \frac{1}{2} OB.OC \sin B\hat{O}C - \frac{1}{2} OC.OA \sin C\hat{O}A \\ &= \frac{1}{2}[r_1 r_2 \sin(\theta_1 - \theta_2) + r_2 r_3 \sin(\theta_2 - \theta_3) \\ &\quad - r_3 r_1 \sin(\theta_1 - \theta_3)] \\ &= \frac{1}{2}[r_1 r_2 \sin(\theta_1 - \theta_2) + r_2 r_3 \sin(\theta_2 - \theta_3) \\ &\quad + r_3 r_1 \sin(\theta_3 - \theta_1)]\end{aligned}$$

Note. If the area is zero, the points are collinear.

8.4 Polar Equation of a Straight Line

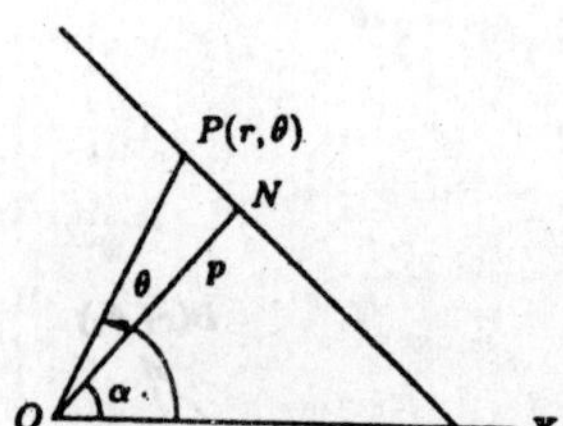

Fig. 8.4

Let (r, θ) be the co-ordinates of a point P on the line PN w.r.t. the pole O and the initial line OX.

ON is perpendicular to the line. Let $ON = p$ and $\angle XON = \alpha$.

Now $ON = OP \cos(\theta - \alpha)$, or $p = r \cos(\theta - \alpha)$. It is the equation of the line.

Corollary 1. If $\alpha = 0, p = r\cos\theta$ is the equation of the line. It is a straight line perpendicular to OX.

Corollary 2. If $\alpha = \frac{\pi}{2}, p = r \sin\theta$ is the equation of the line. It is parallel to OX.

Corollary 3. If $p = 0, \cos(\theta - \alpha) = 0$, or $\theta - \alpha = \frac{\pi}{2}$, or $\theta =$ constant. It is the equation of a line passing through the pole.

Corollary 4. If the line passes through (r_1, θ_1) and makes an angle β with the initial line then $\beta = \frac{\pi}{2} + \alpha$ and $p = r_1 \cos\left\{\frac{\pi}{2} - (\beta - \theta_1)\right\}$, or $p = r_1 \sin(\beta - \theta_1)$.

The equation of the line is $r \sin(\beta - \theta) = r_1 \sin(\beta - \theta_1)$.

Corollary 5. The equation $r\cos(\theta - \alpha) = p$ can be written as

$$r\cos\theta\cos\alpha + r\sin\theta\sin\alpha = p, \quad \text{or} \quad \frac{1}{r} = \frac{\cos\alpha}{p}\cos\theta + \frac{\sin\alpha}{p}\sin\theta$$

which gives another form of the polar equation of the line as

$$\frac{1}{r} = A\cos\theta + B\sin\theta,$$

where A and B are constants. It is the general form of the polar equation of a straight line.

Note. The polar equations of two parallel lines are of the form

$$r\cos(\theta-\alpha) = p \quad \text{and} \quad r\cos(\theta-\alpha) = p'.$$

The polar equations of two mutually perpendicular lines are of the form

$$r\cos(\theta-\alpha) = p \quad \text{and} \quad r\cos\left(\theta-\alpha-\frac{\pi}{2}\right) = p', \quad \text{i.e.,} \quad r\sin(\theta-\alpha) = p'.$$

8.5 Polar Equation of the Line passing through (r_1, θ_1) and (r_2, θ_2)

In cartesian system the equation of a line can be written as $ax + by + c = 0$. Putting $x = r\cos\theta$ and $y = r\sin\theta$,

$$a\cos\theta + b\sin\theta + \frac{c}{r} = 0. \tag{8.5.1}$$

If it passes through (r_1, θ_1) and (r_2, θ_2),

$$a\cos\theta_1 + b\sin\theta_1 + \frac{c}{r_1} = 0 \tag{8.5.2}$$

$$\text{and} \quad a\cos\theta_2 + b\sin\theta_2 + \frac{c}{r_2} = 0. \tag{8.5.3}$$

Eliminating a, b, c from (8.5.1), (8.5.2) and (8.5.3) the required equation is

$$\begin{vmatrix} \cos\theta & \sin\theta & 1/r \\ \cos\theta_1 & \sin\theta_1 & 1/r_1 \\ \cos\theta_2 & \sin\theta_2 & 1/r_2 \end{vmatrix} = 0,$$

$$\text{or} \quad \frac{1}{r}\sin(\theta_1-\theta_2) + \frac{1}{r_1}\sin(\theta_2-\theta) + \frac{1}{r_2}\sin(\theta-\theta_1) = 0.$$

Note.

$$a\cos\theta + b\sin\theta = \frac{k}{r} \tag{8.5.4}$$

$$\text{and} \quad a\cos\theta + b\sin\theta = \frac{k_1}{r} \tag{8.5.5}$$

are parallel lines.

$$a\cos\left(\frac{\pi}{2}+\theta\right) + b\sin\left(\frac{\pi}{2}+\theta\right) = \frac{k_2}{r},$$

i.e., $-a\sin\theta + b\cos\theta = \frac{k_2}{r}$ is perpendicular to (8.5.4) and (8.5.5).

8.6 Polar Equation of a Circle

Let (c, α) be the polar co-ordinates of the centre C of the circle of radius a w.r.t. the pole O and the initial line OX. Let $P(r, \theta)$ be a point on the circle.

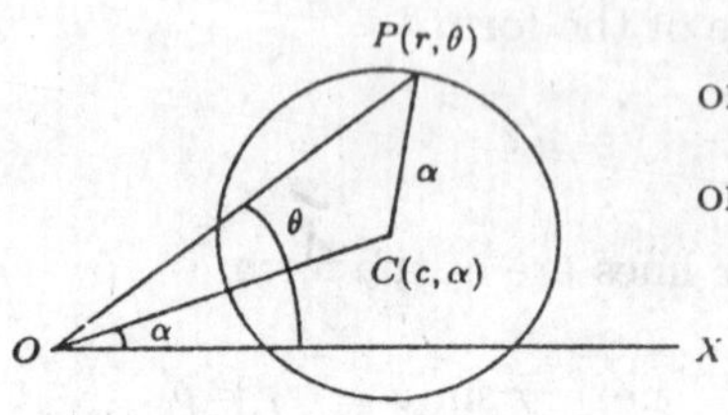

Fig. 8.5

From

$$\Delta OCP, CP^2 = OC^2 + OP^2 - 2OC.OP\cos(\theta - \alpha),$$

$$\text{or} \quad a^2 = c^2 + r^2 - 2cr\cos(\theta - \alpha),$$

$$\text{or} \quad r^2 - 2cr\cos(\theta - \alpha) + c^2 - a^2 = 0. \tag{8.6.1}$$

It is the required equation.

Corollary 1. If the initial line passes through the centre of the circle then $\alpha = 0$. In this case the equation (8.6.1) reduces to

$$r^2 - 2cr\cos\theta + c^2 - a^2 = 0. \tag{8.6.2}$$

Corollary 2. If the pole is on the circle, then $c = a$ and the equation (8.6.1) reduces to

$$r^2 - 2ar\cos(\theta - \alpha) = 0, \quad \text{or} \quad r = 2a\cos(\theta - \alpha). \tag{8.6.3}$$

[This equation is of the form $r = A\cos\theta + B\sin\theta$.]

Corollary 3. If the pole is on the circle and the initial line passes through the centre, then $c = a, \alpha = 0$ and the equation of the circle is

$$r = 2a\cos\theta. \tag{8.6.4}$$

Corollary 4. If the pole coincides with the centre, then the equation of the circle is

$$r = a. \tag{8.6.5}$$

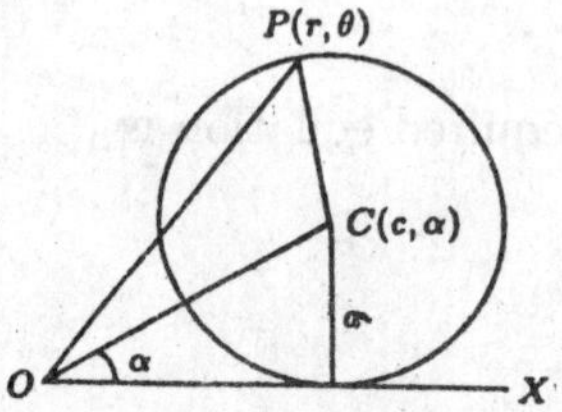

Fig. 8.6

Corollary 5. If the initial line touches the circle then $c = a\operatorname{cosec}\alpha$.

Putting this value of c in (8.6.1)

$$r^2 - 2ar\operatorname{cosec}\alpha\cos(\theta - \alpha) + a^2\operatorname{cosec}^2\alpha - a^2 = 0$$

$$\text{or} \quad r^2 - 2ar\operatorname{cosec}\alpha\cos(\theta - \alpha) + a^2\cot^2\alpha = 0. \tag{8.6.6}$$

It is the equation of the circle.

Corollary 6. If the initial line touches the circle at the pole then $\alpha = 90°$. In this case the equation (8.6.6) reduces to $r^2 - 2ar\cos(\theta - 90°) = 0$,

$$\text{or} \quad r = 2a\sin\theta. \tag{8.6.7}$$

Corollary 7. A secant through the pole, given by the equation $\theta = \theta_1$, meets the circle (8.6.1) in two points whose distances from the pole are given by the roots of $r^2 - 2cr\cos(\theta_1 - \alpha) + c^2 - a^2 = 0$.

If r_1 and r_2 are the roots of this equation, then $r_1r_2 = c^2 - a^2 =$ a constant. It suggests that if the secant meets the circle at P and Q then $OP.OQ =$ a constant for all positions of OPQ.

▶ **Example 1** *Find the equation of the circle when two ends of a diameter are given.* **[B.U. '98]**

Let $P(r,\theta)$ be a point on the circle and the co-ordinates of the end A and B of the diameter AB be (r_1,θ_1) and (r_2,θ_2) respectively w.r.t. the pole O and the initial line OX. From the fig. 8.7.

$$\begin{aligned}AP^2 &= OA^2+OP^2-2OA.OP\cos P\hat{O}A\\ &= r_1^2+r^2-2r_1r\cos(\theta_1-\theta),\\ BP^2 &= OB^2+OP^2-2OB.OP\cos B\hat{O}P\\ &= r_2^2+r^2-2r_2r\cos(\theta-\theta_2),\\ AB^2 &= OA^2+OB^2-2OA.OB\cos A\hat{O}B\\ &= r_1^2+r_2^2-2r_1r_2\cos(\theta_1-\theta_2).\end{aligned}$$

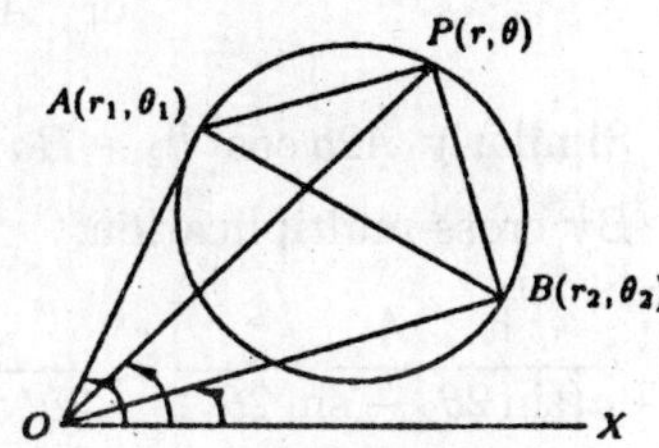

Fig. 8.7

Since AB is a diameter, $\angle APB$ is a right angle.

$$\therefore\ AP^2+BP^2=AB^2,$$

$$\text{or}\quad r_1^2+r^2-2r_1r\cos(\theta_1-\theta)+r_2^2+r^2-2r_2r\cos(\theta-\theta_2),$$

$$=r_1^2+r_2^2-2r_1r_2\cos(\theta_1-\theta_2)$$

$$\text{or}\quad r^2-r\{r_1\cos(\theta-\theta_1)+r_2\cos(\theta-\theta_2)\}+r_1r_2\cos(\theta_1-\theta_2)=0.$$

It is the equation of the circle.

▶ **Example 2** *Find the polar co-ordinates of the centre of the circle* $r=4\cos\theta+3\sin\theta$. **[B.U. '82, '91]**

Let $4=a\cos\alpha, 3=a\sin\alpha$. Then $a=5$ and $\alpha=\tan^{-1}\frac{3}{4}$.

Now $r=4\cos\theta+3\sin\theta=a(\cos\theta\cos\alpha+\sin\theta\sin\alpha)=5\cos(\theta-\alpha)$.

Hence the polar co-ordinates of the centre are $\left(\frac{5}{2},\tan^{-1}\frac{3}{4}\right)$.

▶ **Example 3** *Find the polar equation of the circle which passes through the pole and the two points whose polar co-ordinates are* $(d,0)$ *and* $\left(2d,\frac{\pi}{3}\right)$. *Find also the radius of the circle.* **[C.H. '77]**

Let the equation be $r=A\cos\theta+B\sin\theta$.

By the given conditions $d=A$ and $2d=\frac{A}{2}+B\frac{\sqrt{3}}{2}$, or $B=\sqrt{3}d$.

$\therefore$ the equation is $r=d\cos\theta+\sqrt{3}d\sin\theta=2d\cos\left(\theta-\frac{\pi}{3}\right)$.

Hence the radius of the circle $=d$.

▶ **Example 4** *Prove that the equation of the chord joining the points* $\theta=\theta_1$, $\theta=\theta_2$ *on the circle* $r=2a\cos\theta$ *is* $r\cos(\theta-\theta_1-\theta_2)=2a\cos\theta_1\cos\theta_2$. **[C.H. '68, '76]**

Let the equation of the chord be $\frac{1}{r} = A\cos\theta + B\sin\theta$.

From the equation of the circle the radius vector corresponding to the vectorial angle θ_1 is $2a\cos\theta_1$ and that for θ_2 is $2a\cos\theta_2$.

The ends of the chord are $(2a\cos\theta_1, \theta_1)$ and $(2a\cos\theta_2, \theta_2)$.

$$\therefore \quad \frac{1}{2a\cos\theta_1} = A\cos\theta_1 + B\sin\theta_1$$

$$\text{or} \quad A2a\cos^2\theta_1 + Ba\sin 2\theta_1 - 1 = 0.$$

Similarly $A2a\cos^2\theta_2 + Ba\sin 2\theta_2 - 1 = 0$.

By cross-multiplication,

$$\frac{A}{a(\sin 2\theta_2 - \sin 2\theta_1)} = \frac{B}{2a(\cos^2\theta_1 - \cos^2\theta_2)} = \frac{1}{2a^2(\cos^2\theta_1 \sin 2\theta_2 - \cos^2\theta_2 \sin 2\theta_1)},$$

$$\text{or} \quad \frac{A}{2a\cos(\theta_1+\theta_2)\sin(\theta_2-\theta_1)} = \frac{B}{2a\sin(\theta_1+\theta_2)\sin(\theta_2-\theta_1)} = \frac{1}{4a^2\cos\theta_1\cos\theta_2\sin(\theta_2 -}$$

$$\text{or} \quad A = \frac{\cos(\theta_1+\theta_2)}{2a\cos\theta_1\cos\theta_2}, \quad B = \frac{\sin(\theta_1+\theta_2)}{2a\cos\theta_1\cos\theta_2}.$$

Thus the equation is

$$[\cos\theta\cos(\theta_1+\theta_2) + \sin\theta\sin(\theta_1+\theta_2)] = 2\alpha\cos\theta_1\cos\theta_2 \cdot \frac{1}{r},$$

$$\text{or} \quad r\cos(\theta - \theta_1 - \theta_2) = 2a\cos\theta_1\cos\theta_2.$$

Note. Putting $\theta_1 = \theta_2 = \alpha$, the tangent at $\theta = \alpha$ is $r\cos(\theta - 2\alpha) = 2a\cos^2\alpha$.

8.7 Polar Equation of a Conic with the Focus as the Pole

Let S be the focus, SX the axis and RM the directrix of the conic in the fig. 8.8. Let $P(r, \theta)$ be a point on the conic w.r.t. the pole S and the initial line SX. PM is perpendicular to the directrix, LSL' is the latus rectum. PN and LQ are perpendicular to the initial line and the directrix respectively.

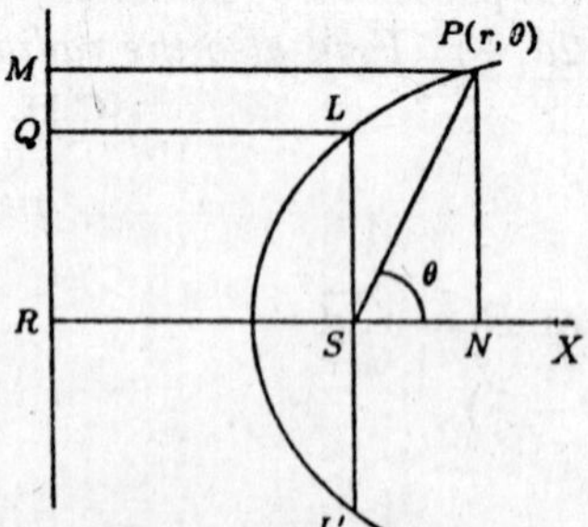

Fig. 8.8

Now the eccentricity $e = \frac{SP}{PM}$,

$$\begin{aligned} \text{or} \quad SP &= e\cdot PM = e\cdot RN \\ &= e\cdot(RS + SN) = e\cdot(LQ + SN) \\ &= e\cdot\left(\frac{SL}{e} + SN\right) \quad \left(\because \frac{SL}{LQ} = e\right) \\ &= SL + e\cdot SP\cos\theta. \end{aligned}$$

If l is the semi-latus rectum then $r = l + er\cos\theta$, or $\frac{l}{r} = 1 - e\cos\theta$ is the equation of the conic.

Corollary 1. If the initial line is taken in the direction SR then the equation of the conic will be $\frac{l}{r} = 1 - e\cos(\pi - \theta)$, i.e., $\frac{l}{r} = 1 + e\cos\theta$.

Corollary 2. If the initial line makes an angle α with the axis then the equation of the conic is $\frac{l}{r} = 1 - e\cos(\theta - \alpha)$.

Corollary 3. If the conic is a parabola then $e = 1$. Thus the equation of a parabola can be expressed as

$$\frac{l}{r} = 1 - \cos\theta, \quad \text{or} \quad \frac{l}{r} = 2\sin^2\frac{\theta}{2} \quad \text{or} \quad r^{1/2}\sin\frac{\theta}{2} = \left(\frac{l}{2}\right)^{1/2}.$$

If the initial line is in the direction SR, then the equation can be written as

$$\frac{l}{r} = 1 + \cos\theta, \quad \text{or} \quad \frac{l}{r} = 2\cos^2\frac{\theta}{2}, \quad \text{or} \quad r^{1/2}\cos\frac{\theta}{2} = \left(\frac{l}{2}\right)^{1/2}.$$

Note. Equation of the directrix: If (r, θ) be a point on the directrix then

$$r\cos(\pi - \theta) = SR, \quad \text{or} \quad -r\cos\theta = \frac{l}{e}, \quad \text{or} \quad \frac{l}{r} = -e\cos\theta.$$

▶ **Example 5** *Find the nature of the following conics:*

(i) $\frac{5}{r} = 2 - 2\cos\theta$, (ii) $\frac{1}{r} = 8 + 5\cos\theta$, (iii) $\frac{5}{r} = 3 - 4\cos\theta$. **[C.U. '92]**

(i) The equation $\frac{5}{r} = 2 - 2\cos\theta$ can be written as $\frac{5/2}{r} = 1 - \cos\theta$. Comparing it with $\frac{l}{r} = 1 - e\cos\theta$, we have $l = 5/2$ and $e = 1$. Since the eccentricity is 1, the given equation is a parabola. Its semi-latus rectum is $5/2$.

(ii) The equation $\frac{1}{r} = 8 + 5\cos\theta$ can be written as $\frac{1/8}{r} = 1 + \frac{5}{8}\cos\theta$.
Comparing it with $\frac{l}{r} = 1 + e\cos\theta$, we have $l = 1/8$ and $e = 5/8$. Since the eccentricity is less than 1 the given equation represents an ellipse whose semi-latus rectum is $1/8$.

(iii) The equation $\frac{5}{r} = 3 - 4\cos\theta$ can be written as $\frac{5/3}{r} = 1 - \frac{4}{3}\cos\theta$.
Comparing it with $\frac{l}{r} = 1 - e\cos\theta$, we have $l = \frac{5}{3}$ and $e = \frac{4}{3}$. Since $e > 1$, the equation represents a hyperbola whose semi-latus rectum is $\frac{5}{3}$.

▶ **Example 6** *Find the points on the conic* $\frac{9}{r} = 2 + \sqrt{2}\cos\theta$ *whose radius vector is* 3.

Putting $r = 3$ in the equation we have

$$3 = 2 + \sqrt{2}\cos\theta, \quad \text{or} \quad \cos\theta = 1/\sqrt{2}.$$

Therefore, $\theta = \pi/4, -\pi/4$.

Thus the required points are $(3, \pi/4)$ and $(3, -\pi/4)$.

► **Example 7** *Find the equation of the directrix of the conic* $r\sin^2\theta/2 = 1$.

$$r\sin^2\theta/2 = 1, \text{ or } r2\sin^2\theta/2 = 2, \text{ or } r(1-\cos\theta) = 2, \text{ or } \frac{2}{r} = 1-\cos\theta.$$

It is a parabola whose semi-latus rectum is 2.

The equation of the directrix is $\frac{2}{r} = -\cos\theta$, or $r\cos\theta + 2 = 0$.

► **Example 8** *Find the polar equation of the left branch of the hyperbola*

$$9x^2 - 16y^2 = 144.$$

The given equation can be written as $\frac{x^2}{16} - \frac{y^2}{9} = 1$.

Comparing it with $\frac{x^2}{a^2} - \frac{y^2}{b^2} = 1$, we get $a^2 = 16$, $b^2 = 9$.

$\therefore$ semi-latus rectum $l = b^2/a = 9/4$ and

$$\text{eccentricity} \quad e = \sqrt{\left(1+\frac{b^2}{a^2}\right)} = \sqrt{\left(1+\frac{9}{16}\right)} = 5/4.$$

The equation of the left branch in polar co-ordinates can be written as

$$\frac{l}{r} = 1 + e\cos\theta, \text{ or } \frac{\frac{9}{4}}{r} = 1 + \frac{5}{4}\cos\theta, \text{ or } \frac{9}{r} = 4 + 5\cos\theta.$$

8.8 Equation of a Chord (joining $\theta = \alpha - \beta$ and $\theta = \alpha + \beta$), a Tangent and a Normal

Chord. Let

$$\frac{l}{r} = 1 - e\cos\theta \tag{8.8.1}$$

$$\text{and} \quad \frac{l}{r} = a\cos\theta + b\sin\theta \tag{8.8.2}$$

be the equations of the conic and the chord respectively.

For the common points of (8.8.1) and (8.8.2)

$$1 - e\cos\theta = a\cos\theta + b\sin\theta.$$

At the common points $\theta = \alpha - \beta$ and $\theta = \alpha + \beta$.

$$\therefore\ 1 - e\cos(\alpha-\beta) = a\cos(\alpha-\beta) + b\sin(\alpha-\beta)$$

$$\text{and } 1 - e\cos(\alpha+\beta) = a\cos(\alpha+\beta) + b\sin(\alpha+\beta).$$

From these two,

$$(a+e)\cos(\alpha-\beta) + b\sin(\alpha-\beta) - 1 = 0$$

$$\text{and } (a+e)\cos(\alpha+\beta) + b\sin(\alpha+\beta) - 1 = 0.$$

By cross-multiplication,

$$\frac{a+e}{-\sin(\alpha-\beta)+\sin(\alpha+\beta)} = \frac{b}{-\cos(\alpha+\beta)+\cos(\alpha-\beta)}$$

$$= \frac{1}{\cos(\alpha-\beta)\sin(\alpha+\beta)-\sin(\alpha-\beta)\cos(\alpha+\beta)}$$

or $$\frac{a+e}{2\cos\alpha\sin\beta} = \frac{b}{2\sin\alpha\sin\beta} = \frac{1}{2\sin\beta\cos\beta}.$$

$$\therefore\ a+e = \frac{\cos\alpha}{\cos\beta}, \quad \text{or} \quad a = \frac{\cos\alpha}{\cos\beta} - e \quad \text{and} \quad b = \frac{\sin\alpha}{\cos\beta}.$$

$\therefore$ the equation of the chord is

$$\frac{l}{r} = \left(\frac{\cos\alpha}{\cos\beta} - e\right)\cos\theta + \frac{\sin\alpha}{\cos\beta}\sin\theta,$$

or $$\frac{l}{r} = -e\cos\theta + \sec\beta\cos(\theta-\alpha). \tag{8.8.3}$$

Corollary. The equation of the chord through the points γ and δ.

Here $\alpha-\beta=\gamma$, $\alpha+\beta=\delta$.

$$\therefore\ \alpha = \frac{\gamma+\delta}{2} \quad \text{and} \quad \beta = \frac{\delta-\gamma}{2}.$$

Putting these values of α and β in (8.8.3) we have

$$\frac{l}{r} = -e\cos\theta + \sec\frac{\delta-\gamma}{2}\cos\left(\theta - \frac{\delta+\gamma}{2}\right). \tag{8.8.4}$$

It is the equation of the chord.

Tangent. The two points corresponding to vectorial angles $\alpha-\beta$ and $\alpha+\beta$ will coincide, if $\beta=0$. In this case the chord will be the tangent at α. Hence the equation of the tangent at α is

$$\frac{l}{r} = -e\cos\theta + \cos(\theta-\alpha). \tag{8.8.5}$$

Normal. Let the equation of the normal at α be

$$\frac{k}{r} = \cos\left(\frac{\pi}{2}+\theta-\alpha\right) - e\cos\left(\frac{\pi}{2}+\theta\right),$$

[$\because$ the normal is perpendicular to the tangent.]

or $$\frac{k}{r} = -\sin(\theta-\alpha) + e\sin\theta.$$

It passes through the point of contact $\left(\frac{l}{1-e\cos\alpha}, \alpha\right)$.

$$\therefore\ \frac{k(1-e\cos\alpha)}{l} = e\sin\alpha, \text{ or } k = \frac{el\sin\alpha}{1-e\cos\alpha}.$$

Putting this value of k, the equation of the normal is

$$\frac{el\sin\alpha}{r(1-e\cos\alpha)} = e\sin\theta - \sin(\theta-\alpha). \tag{8.8.6}$$

Note. If the equation of the conic is $\frac{l}{r} = 1 + e\cos\theta$ then the equations of the tangent and normal at $\theta = \alpha$ are

$$\frac{l}{r} = e\cos\theta + \cos(\theta-\alpha)$$

$$\text{and}\quad \frac{le\sin\alpha}{r(1+e\cos\alpha)} = e\sin\theta + \sin(\theta-\alpha)$$

respectively.

8.9 To find the condition that $\frac{1}{r} = A\cos\theta + B\sin\theta$ may touch the Conic $\frac{l}{r} = 1 - e\cos\theta$

Let the line

$$\frac{1}{r} = A\cos\theta + B\sin\theta \tag{8.9.1}$$

touch the conic $\frac{l}{r} = 1 - e\cos\theta$ at $\theta = \alpha$.

The equation of the tangent at $\theta = \alpha$ to the conic is

$$\frac{l}{r} = -e\cos\theta + \cos(\theta-\alpha), \text{ i.e., } \frac{l}{r} = (\cos\alpha - e)\cos\theta + \sin\alpha\sin\theta.$$

Comparing it with the equation (1), we have

$$\frac{l}{1} = \frac{\cos\alpha - e}{A} = \frac{\sin\alpha}{B}. \qquad \therefore \cos\alpha = Al + e \text{ and } \sin\alpha = Bl.$$

Now $\cos^2\alpha + \sin^2\alpha = (Al+e)^2 + B^2l^2$, or $(Al+e)^2 + B^2l^2 = 1$.

It is the required condition.

8.10 Chord of Contact

Definition 1 *It is a chord joining the points of contact of tangents to a conic from a given point outside the conic.*

Let

$$\frac{l}{r} = 1 - e\cos\theta \tag{8.10.1}$$

be the equation of the conic and (r_1, θ_1) be the point from which tangents are drawn to the conic.

Let $\alpha - \beta$ and $\alpha + \beta$ be the vectorial angles of the two points of contact of the tangents. Then the equation of the chord joining them is

$$\frac{l}{r} = -e\cos\theta + \sec\beta\cos(\theta - \alpha). \tag{8.10.2}$$

[See the equation (8.8.3) of section 8.8.]

The equations of the tangents at the points on the conic having vectorial angles $\alpha - \beta$ and $\alpha + \beta$ are

$$\frac{l}{r} = -e\cos\theta + \cos(\theta - \alpha + \beta) \quad \text{and} \quad \frac{l}{r} = -e\cos\theta + \cos(\theta - \alpha - \beta).$$

These pass through the point (r_1, θ_1) if

$$\frac{l}{r_1} = -e\cos\theta_1 + \cos(\theta_1 - \alpha + \beta) \tag{8.10.3}$$

$$\text{and} \quad \frac{l}{r_1} = -e\cos\theta_1 + \cos(\theta_1 - \alpha - \beta). \tag{8.10.4}$$

Thus $\cos(\theta_1 - \alpha + \beta) = \cos(\theta_1 - \alpha - \beta)$, which is possible if

$$\theta_1 - \alpha + \beta = \pm(\theta_1 - \alpha - \beta)$$

Since $\beta \neq 0$, the upper sign is inadmissible and

$$\theta_1 - \alpha + \beta = -(\theta_1 - \alpha - \beta), \text{ i.e., } \alpha = \theta_1. \tag{8.10.5}$$

Now from (8.8.3), $\frac{l}{r_1} = -e\cos\theta_1 + \cos\beta$ which gives

$$\cos\beta = \frac{l}{r_1} + e\cos\theta_1. \tag{8.10.6}$$

Using (8.8.4) and (8.8.5) in (8.8.2), we get

$$\left(\frac{l}{r} + e\cos\theta\right)\left(\frac{l}{r_1} + e\cos\theta_1\right) = \cos(\theta - \theta_1). \tag{8.10.7}$$

It is the equation of the chord of contact of tangents through the point (r_1, θ_1).

Note. If the equation is $\frac{l}{r} = 1 + e\cos\theta$ then the equation of the chord of contact of the tangents to the conic from the point (r_1, θ_1) is

$$\left(\frac{l}{r} - e\cos\theta\right)\left(\frac{l}{r_1} - e\cos\theta_1\right) = \cos(\theta - \theta_1).$$

8.11 Pole and Polar

Definition 2 *The polar of a point w.r.t. a conic is the locus of the points of intersection of tangents at the extremities of the chords through that point and the point itself is called the pole of its polar.*

Let $\frac{l}{r} = 1 - e\cos\theta$ be the equation of the conic and the point (r_1, θ_1) be the pole.

Let the vectorial angles of the extremities of a chord to the conic through the pole be $\alpha - \beta$ and $\alpha + \beta$. Then the equation of this chord is

$$\frac{l}{r} = -e\cos\theta + \sec\beta\cos(\theta - \alpha).$$

Since it passes through (r_1, θ_1), we have

$$\frac{l}{r_1} = -e\cos\theta_1 + \sec\beta\cos(\theta_1 - \alpha). \tag{8.11.1}$$

The tangents at the extremities are

$$\frac{l}{r} = -e\cos\theta + \cos(\theta - \alpha + \beta)$$

$$\text{and}\quad \frac{l}{r} = -e\cos\theta + \cos(\theta - \alpha - \beta).$$

If these tangents meet at the point (ρ, ϕ), then

$$\frac{l}{\rho} = -e\cos\phi + \cos(\phi - \alpha + \beta)$$

$$\text{and}\quad \frac{l}{\rho} = -e\cos\phi + \cos(\phi - \alpha - \beta).$$

Thus $\cos(\phi - \alpha + \beta) = \cos(\phi - \alpha - \beta)$ which is possible if

$$\phi - \alpha + \beta = \pm(\phi - \alpha - \beta).$$

Since $\beta \neq 0$, the upper sign is inadmissible and

$$\phi - \alpha + \beta = -(\phi - \alpha - \beta), \quad \text{i.e.}\quad \phi = \alpha. \tag{8.11.2}$$

$$\therefore\quad \frac{l}{\rho} = -e\cos\phi + \cos\beta, \quad\text{or}\quad \frac{l}{\rho} + e\cos\phi = \cos\beta. \tag{8.11.3}$$

From (8.8.1)

$$\frac{l}{r_1} + e\cos\theta_1 = \sec\beta\cos(\theta_1 - \phi). \tag{8.11.4}$$

Eliminating β between (8.8.3) and (8.8.4),

$$\left(\frac{l}{\rho} + e\cos\phi\right)\left(\frac{l}{r_1} + e\cos\theta_1\right) = \cos(\phi - \theta_1).$$

Hence the locus of (ρ, ϕ) is

$$\left(\frac{l}{r} + e\cos\theta\right)\left(\frac{l}{r_1} + e\cos\theta_1\right) = \cos(\theta - \theta_1). \tag{8.11.5}$$

It is the equation of the polar of the pole (r_1, θ_1).

Note. The polar of a point w.r.t. a conic coincides with the chord of contact of tangents from that point to the conic when the point is outside the conic.

Worked Out Examples

1. Transform the equations

(a) $(x^2+y^2)^2 = ax^2y$ to polar form; and

(b) $\frac{1}{r} = 1+\cos\theta$ to cartesian form. **[B.U. '82, '84, '96, '99]**

Let (r,θ) be the polar co-ordinates of the point whose cartesian co-ordinates are (x,y). Then $x = r\cos\theta, y = r\sin\theta, r = \sqrt{x^2+y^2}$.

(a) Now $(x^2+y^2)^2 = ax^2y$ transforms to

$$r^4 = ar^2\cos^2\theta\cdot r\sin\theta, \quad \text{or} \quad r = a\cos^2\theta\sin\theta.$$

It is the equation in polar form.

(b) $\frac{1}{r} = 1+\cos\theta$, or $1 = r + r\cos\theta$, or $1 = \sqrt{x^2+y^2}+x$, or $(1-x)^2 = x^2+y^2$, or $y^2 = 1-2x$.

It is the equation in cartesian form.

2. Show that for the conic $\frac{l}{r} = 1+e\cos\theta$, the equation to the directrix corresponding to the focus other than the pole is

$$\frac{l}{r} = -\frac{1-e^2}{1+e^2}e\cos\theta.$$

Let $P(r,\theta)$ be a point on the directrix corresponding to the focus S' (other than the pole S). SR' is perpendicular to the directrix.

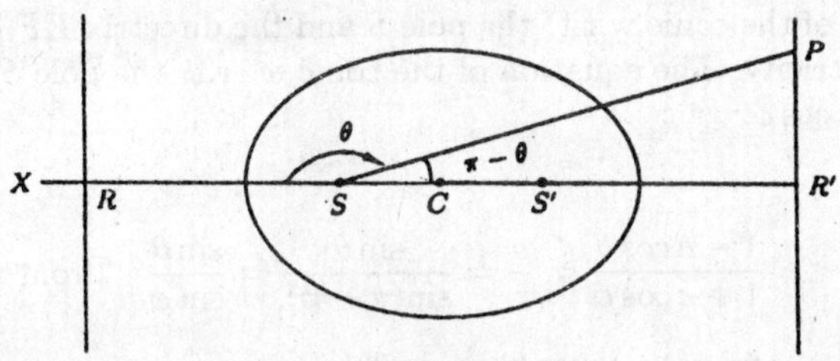

Fig. 8.9

From the fig. 8.9

$$\frac{SR'}{SP} = \cos(\pi-\theta),$$

$$\text{or} \quad r\cos\theta = -SR' = -(RR'-SR) = -\left(\frac{2a}{e}-\frac{l}{e}\right),$$

where l = semi-latus rectum $= \frac{b^2}{a} = \frac{a^2(1-e^2)}{a} = a(1-e^2)$, a = semi-major axis.

$$\therefore \quad r\cos\theta = -\left[\frac{2}{e}\cdot\frac{l}{1-e^2}-\frac{l}{e}\right] = -\frac{l}{e}\,\frac{1+e^2}{1-e^2},$$

$$\text{or} \quad \frac{l}{r} = -\frac{1-e^2}{1+e^2}e\cos\theta.$$

It is the equation of the required directrix.

3. Show that the tangent at $\theta = \alpha$ to

$$\frac{2a}{r} = 1 + \cos\theta \text{ is } r = a\sec\frac{\alpha}{2}\sec\left(\theta - \frac{\alpha}{2}\right).$$

The equation of the tangent at $\theta = \alpha$ is

$$\frac{2a}{r} = \cos\theta + \cos(\theta - \alpha) = 2\cos\left(\theta - \frac{\alpha}{2}\right)\cdot\cos\frac{\alpha}{2},$$

$$\text{or} \quad r = a\sec\frac{\alpha}{2}\sec\left(\theta - \frac{\alpha}{2}\right).$$

4. If θ and ϕ are vectorial angles of a point on a conic referred to two focii and they are measured in the same sense w.r.t. the initial line joining the focii, show that the ratio of $\tan\theta/2$ and $\tan\phi/2$ is constant.

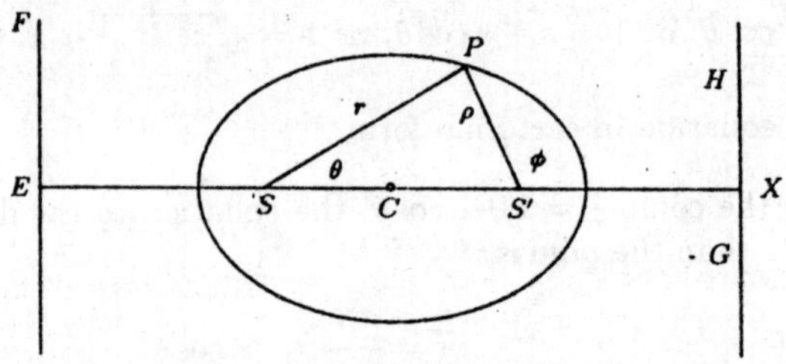

Fig. 8.10

Let S and S' be the focii and SX be the initial line. Let (r, θ) be the co-ordinates of P referred to the pole S and (ρ, ϕ) be those of P referred to the pole S'.

The equation of the conic w.r.t. the pole S and the directrix EF is $\frac{l}{r} = 1 - e\cos\theta$, where e is the eccentricity. The equation of the conic w. r.t. the pole S' and the directrix GH is $l/\rho = 1 + e\cos\phi$.

$$\therefore \quad \frac{1 - e\cos\theta}{1 + e\cos\phi} = \frac{\rho}{r} = \frac{\sin\theta}{\sin(\pi - \phi)} = \frac{\sin\theta}{\sin\phi} \quad [\text{from } \Delta SS'P].$$

$$\therefore \quad \sin\phi(1 - e\cos\theta) = \sin\theta(1 + e\cos\phi),$$

$$\text{or} \quad \sin\phi - \sin\theta = e(\sin\phi\cos\theta + \cos\phi\sin\theta) = e\sin(\phi + \theta),$$

$$\text{or} \quad 2\cos\frac{\phi+\theta}{2}\sin\frac{\phi-\theta}{2} = e\cdot 2\sin\frac{\phi+\theta}{2}\cos\frac{\phi+\theta}{2},$$

$$\text{or} \quad \frac{\sin\frac{\phi-\theta}{2}}{\sin\frac{\phi+\theta}{2}} = e, \quad \text{or} \quad \frac{\sin\frac{\phi+\theta}{2} - \sin\frac{\phi-\theta}{2}}{\sin\frac{\phi+\theta}{2} + \sin\frac{\phi-\theta}{2}} = \frac{1-e}{1+e},$$

$$\text{or} \quad \frac{2\cos\frac{\phi}{2}\sin\frac{\theta}{2}}{2\sin\frac{\phi}{2}\cos\frac{\theta}{2}} = \frac{1-e}{1+e}, \quad \text{or} \quad \frac{\tan\frac{\theta}{2}}{\tan\frac{\phi}{2}} = \frac{1-e}{1+e} = \text{constant}.$$

5. Show that the semi-latus rectum of a conic is the harmonic mean between the segments of a focal chord. **[C.U. '61, '81, '97; N.B. '92; B.H. '83, '85, '97]**

Let the equation of the conic be $\frac{l}{r} = 1 - e\cos\theta$ and PSP' be a focal chord. If $\angle XSP = \alpha$, then the vectorial angle of P' is $\pi + \alpha$.

Now

$$\frac{l}{SP} = 1 - e\cos\alpha$$

and $$\frac{l}{SP'} = 1 - e\cos(\pi + \alpha) = 1 + e\cos\alpha.$$

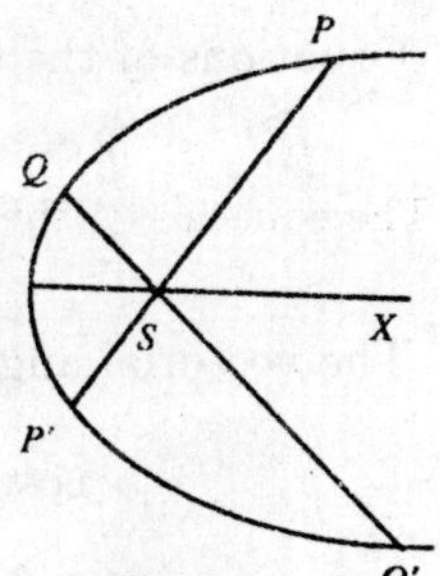

Fig. 8.11

Adding these two, we have

$$\frac{l}{SP} + \frac{l}{SP'} = 2, \quad \text{or} \quad \frac{1}{SP} + \frac{1}{SP'} = \frac{2}{l}.$$

Hence the result follows.

6. In any conic, prove that the sum of the reciprocals of two perpendicular focal chords is constant. **[B.U. '89, '92. '99; C.U. '80, '85, 2000; N.B. '91, '97]**

Let

$$\frac{l}{r} = 1 - e\cos\theta \tag{1}$$

be the equation of the conic and PSP' and QSQ' be the two perpendicular focal chords. [See the fig. 8.11.]

If α be the vectorial angle of P, then the vectorial angles of Q, P' and Q' are $\frac{\pi}{2} + \alpha, \pi + \alpha$ and $\frac{3\pi}{2} + \alpha$ respectively.

From the equation (1), $\frac{l}{SP} = 1 - e\cos\alpha$ and $\frac{l}{SP'} = 1 + e\cos\alpha$.

$$\therefore \quad SP + SP' = \frac{l}{1 - e\cos\alpha} + \frac{l}{1 + e\cos\alpha} = \frac{2l}{1 - e^2\cos^2\alpha},$$

$$\text{or} \quad PP' = \frac{2l}{1 - e^2\cos^2\alpha}, \quad \text{or} \quad \frac{1}{PP'} = \frac{1 - e^2\cos^2\alpha}{2l}. \tag{2}$$

Again

$$\frac{l}{SQ} = 1 - e\cos\left(\frac{\pi}{2} + \alpha\right) = 1 + e\sin\alpha,$$

$$\frac{l}{SQ'} = 1 - e\cos\left(\frac{3\pi}{2} + \alpha\right) = 1 - e\sin\alpha.$$

$$\therefore \quad SQ + SQ' = \frac{l}{1 + e\sin\alpha} + \frac{l}{1 - e\sin\alpha} = \frac{2l}{1 - e^2\sin^2\alpha},$$

$$\text{or} \quad QQ' = \frac{2l}{1 - e^2\sin^2\alpha}, \quad \text{or} \quad \frac{1}{QQ'} = \frac{1 - e^2\sin^2\alpha}{2l}. \tag{3}$$

Thus by (2) and (3)

$$\frac{1}{PP'} + \frac{1}{QQ'} = \frac{1 - e^2\cos^2\alpha}{2l} + \frac{1 - e^2\sin^2\alpha}{2l} = \frac{2 - e^2}{2l} = \text{constant}.$$

7. If PA and PB be the two tangents to the conic $\frac{l}{r} = 1 - e\cos\theta$ at α and β respectively, then show that PS (S is the focus) bisects the angle ASB.

Equations of the tangents at α and β are

$$\frac{l}{r} = \cos(\theta - \alpha) - e\cos\theta \quad \text{and} \quad \frac{l}{r} = \cos(\theta - \beta) - e\cos\theta.$$

The vectorial angle of P is obtained from

$$\cos(\theta - \alpha) - e\cos\theta = \cos(\theta - \beta) - e\cos\theta,$$

$$\text{or} \quad \cos(\theta - \alpha) - \cos(\theta - \beta) = 0,$$

$$\text{or} \quad 2\sin\left(\theta - \frac{\alpha - \beta}{2}\right)\sin\frac{\alpha - \beta}{2} = 0. \quad \therefore\ \alpha \neq \beta, \theta = \frac{\alpha + \beta}{2}.$$

Thus PS bisects the angle ASB.

8. PSP' is a focal chord of a conic; prove that the angle between the tangents at P and P' is $\tan^{-1}\frac{2e\sin\alpha}{1-e^2}$, where α is the angle between the chord and the major axis.

Let $\frac{l}{r} = 1 - e\cos\theta$ be the equation of the conic.

If α be the vectorial angle of P, then the vectorial angle of P' is $\pi + \alpha$.

Tangents at P and P' are

$$\frac{l}{r} = \cos(\theta - \alpha) - e\cos\theta \tag{1}$$

$$\text{and} \quad \frac{l}{r} = \cos(\theta - \pi - \alpha) - e\cos\theta. \tag{2}$$

The equation (1) can be written as

$$l = r\cos\theta(\cos\alpha - e) + r\sin\theta\sin\alpha,$$

$$\text{or} \quad l = x(\cos\alpha - e) + y\sin\alpha.$$

$\therefore$ The slope of this line $= \dfrac{e - \cos\alpha}{\sin\alpha}$.

Similarly the slope of (2) $= \dfrac{e - \cos(\pi + \alpha)}{\sin(\pi + \alpha)} = \dfrac{e + \cos\alpha}{-\sin\alpha}$.

Hence the angle between the tangents

$$= \tan^{-1}\frac{\dfrac{e - \cos\alpha}{\sin\alpha} + \dfrac{e + \cos\alpha}{\sin\alpha}}{1 - \dfrac{(e - \cos\alpha)(e + \cos\alpha)}{\sin^2\alpha}}$$

$$= \tan^{-1}\frac{2e\sin\alpha}{\sin^2\alpha - e^2 + \cos^2\alpha} = \tan^{-1}\frac{2e\sin\alpha}{1 - e^2}.$$

9. Show that the triangle formed by the pole and the points of intersection of the circle $r = 4\cos\theta$ with the line $r\cos\theta = 3$ is an equilateral triangle.

The vectorial angles of the common points of the line and the circle are obtained by eliminating r from the given equations. Thus at the common points

$$4\cos\theta\cdot\cos\theta = 3, \quad \text{or} \quad \cos^2\theta = \frac{3}{4},$$

$$\text{or} \quad \cos\theta = \pm\frac{\sqrt{3}}{2}, \quad \therefore \quad \theta = \pm\pi/6.$$

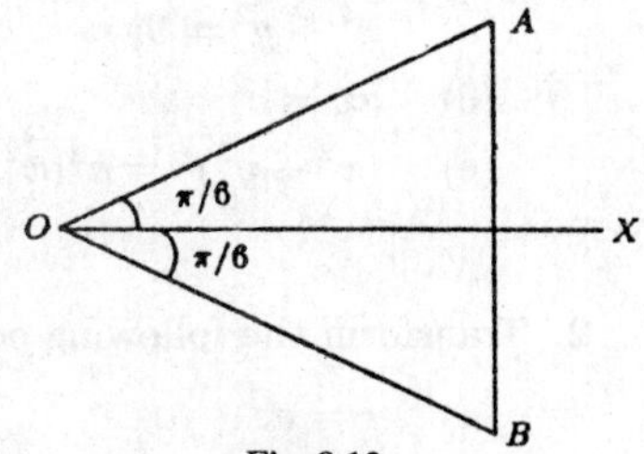

Fig. 8.12

If the line meets the circle at A and B and OX is the initial line with the point O as pole then $OA = OB = 4\cos(\pm\pi/6) = 2\sqrt{3}$.

Again $\angle XOA = \angle BOX = \pi/6$. It follows that $\angle OAB = \angle OBA = \pi/3$. Hence ΔOAB is equilateral.

10. If the two conics $\frac{l_1}{r} = 1 - e_1\cos\theta$ and $\frac{l_2}{r} = 1 - e_2\cos(\theta - \alpha)$ touch one another, show that

$$l_1^2(1 - e_2^2) + l_2^2(1 - e_1^2) = 2l_1l_2(1 - e_1e_2\cos\alpha).$$

[C.H. '71]

Let the conics touch at $\theta = \beta$.

The equations of the tangents at this point are

$$\frac{l_1}{r} = \cos(\theta - \beta) - e_1\cos\theta,$$

$$\text{or} \quad (\cos\beta - e_1)r\cos\theta + \sin\beta.r\sin\theta = l_1 \tag{1}$$

$$\text{and} \quad \frac{l_2}{r} = \cos(\theta - \beta) - e_2\cos(\theta - \alpha),$$

$$\text{or} \quad (\cos\beta - e_2\cos\alpha)r\cos\theta + (\sin\beta - e_2\sin\alpha)r\sin\theta = l_2. \tag{2}$$

Since (1) and (2) are identical

$$\frac{l_2}{l_1} = \frac{\cos\beta - e_2\cos\alpha}{\cos\beta - e_1} = \frac{\sin\beta - e_2\sin\alpha}{\sin\beta}.$$

$$\therefore \quad (l_1 - l_2)\cos\beta = l_1e_2\cos\alpha - l_2e_1$$

$$\text{and} \quad (l_1 - l_2)\sin\beta = l_1e_2\sin\alpha.$$

Now squaring and adding,

$$(l_1 - l_2)^2 = l_1^2e_2^2 + l_2^2e_1^2 - 2l_1l_2e_1e_2\cos\alpha,$$

$$\text{or} \quad l_1^2(1 - e_2^2) + l_2^2(1 - e_1^2) = 2l_1l_2(1 - e_1e_2\cos\alpha).$$

Exercise

1. Transform the following equations into polar form:

   (a) $x^2 + y^2 = 9$; (b) $x^2 + y^2 - ax - by = 0$;
   (c) $xy = c^2$; (d) $y^3 + yx^2 - 2ax^2 = 0$;
   (e) $(x^2 + y^2)^2 = a^2(x^2 - y^2)$.

   [B.U. '87, '90]

2. Transform the following equations into cartesian form:

   (a) $r = a$; (b) $\theta = \pi/3$;
   (c) $\frac{1}{r} = 1 - \cos\theta$; (d) $r = \sec\theta\tan^2\theta$;
   (e) $r = 2a\cos(\theta - \alpha)$; (f) $r = a\sin 3\theta$;
   (g) $r = a - b\cos\theta$; (h) $r(\cos 3\theta + \sin 3\theta) = 2a\sin\theta\cos\theta$.

3. Show that the points $(a\cos\theta, a\sin\theta)$,

$$\left\{a\cos\left(\theta + \frac{2\pi}{3}\right), a\sin\left(\theta + \frac{2\pi}{3}\right)\right\}$$

   and $\{a\cos(\theta - 2\pi/3), a\sin(\theta - 2\pi/3)\}$ form an equilateral triangle whose circumcentre is the pole.

4. (a) Find the equation of the line joining the points $(3, \pi/6)$ and $(4, \pi/3)$.

   (b) Show that (i) the lines $r\cos(\theta - \alpha) = a$ and $r\cos(\theta - \alpha) = b$ are parallel and (ii) the lines $r\cos(\theta - \alpha) = a$ and $r\sin(\theta - \alpha) = b$ are perpendicular.

   (c) Show that the polar equations of the bisectors of the angles between the lines $\theta = \alpha$ and $\theta = \beta$ are $\theta = \frac{1}{2}(\alpha + \beta)$ and $\theta = \frac{\pi}{2} + \frac{1}{2}(\alpha + \beta)$.

   (d) Find the polar co-ordinates of the foot of the perpendicular from the pole on the line joining the points (r_1, θ_1) and (r_2, θ_2).

   (e) Show that the perpendicular distance of the point (r_1, θ_1) from $r\cos(\theta - \alpha) = p$ is $r_1\cos(\theta_1 - \alpha) - p$.

   (f) Show that the four lines $r(\cos\theta + \sin\theta) = \pm 1$ and $r(\cos\theta - \sin\theta) = \pm 1$ form a square whose side is of length $\sqrt{2}$.

   (g) Show that the condition for the concurrence of the lines $r\cos(\theta - \alpha) = a$, $r\cos(\theta - \beta) = b$ and $r\cos(\theta - \gamma) = c$ is

$$\begin{vmatrix} \cos\alpha & \sin\alpha & a \\ \cos\beta & \sin\beta & b \\ \cos\gamma & \sin\gamma & c \end{vmatrix} = 0.$$

   (h) Show that the lines $r\cos(\theta - \alpha) = p$ and $r\cos(\theta - \beta) = p$ intersect at $\{p\sec\frac{1}{2}(\alpha - \beta), \frac{1}{2}(\alpha + \beta)\}$.

5. Find the equations of the circles under the following conditions.

   (a) Centre $(4, 0)$, radius 4.
   (b) Centre $(5, \pi/2)$, radius 5.
   (c) Centre $\left(5, \frac{\pi}{3}\right)$, radius 2.
   (d) Centre $(4, 60^\circ)$, passing through $(7, 45^\circ)$.

(e) Centre on the line $\theta = \pi/3$, passing through pole and $\left(5, \frac{\pi}{2}\right)$.

(f) Extremities of one diameter are $(2, \frac{2\pi}{3})$ and $(-2, \pi/3)$.

6. Find the centre and radius of the following:

(a) $r = 3\sin\theta + 3\sqrt{3}\cos\theta$;

(b) $r^2 - 5\sqrt{2}r(\cos\theta + \sin\theta) + 16 = 0$.

7. Show that the equation $r^2\cos\theta - ar\cos 2\theta - 2a^2\cos\theta = 0$ represents a straight line and a circle. **[C.H. '73]**

8. Show that the line $\frac{l}{r} = A\cos\theta + B\sin\theta$ touches the circle $r = 2a\cos\theta$, if $a^2B^2 + 2Aal = l^2$.

[HINTS. The equation of the line can be written as

$$r\cos(\theta - \alpha) = \frac{l}{\sqrt{A^2 + B^2}}, \quad \text{where} \quad \tan\alpha = \frac{B}{A}.$$

The centre of the circle is $(a, 0)$ and its radius is a. If the line touches the circle then its distance from $(a, 0)$ is a.

$$\therefore \quad a\cos\alpha - \frac{l}{\sqrt{A^2 + B^2}} = a, \text{ or } Aa - l = a\sqrt{A^2 + B^2},$$

$$\text{or} \quad (Aa - l)^2 = a^2(A^2 + B^2), \text{ or } a^2B^2 + 2Aal = l^2.]$$

9. Find the points on

(a) $\frac{8}{r} = 3 - \sqrt{2}\cos\theta$ whose radius vector is 4;

(b) $\frac{5}{r} = 1 + 2\cos\theta$ whose radius vector is 5;

(c) $\frac{l}{r} = 1 - \cos\theta$ which has the smallest radius vector.

[N.B. '80; C.U. '81, '87; B.U. '93]

10. Find the equation of the conic with focus at the pole and having the given e, l or directrix:

(a) $e = 1, l = 4$;

(b) $e = \sqrt{3}/2, l = 6$,

(c) $e = 2$, directrix $r\cos\theta = -5$,

(d) $e = 1$, directrix $r\cos(\theta - \pi/4) = 2\sqrt{2}$.

[HINTS. If $r\cos\theta = -p$ is the directrix, the conic is $r = ep/(1 - e\cos\theta)$.]

11. Find the eccentricity, latus rectum and the directrix of the following conics:

(a) $\frac{3}{r} = 4 - 2\cos\theta$;

(b) $r = \frac{6}{1-\cos\theta}$;

(c) $r = \frac{9}{2+\cos\theta}$;

(d) $\frac{4}{r} = 3 - 5\cos(\theta - \pi/4)$;

(e) $\frac{6}{r} = 1 - \sin\theta$;

(f) $\frac{3}{r} = 1 + 2\sin\theta$.

12. (a) Find the equation of $\frac{x^2}{25}+\frac{y^2}{16}=1$ in the polar form with the right hand focus as the pole and the positive direction of x-axis as the positive direction of the initial line.

(b) Find the equation of $\frac{x^2}{16}-\frac{y^2}{9}=1$ in the polar form with the left hand focus as the pole and the positive direction of the x-axis as the positive direction of the initial line.

(c) Find the equation of $y^2=8x$ in the polar form with the vertex as the pole and the positive direction of the x-axis as the positive direction of the initial line.

13. (a) The latus rectum of a conic is 6 and its eccentricity is $\frac{1}{2}$. Find the length of the focal chord making an angle 45° with the major axis.

[C.U. '62; B.U. '82]

(b) Show that the length of the focal chord of the conic $\frac{l}{r}=1-e\cos\theta$ which is inclined to the initial line at an angle α is $2l/(1-e^2\cos^2\alpha)$ and the locus of mid-points of focal chords is $r(1-e^2\cos^2\theta)=el\cos\theta$. **[N.B. '83, '90, '97; B.U. '99]**

14. Find the points of intersection of the parabolas $\frac{1}{3r}=1-\cos\theta$ and $\frac{1}{r}=1+\cos\theta$.

15. If PSP' and QSQ' are two perpendicular focal chords of a conic, prove that $\frac{1}{PS.SP'}+\frac{1}{QS.SQ'}=a$ constant. **[B.U. '87; C.U. '69, '70, '80]**

16. If r_1 and r_2 are two mutually perpendicular radius vectors of the ellipse $r^2=\frac{b^2}{1-e^2\cos^2\theta}$, prove that $\frac{1}{r_1^2}+\frac{1}{r_2^2}=\frac{1}{a^2}+\frac{1}{b^2}$. **[C.U. '87, B.U. '92]**

[HINTS. The vectorial angles of the points corresponding to r_1 and r_2 are α and $90°+\alpha$.]

17. Show that the perpendicular focal chords of a rectangular hyperbola are equal. $[e=\sqrt{2}]$

18. Show that $r=a\cos(\theta-\alpha)$ and $r=b\sin(\theta-\alpha)$ represent the equation of orthogonal circles. **[B.U. '89]**

19. A circle passes through the focus S of a conic and meets it in four points whose distances from S are r_1, r_2, r_3 and r_4. Prove that

(a) $r_1r_2r_3r_4=d^2l^2/e^2$, where $2l$ and e are the latus rectum and eccentricity of the conic and d is the diameter of the circle; and **[B.U. '92]**

(b) $\frac{1}{r_1}+\frac{1}{r_2}+\frac{1}{r_3}+\frac{1}{r_4}=\frac{2}{l}$. **[C.U. 83]**

[HINTS. Let the equation of the conic be

$$\frac{l}{r}=1-e\cos\theta \tag{1}$$

and that of the circle be

$$r=2a\cos(\theta-\alpha). \tag{2}$$

From (2)

$$(r-2a\cos\alpha\cos\theta)^2=4a^2\sin^2\alpha(1-\cos^2\theta). \tag{3}$$

From (1)

$$\cos\theta=\frac{1}{e}\left(1-\frac{l}{r}\right).$$

Putting it in (3), we have

$$e^2r^4+4ae\cos\alpha r^3+(4a^2-4ael\cos\alpha-4a^2e^2\sin^2\alpha)r^2-8a^2lr+4a^2l^2=0.$$

r_1, r_2, r_3, r_4 are the roots of this equation in r.

$$\therefore \quad r_1r_2r_3 + r_1r_2r_4 + r_1r_3r_4 + r_2r_3r_4 = \frac{8a^2l}{e^2} \tag{4}$$

$$\text{and} \quad r_1r_2r_3r_4 = \frac{4a^2l^2}{e^2} = \frac{d^2l^2}{e^2} \quad (\because\ 2a = d). \tag{5}$$

Dividing (4) by (5),

$$\frac{1}{r_1} + \frac{1}{r_2} + \frac{1}{r_3} + \frac{1}{r_4} = \frac{2}{l}.]$$

20. The tangent at two points P and Q of the parabola $\frac{l}{r} = 1 + \cos\theta$ meet at T. Show that $SP.SQ = ST^2$, where S is the focus. **[C.H. '92]**

[**Note.** If α and β are the vectorial angles of P and Q then the vectorial angle of T is $\frac{1}{2}(\alpha + \beta)$.

Now

$$SP.SQ = \frac{l}{1+\cos\alpha} \cdot \frac{l}{1+\cos\beta} = \left(\frac{l}{2\cos\frac{\alpha}{2} \cdot \cos\frac{\beta}{2}}\right)^2.$$

The equation of the tangent at α is $\frac{l}{r} = \cos\theta + \cos(\theta - \alpha)$.

Since the point T lies on it

$$ST = \frac{l}{\cos\frac{1}{2}(\beta+\alpha) + \cos\frac{1}{2}(\beta - \alpha)} = \frac{l}{2\cos\frac{\alpha}{2}\cos\frac{\beta}{2}}. \quad \therefore\ SP.SQ = ST^2.]$$

21. Show that if the line $\frac{l}{r} = A\cos\theta + B\sin\theta$ touches $\frac{l}{r} = 1 + e\cos\theta$ then $(A-e)^2 + B^2 = 1$. **[C.U. '96, '97]**

[HINTS.

$$\frac{l}{r} = A\cos\theta + B\sin\theta, \tag{1}$$

$$\frac{l}{r} = 1 + e\cos\theta \tag{2}$$

To find the point of intersection between (1) and (2),

$$A\cos\theta + B\sin\theta = 1 + e\cos\theta,$$

$$\text{or} \quad (A-e)\cos\theta + B\sin\theta - 1 = 0,$$

$$\text{or} \quad (A-e)\frac{1-\tan^2\frac{\theta}{2}}{1+\tan^2\frac{\theta}{2}} + 2B\frac{\tan\frac{\theta}{2}}{1+\tan^2\frac{\theta}{2}} - 1 = 0,$$

$$\text{or} \quad (A-e)\left(1-\tan^2\frac{\theta}{2}\right) + 2B\tan\frac{\theta}{2} - \left(1+\tan^2\frac{\theta}{2}\right) = 0,$$

$$\text{or} \quad (A-e+1)\tan^2\frac{\theta}{2} - 2B\tan\frac{\theta}{2} - (A-e-1) = 0. \tag{3}$$

It is a quadratic equation in $\tan\frac{\theta}{2}$.

If the line (1) touches the conic (2) then the equation (3) must have equal roots. By the condition of equal roots

$$4B^2 + 4(A-e+1)(A-e-1) = 0, \quad \text{or} \quad (A-e)^2 + B^2 = 1.]$$

22. Prove that the line $\frac{l}{r} = A\cos\theta + B\sin\theta$ will touch the conic

$$\frac{l}{r} = 1 - e\cos(\theta - \gamma)$$

if $A^2 + B^2 - 2e(A\cos\gamma + B\sin\gamma) + e^2 - 1 = 0$.

[HINTS. Let the given line touch the conic at $\theta = \alpha$.

The equation of the tangent at this point is

$$\frac{l}{r} = e\cos(\theta - \gamma) + \cos(\theta - \alpha)$$

$$= (e\cos\gamma + \cos\alpha)\cos\theta + (e\sin\gamma + \sin\alpha)\sin\theta.$$

It is identical with $\frac{l}{r} = A\cos\theta + B\sin\theta$.

$$\therefore\ A = e\cos\gamma + \cos\alpha, \text{ or } A - e\cos\gamma = \cos\alpha$$

$$\text{and } B = e\sin\gamma + \sin\alpha, \text{ or } B - e\sin\gamma = \sin\alpha.$$

Squaring and adding

$$(A - e\cos\gamma)^2 + (B - e\sin\gamma)^2 = 1,$$

$$\text{or } A^2 + B^2 - 2e(A\cos\gamma + B\sin\gamma) + e^2 - 1 = 0.]$$

Answers

1. (a) $r = 3$, (b) $r - (a\cos\theta + b\sin\theta) = 0$, (c) $r^2\sin 2\theta = 2c^2$, (d) $r = 2a\cos\theta\cot\theta$, (e) $r^2 = a^2\cos 2\theta$. **2.** (a) $x^2 + y^2 = a^2$, (b) $y = \sqrt{3}x$, (c) $1 + 2x = y^2$, (d) $y^2 = x^3$, (e) $x^2 + y^2 = 2a(x\cos\alpha + y\sin\alpha)$, (f) $(x^2 + y^2)^2 = a(3x^2y - y^3)$, (g) $(x^2 + y^2 + bx)^2 = a^2(x^2 + y^2)$, (h) $x^3 - y^3 + 3xy(x - y) = 2axy$. **4.** (a) $\frac{12}{r} = (4\sqrt{3} - 3)\cos\theta - (4 - 3\sqrt{3})\sin\theta$, (d) (p, α), where $p = \frac{r_1 r_2 \sin(\theta_2 - \theta_1)}{AB}$, $\alpha = \tan^{-1}\frac{r_2\cos\theta_2 - r_1\cos\theta_1}{r_1\sin\theta_1 - r_2\sin\theta_2}$. **5.** (a) $r = 8\cos\theta$, (b) $r = 10\sin\theta$, (c) $r^2 - 10r\cos(\theta - \pi/3) + 21 = 0$, (d) $r^2 - 8r\cos(\theta - \pi/3) - 49 + 14(\sqrt{2} + \sqrt{6}) = 0$, (e) $r = \frac{10}{\sqrt{3}}\cos(\theta - \pi/3)$, (f) $r^2 + 2r\cos\theta - 2 = 0$. **6.** (a) $\left(3, \frac{\pi}{6}\right)$, (b) $\left(3, \frac{\pi}{4}\right)$, 7. $r = 2a\cos\theta, r\cos\theta + a = 0$. **9.** (a) $\left(4, \pm\frac{\pi}{4}\right)$, (b) $\left(5, \pm\frac{\pi}{2}\right)$, (c) $\left(\frac{1}{2}, \pi\right)$. **10.** (a) $\frac{4}{r} = 1 \pm \cos\theta$, (b) $\frac{12}{r} = 2 \pm \sqrt{3}\cos\theta$, (c) $\frac{10}{r} = 1 - 2\cos\theta$, (d) $r = \frac{-4}{\sqrt{2} - (\cos\theta + \sin\theta)}$, **11.** (a) $e = \frac{1}{2}, 2l = \frac{3}{2}, r\cos\theta = -\frac{3}{2}$, (b) $e = 1, l = 6, r\cos\theta = -6$, (c) $e = \frac{1}{2}, l = \frac{9}{2}, r\cos\theta = 9$, (d) $e = \frac{5}{3}, l = \frac{4}{3}, r\cos(\theta - \pi/4) = -\frac{4}{5}$, (e) $e = 1, l = 6, r\sin\theta = -6$, (f) $e = 2, l = 3, r\sin\theta = \frac{3}{2}$. **12.** (a) $\frac{16}{r} = 5 + 3\cos\theta$, (b) $9/r = 4 + 5\cos\theta$, (c) $r\sin^2\theta = 8\cos\theta$. **13.** (a) $6\frac{6}{7}$. **14.** $\left(\frac{2}{3}, \pm\pi/3\right)$.

Part One

Short Questions

1. Find the relation between two points when the distance between them is zero.

2. If the points $(5, 3), (-2, 7)$ and $(3, y)$ are collinear, find the value of y.

3. Find the cartesian co-ordinates of the projection of the point $P(-2, 3)$ on the x-axis and the cartesian co-ordinates of the projection of $Q(5, 7)$ on the y-axis.

4. Determine the quadrants in which the point $P(x, y)$ may lie when

   (a) $x + y = 0$; [C.U. '83]

   (b) $xy > 0$; [C.U. '83]

   (c) $x + y > 0$; [C.U. '83]

   (d) $x - y < 0$;

   (e) $xy < 0$.

5. The projections of the line segment $\overline{AB}$ on the x-axis and y-axis are -3 and 5 respectively. If the co-ordinates of B are $(2, 1)$, find those of A.

6. The projections of a line segment on the x-axis and y-axis are -6 and 8 respectively. Find the length of the segment.

7. If $lx + mx + n = 0$ satisfies the relation $al + bm + n = 0$, show that the line passes through (a, b).

8. Show that

$$\left.\begin{aligned} x &= lx + my + n \\ y &= mx - ly + p \end{aligned}\right\} l^2 + m^2 = 1$$

represent an orthogonal transformation. Can it be translation or rotation for different values of l, m, n and p?

9. Interpret the relations

$$\begin{bmatrix} x - h \\ y - k \end{bmatrix} = \begin{bmatrix} \cos\theta & -\sin\theta \\ \sin\theta & \cos\theta \end{bmatrix} \begin{bmatrix} x' \\ y' \end{bmatrix}.$$

10. Find the transformation for translation if the points $(3, -4)$ and $(1, -6)$ lie on the new x-axis and y-axis respectively.

11. The co-ordinate axes are only rotated through an angle of $60°$. If the transformed co-ordinates of a point are $(2\sqrt{3}, -6)$, find its original co-ordinates.

12. If $ax + by$ transforms to $a'x + b'y$ under rotation of axes then show that $a^2 + b^2 = a'^2 + b'^2$. [C.U. '64]

13. Define a conic and the terms focus, eccentricity, directrix of it.

14. What is the eccentricity of a circle?

15. Show that the locus of the point of intersection of the lines $x - y = at$ and $x + y = a/t$, where t is a variable parameter is a rectangular hyperbola.

16. Write down the general equation of second degree in x and y. When it will be non-degenerate? What are the conditions for different non-degenerate conics?

17. Deduce the condition so that the conic $ax^2 + 2hxy + by^2 + 2gx + 2fy + c = 0$ may represent a central conic.

18. Prove that the angle between the lines joining the origin and the points of intersection of the line $y = 3x + 2$ with the curve $x^2 + 2xy + 3y^2 + 4x + 8y - 11 = 0$ is $\tan^{-1} \frac{2\sqrt{2}}{3}$.

[B.U. '77]

19. What is represented by the equation

$$2x^2 - y^2 - x(x-1) + 2y(x+y) - 2xy - 1 = 0?$$

[C.U. '82]

20. Find the values of a and f for which the equation

$$ax^2 + 12xy + 9y^2 + 4x + 2fy - 13 = 0$$

represents (a) a central conic, (b) a non-central conic.

21. Tangents are drawn from (a, b) to the circle $x^2 + y^2 = a^2$. Show that the area of the triangle formed by them and the straight line joining their points of contact is $\frac{ab^3}{a^2+b^2}$.

[C.U. '67]

22. Prove that the locus of the middle points of chords of a parabola which pass through a fixed point is a parabola.

23. What are the position of the pole and initial line when the polar equation of the circle is $r = 2a\cos\theta$?

24. What is the equation of a line passing through the pole?

25. Find the points on the conic $\frac{4}{r} = 1 - \frac{\sqrt{2}}{3}\cos\theta$, where the radius vector is 6.

26. Determine the nature of the conic $\frac{8}{r} = 4 - 5\cos\theta$. [C.U. '82, '90, '93, N.B. '92]

27. On the conic $r = \frac{6}{1-\cos\theta}$, find the point with the least radius vector.

[C.U. '83, 2000]

28. Find the equation of the directrix of the conic $r\sin^2\theta/2 = 1$.

29. What is represented by the equation $p = r\cos(\theta - \alpha)$?

30. Transform the equations

(a) $x^2 - y^2 = 4axy$ to polar form,

(b) $r = \frac{2}{1+\cos\theta}$ to certesian form and

(c) $r^{1/2}\cos\frac{\theta}{2} = a^{1/2}$ to certesian form. [B.U. '87, '98]

31. Find the vertex and length of latus rectum of $\frac{21}{r} = 5 - 2\cos\theta$.

32. PSP' is a focal chord of a conic; prove that the angle between the tangents at P and P' is $\tan^{-1}\frac{2e\sin\alpha}{1-e^2}$, where α is the angle between the chord and the major axis.

33. Show that the line $5/r = \cos(\theta - \pi/4) - \cos\theta$ is a tangent to the parabola $5/r = 1 - \cos\theta$.

34. If a circle cuts the parabola $y^2 = 4ax$ at four points $(at_i^2, 2at_i), (i = 1, 2, 3, 4)$, show that $t_1 + t_2 + t_3 + t_4 = 0$.

35. Show that the area of the ellipse given by the equation $ax^2 + 2hxy + by^2 = 1$ is $\frac{\pi}{\sqrt{ab-h^2}}$.

36. Find the equation of the pair of bisectors of the angles between the co-ordinate axes $x = 0, y = 0$.

37. For what value of h does the equation $3x^2 + 2hxy - 3y^2 - 40x + 30y - 75 = 0$ represents two straight lines?

38. Find the values of a and h show that the equation $ax^2 - hxy - 2y^2 + 5x - 5y + 1 = 0$ represents a pair of perpendicular lines.

39. Find the equation of the circle which passes through $(1, 1)$ and the points of intersection of the circles $x^2 + y^2 + 13x - 3y = 0$ and $2x^2 + 2y^2 + 4x - 7y - 25 = 0$.

40. Show that the centre of a circle passing through the point $(0, 1)$ and touching the curve $y = x^2$ at $(2, 4)$ is at $\left(-\frac{16}{5}, \frac{53}{10}\right)$.

41. Find the co-ordinate transformation which transform the equation $x^2 + y^2 - 2x + 14y + 20 = 0$ into $x'^2 + y'^2 - 30 = 0$. [C.U. '83]

42. Find the centre of the conic $x^2 + y^2 - 4xy - 2x - 20y - 11 = 0$.

43. Reduce the equation $8x^2 + 4xy + 5y^2 - 3 = 0$ into a form $ax'^2 + by'^2 = c$.

44. What is the nature of the conic $4x^2 - 4xy + y^2 - 3x - 1 = 0$.

45. Show that the equations $\frac{l}{r} = 1 - e\cos\theta$ and $\frac{l}{r} = -1 - e\cos\theta$ represents the same conic.

[**Hints.** (r, θ) can be replaced by $(-r, \pi + \theta)$.]

46. Find the equation of the tangent to the conic $2/r = 1 - \cos\theta$ at $\theta = \pi/2$.

47. Find the co-ordinate transformation which transforms the equation $14x^2 - 4xy + 11y^2 = 60$ to $4x'^2 + 6y'^2 = 24$.

48. If by co-ordinate transformation the equation $5x^2 - 2xy + 5y^2 + 2x - 10y - 7 = 0$ transform to $ax'^2 + by'^2 = 1$, find $a + b$ and ab.

49. Show that the circle described on the chord $x\cos\alpha + y\sin\alpha = p$ of the circle $x^2 + y^2 = a^2$ as diameter is $x^2 + y^2 - 2p(x\cos\alpha + y\sin\alpha - p) = 0$.

50. Show that the latus rectum of the parabola

$$(a^2 + b^2)(x^2 + y^2) = (bx + ay - ab)^2 \text{ is } \frac{2b}{\sqrt{(a^2 + b^2)}}.$$

51. Find the equation of the pair of lines intersecting at the origin and parallel to the pair of lines $x^2 + xy - y - 1 = 0$. [B.U. '99]

52. By shifting the origin to the point (α, β) without changing the directions of axes, each of the equations $x - y + 3 = 0$ and $2x - y + 1 = 0$ is reduced to the form $ax' + by' = 0$. Find α and β.

[**Hints.** Here $\alpha - \beta + 3 = 0$ and $2\alpha - \beta + 1 = 0$. From these two equations $\alpha = 2, \beta = 5$.]

53. What is the range of values of e for which the equation $\frac{4}{r} = 8 + e\cos\theta$ will represent an ellipse? **[N.B. '97]**

54. Which of the following statements are true? Justify your answer.

    (a) There is a unique circle through two given points.

    (b) The eccentricity of a pair of lines is one.

    (c) Two conics may intersect in four real points.

    (d) A parabola is a central conic.

    (e) If P is a point on the ellipse with foci and S and S' then $PS = PS'$.

    (f) A pair of straight lines is a conic section.

    (g) The asymptotes of $x^2 - y^2 = 16$ and $x^2 - y^2 = 25$ are same.

    (h) The distance of a point on the ellipse $\frac{x^2}{25} + \frac{y^2}{9} = 1$ from (4, 0) is more than its distance from the line $x = \frac{25}{4}$.

Answers

1. Indentical; **2.** $\frac{9}{2}$; **3.** $(-2, 0), (0, 7)$; **4.** (a) 2nd or 4th (b) 1st or 3rd (c) 1st, 2nd or 3rd (d) 1st, 2nd or 3rd (e) 2nd or 4th; **5.** $(5, -4)$; **6.** 10; **10.** $x = x' + 1, y = y' - 4$; **11.** $(4\sqrt{3}, 0)$; **14.** 0; **19.** circle; **20.** (a) $a \neq 4$; (b) $a = 4, f \neq 3$; **24.** $\theta =$ constant; **25.** $(6, \pi/4), (6, -\pi/4)$; **26.** Hyperbola; **27.** $(6, \pi/2)$; **28.** $r\cos\theta + 2 = 0$; **29.** straight line; **30.** (a) $\cot 2\theta = 2a$; (b) $y^2 = 4(x+1)$; (c) $x^2 + y^2 = (2a - x)^2$; **31.** $(7, 0°), (3, 180°)$, $\frac{42}{5}$; **36.** $y = \pm x$; **37.** 4; **38.** $a = 2, h = \frac{25 \pm \sqrt{56}}{2}$; **39.** $2x^2 + 2y^2 + 15x - 3y - 16 = 0$; **42.** $(-7, -4)$; **44.** Parabola. **53.** $0 < e < 8$. **54.** (a) False, (b) False, (c) True, (d) False, (e) False, (f) True, (g) True, (h) False.

Part II

Geometry of Three Dimensions

Chapter 1

Rectangular Cartesian Co-ordinates

1.1 Introduction

A point can be defined by the plane geometry, if it lies in the plane of the reference to two axes. If the point does not lie on this plane, then it is not possible to define the point by the plane geometry. Moreover the bodies, in general, are of three-dimensions, length, breadth and height. Thus the geometry, which can define a point in space or a body of three-dimensions, is known as the geometry of three-dimensions or solid geometry.

1.2 Origin, Axes and Co-ordinate Planes, Co-ordinates of a Point, Octants

Let $X'OX, Y'OY$ and $Z'OZ$ be three mutually perpendicular lines in space, which intersects at O. O is called the origin and the lines are called rectangular co-ordinate axes, namely x, y and z axes respectively. OX, OY and OZ whose directions are right-handed are taken as positive $(+ve)$ directions of the axes. The phanes XOY, YOZ and ZOX are known as xy, yz and zx co-ordinate planes respectively.

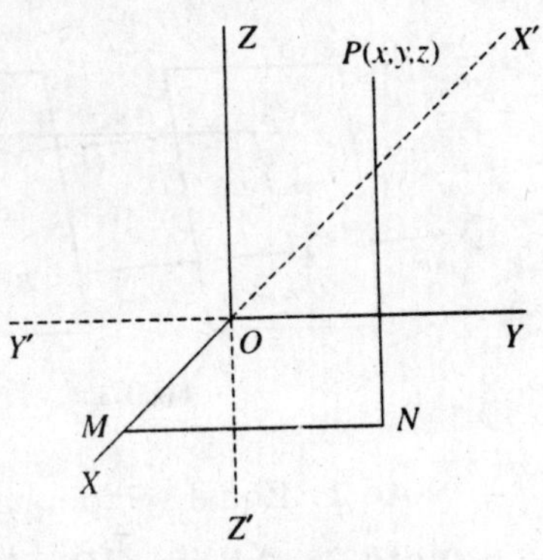

Fig. 1.1

It is obvious that

$$z = 0 \text{ for all points in the } xy \text{ plane,}$$

$$y = 0 \text{ for all points in the } zx \text{ plane,}$$

$$\text{and} \quad x = 0 \text{ for all points in the } yz \text{ plane.}$$

Let P be a point in space, PN is drawn perpendicular to XOY plane. NM is perpendicular to OX, or MN is parallel to OY. If $OM = x, MN = y$ and $NP = z$, then (x, y, z) are the co-ordinates of P. These co-ordinates are $+ve$ or $-ve$ as they are measured in the positive directions of axes or $-ve$ directions of axes.

The co-ordinate planes divide space into eight parts called octants and the signs of co-ordinates of a point lying in the octants are given below:

Octant	$OXYZ$	$OX'YZ$	$OXY'Z$	$OXYZ'$	$OX'Y'Z$	$OXY'Z'$	$OX'Y'Z$	$OX'Y'Z'$
x	+	−	+	+	−	+	−	−
y	+	+	−	+	−	−	−	−
z	+	+	+	−	+	−	+	−

1.3 Geometric Vector

Let P and Q be two points in space. The line-segment going from P to Q is called a *directed line segment* or a *geometric vector* or simple a *vector* and it is designated as $\overline{PQ}$. The point P is the *base* or the *initial point* and the point Q is the *head* or the *terminal point.* The base is also known as the point of application. The *modulus* or *magnitude* of the vector $\overline{PQ}$ is the positive number, which gives the measure of its length and is denoted by $|\overline{PQ}|$. $\overline{PQ}$ and $\overline{QP}$ have the same extremities, the same length but opposite directions.

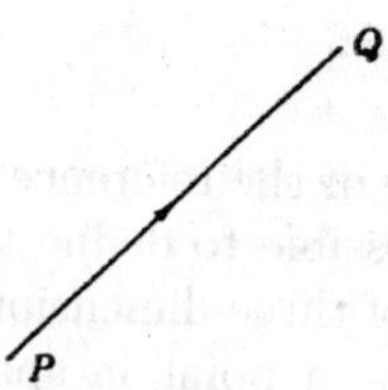

Fig. 1.2

Null Vector. A vector whose base and head coincide is called a *null-vector* or *zero-vector.* It is denoted by $\bar{0}$.

Unit Vector. A vector whose modulus is unity is called a *unit vector.*

Equal Vectors. Two vectors are said to be equal, if they have the same magnitude and the same direction. For this the point of application of a vector may be arbitrary and the geometric vectors are said to be *free vectors.*

Projection of a vector on an axis. Let l be an axis (a directed line) and $\overline{PQ}$ be a given vector. Through the points P and Q two planes α and β are drawn perpendicular to l. These planes intersect l at A and B. A and B are the projections of P and Q on l. The value of the directed segment $\overline{AB}$ is called the projection of $\overline{PQ}$ on the axis l. If θ is the angle between $\overline{PQ}$ and the axis l then

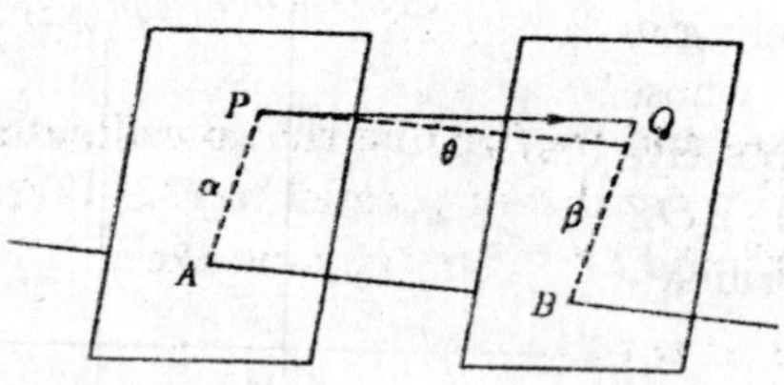

Fig. 1.3

$$|\overline{AB}| = |\overline{PQ}| \cos\theta.$$

Note 1. Equal vectors have equal projections on the same axis.

Note 2. PA and QB are perpendicular to l.

Co-ordinates of a vector. The projections of a vector on the co-ordinate axes are called the co-ordinates of this vector with respect to the co-ordinate system. If X, Y, Z are the projections of $\overline{PQ}$ on the x-axis, y-axis and z-axis respectively then it is written as $\overline{PQ} = (X, Y, Z)$.

Note. The projections of the point (x, y, z) on the xy, yz and zx planes are $(x, y, 0)$, $(0, y, z)$ and $(x, 0, z)$ respectively. Its projections on the x, y and z axes are $(x, 0, 0)$, $(0, y, 0)$, $(0, 0, z)$ respectively.

Theorem 1 *If $P(x_1, y_1, z_1)$ and $Q(x_2, y_2, z_2)$ be any two points then the vector $\overline{PQ} = (x_2 - x_1, y_2 - y_1, z_2 - z_1)$.*

Proof. Let $PQ = (X, Y, Z)$.

Let the planes through P and Q perpendicular to the x-axis intersect it at P' and Q' respectively. Obviously the x-co-ordinates of P' and Q' are x_1 and x_2. Therefore, $P'Q' = x_2 - x_1$. $P'Q'$ is the projection of $\overline{PQ}$ on the x-axis.

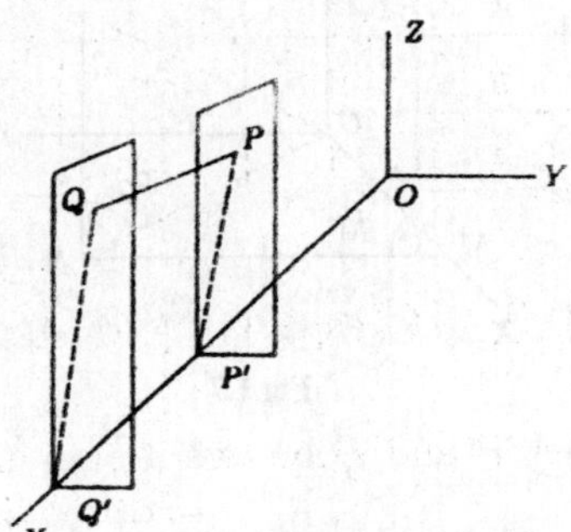

Fig. 1.4

$\therefore$ $P'Q' = X$. Consequently $X = x_2 - x_1$.

Similarly $Y = y_2 - y_1$ and $Z = z_2 - z_1$.

Thus $\overline{PQ} = (x_2 - x_1, y_2 - y_1, z_2 - z_1)$.

Theorem 2 *$P, Q, R, \ldots, T, U$ are any n points in space, the sum of projections of $\overline{PQ}, \overline{QR}, \ldots, \overline{TU}$ on an axis AB is equal to the projection of $\overline{PU}$ on AB.*

Let $P', Q', R', \ldots, T', U'$ be the feet of the perpendiculars from $P, Q, R, \ldots, T, U$ on AB respectively.

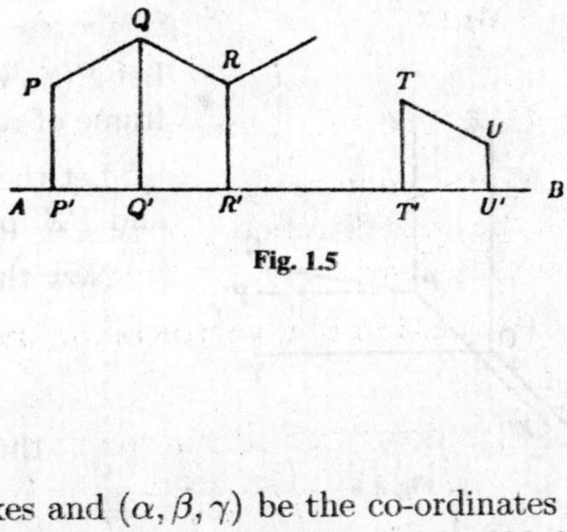

Fig. 1.5

Now $P'Q' + Q'R' + \ldots + T'U' = P'U'$.

$\therefore$ proj. of $\overline{PQ}$+ proj. of $\overline{QR} + \ldots +$ proj. of $\overline{TU}$ on AB = proj. of $\overline{PU}$ on AB.

1.4 Change of Origin

Let OX, OY, OZ be the three rectangular axes and (α, β, γ) be the co-ordinates of a point O' w.r.t. these axes. $O'X', O'Y'$ and $O'Z'$ are drawn parallel to OX, OY, OZ respectively. O' is the new origin and $O'X', O'Y'$ and $O'Z'$ are the new axes.

Let (x, y, z) be the co-ordinates of the point P w.r.t. OX, OY and OZ and (x', y', z') be those of this point w.r.t. new axes. PN is perpendicular to XOY plane. It meets the $X'O'Y'$ plane at N'. NP is parallel to OZ and $O'Z'$. NN' is the distance between two parallel planes XOY and $X'O'Y'$. $O'M$ is perpendicular to XOY plane.

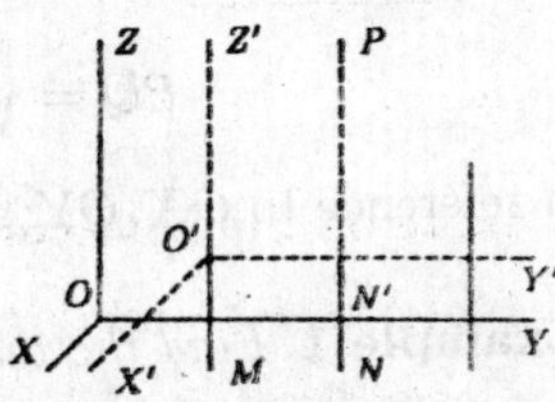

Fig. 1.6

Now $NN' = MO' = \gamma$.

Again $z = NP = NN' + N'P = \gamma + z'$, or $z = z' + \gamma$.

Similarly it can be shown that $x = x' + \alpha, y = y' + \beta$.

Note. Co-ordinates of P w.r.t. new axes are $(x - \alpha, y - \beta, z - \gamma)$.

1.5 Distance of a Point from the Origin

Let P be a point whose co-ordinates are (x, y, z) w.r.t. three rectangular axes OX, OY and OZ.

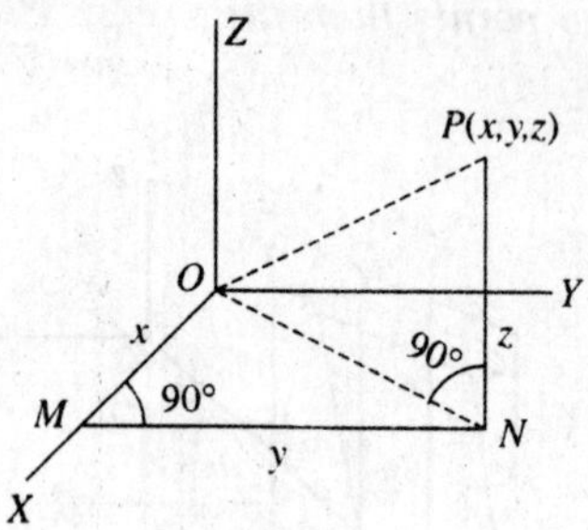

Fig 1.7

In the fig. 1.7, PN is perpendicular to XOY plane and MN is parallel to OY. $\therefore$ $OM = x, MN = y, NP = z$. ON and OP are joined.

From the right-angled triangle OMN,

$$ON^2 = OM^2 + MN^2 = x^2 + y^2.$$

From the right-angled triangle ONP,

$$OP^2 = ON^2 + NP^2 = x^2 + y^2 + z^2.$$

$$\therefore\ OP = \sqrt{x^2 + y^2 + z^2}.$$

Note. $\overline{OP} = (x, y, z)$.

$$\therefore\quad |\overline{OP}| = \sqrt{x^2 + y^2 + z^2}.$$

1.6 Distance between Two Points

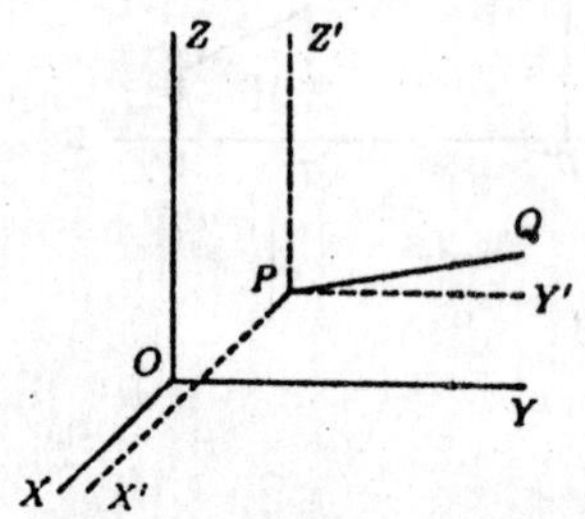

Fig. 1.8

Let $P(x_1, y_1, z_1)$ and $Q(x_2, y_2, z_2)$ be the two points in the frame of reference of rectangular axes OX, OY and OZ.

Let the origin be transferred to P with the PX', PY' and PZ' parallel to OX, OY and OZ respectively.

Now the co-ordinates of Q w.r.t. PX', PY', PZ' are

$$(x_2 - x_1, y_2 - y_1, z_2 - z_1).$$

$\therefore$ the distance of Q from P

$$= \sqrt{\{(x_2 - x_1)^2 + (y_2 - y_1)^2 + (z_2 - z_1)^2\}}.$$

Since the distance between two points is invariant,

$$PQ = \sqrt{\{(x_2 - x_1)^2 + (y_2 - y_1)^2 + (z_2 - z_1)^2\}}$$

with reference to OX, OY and OZ.

► **Example 1** *Find the distance between the points* $(4, 3, -6)$ *and* $(-2, 1, -3)$.

$$\textit{Distance} = \sqrt{\{(-2-4)^2 + (1-3)^2 + (-3+6)^2\}}$$

$$= \sqrt{(36 + 4 + 9)} = \sqrt{49} = 7.$$

1.7 To find the Co-ordinates of the Point which divides the join of $P(x_1, y_1, z_1)$ and $Q(x_2, y_2, z_2)$ in a given ratio $m : n$

Let $R(x, y, z)$ divide the distance between the points $P(x_1, y_1, z_1)$ and $Q(x_2, y_2, z_2)$ in the ratio $m : n$. $(PR : RQ = m : n)$.

PL, QM and RN are drawn perpendicular to XOY plane. The lines PL, QM and RN lie in the same plane, so the points L, M, N lie on a straight line in the XOY plane. $PN'M'$ is drawn parallel to LMN. It is perpendicular to RN and QM.

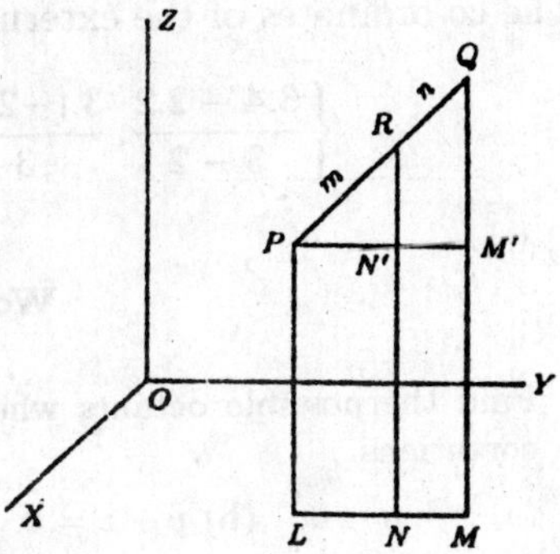

Fig. 1.9

Here

$$LP = MM' = NN' = z_1,$$

$$N'R = NR - NN' = z - z_1,$$

$$M'Q = MQ - MM' = z_2 - z_1.$$

$\Delta PN'R$ and $\Delta PM'Q$ are similar.

$$\therefore \quad \frac{PR}{PQ} = \frac{N'R}{M'Q}, \quad \text{or} \quad \frac{m}{m+n} = \frac{z - z_1}{z_2 - z_1}, \quad \text{or} \quad z = \frac{mz_2 + nz_1}{m+n}.$$

Similarly drawing perpendiculars to YOZ and ZOX planes from P, Q and R, it can be shown that

$$x = \frac{mx_2 + nx_1}{m+n}, \quad y = \frac{my_2 + ny_1}{m+n}.$$

$\therefore$ the co-ordinates of R are

$$\left(\frac{mx_2 + nx_1}{m+n}, \frac{my_2 + ny_1}{m+n}, \frac{mz_2 + nz_1}{m+n} \right).$$

Corollary 1. Mid-point: The co-ordinates of the middle point are

$$\left(\frac{x_1 + x_2}{2}, \frac{y_1 + y_2}{2}, \frac{z_1 + z_2}{2} \right).$$

Corollary 2. If R divides PQ externally in the ratio $m : n$, then the co-ordinates of R are

$$\left(\frac{mx_2 - nx_1}{m-n}, \frac{my_2 - ny_1}{m-n}, \frac{mz_2 - nz_1}{m-n} \right).$$

Corollary 3. If $\frac{m}{n} = \lambda$, i.e., R divides PQ in the ratio $\lambda : 1$, then

$$x = \frac{x_1 + \lambda x_2}{1+\lambda}, \quad y = \frac{y_1 + \lambda y_2}{1+\lambda}, \quad z = \frac{z_1 + \lambda z_2}{1+\lambda}.$$

► **Example 2** *Find the co-ordinates of the points, which divide the distance between two points* $(2,0,1)$ *and* $(4,-2,5)$ *internally and externally in the ratio* $3:2$.

The co-ordinates of the internal point are

$$\left\{\frac{3.4+2.2}{3+2}, \frac{3(-2)+2.0}{3+2}, \frac{3.5+2.1}{3+2}\right\}, \text{ i.e., } \left\{\frac{16}{5}, \frac{-6}{5}, \frac{17}{5}\right\}.$$

The co-ordinates of the external point are

$$\left\{\frac{3.4-2.2}{3-2}, \frac{3.(-2)-2.0}{3-2}, \frac{3.5-2.1}{3-2}\right\}, \text{ i.e., } (8,-6,13).$$

Worked Out Examples

1. Find the possible octants where the point (x,y,z) may lie for each of the following conditions.

   (a) $x-y=0$; (b) $y+z=0$.

   (a) In this case x and y have the same sign. The same sign of these two co-ordinates occur in the octants $OXYZ$, $OXYZ'$, $OX'Y'Z$ and $OX'Y'Z'$. Hence the point may lie in any one of the above four octants.

   (b) Here y and z have opposite signs. This happens in the octants $OXY'Z$, $OXYZ'$, $OX'Y'Z$ and $OX'YZ'$. Therefore, the point may lie in any one of the above octants.

2. Show that the points $(0,7,10), (-1,6,6), (-4,9,6)$ form an isosceles right-angled triangle.

   Let the points be denoted by A, B, C respectively.

   Now

$$AB = \sqrt{\{(-1)^2+(6-7)^2+(6-10)^2\}} = \sqrt{(1+1+16)} = \sqrt{18},$$

$$BC = \sqrt{\{(-4+1)^2+(9-6)^2+(6-6)^2\}} = \sqrt{(9+9+0)} = \sqrt{18},$$

$$CA = \sqrt{\{(4)^2+(7-9)^2+(10-6)^2\}} = \sqrt{(16+4+16)} = \sqrt{36}.$$

   $\because$ $AB = BC$, the triangle is isosceles.

   Again $AB^2 + BC^2 = 18 + 18 = 36 = CA^2$.

   $\therefore$ the triangle is right-angled.

3. A, B, C are the points on the x, y and z axes respectively. The distance of A, B, C from the origin O are a, b, c. Find the co-ordinates of the point, which is equidistant from O, A, B and C.

   Let $P(x,y,z)$ be the point, which is equidistant from O, A, B and C. The co-ordinates of O, A, B, and C are $(0,0,0), (a,0,0), (0,b,0)$ and $(0,0,c)$ respectively.

   Since $OP = PA, x^2+y^2+z^2 = (x-a)^2+y^2+z^2$, or $0 = a^2 - 2ax$, or $x = \frac{a}{2}$.

   Similarly from $OP = PB, y = \frac{b}{2}$ and from $OP = PC, z = \frac{c}{2}$.

   $\therefore$ the required point is $\left(\frac{a}{2}, \frac{b}{2}, \frac{c}{2}\right)$.

4. Find the point where the line joining $(5, -2, 3)$ and $(3, 0, 1)$ pierces the xy plane.

[C.U. '65]

Let the point divide the join of $(5, -2, 3)$ and $(3, 0, 1)$ in the ratio $\lambda : 1$. On the XOY plane, z-co-ordinate is zero.

$$\therefore \quad 0 = \frac{\lambda.1 + 1.3}{\lambda + 1}, \quad \text{or} \quad \lambda + 3 = 0, \quad \text{or} \quad \lambda = -3.$$

Thus the x-co-ordinate of the point

$$= \frac{-3.3 + 1.5}{-3 + 1} = \frac{4}{2} = 2$$

and the y-co-ordinate of the point

$$\frac{-3.0 + 1.(-2)}{-3 + 1} = \frac{2}{2} = 1.$$

Hence the required point is $(2, 1, 0)$.

5. Find the co-ordinates of the centroid of the triangle whose vertices are (x_1, y_1, z_1), (x_2, y_2, z_2) and (x_3, y_3, z_3).

Let $(x_1, y_1, z_1), (x_2, y_2, z_2)$ and (x_3, y_3, z_3) be the co-ordinates of A, B, C respectively of ΔABC and AD be a median of it.

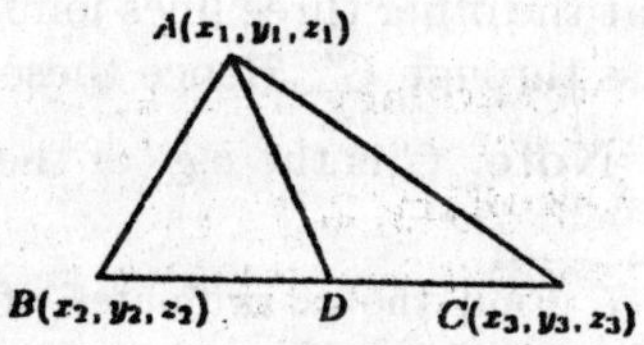

Fig. 1.10

Since D is the middle point of BC, the co-ordinates of D are

$$\left\{\frac{x_2 + x_3}{2}, \quad \frac{y_2 + y_3}{2}, \quad \frac{z_2 + z_3}{2}\right\}.$$

Let G be the centroid. G divides AD in the ratio $2 : 1$.

$\therefore$ the co-ordinates of G are

$$\left(\frac{2 \cdot \frac{x_2+x_3}{2} + 1.x_1}{2+1}, \quad \frac{2 \cdot \frac{y_2+y_3}{2} + 1.y_1}{2+1}, \quad \frac{2 \cdot \frac{z_2+z_3}{2} + 1.z_1}{2+1}\right),$$

$$\text{i.e.,} \quad \left\{\frac{x_1 + x_2 + x_3}{3}, \frac{y_1 + y_2 + y_3}{3}, \frac{z_1 + z_2 + z_3}{3}\right\}.$$

6. Show that the lines joining the mid-points of the opposite edges of a tetrahedron and lines joining the vertices to centroids of the opposite faces of the tetrahedron are concurrent.

Let $ABCD$ be a tetrahedron and $(x_1, y_1, z_1), (x_2, y_2, z_2)$, (x_3, y_3, z_3) and (x_4, y_4, z_4) be the co-ordinates of A, B, C, D respectively. If E and F are the mid-points of the opposite edges AB and CD, then the co-ordinates of E and F are

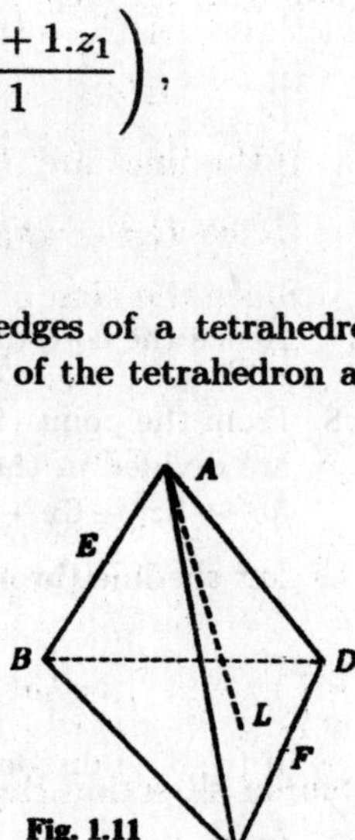

Fig. 1.11

$$\left(\frac{x_1 + x_2}{2}, \frac{y_1 + y_2}{2}, \frac{z_1 + z_2}{2}\right)$$

$$\text{and} \quad \left(\frac{x_3 + x_4}{2}, \frac{y_3 + y_4}{2}, \frac{z_3 + z_4}{2}\right)$$

respectively.

If G be the mid-point of EF, the co-ordinates of G are

$$\left(\frac{x_1+x_2+x_3+x_4}{4}, \frac{y_1+y_2+y_3+y_4}{4}, \frac{z_1+z_2+z_3+z_4}{4}\right).$$

The symmetry of the co-ordinates of G suggests that the other two lines joining the mid-points of the opposite edges will pass through G.

Let L be the centroid of ΔBCD. Then the co-ordinates of L are

$$\left(\frac{x_2+x_3+x_4}{3}, \frac{y_2+y_3+y_4}{3}, \frac{z_2+z_3+z_4}{3}\right).$$

Let G' be a point on AL, which divides AL in the ratio $3:1$. The co-ordinates of G' are

$$\left(\frac{x_1+x_2+x_3+x_4}{4}, \frac{y_1+y_2+y_3+y_4}{4}, \frac{z_1+z_2+z_3+z_4}{4}\right).$$

$\therefore$ G' coincides with G. Again the symmetry of the co-ordinates of G' suggests that the other three lines joining the vertices to the centroids of the opposite faces must pass through G'. Hence these seven lines are concurrent.

Note. G is the c.g. of the tetrahedron.

7. Find the co-ordinates of a vector whose magnitude is 4 and which makes angles $60°, 120°, 45°$ with the x, y and z axes respectively.

Let X, Y, Z be the projections of the vector on the x, y and z axes respectively.

Then

$$X = 4\cos 60° = 4\cdot\frac{1}{2} = 2,$$

$$Y = 4\cos 120° = 4\cdot\left(-\frac{1}{2}\right) = -2,$$

$$Z = 4\cos 45° = 4\cdot\frac{1}{\sqrt{2}} = 2\sqrt{2}.$$

Hence the co-ordinates of the vector are $(2, -2, 2\sqrt{2})$.

8. From the point $(1, -2, 3)$ lines are drawn to meet the sphere $x^2+y^2+z^2 = 4$ and they are divided in the ratio $2:3$. Prove that the points of section lie on the sphere $5x^2 + 5y^2 + 5z^2 - 6x + 12y - 18z + 22 = 0$.

Let the line through $(1, -2, 3)$ meet the sphere at (x_1, y_1, z_1).

$$\therefore \quad x_1^2 + y_1^2 + z_1^2 = 4. \tag{1}$$

If (α, β, γ) divides $(1, -2, 3)$ and (x_1, y_1, z_1) in the ratio $2:3$, then

$$\alpha = \frac{2x_1+3}{5}, \quad \beta = \frac{2y_1-6}{5}, \quad \gamma = \frac{2z_1+9}{5},$$

$$\text{or} \quad x_1 = \frac{5\alpha-3}{2}, \quad y_1 = \frac{5\beta+6}{2}, \quad z_1 = \frac{5\gamma-9}{2}.$$

From (1)

$$\left(\frac{5\alpha-3}{2}\right)^2+\left(\frac{5\beta+6}{2}\right)^2+\left(\frac{5\gamma-9}{2}\right)^2=4,$$

$$\text{or} \quad 5\alpha^2+5\beta^2+5\gamma^2-6\alpha+12\beta-18\gamma+22=0.$$

Hence (α,β,γ) lies on the sphere

$$5x^2+5y^2+5z^2-6x+12y-18z+22=0.$$

Exercise

1. Find the co-ordinates of the projections of the points $(2,1,3),(-1,3,5),(-3,0,6)$ and $(0,0,4)$ on the co-ordinates planes and on the co-ordinate axes.

2. Find the possible octants where the point (x,y,z) may lie for each of the following conditions.

   (a) $x+y=0$; (b) $z-x=0$; (c) $z-y=0$.

3. Find the octants, which contain the points whose co-ordinates (x,y,z) satisfy each of the following conditions.

   (a) $xy>0$, (b) $xyz>0$.

4. (a) If P and Q are points $(3,-2,5)$ and $(1,0,-7)$, find the co-ordinates of $\overline{PQ}$ and $\overline{QP}$.

   (b) If a vector of modulus 6 makes angles $120°, 45°$ and $60°$ with the x, y and z axes respectively, find the co-ordinates of it.

5. Find the distance between the points.

   (a) $(4,-1,5),(1,2,-2)$; (b) $(0,7,10),(5,4,-7)$.

6. Find the centre of the sphere passing through $(0,0,0),(2,0,0),(0,3,0)$ and $(0,0,4)$.

7. Show that the points $(1,1,1),(-2,4,1),(-1,5,5,)$ and $(2,2,5)$ form a square.

8. Show that the following points are collinear.

   (a) $(3,-2,4),(1,1,1),(-1,4,-2)$.

   [HINTS. Show that $(1,1,1)$ divides the join of the other two in a certain ratio.]

   (b) $(2,5,-4),(1,4,-3),(4,7,-6)$.

9. (a) Find the co-ordinates of the point, which divides the segment joining the points $(1,3,-9)$ and $(6,-2,-4)$ in the ratio $2:3$.

   (b) Find the co-ordinates of the points which divide the segment joining the points $A(2,7,1)$ and $B(8,-2,5)$ into three equal parts.

10. (a) Find the ratio in which the line joining $(4,6,7)$ and $(-1,2,5)$ is divided by the yz plane. Find also the co-ordinates of the point on the yz plane.

    (b) A point P lies on the line passing through the points $(2,-3,1)$ and $(3,4,-5)$. If the x-co-ordinates of P is 5, find the other co-ordinates of it.

    (c) Find the co-ordinates of the point on the y-axis that is equidistant from $A(4,3.1)$ and $B(-2,-6,-2)$.

11. Find the point where the line joining $(2, -3, 1)$ and $(3, -4, -5)$ cuts the plane $2x + y + z = 7$.

12. The co-ordinates of A, B, C, D are $(1, 1, 1), (-1, 3, -3), (3, -1, 2)$ and $(-3, 5, -4)$ respectively. Show that the lines AB and CD intersect and find the point of intersection.

[HINTS. Let the co-ordinates of a point P on AB be

$$\left(\frac{-\lambda+1}{\lambda+1}, \frac{3\lambda+1}{\lambda+1}, \frac{-3\lambda+1}{\lambda+1}\right)$$

and those of a point Q on CD be

$$\left(\frac{-3\lambda'+3}{\lambda'+1}, \frac{5\lambda'-1}{\lambda'+1}, \frac{-4\lambda'+2}{\lambda'+1}\right).$$

The lines will intersect if there exist values of λ and λ' such that

$$\frac{-\lambda+1}{\lambda+1} = \frac{-3\lambda'+3}{\lambda'+1}, \quad \frac{3\lambda+1}{\lambda+1} = \frac{5\lambda'-1}{\lambda'+1}, \quad \frac{-3\lambda+1}{\lambda+1} = \frac{-4\lambda'+2}{\lambda'+1}.$$

From these relations it is found that $\lambda = \lambda' = 1$ and the point of intersection is $(0, 2, -1)$.]

13. Find the locus of the point, which is equidistant from $(0, 2, 3)$ and $(2, 0, 4)$,

14. Find the locus of a point whose distance from the y-axis is always 3.

15. Find the ratio in which the sphere $x^2 + y^2 + z^2 = 504$ divides the line joining the points $(12, -4, 8)$ and $(27, -9, 18)$.

16. Show that the points where the line joining the points $(3, 0, -2)$ and $(6, 6, 7)$ cuts the surface represented by $x^2 + y^2 + z^2 - 4x + 4y + 10z - 23 = 0$ are $(4, 2, 1)$ and $(0, -6, -11)$.

Answers

1. Co-ordinates on the xy plane $(2, 1, 0), (-1, 3, 0), (-3, 0, 0), (0, 0, 0)$; Co-ordinates on the yz plane $(0, 1, 3), (0, 3, 5), (0, 0, 6), (0, 0, 4)$; Co-ordinates on the zx plane $(2, 0, 3), (-1, 0, 5), (-3, 0, 6), (0, 0, 4)$; Co-ordinates on the x-axis $(2, 0, 0), (-1, 0, 0), (-3, 0, 0), (0, 0, 0)$; Co-ordinates on the y-axis $(0, 1, 0), (0, 3, 0), (0, 0, 0), (0, 0, 0)$; Co-ordinates on the z-axis $(0, 0, 3), (0, 0, 5), (0, 0, 6), (0, 0, 4)$. **2.** (a) $OX'YZ, OXY'Z, OXY'Z', OX'YZ'$. (b) $OXYZ, OXY'Z, OX'YZ', OX'Y'Z'$ (c) $OXYZ, OX'YZ, OXYZ', OX'Y'Z'$ **3.** (a) $OXYZ$, $OXYZ'$, $OX'Y'Z'$, $OX'Y'Z$, (b) $OXYZ$, $OXY'Z'$, $OX'Y'Z$, $OX'YZ'$. **4.** (a) $\overline{PQ} = (-2, 2, -12)$, $\overline{QP} = (2, -2, 12)$ (b) $(-3, 3\sqrt{2}, 3)$. **5.** (a) $\sqrt{67}$, (b) $\sqrt{323}$. **6.** (a) $\left(1, \frac{3}{2}, 2\right)$. **9.** (a) $(3, 1, -7)$, (b) $\left(4, 4, \frac{7}{3}\right)$, $\left(6, 1, \frac{11}{3}\right)$. **10.** (a) $4 : 1$, $\left(0, \frac{14}{5}, \frac{27}{5}\right)$, (b) $(5, 18, -17)$, (c) $(0, -1, 0)$. **11.** $(1, -2, 7)$. **13.** $4x - 4y + 2z - 7 = 0$. **14.** $z^2 + x^2 = 9$. **15.** $2 : 3, -2 : 3$.

Chapter 2

Direction Cosines and Angle Between Two Lines

2.1 Angle between Two Non-Coplanar Lines

We have a knowledge about the angle between two lines lying on the same plane. If the lines are non-coplaner, then we define that the angle between them is equal to the angle between the two coplaner straight lines parallel to the non-coplanar straight lines.

2.2 Direction Cosines of a Line

Definition 1 ***If a directed line makes angles α, β, γ with the +ve directions of x, y and z axes respectively, then $\cos\alpha, \cos\beta, \cos\gamma$ are called the direction cosines (d. cs.) of this line. These numbers are generally denoted by l, m, n.***

In the fig. 2.1, $\overline{OP}$ makes angles α, β, γ with the co-ordinate axes. If (x, y, z) are the co-ordinates of P and we draw PM perpendicular to x-axis, then we have

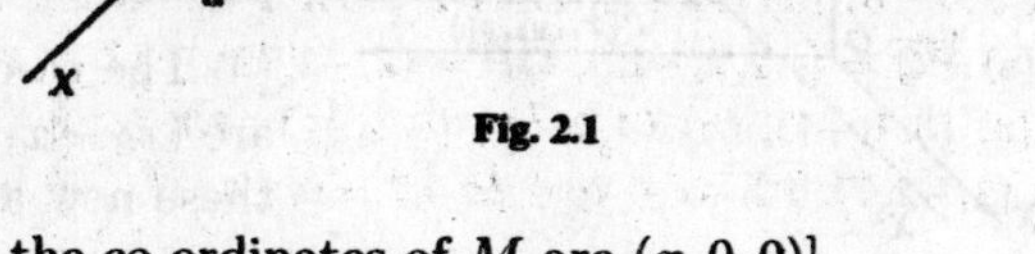

Fig. 2.1

$$\cos\alpha = \frac{OM}{|\overline{OP}|} = \frac{x}{|\overline{OP}|} \quad [\because \text{ the co-ordinates of } M \text{ are } (x, 0, 0)]$$

Similarly $$\cos\beta = \frac{y}{|\overline{OP}|}, \quad \cos\gamma = \frac{z}{|\overline{OP}|}.$$

$$\therefore \quad \cos^2\alpha + \cos^2\beta + \cos^2\gamma = \frac{x^2 + y^2 + z^2}{OP^2} = \frac{OP^2}{OP^2} = 1,$$

i.e.,

$$l^2 + m^2 + n^2 = 1. \tag{2.2.1}$$

Again

$$\frac{x}{l} = \frac{y}{m} = \frac{z}{n}. \tag{2.2.2}$$

x, y, z are proportional to the direction cosines. *Any three numbers, which are proportional to the direction cosines of a line are called the direction ratios of the line.*

Note 1. Any three numbers cannot be d. cs. of a line. They will be d. cs. only when the sum of the squares of them is equal to 1.

Note 2. If l, m, n are the d. cs. of $\overline{OP}$, then $-l, -m, -n$ are the d. cs. of $\overline{PO}$.

Corollary 1. From (2)

$$\frac{x}{l} = \frac{y}{m} = \frac{z}{n} = \pm\sqrt{x^2 + y^2 + z^2}.$$

$\therefore$ the d. cs. of $\overline{OP}$ are

$$\frac{x}{\sqrt{x^2+y^2+z^2}}, \frac{y}{\sqrt{x^2+y^2+z^2}}, \frac{z}{\sqrt{x^2+y^2+z^2}}$$

and those of $\overline{PO}$ are

$$-\frac{x}{\sqrt{x^2+y^2+z^2}}, -\frac{y}{\sqrt{x^2+y^2+z^2}}, -\frac{z}{\sqrt{x^2+y^2+z^2}}.$$

Thus the d. cs. can be found out when d. rs. (direction ratios) are given.

Corollary 2.

$$\sin^2\alpha + \sin^2\beta + \sin^2\gamma = (1 - \cos^2\alpha) + (1 - \cos^2\beta) + (1 - \cos^2\gamma)$$
$$= 3 - (\cos^2\alpha + \cos^2\beta + \cos^2\gamma) = 3 - 1 = 2.$$

2.3 Direction Cosines of a Line Joining Two Points

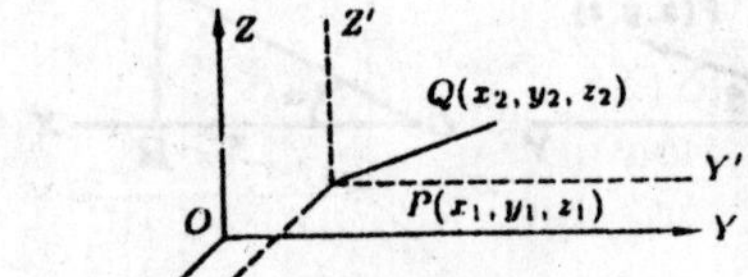

Fig. 2.2

Let $P(x_1, y_1, z_1)$ and $Q(x_2, y_2, z_2)$ be the two points in the frame of references OX, OY and OZ. Let the origin O be transferred to P with the axes remaining parallel to the original axes.

The co-ordinates of Q w.r.t. the new set of axes are $(x_2 - x_1, y_2 - y_1, z_2 - z_1)$. The d. cs. of $\overline{PQ}$ w.r.t. these new axes are

$$\frac{x_2 - x_1}{|\overline{PQ}|}, \quad \frac{y_2 - y_1}{|\overline{PQ}|}, \quad \frac{z_2 - z_1}{|\overline{PQ}|}.$$

Since the old set of axes are parallel to the new set of axes, the angles made by $\overline{PQ}$ with the new set of axes are equal to those made by it with OX, OY and OZ. Hence the d. cs. of $\overline{PQ}$ w.r.t. OX, OY and OZ are

$$\frac{x_2 - x_1}{|\overline{PQ}|}, \frac{y_2 - y_1}{|\overline{PQ}|}, \frac{z_2 - z_1}{|\overline{PQ}|}$$

where $|\overline{PQ}| = \sqrt{\{(x_2 - x_1)^2 + (y_2 - y_1)^2 + (z_2 - z_1)^2\}}$.

► **Example 1** *Find the d. cs. of the line joining the points* $(2,3,4)$ *and* $(-1,4,6)$.

Distance between the points $= \sqrt{\{(-1-2)^2 + (4-3)^2 + (6-4)^2\}} = \sqrt{14}$.

The d. cs. are

$$\frac{-3}{\sqrt{14}}, \quad \frac{1}{\sqrt{14}}, \quad \frac{2}{\sqrt{14}}.$$

2.4 If $P(x_1, y_1, z_1)$ and $Q(x_2, y_2, z_2)$ be the two given points then the projection of $\overline{PQ}$ on a line AB whose d. cs. are l, m, n is

$$l(x_2 - x_1) + m(y_2 - y_1) + n(z_2 - z_1).$$

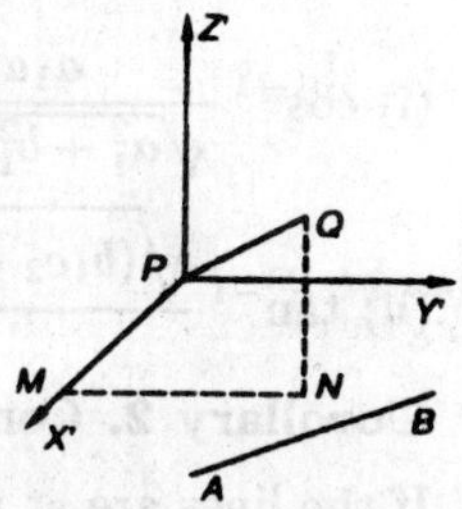

Fig. 2.3

If the origin is transferred to P with the axes remaining parallel, the co-ordinates of Q w.r.t. this new set of axes are

$$(x_2 - x_1, y_2 - y_1, z_2 - z_1).$$

In the fig. 2.3. QN is perpendicular to $X'PY'$ plane and MN is parallel to PY'.

$\therefore\ PM = x_2 - x_1, MN = y_2 - y_1$ and $NQ = z_2 - z_1$.

Now proj. of $\overline{PQ}$ on $AB =$ proj. of PM on $AB+$ proj. of MN on $AB+$ proj. of NQ on AB

$$= l(x_2 - x_1) + m(y_2 - y_1) + n(z_2 - z_1).$$

[$\because$ cosines of the angles between AB and PX', PY', PZ' are l, m, n respectively.]

2.5 Angle between Two Lines

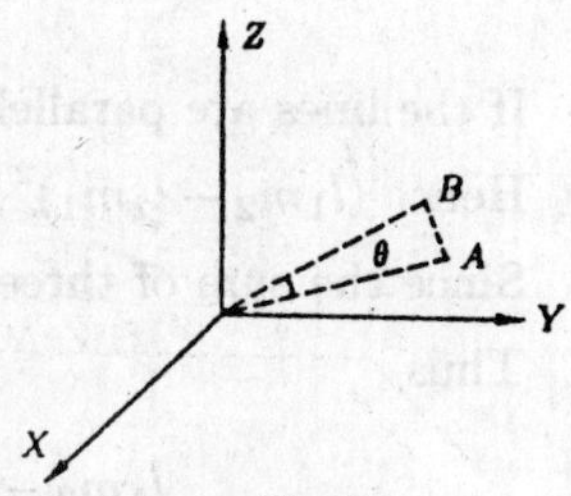

Fig. 2.4

Let the d. cs. of two lines OA and OB be (l_1, m_1, n_1) and (l_2, m_2, n_2). If O is not the origin, the origin is shifted to O with the axes remaining parallel. For this transformation the d. cs. of the lines will not be changed.

Let us choose $OA = OB = 1$. Then the co-ordinates of A and B are (l_1, m_1, n_1) and (l_2, m_2, n_2).

Method I. If θ be the angle between OA and OB, then from

$$\Delta OAB, AB^2 = OA^2 + OB^2 - 2.OA.OB.\cos\theta,$$

or $\quad (l_2 - l_1)^2 + (m_2 - m_1)^2 + (n_2 - n_1)^2 = 1 + 1 - 2\cos\theta,$

or $\quad (l_1^2 + m_1^2 + n_1^2) + (l_2^2 + m_2^2 + n_2^2) - 2(l_1 l_2 + m_1 m_2 + n_1 n_2) = 2 - 2\cos\theta,$

or $\quad 1 + 1 - 2(l_1 l_2 + m_1 m_2 + n_1 n_2) = 2 - 2\cos\theta,$

or $\quad \cos\theta = l_1 l_2 + m_1 m_2 + n_1 n_2,$

or $\quad \theta = \cos^{-1}(l_1 l_2 + m_1 m_2 + n_1 n_2).$

Method II. Considering projection of OB on OA we have

$$OB\cos\theta = l_1l_2 + m_1m_2 + n_1n_2,$$

[$\because$ $(0,0,0)$ and (l_2, m_2, n_2) are the co-ordinates of O and B and the d. cs. of OA are l_1, m_1, n_1]

$$\text{or}\quad \cos\theta = l_1l_2 + m_1m_2 + n_1n_2,$$

$$\text{or}\quad \theta = \cos^{-1}(l_1l_2 + m_1m_2 + n_1n_2).$$

Corollary 1. Angle with direction ratios:

If (a_1, b_1, c_1) and (a_2, b_2, c_2) are the d. rs. of two lines, then the angle between them is

(i) $\cos^{-1}\dfrac{a_1a_2 + b_1b_2 + c_1c_2}{\sqrt{a_1^2+b_1^2+c_1^2}\cdot\sqrt{a_2^2+b_2^2+c_2^2}}$, or

(ii) $\tan^{-1}\dfrac{\sqrt{(b_1c_2-b_2c_1)^2+(c_1a_2-c_2a_1)^2+(a_1b_2-a_2b_1)^2}}{a_1a_2+b_1b_2+c_1c_2}$.

Corollary 2. Condition for perpendicularity:

If the lines are at right angle, then $\theta = 90°$. $\therefore$ $\cos\theta = 0$.

Hence the condition is $l_1l_2 + m_1m_2 + n_1n_2 = 0$.

From Corollary 1. $a_1a_2 + b_1b_2 + c_1c_2 = 0$.

Corollary 3. Condition for parallel lines:

$$\sin^2\theta = 1 - \cos^2\theta = (l_1^2+m_1^2+n_1^2)(l_2^2+m_2^2+n_2^2) - (l_1l_2+m_1m_2+n_1n_2)^2$$
$$= (l_1m_2 - l_2m_1)^2 + (m_1n_2 - m_2n_1)^2 + (n_1l_2 - n_2l_1)^2.$$

[Lagrange's identity]

If the lines are parallel, then $\theta = 0$. $\therefore$ $\sin\theta = 0$.

Hence $(l_1m_2 - l_2m_1)^2 + (m_1n_2 - m_2n_1)^2 + (n_1l_2 - n_1l_1)^2 = 0$.

Since the sum of three squares is zero, each of them is zero.

Thus

$$l_1m_2 - l_2m_1 = 0, m_1n_2 - m_2n_1 = 0, n_1l_2 - n_2l_1 = 0,$$

$$\text{or}\quad \frac{l_1}{l_2} = \frac{m_1}{m_2} = \frac{n_1}{n_2} = \frac{\sqrt{l_1^2+m_1^2+n_1^2}}{\sqrt{l_2^2+m_2^2+n_2^2}} = 1.$$

$\therefore$ $l_1 = l_2, m_1 = m_2, n_1 = n_2$. [d. cs. are same.]

For direction ratios

$$\frac{a_1}{a_2} = \frac{b_1}{b_2} = \frac{c_1}{c_2}.$$ [d. rs. are proportional.]

Worked Out Examples

1. Can numbers $\frac{1}{\sqrt{2}}, \frac{1}{\sqrt{2}}, \frac{1}{\sqrt{2}}$ be the d. cs. of any directed line?

[C.U. '61, '85; B.U. '82, '87]

Since

$$\left(\frac{1}{\sqrt{2}}\right)^2 + \left(\frac{1}{\sqrt{2}}\right)^2 + \left(\frac{1}{\sqrt{2}}\right)^2 = \frac{1}{2} + \frac{1}{2} + \frac{1}{2} = \frac{3}{2} \neq 1,$$

the given numbers cannot be the d. cs. of a line.

2. Find the d. cs. of the line, which is equally inclined to the axes.

Let the line make an angle θ with each of the axes. Therefore, the d. cs. l, m, n are equal and $l = m = n = \cos\theta$.

$$\because \quad l^2 + m^2 + n^2 = 1, 3l^2 = 1, \quad \text{or} \quad l = \pm\frac{1}{\sqrt{3}}.$$

Hence the d. cs. are

$$\left(\frac{1}{\sqrt{3}}, \frac{1}{\sqrt{3}}, \frac{1}{\sqrt{3}}\right)$$

whence the acute angles made by the line are considered.

3. A directed line makes angles 60° and 45° with the axes of x and y respectively. What angle does it make with the axis of z? **[C.U. '97]**

Here x-d.c. $= \cos 60° = \frac{1}{2}$, y-d.c. $= \cos 45° = \frac{1}{\sqrt{2}}$.

If $n = z$-d.c., then $\left(\frac{1}{2}\right)^2 + \left(\frac{1}{\sqrt{2}}\right)^2 + n^2 = 1$, or $n^2 = 1 - \frac{1}{4} - \frac{1}{2} = \frac{1}{4}$.

$\therefore \quad n = \pm\frac{1}{2}$.

Thus the acute angle made by the line with the z-axis is 60°.

4. If l_1, m_1, n_1 and l_2, m_2, n_2 are d. cs. of two mutually perpendicular lines, show that the line whose d. r.s. are $m_1n_2 - m_2n_1, n_1l_2 - n_2l_1, l_1m_2 - l_2m_1$ is perpendicular to both of them. Find also the d. cs. of this line. **[C.U. '90, B.U. '85]**

Let l, m, n be the d. cs. of the line, which is perpendicular to both of the given lines.

Then $ll_1 + mm_1 + nn_1 = 0$ and $ll_2 + mm_2 + nn_2 = 0$.

By cross-multiplication

$$\frac{l}{m_1n_2 - m_2n_1} = \frac{m}{n_1l_2 - n_2l_1} = \frac{n}{l_1m_2 - l_2m_1} = \pm\frac{1}{\sqrt{\Sigma(m_1n_2 - m_2n_1)^2}}.$$

If θ be the angle between the given lines, then $\sin^2\theta = \Sigma(m_1n_2 - m_2n_1)^2$.

Here $\theta = 90°$. $1 = \Sigma(m_1n_2 - m_2n_1)^2$.

$$\therefore \quad \frac{l}{m_1n_2 - m_2n_1} = \frac{m}{n_1l_2 - n_2l_1} = \frac{n}{l_1m_2 - l_2m_1} = \pm 1.$$

Hence $m_1n_2 - m_2n_1, n_1l_2 - n_2l_1$ and $l_1m_2 - l_2m_1$ are the d. rs. of the line. The d. cs. are

$$\pm(m_1n_2 - m_2n_1), \pm(n_1l_2 - n_2l_1), \pm(l_1m_2 - l_2m_1).$$

5. Prove that the acute angle between the lines whose d. cs. are given by the relations $l+m+n=0$ and $l^2+m^2-n^2=0$ is $\pi/3$.

$$l+m+n=0, \tag{1}$$

$$l^2+m^2-n^2=0. \tag{2}$$

From (1) $n=-(l+m)$. Putting in (2)

$$l^2+m^2-(l+m)^2=0, \quad \text{or} \quad lm=0.$$

Either $l=0$, or $m=0$.

Taking $l=0, m+n=0$ and $m^2-n^2=0$.

Again

$$l^2+m^2+n^2=1. \tag{3}$$

By (2) and (3), $2n^2=1, n^2=\frac{1}{2}, n=\pm\frac{1}{\sqrt{2}}$.

But $n=-m$. $\therefore$ $m=\mp\frac{1}{\sqrt{2}}$.

Considering only one sign, the d. cs. of one line are $0, \frac{1}{\sqrt{2}}, -\frac{1}{\sqrt{2}}$.

Taking $m=0$, the d. cs. of the other line are $\frac{1}{\sqrt{2}}, 0, -\frac{1}{\sqrt{2}}$.

If θ be the angle between the lines

$$\cos\theta = 0\cdot\frac{1}{\sqrt{2}}+\frac{1}{\sqrt{2}}\cdot 0+\left(-\frac{1}{\sqrt{2}}\right)\cdot\left(-\frac{1}{\sqrt{2}}\right)=\frac{1}{2}. \quad \therefore\ \theta=\frac{\pi}{3}.$$

6. Show that the straight lines whose d. cs. are given by $al+bm+cn=0$, $fmn+gnl+hlm=0$ are perpendicular if $\frac{f}{a}+\frac{g}{b}+\frac{h}{c}=0$ and parallel if $\sqrt{af}\pm\sqrt{bg}\pm\sqrt{ch}=0$.

$$al+bm+cn=0, \tag{1}$$

$$fmn+gnl+hlm=0. \tag{2}$$

From (1) $l=-\frac{bm+cn}{a}$.

Putting in (2)

$$fmn+(gn+hm)\left(-\frac{bm+cn}{a}\right)=0,$$

$$\text{or} \quad bhm^2+(ch+bg-af)mn+cgn^2=0,$$

$$\text{or} \quad bh\left(\frac{m}{n}\right)^2+(ch+bg-af)\frac{m}{n}+cg=0. \tag{3}$$

If l_1, m_1, n_1 and l_2, m_2, n_2 are the d. cs. of the lines, then $\frac{m_1}{n_1}$ and $\frac{m_2}{n_2}$ are the roots of the equation (3). If the lines are parallel, then $\frac{m_1}{n_1}=\frac{m_2}{n_2}$.

From the condition of equal roots

$$(ch+bg-af)^2=4.bh.cg,$$

$$\text{or} \quad ch+bg-af=\pm2\sqrt{bcgh},$$

$$\text{or} \quad af=(\sqrt{ch})^2+(\sqrt{bg})^2\pm2\sqrt{bg.ch}=(\sqrt{ch}\pm\sqrt{bg})^2.$$

$$\therefore\ \sqrt{af}=\pm(\sqrt{ch}\pm\sqrt{bg}), \quad \text{or} \quad \sqrt{af}\pm\sqrt{bg}\pm\sqrt{ch}=0.$$

Hence the condition of parallelism is followed.

From (3)

$$\frac{m_1}{n_1}\cdot\frac{m_2}{n_2}=\frac{cg}{bh}, \quad \text{or} \quad \frac{m_1m_2}{cg}=\frac{n_1n_2}{bh}, \quad \text{or} \quad \frac{m_1m_2}{g/b}=\frac{n_1n_2}{h/c}.$$

Similarly eliminating n from (1) and (2), we get

$$\frac{l_1l_2}{f/a}=\frac{m_1m_2}{g/b}.$$

If the lines are perpendicular to each other, then

$$l_1l_2+m_1m_2+n_1n_2=0. \quad \therefore \frac{f}{a}+\frac{g}{b}+\frac{h}{c}=0.$$

7. If three concurrent lines with d. cs. $l_1, m_1, n_1; l_2, m_2, n_2$ and l_3, m_3, n_3 are coplanar, prove that

$$\begin{vmatrix} l_1 & m_1 & n_1 \\ l_2 & m_2 & n_2 \\ l_3 & m_3 & n_3 \end{vmatrix}=0.$$

[B.U. '80: C.U. '89, '92, 2000]

Since the lines are coplanar, they are perpendicular to the normal to the plane of them. If l, m, n be the d. cs. of the normal to the plane, then

$$ll_1+mm_1+nn_1=0, \tag{1}$$
$$ll_2+mm_2+nn_2=0, \tag{2}$$
$$ll_3+mm_3+nn_3=0. \tag{3}$$

Eliminating l, m, n from (1), (2) and (3) we have

$$\begin{vmatrix} l_1 & m_1 & n_1 \\ l_2 & m_2 & n_2 \\ l_3 & m_3 & n_3 \end{vmatrix}=0.$$

8. If a line makes angles $\alpha, \beta, \gamma, \delta$ with the four diagonals of a cube, prove that $\cos^2\alpha+\cos^2\beta+\cos^2\gamma+\cos^2\delta=\frac{4}{3}$.

Let O, A, B, C, D, E, F, G be the corners of a cube whose length of each edge is unity. Let the three mutually perpendicular edges OA, OC and OE be take as x, y and z-axes respectively.

Now the co-ordinates of the corners are as the following:

$O(0,0,0)$, $A(1,0,0)$, $B(1,1,0)$, $C(0,1,0)$,
$D(0,1,1)$, $E(0,0,1)$, $F(1,0,1)$, $G(1,1,1)$.

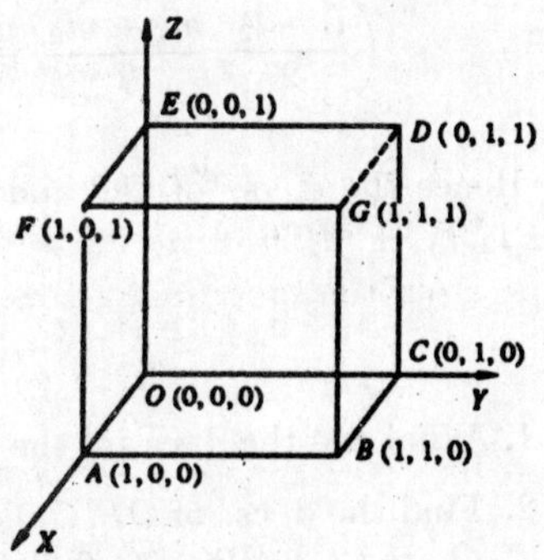

Fig. 2.5

The diagonals of the cube are OG, AD, BE and CF.

The d. cs. of OG are $\frac{1}{\sqrt{3}}, \frac{1}{\sqrt{3}}, \frac{1}{\sqrt{3}}$.

The d. cs. of AD are $-\frac{1}{\sqrt{3}}, \frac{1}{\sqrt{3}}, \frac{1}{\sqrt{3}}$.

The d. cs. of BE are $-\frac{1}{\sqrt{3}}, -\frac{1}{\sqrt{3}}, \frac{1}{\sqrt{3}}$.

The d. cs. of CF are $\frac{1}{\sqrt{3}}, -\frac{1}{\sqrt{3}}, \frac{1}{\sqrt{3}}$.

Let l, m, n be the d. cs. of the line which makes angles $\alpha, \beta, \gamma, \delta$ with OG, AD, BE and CF respectively.

Then

$$\cos\alpha = \frac{1}{\sqrt{3}}(l+m+n), \quad \cos\beta = \frac{1}{\sqrt{3}}(-l+m+n),$$

$$\cos\gamma = \frac{1}{\sqrt{3}}(-l-m+n), \quad \cos\delta = \frac{1}{\sqrt{3}}(l-m+n).$$

$$\therefore \cos^2\alpha + \cos^2\beta + \cos^2\gamma + \cos^2\delta$$
$$= \frac{1}{3}[(l+m+n)^2 + (-l+m+n)^2 + (-l-m+n)^2 + (l-m+n)^2]$$
$$= \frac{4}{3}(l^2+m^2+n^2) = \frac{4}{3}.$$

9. If l_1, m_1, n_1 and l_2, m_2, n_2 are the d. cs. of two concurrent lines, show that $l_1 \pm l_2, m_1 \pm m_2, n_1 \pm n_2$ are the d. rs. of the bisectors of the angles between the lines.

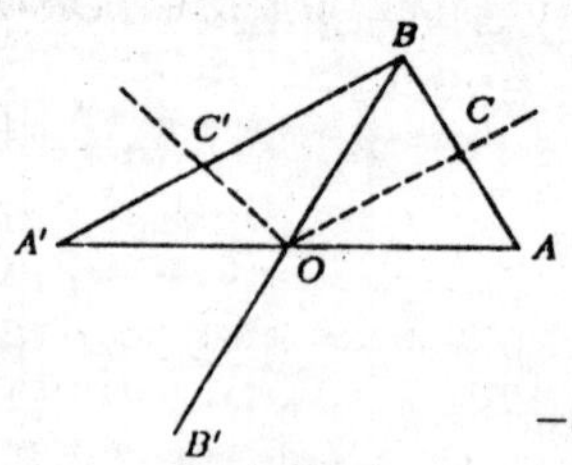

Fig. 2.6

Let AOA' and BOB' be the given lines.

Let us take O as origin and

$$|OA| = |OA'| = |OB| = 1.$$

The d. cs. of OA, OB and OA' are $l_1, m, n_1; l_2, m_2, n_2$ and $-l_1, -m_1, -n$.

$\therefore$ the co-ordinates of A, B and A' are (l_1, m_1, n_1) (l_2, m_2, n_2) and $(-l_1, -m_1, -n_1)$ respectively.

Let OC and OC' be the bisectors of the angles $\angle AOB$ and $\angle BOA'$. According to constructions C and C' are the mid-points of AB and BA'.

$\therefore$ the co-ordinates of C and C' are

$$\left(\frac{l_1+l_2}{2}, \frac{m_1+m_2}{2}, \frac{n_1+n_2}{2}\right) \quad \text{and} \quad \left(\frac{l_2-l_1}{2}, \frac{m_2-m_1}{2}, \frac{n_2-n_1}{2}\right).$$

Hence the d. rs. of OC and OC', i.e., the bisectors of the angles between the lines are $l_1 \pm l_2, m_1 \pm m_2, n_1 \pm n_2$.

Exercise

1. What are the d. cs. of the co-ordinates axes?
2. Find the d. cs. of OP, OQ and PO where O is the origin and the co-ordinates of P and Q are $(2, 3, 4)$ and $(-6, 5, -1)$.
3. Find the d. cs. of the line joining the points $(3, 2, 5)$ and $(-1, 3.2)$.

4. (a) A line in the XOY plane makes an angle $30°$ with the x-axis. Find the d. cs. of this line.

   (b) A line makes angles $60°$ and $45°$ with the y-axis and z-axis respectively. Find the d. cs. of the line.

5. If A, B, C, D are the points $(2,3,-1), (3,5,-3), (1,2,3)$ and $(3,5,7)$ respectively, prove by projections that AB is perpendicular to CD.

6. Find the angle between the lines whose d. rs. are $1, 1, 2$ and $\sqrt{3}-1,\ -\sqrt{3}-1, 1$.

7. (a) A, B, C are $(2,3,5), (-1,3,2)$ and $(3,5,-2)$. Find the angles of ΔABC.

   (b) Show that the points $(4,5,0), (2,6,2), (2,3,-1)$ are the vertices of an isosseles triangle. Find equal angles of the triangle. **[B.U. '81]**

8. Prove by d. cs. that the points $(3,2,4), (4,5,2), (5,8,0)$ and $(2,-1,6)$ are collinear.

9. Find the d. cs. of the line which is perpendicular to each of the line with d. rs. $(1,-2,-2)$ and $(0,2,1)$.

10. Prove that the lines drawn from the origin whose d. cs. are proportional to $(1,-1,1), (2,-3,0)$ and $(1,0,3)$ are coplanar. **[C.U. '91]**

11. Find the point where the line through the origin meets the join of $(-9,4,5)$ and $(11,0,-1)$ perpendicularly.

12. (a) Prove that the angle between two diagonals of a cube is $\cos^{-1}\frac{1}{3}$. **[B.U. '96]**

    (b) If the edges of a rectangular parallelopiped be a, b, c, show that the angles between the four diagonals are given by

$$\cos^{-1}\frac{\pm a^2 \pm b^2 \pm c^2}{a^2+b^2+c^2}.$$

    (c) If two pairs of opposite edges of a tetrahedron are at right angles, then show that the third pair is also at right angles.

    [HINTS. Let $OABC$ be the tetrahedron. Taking O as origin, let the co-ordinates of A, B, C be $(x_i, y_i, z_i), i = 1, 2, 3$. Let the opposite edges OA, BC and OB, CA be perpendicular. D. rs. of OA and BC are x_1, y_1, z_1 and $x_3 - x_2, y_3 - y_2, z_3 - z_2$. Since $OA \perp BC$, we have

$$x_1(x_3 - x_2) + y_1(y_3 - y_2) + z_1(z_3 - z_2) = 0. \tag{1}$$

    Similarly

$$x_2(x_1 - x_3) + y_2(y_1 - y_3) + z_2(z_1 - z_3) = 0 \tag{2}$$

    as $OB \perp CA$.

    Adding (1) and (2), $x_3(x_1 - x_2) + y_3(y_1 - y_2) + z_3(z_1 - z_2) = 0$.

    $\therefore$ OC is perpendicular to AB.]

13. Show that the figure formed by the points $(5,-1,1), (7,-4,7), (1,-6,10)$ and $(-1,-3,4)$ as vertices is a rhombus.

14. (a) Prove that the two lines whose d. cs. are given by the relations $2l + 2m - n = 0$ and $lm + mn + nl = 0$ are perpendicular to each other. **[C. U. '88, '96]**

    (b) The d. cs. of two lines are connected by the relations $l - 5m + 3n = 0$ and $7l^2 + 5m^2 - 3n^2 = 0$. Find the angle between them.

    (c) Show that the lines whose d. cs. are given by the relations $a^2l + b^2m + c^2n = 0$ and $mn + nl + lm = 0$ will be parallel, if $a + b + c = 0$. **[N.B. '87]**

(d) Show that the straight lines whose d. cs. are given by the equations $al + bm + cn = 0, ul^2 + vm^2 + wn^2 = 0$ are perpendicular or parallel according as $(v+w)a^2 + (w+u)b^2 + (u+v)c^2 = 0$,

$$\text{or} \quad \frac{a^2}{u} + \frac{b^2}{v} + \frac{c^2}{w} = 0.$$

15. If $l_1, m_1, n_1; l_2, m_2, n_2; l_3, m_3, n_3$ are the d. cs. of three mutually perpendicular lines, show that the line whose d. rs. are $l_1 + l_2 + l_3, m_1 + m_2 + m_3, n_1 + n_2 + n_3$ makes equal angles with them.

16. The d. cs. of two intersecting lines are l_1, m_1, n_1 and l_2, m_2, n_2. Prove that all lines passing through the intersection of these two and having d. cs. proportional to $l_1 + \lambda l_2, m_1 + \lambda m_2, n_1 + \lambda n_2$ are coplanar with them.

Answers

1. 1, 0, 0; 0, 1, 0; 0, 0, 1; **2.** $\frac{2}{\sqrt{29}}, \frac{3}{\sqrt{29}}, \frac{4}{\sqrt{29}}; \frac{-6}{\sqrt{62}}, \frac{5}{\sqrt{62}}, \frac{-1}{\sqrt{62}}; \frac{-2}{\sqrt{29}}, \frac{-3}{\sqrt{29}}, \frac{-4}{\sqrt{29}}$; **3.** $\frac{-4}{\sqrt{26}}, \frac{1}{\sqrt{26}}, \frac{-3}{\sqrt{26}}$; **4.** (a) $\frac{\sqrt{3}}{2}, \frac{1}{2}, 0$; (b) $\pm\frac{1}{2}, \frac{1}{2}. \frac{1}{\sqrt{2}}$. **6.** $\frac{\pi}{2}$. **7.** (a) $\cos^{-1}\frac{1}{\sqrt{3}}, \frac{\pi}{2}, \cos^{-1}\sqrt{\frac{2}{3}}$; (b) $\frac{\pi}{4}$; **9.** $\frac{2}{3}, \frac{-1}{3}, \frac{2}{3}$; **11.** (1, 2, 2); **14.** (b) $\cos^{-1}\frac{7}{\sqrt{84}}$.

Chapter 3

Plane

3.1 Definition of a Plane

A plane is a flat surface such that if any two points are taken on the surface and joined by a straight line, the straight line lies wholly on the surface.

3.2 Every Linear (i.e., First Degree) Equation Represents a Plane

The most general equation of the first degree in x, y, z is of the form

$$ax + by + cz + d = 0, \tag{3.2.1}$$

where a, b, c, d are constants and a, b, c are not all zero.

Let (x_1, y_1, z_1) and (x_2, y_2, z_2) be any two points on the surface represented by the equation (3.2.1).

Then

$$ax_1 + by_1 + cz_1 + d = 0 \tag{3.2.2}$$

$$\text{and} \quad ax_2 + by_2 + cz_2 + d = 0. \tag{3.2.3}$$

Multiplying (3.2.3) by λ and then adding to (3.2.2)

$$a(x_1 + \lambda x_2) + b(y_1 + \lambda y_2) + c(z_1 + \lambda z_2) + d(1 + \lambda) = 0.$$

Dividing by $1 + \lambda$,

$$a\frac{x_1 + \lambda x_2}{1 + \lambda} + b\frac{y_1 + \lambda y_2}{1 + \lambda} + c\frac{z_1 + \lambda z_2}{1 + \lambda} + d = 0.$$

It shows that the point

$$\left(\frac{x_1 + \lambda x_2}{1 + \lambda}, \quad \frac{y_1 + \lambda y_2}{1 + \lambda}, \quad \frac{z_1 + \lambda z_2}{1 + \lambda}\right)$$

lies on the surface.

Again this point is on the line joining (x_1, y_1, z_1) and (x_2, y_2, z_2) and divides this line in the ratio $\lambda : 1$. λ may have any value. Therefore, all the points on the line lie on the surface, i.e., the line lies wholly on the surface.

Hence the equation (1) represents a plane.

Note. $ax + by + cz + d = 0$ is known as the *general equation* of a plane. It can be written as $\frac{a}{d}x + \frac{b}{d}y + \frac{c}{d}z + 1 = 0$.

Therefore, there are three arbitrary constants in the general equation of a plane.

Corollary 1. $ax + by + cz = 0$ represents a plane passing through the origin.

Corollary 2. $a(x - x_1) + b(y - y_1) + c(z - z_1) = 0$ represents a plane passing through (x_1, y_1, z_1).

3.3 Equation of a Plane in the Intercept Form

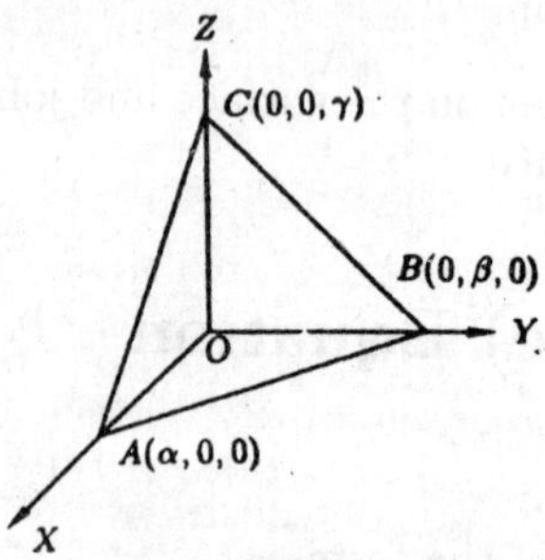

Fig. 3.1

Let $ax + by + cz + d = 0$ be the equation of a plane which meets the co-ordinate axes at A, B, C. If the co-ordinates of A, B, C be $(\alpha, 0, 0), (0, \beta, 0)$ and $(0, 0, \gamma)$ respectively, then $a\alpha + d = 0$, or $\alpha = -\frac{d}{a}$.

Similarly $\beta = -\frac{d}{b}, \gamma = -\frac{d}{c}$.

The equation of the plane can be written as $ax + by + cz = -d$,

$$\text{or} \quad \frac{x}{-\frac{d}{a}} + \frac{y}{\frac{-d}{b}} + \frac{z}{-\frac{d}{c}} = 1, \quad \text{or} \quad \frac{x}{\alpha} + \frac{y}{\beta} + \frac{z}{\gamma} = 1.$$

It is known as the equation of a plane in the intercept form. α, β, γ are the *intercepts* made by the plane on the x, y and z-axes respectively.

► **Example 1** *Find the equation of the plane which makes intercepts* $2, 3, -4$ *on the axes.*

The equation of the plane is $\frac{x}{2} + \frac{y}{3} - \frac{z}{4} = 1$.

► **Example 2** *Find the intercepts made by the plane* $x - 2y + 3z - 18 = 0$ *on the axes.*

The equation of the plane can be written as $\frac{x}{18} + \frac{y}{-9} + \frac{z}{6} = 1$.
Therefore, the intercepts on the x, y and z-axes are $18, -9, 6$ respectively.

► **Example 3** *If a plane meets the axes in* A, B, C *and the centroid of* ΔABC *is* (α, β, γ), *show that the equation of the plane is* $\frac{x}{\alpha} + \frac{y}{\beta} + \frac{z}{\gamma} = 3$. **[B.U. '99; C.U. '99]**

Let $\frac{x}{a} + \frac{y}{b} + \frac{z}{c} = 1$ be the equation of the plane.

The co-ordinates of A, B, C are $(a, 0, 0), (0, b, 0)$ and $(0, 0, c)$ respectively. The centroid of ΔABC is $\left(\frac{a}{3}, \frac{b}{3}, \frac{c}{3}\right)$. By the given co-ordinates of the centroid we have $a/3 = \alpha$, or $a = 3\alpha, b = 3\beta$ and $c = 3\gamma$.

$\therefore$ the equation of the plane is

$$\frac{x}{3\alpha} + \frac{y}{3\beta} + \frac{z}{3\gamma} = 1, \quad \text{or} \quad \frac{x}{\alpha} + \frac{y}{\beta} + \frac{z}{\gamma} = 3.$$

3.4 Equation of a Plane Passing through Three given points

Let

$$ax + by + cz + d = 0 \qquad (3.4.1)$$

represent a plane passing through $(x_1, y_1, z_1), (x_2, y_2, z_2)$ and (x_3, y_3, z_3).

Since the plane passes through these points

$$ax_1 + by_1 + cz_1 + d = 0, \qquad (3.4.2)$$

$$ax_2 + by_2 + cz_2 + d = 0 \qquad (3.4.3)$$

$$\text{and} \quad ax_3 + by_3 + cz_3 + d = 0. \qquad (3.4.4)$$

Eliminating a, b, c, d from (3.4.1), (3.4.2), (3.4.3) and (3.4.4) we have

$$\begin{vmatrix} x & y & z & 1 \\ x_1 & y_1 & z_1 & 1 \\ x_2 & y_2 & z_2 & 1 \\ x_3 & y_3 & z_3 & 1 \end{vmatrix} = 0.$$

It is the required equation.

▶ **Example 4** *Find the equation of the plane passing through* $(0, 2, 4), (3, 1, 1)$ *and* $(2, 0, -1)$.

The equation of the plane is

$$\begin{vmatrix} x & y & z & 1 \\ 0 & 2 & 4 & 1 \\ 3 & 1 & 1 & 1 \\ 2 & 0 & -1 & 1 \end{vmatrix} = 0,$$

$$\text{or} \quad x\begin{vmatrix} 2 & 4 & 1 \\ 1 & 1 & 1 \\ 0 & -1 & 1 \end{vmatrix} - y\begin{vmatrix} 0 & 4 & 1 \\ 3 & 1 & 1 \\ 2 & -1 & 1 \end{vmatrix} + z\begin{vmatrix} 0 & 2 & 1 \\ 3 & 1 & 1 \\ 2 & 0 & 1 \end{vmatrix} - \begin{vmatrix} 0 & 2 & 4 \\ 3 & 1 & 1 \\ 2 & 0 & -1 \end{vmatrix} = 0,$$

or $x + 9y + 4z + 2 = 0$.

3.5 Equation of the Plane in the Normal Form

Let ABC be a plane in the co-ordinate system OX, OY and OZ. ON is perpendicular to the plane. ON is called the normal to the plane. Let l, m, n be the d. cs. of ON and $ON = p$.

$P(x, y, z)$ is a point on the plane. PM is perpendicular to XOY plane. LM is parallel to OY.

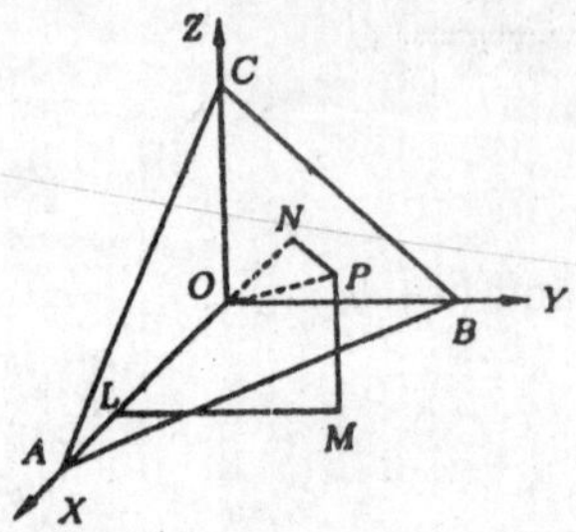

Fig. 3.2

$$\therefore \quad OL = x, LM = y, MP = z.$$

By the property of projection, proj. of OP on $ON =$ proj. of OL on $ON+$ proj. of LM on $ON+$ proj. of MP on ON.

Since ON is perpendicular to the plane ABC, ON is perpendicular to NP.

$$\therefore \quad \text{proj. of } OP \text{ on } ON = ON = p.$$

Again proj. of OL on $ON = lx$,

proj. of LM on $ON = my$,

proj. of MP on $ON = nz$.

$$\therefore \quad lx + my + nz = p.$$

P is any point on the plane. Hence it is the equation of the plane.

Note 1. $p =$ perpendicular distance from the origin to the plane. It is always positive.

Note 2. l, m, n are the d. cs. of the normal to the plane. If the normal makes angles α, β, γ with the co-ordinate axes then the equation to the plane is $x \cos\alpha + y \cos\beta + z \cos\gamma = p$.

3.6 Reduction of $ax + by + cz + d = 0$ in the Normal Form

Let $lx + my + nz = p$ be the normal form of the equation. Comparing these two equations, $\frac{l}{a} = \frac{m}{b} = \frac{n}{c} = \frac{-p}{d}$.

$$l = -\frac{ap}{d}, \quad m = -\frac{bp}{d}, \quad n = -\frac{cp}{d}.$$

$$l^2 + m^2 + n^2 = \frac{a^2 + b^2 + c^2}{d^2} p^2, \quad \text{or} \quad 1 = \frac{a^2 + b^2 + c^2}{d^2} \cdot p^2.$$

$$\text{or} \quad p^2 = \frac{d^2}{a^2 + b^2 + c^2}, \quad \text{or} \quad p = \pm\frac{d}{\sqrt{a^2 + b^2 + c^2}}.$$

Since p is always +ve,

$$p = \frac{d}{\sqrt{a^2 + b^2 + c^2}}$$

when d is +ve;

$$p = \frac{-d}{\sqrt{a^2 + b^2 + c^2}}$$

when d is −ve.

Now

$$l = \mp\frac{a}{\sqrt{a^2+b^2+c^2}}, \quad m = \mp\frac{b}{\sqrt{a^2+b^2+c^2}},$$
$$n = \mp\frac{c}{\sqrt{a^2+b^2+c^2}}$$

according as d is +ve or −ve.

Thus the normal form of the general equation is

$$\frac{a}{\sqrt{a^2+b^2+c^2}}x + \frac{b}{\sqrt{a^2+b^2+c^2}}y + \frac{c}{\sqrt{a^2+b^2+c^2}}z = \mp\frac{d}{\sqrt{a^2+b^2+c^2}}$$

according as d is +ve or −ve.

Note 1. Perpendicular distance from the origin to $ax+by+cz+d=0$ is

$$\pm\frac{d}{\sqrt{a^2+b^2+c^2}}$$

according as d is +ve or −ve.

Note 2. The d. cs. of the normal to the plane $ax+by+cz+d=0$ are

$$\mp\left(\frac{a}{\sqrt{a^2+b^2+c^2}}, \frac{b}{\sqrt{a^2+b^2+c^2}}, \frac{c}{\sqrt{a^2+b^2+c^2}}\right)$$

according as d is +ve or −ve. a, b, c are the d. rs. of the normal to the plane.

Corollary 1. Equations of the co-ordinate planes.

The XOY plane passes through the origin and z-axis is normal to the plane. The d. cs. of z-axis are $0, 0, 1$. Hence the equation of the XOY plane is $z=0$. Similarly the equations of YOZ and ZOX are $x=0$ and $y=0$.

Corollary 2. Equations of planes parallel to the axes.

If a plane is parallel to z-axis, then the normal to the plane is perpendicular to z-axis. Therefore, the z-d.c. of this normal is zero. Hence the equation of this plane is $ax+by+d=0$. Similarly the equations of the planes parallel to x and y axes are $by+cz+d=0$ and $ax+cz+d=0$ respectively.

► **Example 5** *Express the equation of the plane $2x+6y-3z+5=0$ in the normal form and hence obtain the length of perpendicular from the origin upon the plane.*

[B.U. '81, 85]

Here $\sqrt{2^2+6^2+(-3)^2} = 7$ and the constant term $= 5$.

Dividing each term of the given equation by −7, the required normal form is

$$-\frac{2}{7}x - \frac{6}{7}y + \frac{3}{7}z = \frac{5}{7}.$$

Perpendicular distance from the origin $= \frac{5}{7}$.

► **Example 6** *Find the equations of three planes through the points* $(3,1,1)$, $(1,-2,3)$, *parallel to the co-ordinate axes.* **[C.H. '84]**

(i) Let the equation of the plane parallel to the x-axis be $by+cz+d=0$. If it passes through the given points then $b+c+d=0$ and $-2b+3c+d=0$. From these relations $b=-\frac{2}{5}d, c=-\frac{3}{5}d$.

$\therefore$ the required equation is $2y+3z-5=0$.

(ii) Let the equation of the plane parallel to the y-axis be $ax+cz+d=0$. If it passes through the given points then $3a+c+d=0$ and $a+3c+d=0$.

From these $a=-\frac{d}{4}, c=-\frac{d}{4}$.

$\therefore$ the required equation is $x+z-4=0$.

(iii) Let the equation of the plane parallel to the z-axis be $ax+by+d=0$. If it passes through the given points then $3a+b+d=0$ and $a-2b+d=0$.

From these $a=-\frac{3d}{7}, b=\frac{2d}{7}$.

$\therefore$ the required equation is $3x-2y-7=0$.

3.7 Perpendicular Distance of a Point from a Plane

Let

$$ax+by+cz+d=0 \tag{3.7.1}$$

be the equation of a plane and (x_1, y_1, z_1) be a given point.

The equation of the plane parallel to (3.7.1) and passing through (x_1, y_1, z_1) is

$$(x_1, y_1, z_1) \text{ is } a(x-x_1)+b(y-y_1)+c(z-z_1)=0, \tag{3.7.2}$$

$$\text{or} \quad ax+by+cz-(ax_1+by_1+cz_1)=0. \tag{3.7.3}$$

The distances of (3.7.1) and (3.7.2) from the origin are

$$\pm\frac{d}{\sqrt{a^2+b^2+c^2}} \quad \text{and} \quad \mp\frac{ax_1+by_1+cz_1}{\sqrt{a^2+b^2+c^2}}.$$

$\therefore$ the required distance

$$=\pm\frac{d}{\sqrt{a^2+b^2+c^2}}-\left(\mp\frac{ax_1+by_1+cz_1}{\sqrt{a^2+b^2+c^2}}\right)=\frac{ax_1+by_1+cz_1+d}{\pm\sqrt{a^2+b^2+c^2}}$$

according as d is +ve or −ve.

Corollary. If d is +ve, the expression $ax_1+by_1+cz_1+d$ will be +ve when (x_1, y_1, z_1) and the origin are on the same side of the plane $ax+by+cz+d=0$ and −ve when they are on opposite sides of the plane.

Note. Perpendicular distance of a point (x_1, y_1, z_1) from the plane $x\cos\alpha+y\cos\beta+z\cos\gamma=p$ is $p-x_1\cos\alpha-y_1\cos\beta-z_1\cos\gamma$.

► **Example 7** *Find the distances of* $(2,3,4)$ *from the planes* $3x+4y+5z+7=0$ *and* $2x-3y+5z-8=0$.

The distance of $(2,3,4)$ from $3x+4y+5z+7=0$ is

$$\frac{3.2+4.3+5.4+7}{\sqrt{3^2+4^2+5^2}}=\frac{45}{\sqrt{50}}=\frac{9}{\sqrt{2}}.$$

The distance of $(2,3,4)$ from $2x-3y+5z-8=0$ is.

$$\frac{2.2-3.3+5.4-8}{\sqrt{2^2+3^2+5^2}}=\frac{7}{\sqrt{38}}.$$

3.8 Angle between Two Planes

Definition 1 *Angle between two planes is defined as the angle between the normals to the planes drawn through a point.*

Let

$$a_1x+b_1y+c_1z+d_1=0 \tag{3.8.1}$$

$$\text{and}\quad a_2x+b_2y+c_2z+d_2=0 \tag{3.8.2}$$

be the equations of two planes. The d. rs. of the normals to the planes are a_1, b_1, c_1 and a_2, b_2, c_2.

If θ be the angle between them, then

$$\cos\theta=\frac{a_1a_2+b_1b_2+c_1c_2}{\sqrt{a_1^2+b_1^2+c_1^2}\cdot\sqrt{a_2^2+b_2^2+c_2^2}}.$$

Corollary 1. Condition for perpendicular planes $a_1a_2+b_1b_2+c_1c_2=0$.

Corollary 2. Conditions for parallel planes $\frac{a_1}{a_2}=\frac{b_1}{b_2}=\frac{c_1}{c_2}$.
Hence $ax+by+cz+d=0$ is parallel to $ax+by+cz+d'=0$.

▶ **Example 8** *Find the angle between the planes*

$$2x-y+3z+7=0$$

$$\text{and}\quad x-2y-3z+8=0.$$

Angle between the planes

$$=\cos^{-1}\frac{2.1+(-1)\cdot(-2)+3\cdot(-3)}{\sqrt{2^2+1^2+3^2}\cdot\sqrt{1^2+2^2+3^2}}=\cos^{-1}\frac{-5}{14}.$$

3.9 Planes Bisecting the Angle between Two Planes

Let the equations of the planes be

$$a_1x+b_1y+c_1z+d_1=0 \tag{3.9.1}$$

$$\text{and}\quad a_2x+b_2y+c_2z+d_2=0. \tag{3.9.2}$$

If $P(x', y', z')$ be any point on either of the two bisecting planes, then P is equidistant from the two planes.

$$\therefore \quad \frac{a_1x' + b_1y' + c_1z' + d_1}{\sqrt{a_1^2 + b_1^2 + c_1^2}} = \pm \frac{a_2x' + b_2y' + c_2z' + d_2}{\sqrt{a_2^2 + b_2^2 + c_2^2}}.$$

Hence the equations of the planes are

$$\frac{a_1x + b_1y + c_1z + d_1}{\sqrt{a_1^2 + b_1^2 + c_1^2}} = \pm \frac{a_2x + b_2y + c_2z + d_2}{\sqrt{a_2^2 + b_2^2 + c_2}}.$$

Note. To distinguish the two bisecting planes

First write the equations of the planes with the same sign of the constants. Now calculate the angle between the planes. If $\cos\theta$ is $-$ve, i.e., $a_1a_2 + b_1b_2 + c_1c_2 = -$ve, then the origin is within the acute angle between the planes. If $a_1a_2 + b_1b_2 + c_1c_2 =$ +ve, then the origin is within the obtuse angle between the planes.

Writing the equations with the constants d_1 and d_2 as +ve numbers the equation

$$\frac{a_1x + b_1y + c_1z + d_1}{\sqrt{a_1^2 + b_1^2 + c_1^2}} = \frac{a_2x + b_2y + c_2z + d_2}{\sqrt{a_2^2 + b_2^2 + c_2^2}}$$

represents the plane bisecting that angle between the planes which contains the origin. If the origin lies in the acute angle, then the above plane bisects the acute angle, otherwise it will bisect the obtuse angle.

▶ **Example 9** *Find the equation of the plane which bisects the acute angle between the planes $x + 2y + 2z = 9$ and $4x - 3y + 12z + 13 = 0$.* **[C.U. '89]**

First we write the equations with the positive sign of the constants in the following way.

$$4x - 3y + 12z + 13 = 0 \tag{1}$$

$$-x - 2y - 2z + 9 = 0. \tag{2}$$

Here $4\cdot(-1) - 3\cdot(-2) + 12\cdot(-2) = 4 + 6 - 24 = -$ve.

Therefore, the origin is within the acute angle between the planes. Hence the equation of the plane bisecting the acute angle between the planes is

$$\frac{4x - 3y + 12z + 13}{\sqrt{4^2 + 3^2 + 12^2}} = \frac{-x - 2y - 2z + 9}{\sqrt{1^2 + 2^2 + 2^2}},$$

$$\text{or} \quad \frac{4x - 3y + 12z + 13}{13} = \frac{-x - 2y - 2z + 9}{3},$$

$$\text{or} \quad 25x + 17y + 62z - 78 = 0.$$

3.10 Any Linear Combination of Two Linear Equations represents a Plane passing through the Common Line of the Planes

Let

$$a_1x + b_1y + c_1z + d_1 = 0 \quad (3.10.1)$$

$$\text{and} \quad a_2x + b_2x + c_2z + d_2 = 0 \quad (3.10.2)$$

be equations of two planes.

The eqaution

$$a_1x + b_1y + c_1z + d_1 + \lambda(a_2x + b_2y + c_2z + d_2) = 0 \quad (3.10.3)$$

represents a plane, since it is of first degree. Here λ is a variable parameter. Again any point satisfying the equations (3.10.1) and (3.10.2) must satisfy the equation (3.10.3). Therefore, the equation represents a plan passing through the line of intersection of the planes (3.10.1) and (3.10.2).

► **Example 10** *Find the equation of the plane passing through the point* $(-1, 0, 1)$ *and the line of intersection of the planes* $4x - 3y + 1 = 0$ *and* $y - 4z + 13 = 0$. **[C.U. '64]**

The equation of a plane passing through the line of intersection of the two planes can be written as

$$4x - 3y + 1 + \lambda(y - 4z + 13) = 0,$$

where λ is a constant.

As it passes through $(-1, 0, 1), -4 + 1 + \lambda(-4 + 13) = 0$ or $\lambda = \frac{1}{3}$.

$\therefore$ the required equation is

$$4x - 3y + 1 + \frac{1}{3}(y - 4z + 13) = 0,$$

$$\text{or} \quad 12x - 8y - 4z + 16 = 0,$$

$$\text{or} \quad 3x - 2y - z + 4 = 0.$$

Worked Out Examples

1. Find the equation to the plane through the point $(1, 2, -3)$ and normal to the straight line joining the points $(-1, 3, 4)$ and $(5, 2, -1)$. **[C.U. '66]**

   Let the equation of the plane through the point $(1, 2, -3)$ be $a(x-1) + b(y-2) + c(z+3) = 0$.

   The d. rs. of the line joining the given points are $6, -1, -5$.

   Since this line is normal to the plane $\frac{a}{6} = \frac{b}{-1} = \frac{c}{-5}$.

   $\therefore$ the equation of the plane is $6(x-1) - (y-2) - 5(z+3) = 0$, i.e., $6x - y - 5z - 19 = 0$.

2. Find the equation of the plane which is perpendicular to the plane $x + 2y - z + 1 = 0$ and which contains the line of intersection of the planes $x + 2y + 3z - 4 = 0$ and $2x + y + z + 2 = 0$. **[B.U. '66]**

   The equation of a plane through the line of intersection of the given planes can be written as

$$x + 2y + 3z - 4 + \lambda(2x + y + z + 2) = 0,$$

$$\text{or} \quad (1 + 2\lambda)x + (2 + \lambda)y + (3 + \lambda)z + 2\lambda - 4 = 0,$$

   where λ is an arbitrary constant.

   Since it is perpendicular to $x + 2y - z + 1 = 0$,

$$(1 + 2\lambda)\cdot 1 + (2 + \lambda)\cdot 2 + (3 + \lambda)\cdot(-1) = 0, \quad \text{or} \quad \lambda = -\frac{2}{3}.$$

   $\therefore$ the required equation is $x - 4y - 7z + 16 = 0$.

3. Find the equation of the plane passing through the point $(2, 5, -8)$ and perpendicular to each of the planes $2x - 3y + 4z + 1 = 0$ and $4x + y - 2z + 6 = 0$.

   [C.U: '92; N.B. '85]

   Let the equation of the plane through $(2, 5, -8)$ be

$$a(x-2) + b(y-5) + c(z+8) = 0.$$

   Since it is perpendicular to each of the given planes,

$$2a - 3b + 4c = 0 \quad \text{and} \quad 4a + b - 2c = 0.$$

   By cross-multiplication

$$\frac{a}{2} = \frac{b}{20} = \frac{c}{14}, \quad \text{i.e.,} \quad \frac{a}{1} = \frac{b}{10} = \frac{c}{7}.$$

   $\therefore$ the equation of the plane is

$$1\cdot(x-2) + 10(y-5) + 7\cdot(z+8) = 0, \quad \text{i.e.,} \quad x + 10y + 7z + 4 = 0.$$

4. **Find the equation of the plane which bisects the line joining the points $(1, 2, 3)$ and $(3, 4, 5)$ at right angles and show that it makes equal intercepts on the axes.**

The d. rs. of the line are $2, 2, 2$.

The equation of a plane through $(2, 3, 4)$ can be written as

$$a(x-2)+b(y-3)+c(z-4)=0.$$

If it is perpendicular to the line of d. rs. $2, 2, 2$ then $\frac{a}{2}=\frac{b}{2}=\frac{c}{2}$.

$\therefore$ the equation of the plane is

$$2(x-2)+2(y-3)+2(z-4)=0, \quad \text{i.e.,} \quad x+y+z=9.$$

Obviously the plane makes equal intercepts on the axes and each of them is equal to 9.

5. **Show that the points $(2, 3, -5)$ and $(3, 4, 7)$ lie on the opposite sides of the plane $x+2y-2z=9$.**

Perpendicular distances of the plane from the points $(2, 3, -5)$ and $(3, 4, 7)$ are

$$\frac{2+2.3+2.5-9}{\sqrt{1+4+4}}=\frac{9}{3}=3 \quad \text{and} \quad \frac{3+2.4-2.7-9}{\sqrt{1+4+4}}=\frac{-12}{3}=-4.$$

Since these are of opposite signs the points lie on the opposite sides of the plane.

6. **The plane $x-2y+3z-4=0$ is rotated through a right angle about its line of intersection with the plane $2x+3y-4z-5=0$. Find the equation of the plane in its new position.**

The plane

$$x-2y+3z-4=0 \tag{1}$$

when rotated through a right-angle about the line of intersection with

$$2x+3y-4z-5=0 \tag{2}$$

is the plane which is perpendicular to the plane through the line of intersection of (1) and (2).

The equation of the plane through the line of intersection of (1) and (2) is

$$x-2y+3z-4+\lambda(2x+3y-4z-5)=0, \tag{3}$$

$$\text{i.e.,} \quad (1+2\lambda)x+(3\lambda-2)y+(3-4\lambda)z-(4+5\lambda)=0. \tag{4}$$

Since it is perpendicular to (1)

$$(1+2\lambda)-2(3\lambda-2)+3(3-4\lambda)=0, \quad \text{i.e.,} \quad \lambda=\frac{7}{8}.$$

Therefore the required equation is

$$\left(1+\frac{7}{4}\right)x+\left(\frac{21}{8}-2\right)y+\left(3-\frac{7}{2}\right)z-\left(4+\frac{35}{8}\right)=0,$$

$$\text{or} \quad 22x+5y-4z-67=0.$$

7. The sum of the squares of the distances of a point from the planes $x+y+z=0$ and $x-2y+z=0$ is equal to the square of its distance from the plane $x-z=0$. Find the locus of the point.

The planes are

$$x+y+z=0, \tag{1}$$

$$x-2y+z=0, \tag{2}$$

$$\text{and} \quad x-z=0. \tag{3}$$

Let the point be (x_1, y_1, z_1).

The distances of the planes from this point are

$$\frac{x_1+y_1+z_1}{\sqrt{1+1+1}}, \quad \frac{x_1-2y_1+z_1}{\sqrt{1+4+1}}, \quad \frac{x_1-z_1}{\sqrt{1+1}}.$$

By the given condition, we have

$$\frac{(x_1+y_1+z_1)^2}{3}+\frac{(x_1-2y_1+z_1)^2}{6}=\frac{(x_1-z_1)^2}{2}, \quad \text{or} \quad y_1^2+2z_1x_1=0.$$

Hence the required locus is $y^2+2zx=0$.

8. Find the equations of the planes parallel to the plane $4x-3y+12z-39=0$ and at a distance 3 from it.

Let the plane be $4x-3y+12z+c=0$, where c is a constant.

Now distances of the above plane and the given plane from the origin are

$$\frac{c}{\sqrt{4^2+3^2+12^2}}, \quad \text{i.e.,} \quad \frac{c}{13} \quad \text{and} \quad \frac{-39}{\sqrt{4^2+3^2+12^2}}=\frac{-39}{13}=-3.$$

By the given condition, $c-(-3)=\pm 3 \quad \therefore \quad c=0$ or -6.

Hence the equations of the planes are $4x-3y+12z=0$ and $4x-3y+12z-6=0$.

9. A variable plane passes through a fixed point (α, β, γ) and meets the axes of reference in A, B and C. Show that the locus of the point of intersection of the planes through A, B and C parallel to the co-ordinate planes is $\frac{\alpha}{x}+\frac{\beta}{y}+\frac{\gamma}{z}=1$.

Let the the variable plane be $\frac{x}{a}+\frac{y}{b}+\frac{z}{c}=1$.

As it passes through

$$(\alpha, \beta, \gamma), \quad \frac{\alpha}{a}+\frac{\beta}{b}+\frac{\gamma}{c}=1. \tag{1}$$

The plane meets the axes at $A(a,0,0)$, $B(0,b,0)$ and $C(0,0,c)$. The plane through A, B and C parallel to the co-ordinate planes are $x=a, y=b$ and $z=c$. These three planes meet at $(a,b,c)\cdot(a,b,c)$ satisfies the relation (1). Hence the required locus is

$$\frac{\alpha}{x}+\frac{\beta}{y}+\frac{\gamma}{z}=1.$$

10. A variable plane at a constant distance p from the origin meets the axes at A, B, C. Show that the locus of the centroid of the tetrahedron $OABC$ is

$$\frac{1}{x^2}+\frac{1}{y^2}+\frac{1}{z^2}=\frac{16}{p^2}.$$

[C.U. '90]

Let the plane be $\frac{x}{a}+\frac{y}{b}+\frac{z}{c}=1$.

Since its distance from the origin is p,

$$p=\frac{1}{\sqrt{\left(\frac{1}{a^2}+\frac{1}{b^2}+\frac{1}{c^2}\right)}}, \quad \text{or} \quad \frac{1}{a^2}+\frac{1}{b^2}+\frac{1}{c^2}=\frac{1}{p^2}.$$

The plane meets the axes $A(a,0,0), B(0,b,0)$ and $C(0,0,c)$. If (α,β,γ) be the centroid of the tetrahedron $OABC$, (O the origin), then

$$\alpha=\frac{a}{4}, \quad \beta=\frac{b}{4}, \quad \gamma=\frac{c}{4}.$$

From (1)

$$\frac{1}{16\alpha^2}+\frac{1}{16\beta^2}+\frac{1}{16\gamma^2}=\frac{1}{p^2}.$$

Hence the locus of (α,β,γ) is $\frac{1}{x^2}+\frac{1}{y^2}+\frac{1}{z^2}=\frac{16}{p^2}$.

Exercise

1. Find the intercepts made by (i) $x-y+z=2$, (ii) $2x+3y+6z-6=0$ on the axes.
2. Calculate the d. cs. and the length of the directed normal from the origin to the plane: (i) $2x-3y+6z=7$, (ii) $4x+3y+2z+12=0$.
3. Express the following in the normal form:
   (a) $2x-3y+4z-5=0$, (b) $x+5=0$.
4. Find the equation of the plane
   (a) passing through $(3,5,1),(2,3,0)$ and $(0,6,0)$;
   (b) through $(4,1,1)$ and parallel to $3x-4y+7z+5=0$;
   (c) through $(2,0,-1)$ and perpendicular to the line whose d. rs. are $3,4,-2$;
   (d) passing through the point $(1,2,3)$ and parallel to the zx plane;
   (e) which contains the line of intersection of the planes $x+2y+3z-4=0$ and $2x+y-z+5=0$ and which is perpendicular to the plane $5x+3y+6z+8=0$; **[C.U. 2000; N.B. '92; B.U. '87, '99]**
   (f) through the line $x+y-2z+4=0=3x-y+2z+1$ and parallel to the line with d. rs. $2,3,-1$; **[B.U. '67; N.B. '70]**
   (g) parallel to the plane containing OZ and the point $(2,2,1)$ and passing through $(1,-1,1)$;
   (h) passing through the points $(2,3,-4)$, $(1,-1,3)$ and perpendicular to YOZ plane;
   (i) passing through $(2,1,4)$ and perpendicular to each of the planes $x+y+2z-4=0$ and $2x-3y+z+5=0$; **[C.U. '85]**
   (j) through the points $(2,,2,1)$ and $(9,3,6)$ and perpendicular to the plane $2x+6y+6z=9$; **[N.B. '84]**
   (k) bisecting the line joining the points $(3,4,-1)$ and $(3,-1,5)$ normally; **[C.U. '89]**
   (l) through the points $(4,3,1)$ and $(1,-3,4)$ and parallel to the y-axis.

5. (a) The x and y intercepts of a plane are 3 and -5 respectively and the plane passes through $(2, 1, 8)$. Find its equation.

   (b) A plane passes through $(5, 1, 5)$ and $(-1, -2, 4)$ and has its y intercept equal to -4. Find its equation.

6. Show that the following points are coplanar.

   (a) $(0, -1, 0), (2, 1, -1), (1, 1, 1), (3, 3, 0)$; **[B.U. '98]**

   (b) $(1, 1, -1), (-1, -5, -3), (0, 4, 3), (-2, -2, 1)$.

   Obtain the equation of the plane.

7. Find the angle between the planes

   (a) $2x + y + z = 6, x - y + 2z = 3$;

   (b) $x + 2y + 3z - 4 = 0, 2x + 5y - 4z = 3$.

8. (a) Find the distances of the points $(2, 0, 1)$ and $(3, -3, 2)$ from the plane $x - 2y + z - 6 = 0$. Do the points lie on the same side or opposite sides of the plane?

   (b) Find a point on the y-axis such that it is equidistant from $6x + 12y - 4z + 2 = 0$ and $x - 2y + 2z + 5 = 0$.

   (c) Find the distance between the parallel planes $x - 4y + 8z - 9 = 0$ and $x - 4y + 8z + 18 = 0$.

   [HINTS. Distances of the planes from the origin are $\frac{-9}{9} = -1$ and $\frac{18}{9} = 2$. $\therefore$ the required distance $= 2 - (-1) = 3$.]

9. Find the locus of a point the sum of squares of whose distance from the planes $x + y + z = 0, x - z = 0, x - 2y + z = 0$ is 9. **[B.U. '09]**

10. The foot of the perpendicular drawn from the origin to a plane is $(12, -4, -3)$, find the equation of the plane.

11. Find the equations of the two planes through the points $(0, 4, -3)$ and $(6, -4, 3)$ other than the plane through the origin which cut off from the axes intercepts whose sum is zero.

12. (a) Find the equation of the planes bisecting the angle between the planes $3x - 4y + 12z = 26, x + 2y - 2z = 9$ and distinguish them.

    (b) Find the equation of the plane which bisects the obtuse angle between the planes $4x - 3y + 12z + 13 = 0$ and $x + 2y + 2z = 9$.

13. Show that the equation of the plane through the point (x_1, y_1, z_1) and perpendicular to both the planes $a_1x + b_1y + c_1z + d_1 = 0$ and $a_2x + b_2y + c_2z + d_2 = 0$ can be expressed as

$$\begin{vmatrix} x - x_1 & y - y_1 & z - z_1 \\ a_1 & b_1 & c_1 \\ a_2 & b_2 & c_2 \end{vmatrix} = 0.$$

14. Find the equations of the planes parallel to the plane $16x + 12y - 15z + 75 = 0$ and at distance 4 from it.

15. The plane $2x - 3y - 4z + 3 = 0$ is rotated through a right-angle about its line of intersection with the plane $x + 5y - 6z - 3 = 0$. Find the equation of the plane in its new position.

16. Find a point on the y-axis which is equidistant from the planes $6x - 12y + 4z - 5 = 0$ and $x + 2y + 2z - 3 = 0$.

[HINTS. Let $(0, b, 0)$ be the point.

Then

$$\frac{-12b-5}{\sqrt{6^2+12^2+4^2}} = \pm\frac{2b-3}{\sqrt{1^2+2^2+2^2}}, \quad \text{or} \quad \frac{-12b-5}{14} = \pm\frac{2b-3}{3},$$

$$\text{or} \quad -36b - 15 = \pm(28b - 42), \quad \text{or} \quad b = \frac{27}{64}, \ -\frac{57}{8}.]$$

17. A plane meets the co-ordinate axes at A, B, C such that the centroid of the triangle ABC is the point (α, β, γ); show that the equation of the plane is $\frac{x}{\alpha} + \frac{y}{\beta} + \frac{z}{\gamma} = 3$.

[B.U. '99; C.U. '99]

[HINTS. Let the equation of ABC be $\frac{x}{a} + \frac{y}{b} + \frac{z}{c} = 1$. Then the co-ordinates of A, B, C are $(a, 0, 0)$, $(0, b, 0)$ and $(0, 0, c)$.

If (α, β, γ) is the centroid of ΔABC then $\alpha = \frac{a}{3}, \beta = \frac{b}{3}, \gamma = \frac{c}{3}$.

$\therefore$ the equation of the plane is $\frac{x}{3\alpha} + \frac{y}{3\beta} + \frac{z}{3\gamma} = 1$, or, $\frac{x}{\alpha} + \frac{y}{\beta} + \frac{z}{\gamma} = 3$.]

18. A variable plane is at a constant distance $3p$ from the origin and meets the axes in A, B, C, show that the locus of the centroid of ΔABC is $\frac{1}{x^2} + \frac{1}{y^2} + \frac{1}{z^2} = \frac{1}{p^2}$. [B.U. '97]

[HINTS. Let the plane be $\frac{x}{a} + \frac{y}{b} + \frac{z}{c} = 1$. Then the co-ordinates of A, B, C are $(a, 0, 0)$, $(0, b, 0)$ and $(0, 0, c)$. If (α, β, γ) is the centroid of ΔABC then $\alpha = \frac{a}{3}, \beta = \frac{b}{3}, \gamma = \frac{c}{3}$.

Here

$$\frac{1}{\sqrt{\frac{1}{a^2}+\frac{1}{b^2}+\frac{1}{c^2}}} = 3p, \quad \text{or} \quad \frac{1}{a^2}+\frac{1}{b^2}+\frac{1}{c^2} = \frac{1}{9p^2}.$$

$$\therefore \quad \frac{1}{9\alpha^2}+\frac{1}{9\beta^2}+\frac{1}{9\gamma^2} = \frac{1}{9p^2}, \quad \text{or} \quad \frac{1}{\alpha^2}+\frac{1}{\beta^2}+\frac{1}{\gamma^2} = \frac{1}{p^2}.$$

Hence the locus of (α, β, γ) is $\frac{1}{x^2} + \frac{1}{y^2} + \frac{1}{z^2} = \frac{1}{p^2}$.]

19. The plane $ax + by + cz + d = 0$ bisects an angle between a pair of planes of which one is $lx + my + nz + p = 0$. Show that the equation of the other plane of the pair is

$$(lx + my + nz + p)(a^2 + b^2 + c^2) = 2(al + bm + cn)(ax + by + cz + d).$$

[HINTS. Let the other plane be

$$ax + by + cz + d + \lambda(lx + my + nz + p) = 0. \tag{1}$$

$\left(-\frac{d}{a}, 0, 0\right)$ is a point on $ax + by + cz + d = 0$. Distances of the planes $lx + my + nz + p = 0$ and (1) are equal

$$\therefore \quad \frac{\left|p - \frac{ld}{a}\right|}{\sqrt{l^2+m^2+n^2}} = \frac{|\lambda|\left|p - \frac{ld}{a}\right|}{\sqrt{\Sigma(a+l\lambda)^2}}, \quad \text{or} \quad \lambda = -\frac{a^2+b^2+c^2}{2(al+bm+cn)}.$$

Hence the result follows.]

20. From the point $P(\alpha, \beta, \gamma)$ perpendiculars PA, PB, PC are drawn on the co-ordinate planes. Show that the equation of the plane ABC is $\frac{x}{\alpha} + \frac{y}{\beta} + \frac{z}{\gamma} = 2$. [C.U. '89]

[HINTS. The co-ordinates of A, B, C are $(\alpha, \beta, 0)$ $(0, \beta, \gamma)$ and $(\alpha, 0, \gamma)$.

Let the plane through A, B, C be $ax + by + cz + d = 0$. Then

$$a\alpha + b\beta + d = 0, \tag{1}$$

$$b\beta + c\gamma + d = 0, \tag{2}$$

$$\text{and} \quad a\alpha + c\gamma + d = 0. \tag{3}$$

From (1) and (2) $a\alpha = c\gamma$. From (3) $2c\gamma + d = 0$, or $c = -\frac{d}{2\gamma}$.

Similarly $a = -\frac{d}{2\alpha}$ and $b = -\frac{d}{2\beta}$.

Thus the equation of the plane is

$$-\frac{d}{2\alpha}x - \frac{d}{2\beta}y - \frac{d}{2\gamma}z + d = 0, \quad \text{or} \quad \frac{x}{\alpha} + \frac{y}{\beta} + \frac{z}{\gamma} = 2.]$$

21. Find the common point of the planes

$$2x - y + z = 4, \quad 5x + 7y + 2z = 0, \quad 3x + 4y - 2z + 3 = 0.$$

22. A variable plane at a constant distance p from the origin and meets the axes in A, B, C. Through A, B, C planes are drawn parallel to co-ordinate planes. Show that the locus of their point of intersection is $\frac{1}{x^2} + \frac{1}{y^2} + \frac{1}{z^2} = \frac{1}{p^2}$. [C.U. '98]

Answers

1. (a) 2, −2, 2; (b) 3, 2, 1; **2.** (a) $\frac{2}{7}, \frac{-3}{7}, \frac{6}{7}$; 1; (b) $\frac{-4}{\sqrt{29}}, \frac{-3}{\sqrt{29}}, \frac{-2}{\sqrt{29}}$; $\frac{12}{\sqrt{29}}$; **3.** (a) $\frac{2}{\sqrt{29}}x - \frac{3}{\sqrt{29}}y + \frac{4}{\sqrt{29}}z = \frac{5}{\sqrt{29}}$; (b) $-x = 5$; **4.** (a) $3x+2y-7z-12 = 0$; (b) $3x-3y+7z-15 = 0$; (c) $3x+4y-2z-8 = 0$; (d) $y = 2$; (e) $51x + 15y - 50z + 173 = 0$; (f) $20x - 8y + 16z + 3 = 0$; (g) $x - y - 2 = 0$; (h) $7y + 4z - 5 = 0$; (i) $7x + 3y - 5z + 3 = 0$; (j) $3x + 4y - 5z - 9 = 0$; (k) $10y - 12z + 9 = 0$; (l) $x + z - 5 = 0$. **5.** (a) $\frac{x}{3} - \frac{y}{5} + \frac{z}{15} = 1$; (b) $\frac{x}{10} - \frac{y}{4} + \frac{3z}{20} = 1$; **6.** (a) $4x - 3y + 2z - 3 = 0$; (b) $9x - 5y + 6z + 2 = 0$; **7.** (a) $\frac{\pi}{3}$; (b) $\frac{\pi}{2}$; **8.** (a) $\frac{3}{\sqrt{6}}, \frac{5}{\sqrt{6}}$, opposite sides; (b) $(0, 1, 0)$, $\left(0, -\frac{19}{2}, 0\right)$; **9.** $x^2 + y^2 + z^2 = 9$; **10.** $12x - 4y - 3z = 169$; **11.** $6x + 3y - 2z = 18$, $2x - 3y - 6z = 6$; **12.** (a) $4x + 38y - 62z - 39 = 0$ (acute angle), $22x + 14y + 10z - 195 = 0$; (b) $x + 35y - 10z - 156 = 0$; **14.** $16x + 12y - 15z + 175 = 0$, $16x + 12y - 15z - 25 = 0$; **15.** $7x + 178y - 130z - 120 = 0$; **21.** $(1, -1, 1)$.

Chapter 4

Straight Line

4.1 To find the Equations of a Line passing through a given point with the given Direction Cosines

Let (x_1, y_1, z_1) be the given point and l, m, n be proportional to d. cs.. of the line. If (x, y, z) be any point on the line, then d. cs. are proportional to $x - x_1, y - y_1, z - z_1$.

Hence the equations of the line are

$$\frac{x - x_1}{l} = \frac{y - y_1}{m} = \frac{z - z_1}{n}.$$

This is known as the equations of a line in ***symmetrical form*** or the ***canonical equations*** of a line.

Corollary 1. Equation of a line through the origin.

If the line passes through the origin, the equations of the line

$$\frac{x}{l} = \frac{y}{m} = \frac{z}{n}.$$

Corollary 2. Equations of the line through two given points.

If the line passes through two given points (x_1, y_1, z_1) and (x_2, y_2, z_2) then

$$\frac{x_2 - x_1}{l} = \frac{y_2 - y_1}{m} = \frac{z_2 - z_1}{n}.$$

Therefore the equations of the line are

$$\frac{x - x_1}{x_2 - x_1} = \frac{y - y_1}{y_2 - y_1} = \frac{z - z_1}{z_2 - z_1}.$$

Corollary 3. Any point on the line.

Let

$$\frac{x - x_1}{l} = \frac{y - y_1}{m} = \frac{z - z_1}{n} = r.$$

Then $x = lr + x_1, y = mr + y_1, z = nr + z_1$. These are called the *parametric equations* of the line.

Thus any point on the line is $(lr + x_1, mr + y_1, nr + z_1)$.

Note. If (l, m, n) are the actual direction cosines of the line, then the point $(lr + x_1, mr + y_1, nr + z_1)$ is at a distance r from the point (x_1, y_1, z_1) on the line.

4.2 To find the Equation of the Line of Intersection of Two Planes in Symmetrical Form

Let the planes be

$$a_1x + b_1y + c_1z + d_1 = 0 \tag{4.2.1}$$

$$\text{and} \quad a_2x + b_2y + c_2z + d_2 = 0. \tag{4.2.2}$$

The equations (4.2.1) and (4.2.2) together represent the straight line of intersection of the planes, since they are satisfied by any point on the line of intersection of the planes. It is the *unsymmetrical form* of a line.

Since the line lies on both the planes, it must be perpendicular to the normals to the planes. If l, m, n be the d. cs. of the line, then

$$la_1 + mb_1 + nc_1 = 0 \quad \text{and} \quad la_2 + mb_2 + nc_2 = 0.$$

From these two relations,

$$\frac{l}{b_1c_2 - b_2c_1} = \frac{m}{c_1a_2 - c_2a_1} = \frac{n}{a_1b_2 - a_2b_1}.$$

If the line meets the plane $z = 0$ at $(x', y', 0)$, then

$$a_1x' + b_1y' + d_1 = 0 \quad \text{and} \quad a_2x' + b_2y' + d_2 = 0.$$

From these two

$$\frac{x'}{b_1d_2 - b_2d_1} = \frac{y'}{d_1a_2 - d_2a_1} = \frac{1}{a_1b_2 - a_2b_1}.$$

$\therefore$ the d. rs. of the line are

$$b_1c_2 - b_2c_1, c_1a_2 - c_2a_1, a_1b_2 - a_2b_1$$

and the line passes through

$$\left(\frac{b_1d_2 - b_2d_1}{a_1b_2 - a_2b_1}, \frac{d_1a_2 - d_2a_1}{a_1b_2 - a_2b_1}, 0\right).$$

Hence the equation of the common line in symmetrical form is

$$\frac{x - \dfrac{b_1d_2 - b_2d_1}{a_1b_2 - a_2b_1}}{b_1c_2 - b_2c_1} = \frac{y - \dfrac{d_1a_2 - d_2a_1}{a_1b_2 - a_2b_1}}{c_1a_2 - c_2a_1} = \frac{z}{a_1b_2 - a_2b_1}.$$

► **Example 1** *Obtain in the symmetrical form, equations of the line*

$$x - 2y + 3z - 4 = 0 = 2x - 3y + 4z - 5.$$

$$x - 2y + 3z - 4 = 0, \qquad (1)$$

$$2x - 3y + 4z - 5 = 0. \qquad (2)$$

The d. rs. of the normal to the planes (1) and (2) are $1, -2, 3$ and $2, -3, 4$ respectively.

Let l, m, n be the d. rs. of the line. Since it lies on both the planes, it is perpendicular to the normals to both of them.

$$\therefore \quad l - 2m + 3n = 0$$

$$\text{and} \quad 2l - 3m + 4n = 0.$$

By cross-multiplication $\frac{l}{1} = \frac{m}{2} = \frac{n}{1}$.

If the line meets the plane $z = 0$ at $(x', y', 0)$, then

$$x' - 2y' - 4 = 0 \quad \text{and} \quad 2x' - 3y' - 5 = 0.$$

From these, $x' = -2$, $y' = -3$.

Now the d. rs. of the line are 1, 2, 1 and it passes through the point $(-2, -3, 0)$.

Hence the equations of the line in symmetrical form are

$$\frac{x+2}{1} = \frac{y+3}{2} = \frac{z}{1}.$$

► **Example 2** *Show that the line $2x + 2y - z - 6 = 0 = 2x + 3y - z - 8$ is parallel to a co-ordinate plane and find the equation of the plane normal to this line and passing through the point, where this line meets the plane $x = 0$.* **[C.U. '68]**

Let l, m, n be the d. cs. of the line. Then $2l + 2m - n = 0$ and $2l + 3m - n = 0$.

By cross-multiplication $\frac{l}{1} = \frac{m}{0} = \frac{n}{2}$.

Since the y-d.c. is zero, the line is perpendicular to y-axis, or it is parallel to ZOX plane.

Let the line meet the plane $x = 0$ at $(0, y', z')$.

$\therefore \quad 2y' - z' - 6 = 0, 3y' - z' - 8 = 0.$

From these two, $y' = 2, z' = -2$.

Now we find out the equation of the plane passing through $(0, 2, -2)$ and the d. rs. of the normal are 1, 0, 2. The equation of the plane is

$$x + 2(z + 2) = 0, \quad \text{or} \quad x + 2z + 4 = 0.$$

4.3 The Plane and a Straight Line

Let the equations of the plane and the line be

$$ax + by + cz + d = 0 \tag{4.3.1}$$

$$\text{and} \quad \frac{x - x_1}{l} = \frac{y - y_1}{m} = \frac{z - z_1}{n}. \tag{4.3.2}$$

Any point on the line is $(lr + x_1, mr + y_1, nr + z_1)$.

If the line meets the plane at this point, then

$$a(lr + x_1) + b(mr + y_1) + c(nr + z_1) + d = 0,$$

$$\text{or} \quad (al + bm + cn)r + ax_1 + by_1 + cz_1 + d = 0. \tag{4.3.3}$$

r is proportional to the distance of the point from (x_1, y_1, z_1).

Thus the line will be parallel to the plane, if

$$al + bm + cn = 0.$$

If the line lies on the plane, then

$$al + bm + cn = 0 \quad \text{and} \quad ax_1 + by_1 + cz_1 + d = 0.$$

If the line is perpendicular to the plane, then it is parallel to the normal to the plane. In this case, $\frac{l}{a} = \frac{m}{b} = \frac{n}{c}$.

If θ is the angle between the plane and the line then $90° - \theta$ is the angle between the line and a normal to the plane. Thus

$$\cos(90° - \theta) = \frac{al + bm + cn}{\sqrt{l^2 + m^2 + n^2}\sqrt{a^2 + b^2 + c^2}},$$

$$\text{i.e.,} \quad \theta = \sin^{-1}\left[\frac{al + bm + cn}{\sqrt{l^2 + m^2 + n^2}\sqrt{a^2 + b^2 + c^2}}\right].$$

► **Example 3** *Prove that the distance of the point of intersection of the line* $\frac{x-2}{3} = \frac{y+1}{4} = \frac{z-2}{12}$ *and the plane* $x - y + z = 5$ *from the point* $(-1, -5, -10)$ *is* 13.

Any point on the line is $(3r + 2, 4r - 1, 12r + 2)$.

If the line meets the plane at this point, then

$$3r + 2 - (4r - 1) + 12r + 2 = 5, \quad \text{or} \quad 11r = 0, r = 0.$$

$\therefore$ the point of intersection of the line and the plane is $(2, -1, 2)$. Now the distance between $(2, -1, 2)$ and $(-1, -5, -10)$

$$= \sqrt{\{(2+1)^2 + (-1+5)^2 + (2+10)^2\}} = \sqrt{169} = 13.$$

4.4 The necessary and sufficient condition for the Coplanarity of Two non-parallel Straight Lines

$$\frac{x-x_1}{l_1}=\frac{y-y_1}{m_1}=\frac{z-z_1}{n_1} \tag{4.4.1}$$

$$\text{and}\quad \frac{x-x_2}{l_2}=\frac{y-y_2}{m_2}=\frac{z-z_2}{n_2} \tag{4.4.2}$$

is that

$$\begin{vmatrix} x_2-x_1 & y_2-y_1 & z_2-z_1 \\ l_1 & m_1 & n_1 \\ l_2 & m_2 & n_2 \end{vmatrix}=0.$$

(i) *The condition is necessary.*

Let the lines be coplanar and lie on $ax+by+cz+d=0$.

$$\therefore\quad al_1+bm_1+cn_1=0, \tag{4.4.3}$$

$$al_2+bm_2+cn_2=0, \tag{4.4.4}$$

$$ax_1+by_1+cz_1+d=0, \tag{4.4.5}$$

$$\text{and}\quad ax_2+by_2+cz_2+d=0. \tag{4.4.6}$$

From (5) and (6),

$$a(x_2-x_1)+b(y_2-y_1)+c(z_2-z_1)=0, \tag{4.4.7}$$

Eliminating a, b, c from (4.4.7), (4.4.3) and (4.4.4), we get

$$\begin{vmatrix} x_2-x_1 & y_2-y_1 & z_2-z_1 \\ l_1 & m_1 & n_1 \\ l_2 & m_2 & n_2 \end{vmatrix}=0.$$

(ii) *The condition is sufficient.*

If the condition holds then the system of homogeneous equations

$$(x_2-x_1)x+(y_2-y_1)y+(z_2-z_1)z=0, \tag{4.4.8}$$

$$l_1x+m_1y+n_1z=0, \tag{4.4.9}$$

$$\text{and}\quad l_2x+m_2y+n_2x=0. \tag{4.4.10}$$

has non-trivial solutions.

Let (α, β, γ) be a non-zero solution.

Then

$$(x_2-x_1)\alpha+(y_2-y_1)\beta+(z_2-z_1)\gamma=0, \tag{4.4.11}$$

$$l_1\alpha+m_1\beta+n_1\gamma=0, \tag{4.4.12}$$

$$\text{and}\quad l_2\alpha+m_2\beta+n_2\gamma=0. \tag{4.4.13}$$

From (4.4.11) $\alpha x_1 + \beta y_1 + \gamma z_1 = \alpha x_2 + \beta y_2 + \gamma z_2 = k$ (say),

$$\text{i.e.,} \quad \alpha x_1 + \beta y_1 + \gamma z_1 - k = 0, \tag{4.4.14}$$

$$\text{and} \quad \alpha x_2 + \beta y_2 + \gamma z_2 - k = 0. \tag{4.4.15}$$

The conditions (4.4.12) and (4.4.14) suggest that the line (4.4.1) lies on

$$\alpha x + \beta y + \gamma z - k = 0. \tag{4.4.16}$$

By the conditions (4.4.13) and (4.4.15) the line (4.4.2) also lies on the plane (4.4.16). Hence the lines are coplanar.

Corollary. The equation of the plane on which the lines lie is

$$\begin{vmatrix} x - x_1 & y - y_1 & z - z_1 \\ l_1 & m_1 & n_1 \\ l_2 & m_2 & n_2 \end{vmatrix} = 0$$

provided the condition of coplanarity is satisfied.

Note. The condition of coplanarity is also the condition of the intersection of the lines (4.4.1) and (4.4.2).

▶ **Example 4** *Prove that* $\frac{x-1}{2} = \frac{y-2}{3} = \frac{z-3}{4}$ *and* $\frac{x-2}{3} = \frac{y-3}{4} = \frac{z-4}{5}$ *are coplanar. Find also the equation of the plane.*

Since

$$\begin{vmatrix} 2-1 & 3-2 & 4-3 \\ 2 & 3 & 4 \\ 3 & 4 & 5 \end{vmatrix} = \begin{vmatrix} 1 & 1 & 1 \\ 2 & 3 & 4 \\ 3 & 4 & 5 \end{vmatrix} = \begin{vmatrix} 1 & 1 & 1 \\ 3 & 4 & 5 \\ 3 & 4 & 5 \end{vmatrix} = 0,$$

(Adding the elements of the first row with those of the second row, the two rows become identical) the lines are coplanar.

The equation of the plane is

$$\begin{vmatrix} x-1 & y-2 & z-3 \\ 2 & 3 & 4 \\ 3 & 4 & 5 \end{vmatrix} = 0,$$

$$\text{or} \quad -(x-1) + 2(y-2) - (z-3) = 0, \quad \text{or} \quad x - 2y + z = 0.$$

▶ **Example 5** *Show that the lines* $x + 2y - 5z + 9 = 0 = 3x - y + 2z - 5$, $2x + 3y - z - 3 = 0 = 4x - 5y + z + 3$ *are coplanar.*

[B.U. '66]

First we write the line $x + 2y - 5z + 9 = 0 = 3x - y + 2z - 5$ in symmetrical form. Let l, m, n be the d. cs. of the line.

Then $l + 2m - 5n = 0, 3l - m + 2n = 0$.

From these, $\frac{l}{1} = \frac{m}{17} = \frac{n}{7}$.

Let the line meet the plane $z = 0$ at $(\alpha, \beta, 0)$.

Then $\alpha + 2\beta + 9 = 0, 3\alpha - \beta - 5 = 0$.

These two give

$$\frac{\alpha}{-1} = \frac{\beta}{32} = \frac{1}{-7}, \quad \text{or} \quad \alpha = \frac{1}{7}, \quad \beta = -\frac{32}{7}.$$

$\therefore$ the equation of the line is

$$\frac{x - \frac{1}{7}}{1} = \frac{y + \frac{32}{7}}{17} = \frac{z}{7}. \qquad (1)$$

Any plane through the second line is

$$(2 + 4\lambda)x + (3 - 5\lambda)y - (1 - \lambda)z - 3 + 3\lambda = 0,$$

where λ is a parameter.

If (1) be parallel to this plane, then

$$(2 + 4\lambda) \cdot 1 + (3 - 5\lambda) \cdot 17 - (1 - \lambda).7 = 0, \quad \text{or} \quad \lambda = \frac{\mathbf{23}}{\mathbf{37}}.$$

The equation of the plane is $83x - 2y - 7z - 21 = 0$.

Again the point $\left(\frac{1}{7} - \frac{32}{7}, 0\right)$ satisfies this equation.

Therefore, the line (1) lies on this plane. Hence the lines are coplanar.

Point of intersection. Any point on the line (1) is $\left(r + \frac{1}{7}, 17r - \frac{32}{7}, 7r\right)$. As the two lines meet, this point should satisfy the equation of the planes of the second line.

$$\therefore \quad 2\left(r + \frac{1}{7}\right) + 3\left(17r - \frac{32}{7}\right) - 7r - 3 = 0, \quad \text{or} \quad r = \frac{5}{14}$$

$$\text{and} \quad 4\left(r + \frac{1}{7}\right) - 5\left(17r - \frac{32}{7}\right) + 7r + 3 = 0, \quad \text{or} \quad r = \frac{5}{14}.$$

Since these two values of r are equal, the lines intersect at a point and this point is $\left(\frac{1}{2}, \frac{3}{2}, \frac{5}{2}\right)$.

4.5 Distance of a Point from a Line

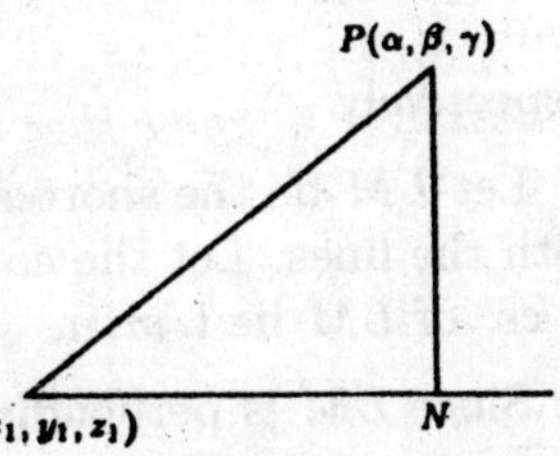

Fig. 4.1

Let $\frac{x - x_1}{l} = \frac{y - y_1}{m} = \frac{z - z_1}{n}$ be the equation of the line AN and $P(\alpha, \beta, \gamma)$ be a given point.

Let PN be perpendicular to the line AN and the co-ordinates of A be (x_1, y_1, z_1).

The d. cs. of AN are

$$\frac{l}{\sqrt{l^2 + m^2 + n^2}}, \frac{m}{\sqrt{l^2 + m^2 + n^2}}, \frac{n}{\sqrt{l^2 + m^2 + n^2}}.$$

The d. cs. of AP are $\frac{\alpha-x_1}{AP}, \frac{\beta-y_1}{AP}, \frac{\gamma-z_1}{AP}$.

If θ be the angle between AN and AP, then

$$PN = AP\sin\theta = AP\sqrt{1-\cos^2\theta}$$

$$= AP\sqrt{\left[1-\frac{\{l(\alpha-x_1)+m(\beta-y_1)+n(\gamma-z_1)\}^2}{(l^2+m^2+n^2)\cdot AP^2}\right]}$$

$$= \sqrt{\left[AP^2-\frac{\{l(\alpha-x_1)+m(\beta-y_1)+n(\gamma-z_1)\}^2}{l^2+m^2+n^2}\right]}$$

$$= \left[(\alpha-x_1)^2+(\beta-y_1)^2+(\gamma-z_1)^2-\frac{\{l(\alpha-x_1)+m(\beta-y_1)+n(\gamma-z_1)\}^2}{l^2+m^2+n^2}\right]^{\frac{1}{2}}.$$

▶ **Example 6** *Find the distance of the point* $(3, 2, 1)$ *from the line* $\frac{x-1}{3}=\frac{y}{4}=\frac{z-2}{1}$.

The required distance

$$= \left[(3-1)^2+2^2+(1-2)^2-\frac{\{3.(3-1)+4.2+1.(1-2)\}^2}{3^2+4^2+1^2}\right]^{\frac{1}{2}}$$

$$= \left[9-\frac{169}{26}\right]^{\frac{1}{2}} = \left[9-\frac{13}{2}\right]^{\frac{1}{2}} = \sqrt{\frac{5}{2}}.$$

4.6 Shortest Distance between Two Skew Lines

Definition 1 *If two non-parallel lines do not intersect, then they are called skew lines. These are non-coplanar lines.*

Let the equation of the skew lines LA and MB be

$$\frac{x-x_1}{l_1}=\frac{y-y_1}{m_1}=\frac{z-z_1}{n_1} \tag{4.6.1}$$

$$\text{and } \frac{x-x_2}{l_2}=\frac{y-y_2}{m_2}=\frac{z-z_2}{n_2} \tag{4.6.2}$$

respectively.

Let LM be the shortest distance between them. Therefore LM is perpendicular to both the lines. Let the co-ordinates of A and B be (x_1, y_1, z_1) and (x_2, y_2, z_2) and the d. cs. of LM be l, m, n.

Since LM is perpendicular to both the lines,

$$ll_1+mm_1+nn_1=0 \quad\text{and}\quad ll_2+mm_2+nn_2=0.$$

By cross-multiplication

$$\frac{l}{m_1n_2 - m_2n_1} = \frac{m}{n_1l_2 - n_2l_1} = \frac{n}{l_1m_2 - l_2m_1}$$
$$= \pm\frac{1}{\sqrt{\Sigma(m_1n_2 - m_2n_1)^2}}. \quad (4.6.3)$$

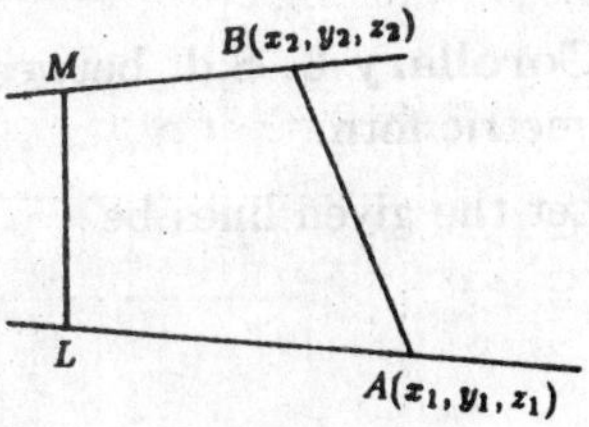

Fig. 4.2

The d. cs. of AB are

$$\frac{x_2 - x_1}{AB}, \frac{y_2 - y_1}{AB}, \frac{z_2 - z_1}{AB}.$$

LM is the projection of AB on LM. If θ be the angle between AB and LM, then $LM = AB\cos\theta$

$$= AB\cdot\frac{l(x_2 - x_1) + m(y_2 - y_1) + n(z_2 - z_1)}{AB}$$
$$= \pm\frac{\Sigma(x_2 - x_1)(m_1n_2 - m_2n_1)}{\sqrt{\Sigma(m_1n_2 - m_2n_1)^2}} \quad \text{[by (4.6.3)]}$$
$$= \pm\frac{\begin{vmatrix} x_2 - x_1 & y_2 - y_1 & z_2 - z_1 \\ l_1 & m_1 & n_1 \\ l_2 & m_2 & n_2 \end{vmatrix}}{\left[\begin{vmatrix} m_1 & n_1 \\ m_2 & n_2 \end{vmatrix}^2 + \begin{vmatrix} n_1 & l_1 \\ n_2 & l_2 \end{vmatrix}^2 + \begin{vmatrix} l_1 & m_1 \\ l_2 & m_2 \end{vmatrix}^2\right]^{1/2}}.$$

(positive value is taken.)

Corollary 1. Equation of the line of s.d.

S.d. is the line of intersection of the planes ALM and BLM. Therefore, the equation of it is

$$\begin{vmatrix} x - x_1 & y - y_1 & z - z_1 \\ l_1 & m_1 & n_1 \\ l & m & n \end{vmatrix} = 0 = \begin{vmatrix} x - x_2 & y - y_2 & z - z_2 \\ l_2 & m_2 & n_2 \\ l & m & n \end{vmatrix}.$$

Corollary 2. Points of intersection of s.d. and the lines.

Let

$$(l_1r_1 + x_1, m_1r_1 + y_1, n_1r_1 + z_1)$$
$$\text{and } (l_2r_2 + x_2, m_2r_2 + y_2, n_2y_2 + z_2)$$

be any two points on the lines. If these points are the points of intersection, then the d. rs. of the s.d. can be calculated by the help of these two points. Applying the condition of perpendicularity of it with the lines two equations of r_1 and r_2 are obtained. From these equations r_1 and r_2 are also obtained. Consequently the length of s.d., the points of intersection and its equation are found out.

Corollary 3. S.d. between two lines when one is in general form and the other in symmetric form.

Let the given lines be

$$u = 0 = v \tag{4.6.4}$$

$$\text{and} \quad \frac{x - x_1}{l} = \frac{y - y_1}{m} = \frac{z - z_1}{n}. \tag{4.6.5}$$

To find the s.d. the plane $\alpha = u + \lambda v = 0$ is determined in such a way that $\alpha = 0$ is parallel to the line (4.6.5). The s.d. is the distance of the point (x_1, y_1, z_1) from the plane $\alpha = 0$. The equations of the line of s.d. are the planes through the given lines and perpendicular to $\alpha = 0$.

Note. Here s.d. can be found out by writing the line $u = 0 = v$ in symmetric form.

Corollary 4. S.d. between two lines when both are in general form.

Let the lines be $u_1 = 0 = v_1$ and $u_2 = 0 = v_2$.

The values of λ_1 and λ_2 are determined in such a way that the planes $u_1 + \lambda_1 v_1 = 0$ and $u_2 + \lambda_2 v_2 = 0$ are parallel. The s.d. is the distance between these two parallel planes. The equations of the line of s.d. are the plannes through the given lines and perpendicular to any one of the parallel planes.

Note. S.d. can also be found out by expressing the given lines in symmetric form.

► **Example 7** *Find the s.d. between the lines*

$$\frac{x-3}{3} = \frac{y-8}{-1} = \frac{z-3}{1}, \tag{1}$$

$$\frac{x+3}{-3} = \frac{y+7}{2} = \frac{z-6}{4}. \tag{2}$$

[B.U. '97]

Find also the equations and the points of intersection in which it meets the lines.

[B.U. '80]

Let the s.d. meet (1) and (2) at $(3r_1 + 3, -r_1 + 8, r_1 + 3)$ and $(-3r_2 - 3, 2r_2 - 7, 4r_2 + 6)$ respectively. The d. rs. of the s.d. are $3r_1 + 3r_2 + 6, -r_1 - 2r_2 + 15, r_1 - 4r_2 - 3$.

Since the s.d. is perpendicular to (1) and (2),

$$3(3r_1 + 3r_2 + 6) - 1\cdot(-r_1 - 2r_2 + 15) + 1\cdot(r_1 - 4r_2 - 3) = 0,$$

$$\text{or} \quad 11r_1 + 7r_2 = 0 \tag{3}$$

$$\text{and} \quad -3\cdot(3r_1 + 3r_2 + 6) + 2\cdot(-r_1 - 2r_2 + 15) + 4\cdot(r_1 - 4r_2 - 3) = 0,$$

$$\text{or} \quad 7r_1 + 29r_2 = 0. \tag{4}$$

From (3) and (4), $r_1 = r_2 = 0$.

$\therefore$ the points of intersection are $(3, 8, 3)$ and $(-3, -7, 6)$

Hence the length of the s.d. $= \sqrt{(3+3)^2 + (8+7)^2 + (3-6)^2} = 3\sqrt{30}.$

The d. rs. of s.d. are $6, 15, -3$ and the equation of s.d. is

$$\frac{x-3}{2} = \frac{y-8}{5} = \frac{z-3}{-1}.$$

▶ **Example 8** *Find the magnitude and position of the line of s.d. between the lines* $\frac{x}{4} = \frac{y+1}{3} = \frac{z-2}{2}$ *and* $5x - 2y - 3z + 6 = 0 = x - 3y + 2z - 3$.

The equation of a plane through the second line is

$$(5x - 2y - 3z + 6) + \lambda(x - 3y + 2z - 3) = 0,$$

$$\text{or} \quad (5+\lambda)x + (-2-3\lambda)y + (-3+2\lambda)z + 6 - 3\lambda = 0.$$

If it is parallel to the first line then

$$4(5+\lambda) - 3(2+3\lambda) - 2(3-2\lambda) = 0, \quad \text{or} \quad \lambda = 8.$$

In this case this equation of the plane is

$$13x - 26y + 13z - 18 = 0. \tag{1}$$

S.d. = distance of $(0, -1, 2)$ from (1)

$$= \frac{26 + 26 - 18}{13\sqrt{1+4+1}} = \frac{34}{13\sqrt{6}} = \frac{17\sqrt{6}}{39}.$$

The plane through the first line and perpendicular to (1) is

$$\begin{vmatrix} x & y+1 & z-2 \\ 4 & 3 & 2 \\ 13 & -26 & 13 \end{vmatrix} = 0, \quad \text{or} \quad 7x - 2y - 11z + 20 = 0. \tag{2}$$

The equation of a plane through the second line is

$$(5+\mu)x + (-2-3\mu)y + (-3+2\mu)z + 6 - 3\mu = 0.$$

If it is perpendicular to (1) then

$$13(5+\mu) + 26(2+3\mu) + 13(-3+2\mu) = 0, \quad \text{or} \quad \mu = -\frac{2}{3}.$$

The equation of the plane is

$$13x - 13z + 24 = 0 \tag{3}$$

Planes (2) and (3) give the equations of s.d.

► **Example 9** *Find the magnitude and the position of the line of s.d. between the lines* $2x + y - z = 0 = x - y + 2z$ *and* $x + 2y - 3z - 4 = 0 = 2x - 3y + 4z - 5$.

Equation of planes through the given lines are

$$2x + y - z + \lambda_1(x - y + 2z) = 0 \quad (1)$$

$$\text{and} \quad x + 2y - 3z - 4 + \lambda_2(2x - 3y + 4z - 5) = 0. \quad (2)$$

If these planes are parallel then

$$\frac{2+\lambda_1}{1+2\lambda_2} = \frac{1-\lambda_1}{2-3\lambda_2} = \frac{-1+2\lambda_1}{-3+4\lambda_2}.$$

Solving for λ_1 and $\lambda_2, \lambda_1 = -1, \lambda_2 = 0$.

The planes are

$$x + 2y - 3z = 0 \quad (3)$$

$$\text{and} \quad x + 2y - 3z - 4 = 0. \quad (4)$$

A point on (3) is $(0, 0, 0)$.

S.d. = distance between the parallel planes (3) and (4).

$$= \frac{4}{\sqrt{1+4+9}} = \frac{4}{\sqrt{14}}. \quad (5)$$

If the planes (1) and (2) are perpendicular to (3) then

$$(2 + \lambda_1).1 + (1 - \lambda_1).2 + (-1 + 2\lambda_1)(-3) = 0, \quad \text{or} \quad \lambda_1 = 1$$

$$\text{and} \quad (1 + 2\lambda_2)\cdot 1 + (2 - 3\lambda_2)\cdot 2 + (-3 + 4\lambda_2)\cdot(-3) = 0, \quad \text{or} \quad \lambda_2 = \frac{7}{8}.$$

For these values of λ_1 and λ_2, the planes are

$$3x + z = 0 \quad \text{and} \quad 22x - 5y + 4z - 67 = 0.$$

These planes give the equation of s.d.

4.7 Equations of Two Skew Lines in simplified form w.r.t. suitable Co-ordinate Axes

Let LM be the s.d. between two skew lines AB and CD. Let $LM = 2c$ and let the angle between the skew lines $= 2\alpha$.

The mid-point of LM is chosen as the origin O. OP and OQ are drawn parallel to AB and CD respectively. The internal and external bisectors of the angle POQ, i.e., OX and OY are chosen as the x-axis and y-axis respectively. OL is taken as the z-axis.

In this system of axes the line OP or AB makes angles $\alpha, \frac{\pi}{2}-\alpha, \frac{\pi}{2}$ with x, y, z axes respectively and the line OQ or CD makes angles $-\alpha, \frac{\pi}{2}+\alpha, \frac{\pi}{2}$ with x, y, z axes respectively. Therefore the d.cs. of AB are $\cos\alpha, \cos\left(\frac{\pi}{2}-\alpha\right), \cos\frac{\pi}{2}$, i.e., $\cos\alpha, \sin\alpha, 0$ and those of CD are $\cos(-\alpha), \cos\left(\frac{\pi}{2}+\alpha\right)$, $\cos\frac{\pi}{2}$, i.e., $\cos\alpha, -\sin\alpha, 0$. The co-ordinates of L and M are $(0,0,c)$ and $(0,0,-c)$. Now the equation of AB and CD are

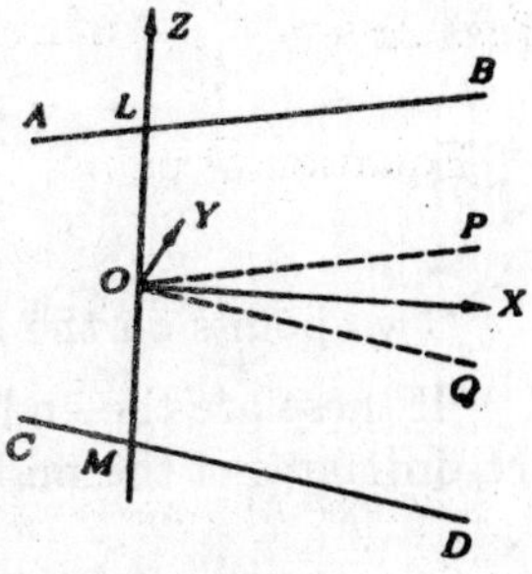

Fig. 4.3

$$\frac{x-0}{\cos\alpha}=\frac{y-0}{\sin\alpha}=\frac{z-c}{0} \quad \text{and} \quad \frac{x-0}{\cos\alpha}=\frac{y-0}{-\sin\alpha}=\frac{z+c}{0},$$

or $y = x\tan\alpha, z = c$ and $y = -x\tan\alpha, z = -c$

respectively.

Note 1. If we put $\tan\alpha = m$, then the equations of the skew lines are $y = mx$, $z = c$; $y = -mx, z = -c$ respectively.

Note 2. The parametric co-ordinates of a point on AB are $(r, r\tan\alpha, c)$ and those of a point on CD and $(r', -r'\tan\alpha, -c)$.

► **Example 10** *Find the surface generated by the lines which intersect the lines $y = mx, z = c; y = -mx, z = -c$ and x-axis.*

The equations of a line intersecting the given skew lines can be written as

$$(y-mx)+\lambda_1(z-c) = 0, \tag{1}$$

$$(y+mx)+\lambda_2(z+c) = 0. \tag{2}$$

Let this line meet the x-axis. On the x-axis $y = 0, z = 0$.

Putting $y = 0, z = 0$ in (1) and (2)

$$-mx-\lambda_1 c = 0 \quad \text{and} \quad mx+\lambda_2 c = 0,$$

$$\text{or} \quad x = -\frac{c\lambda_1}{m} = -\frac{c\lambda_2}{m} \quad \therefore \quad \lambda_1 = \lambda_2. \tag{3}$$

Now eliminating λ_1 and λ_2 from (1), (2) and (3), we have

$$\frac{y-mx}{z-c}=\frac{y+mx}{z+c},$$

$$\text{or} \quad (y-mx)(z+c) = (y+mx)(z-c), \quad \text{or} \quad mzx = cy.$$

It is the equation of the required surface.

► **Example 11** *A line of constant length 2k has its extremities on two fixed lines. Prove that the locus of its mid-point is an ellipse whose axes are equally inclined to the fixed lines.*

Let the fixed lines be

$$\frac{x}{1}=\frac{y}{m}=\frac{z-c}{0}=r \tag{1}$$

$$\text{and}\quad \frac{x}{1}=\frac{y}{-m}=\frac{z+c}{0}=r'. \tag{2}$$

Two points on the lines (1) and (2) are (r, mr, c) and $(r', -mr', -c)$.

If these are the end points of the variable line at one position and (α, β, γ) are the co-ordinates of the mid-point of it then

$$2\alpha = r + r', \quad 2\beta = m(r - r'), \quad 2\gamma = 0. \tag{3}$$

Since the length of the variable line is $2k$,

$$(r-r')^2 + m^2(r+r')^2 + (c+c)^2 = 4k^2,$$

$$\text{or}\quad \frac{4\beta^2}{m^2} + 4m^2\alpha^2 + 4c^2 = 4k^2, \quad \text{or}\quad m^2\alpha^2 + \frac{\beta^2}{m^2} = k^2 - c^2.$$

Hence the required locus is $m^2x^2 + \frac{y^2}{m^2} = k^2 - c^2, z = 0$.

It is an ellipse in the xy plane and its axes (x-axis, y-axis) are equally inclined to the fixed lines.

4.8 Equations of a Line Intersecting Two Lines

Case 1. *Given lines in symmetric form.*

Let the given lines be

$$\frac{x-x_1}{l_1}=\frac{y-y_1}{m_1}=\frac{z-z_1}{n_1}=r_1 \tag{4.8.1}$$

$$\text{and}\quad \frac{x-x_2}{l_2}=\frac{y-y_2}{m_2}=\frac{z-z_2}{n_2}=r_2. \tag{4.8.2}$$

Here r_1 and r_2 are parameters. Any point on (4.8.1) is

$$(l_1r_1 + x_1, m_1r_1 + y_1, n_1r_1 + z_1)$$

and any point on (4.8.2) is

$$(l_2r_2 + x_2, m_2r_2 + y_2, n_2r_2 + z_2).$$

The line passing through these two points intersects the given lines. r_1 and r_2 are obtained by the conditions of the problem.

Case 2. *Given lines in general form.*

Let the given lines be

$$a_1x + b_1y + c_1z + d_1 = 0 = a_2x + b_2y + c_2z + d_2$$

$$\text{and}\quad a_3x + b_3y + c_3z + d_3 = 0 = a_4x + b_4y + c_4z + d_4.$$

The equations of the required line are

$$a_1x + b_1y + c_1z + d_1 + \lambda_1(a_2x + b_2y + c_2y + d_2) = 0,$$

$$a_3x + b_3y + c_3z + d_3 + \lambda_2(a_4x + b_4y + c_4z + d_4) = 0.$$

λ_1 and λ_2 are determined by the conditions of the problem.

► **Example 12** *A line with d. cs. proportional to* $(2, 7, -5)$ *is drawn to intersect the lines,* $\frac{x-5}{3} = \frac{y-7}{-1} = \frac{z+2}{1}$ *and* $\frac{x+3}{-3} = \frac{y-3}{2} = \frac{z-6}{4}$.

Find the co-ordinates of the points of intersection and the length of the intercept between the lines.

Let the line with d. rs. $(2, 7, -5)$ meet the given lines at

$$(3r_1 + 5, -r_1 + 7, r_1 - 2) \quad \text{and} \quad (-3r_2 - 3, 2r_2 + 3, 4r_2 + 6).$$

Then the d. rs. of the line are

$$(-3r_2 - 3r_1 - 8, 2r_2 + r_1 - 4, 4r_2 - r_1 + 8).$$

$$\therefore \quad \frac{-3r_2 - 3r_1 - 8}{2} = \frac{2r_2 + r_1 - 4}{7} = \frac{4r_2 - r_1 + 8}{-5}.$$

$$\therefore \quad -21r_2 - 21r_1 - 56 = 4r_2 + 2r_1 - 8,$$

$$\text{or} \quad 23r_1 + 25r_2 + 48 = 0 \tag{1}$$

$$\text{and} \quad -10r_2 - 5r_1 + 20 = 28r_2 - 7r_1 + 56,$$

$$\text{or} \quad r_1 - 19r_2 - 18 = 0. \tag{2}$$

Form (1) and (2), $r_1 = r_2 = -1$.

Thus the points of intersection are $(2, 8, -3)$ and $(0, 1, 2)$.

Intercepted length $= \sqrt{\{(-2)^2 + (1-8)^2 + (2+3)^2\}} = \sqrt{78}$.

► **Example 13** *Find the equation to the line which intersects the lines* $z = 5x - 6 = 4y + 3$ *and* $z = 2x - 4 = 3y + 5$ *and it is parallel to* $\frac{x}{4} = \frac{y}{1} = \frac{z}{1}$.

Equations of the line which intersects the given lines are

$$z - 5x + 6 + \lambda_1(z - 4y - 3) = 0,$$

$$z - 2x + 4 + \lambda_2(z - 3y - 5) = 0,$$

$$\text{i.e.,} \quad -5x - 4\lambda_1 y + (\lambda_1 + 1)z + 6 - 3\lambda_1 = 0,$$

$$-2x - 3\lambda_2 y + (\lambda_2 + 1)z + 4 - 5\lambda_2 = 0.$$

If it is parallel to $\frac{x}{4} = \frac{y}{1} = \frac{z}{1}$,

$$-5.4 - 4\lambda_1 \cdot 1 + (\lambda_1 + 1) \cdot 1 = 0, \quad \text{or} \quad \lambda_1 = -\frac{19}{3}$$

$$\text{and} \quad -2.4 - 3\lambda_2 \cdot 1 + (\lambda_2 + 1) \cdot 1 = 0, \quad \text{or} \quad \lambda_2 = -\frac{7}{2}.$$

$\therefore$ the equations of the required line are

$$15x - 76y + 16z + 75 = 0, \quad 4x - 21y + 5z - 43 = 0.$$

Worked Out Examples

1. Find the distance of the point $(-1, -5, -10)$ from the point of intersection of the line $\frac{x-2}{3} = \frac{y+1}{4} = \frac{z-2}{12}$ and the plane $x - y + z = 5$.

Let

$$\frac{x-2}{3} = \frac{y+1}{4} = \frac{z-2}{12} = r.$$

Any point on the line is $(3r+2, 4r-1, 12r+2)$.

If this point lies on the plane $x - y + z = 5$, then $(3r+2) - (4r-1) + (12r+2) = 5$, or $r = 0$. Therefore, the line meets the plane at the point $(2, -1, 2)$.

Now the distance between the points $(-1, -5, -10)$ and $(2, -1, 2)$

$$= \sqrt{(2+1)^2 + (-1+5)^2 + (2+10)^2} = 13.$$

2. Determine the equation of the straight line through the point $(3, 1, -6)$ and parallel to each of the planes $x + y + 2z - 4 = 0$ and $2x - 3y + z + 5 = 0$.

[B.U. '92; C.U. '85]

Let the equations of the line be

$$\frac{x-3}{l} = \frac{y-1}{m} = \frac{z+6}{n}. \tag{1}$$

Since it is parallel to the given planes it is perpendicular to the normals to the planes.

$$\therefore \quad l + m + 2n = 0 \quad \text{and} \quad 2l - 3m + n = 0.$$

By cross-multiplication

$$\frac{l}{7} = \frac{m}{3} = \frac{n}{-5}.$$

Thus the equations of the line are

$$\frac{x-3}{7} = \frac{y-1}{3} = \frac{z+6}{-5}.$$

3. Determine the value of k so that the lines

$$\frac{x-1}{2} = \frac{y-4}{1} = \frac{z-5}{2} \quad \text{and} \quad \frac{x-2}{-1} = \frac{y-8}{k} = \frac{z-11}{4}$$

may intersect.

[C.U. '80, N.B. '92]

A point on the first line is $(2r_1 + 1, r_1 + 4, 2r_1 + 5)$ and a point on the second line is $(-r_2 + 2, kr_2 + 8, 4r_2 + 11)$.

The given lines intersect, if the above points are identical.

In this case

$$2r_1 + 1 = -r_2 + 2, \tag{1}$$

$$r_1 + 4 = kr_2 + 8 \tag{2}$$

$$\text{and} \quad 2r_1 + 5 = 4r_2 + 11. \tag{3}$$

By (1) and (3), $r_1 = 1$, $r_2 = -1$.

From (2) $5 = -k + 8$, or $k = 3$.

4. Find the equation of the perpendicular from the point $(2,4,-1)$ to the line

$$x+5=\frac{1}{4}(y+3)=-\frac{1}{9}(z-6).$$

Obtain the perpendicular distance and the foot of the perpendicular.

Let

$$\frac{x+5}{1}=\frac{y+3}{4}=\frac{z-6}{-9}=r. \tag{1}$$

Any point on the line is $(r-5,4r-3,-9r+6)$. This will be the foot of the perpendicular, if the line joining the points $(2,4,-1)$ and $(r-5,4r-3,-9r+6)$ is perpendicular to the given line. The d. rs. of the line joining the points are $r-7,4r-7,-9r+7$.

$\therefore\quad 1\cdot(r-7)+4\cdot(4r-7)-9\cdot(9r+7)=0$, i.e., $r=1$.

$\therefore\quad$ the foot of the perpendicular is $(-4,1,-3)$ and the d. rs. of it are $-6,-3,-2$. Hence the equations of the line are

$$\frac{x+4}{6}=\frac{y-1}{3}=\frac{z+3}{2}.$$

Perpendicular distance = distance between the points (2,4,−1) and (−4,1,−3)

$$=\sqrt{(2+4)^2+(4-1)^2+(1+3)^2}=7.$$

5. Find the equation of the plane which contains the lines

$$\frac{x-1}{2}=-y-1=\frac{z-3}{4}$$

and is perpendicular to the plane $x+2y+z=12$. **[C.U. '93]**

The equation of the plane containing the line is

$$a(x-1)+b(y+1)+c(z-3)=0, \tag{1}$$

where

$$2a-b+4c=0. \tag{2}$$

Since the given plane is perpendicular to the plane (1),

$$a+2b+c=0. \tag{3}$$

From (2) and (3)

$$\frac{a}{-9}=\frac{b}{2}=\frac{c}{5}.$$

Thus the required equation is $-9(x-1)+2(y+1)+5(z-3)=0$, i.e., $9x-2y-5z+4=0$.

6. Find the distance of the point $(1,2,3)$ from the plane $x+y+z=3$ measured in the direction parallel to the line

$$\frac{x}{1}=\frac{y}{2}=\frac{z}{5}.$$

Equations of the line passing through the point $(1,2,3)$ and parallel to the given line are

$$\frac{x-1}{1}=\frac{y-2}{2}=\frac{z-3}{5}=r\quad \text{(say)}.$$

Any point on the line is $(r+1, 2r+2, 5x+3)$. If this point lies on the plane $x+y+z=3$, then $r+1+2r+2+5r+3=3$, or $r=-\frac{3}{8}$.

Hence the line intersects the plane at the point

$$\left(1-\frac{3}{8},\ 2-\frac{3}{4},\ 3-\frac{15}{8}\right),\quad \text{i.e.,}\quad \left(\frac{5}{8},\ \frac{5}{4},\ \frac{9}{8}\right).$$

$$\therefore\ \text{distance} = \sqrt{\left(1-\frac{5}{8}\right)^2+\left(2-\frac{5}{4}\right)^2+\left(3-\frac{9}{8}\right)^2} = \frac{3\sqrt{30}}{8}.$$

7. Find the co-ordinates of the point which is the image of the point $(1,-2,5)$ in the plane $2x-3y+4z+1=0$.

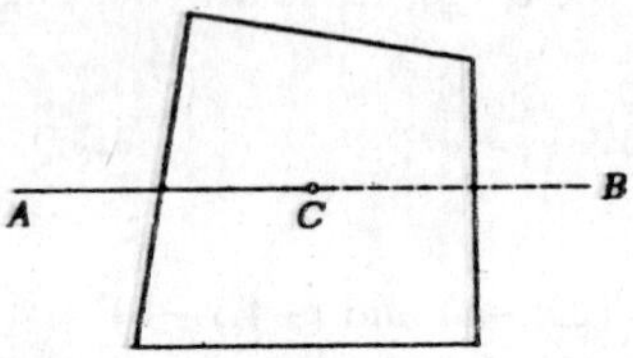

Fig. 4.4

Let $A(1,-2,5)$ be the given point and ACB be the normal to the given plane. B is the image of A.

The equation of ACB is

$$\frac{x-1}{2} = \frac{y+2}{-3} = \frac{z-5}{4}.$$

Let the co-ordinates of B be $(2r+1, -3r-2, 4r+5)$.

The line meets the plane at C which is the midpoint of AB. Therefore, the co-ordinates of C are

$$\left(r+1,\ \frac{-3r-4}{2},\ 2r+5\right).$$

Since this point lies on the plane,

$$2r+2+\frac{9r}{2}+6+8r+20+1=0,\quad \text{or}\quad r=-2.$$

Hence the image point is $(-3,4,-3)$.

8. Prove that the lines $x=ay+b=cz+d$ and $x=\alpha y+\beta=\gamma z+\delta$ are coplanar, if $(\gamma-c)(a\beta-b\alpha)-(\alpha-a)(c\delta-d\gamma)=0$.

The lines in symmetrical form are

$$\frac{x}{1} = \frac{y+\frac{b}{a}}{\frac{1}{a}} = \frac{z+\frac{d}{c}}{\frac{1}{c}}\quad \text{and}\quad \frac{x}{1} = \frac{y+\frac{\beta}{\alpha}}{\frac{1}{\alpha}} = \frac{z+\frac{\delta}{\gamma}}{\frac{1}{\gamma}}.$$

From the condition of coplanarity

$$\begin{vmatrix} 0 & \frac{\beta}{\alpha}-\frac{b}{a} & \frac{\delta}{\gamma}-\frac{d}{c} \\ 1 & \frac{1}{a} & \frac{1}{c} \\ 1 & \frac{1}{\alpha} & \frac{1}{\gamma} \end{vmatrix} = 0,\quad \text{or}\quad \begin{vmatrix} 0 & \frac{\beta}{\alpha}-\frac{b}{a} & \frac{\delta}{\gamma}-\frac{d}{c} \\ 0 & \frac{1}{a}-\frac{1}{\alpha} & \frac{1}{c}-\frac{1}{\gamma} \\ 1 & \frac{1}{\alpha} & \frac{1}{\gamma} \end{vmatrix} = 0,$$

$$\text{or}\quad \left(\frac{\beta}{\alpha}-\frac{b}{a}\right)\left(\frac{1}{c}-\frac{1}{\gamma}\right)-\left(\frac{\delta}{\gamma}-\frac{d}{c}\right)\left(\frac{1}{a}-\frac{1}{\alpha}\right)=0,$$

$$\text{or}\quad (\gamma-c)(a\beta-b\alpha)-(\alpha-a)(c\delta-d\gamma)=0.$$

9. Show that the equation to the plane containing the line $\frac{y}{b}+\frac{z}{c}=1, x=0$ and parallel to the line $\frac{x}{a}-\frac{z}{c}=1, y=0$ is $\frac{x}{a}-\frac{y}{b}-\frac{z}{c}+1=0$ and if $2d$ is the s.d. between the lines, prove that

$$\frac{1}{d^2}=\frac{1}{a^2}+\frac{1}{b^2}+\frac{1}{c^2}.$$

[N.B. '90, B.H. '83, C.U. '83, '88]

The equation of any plane through the first line is

$$\frac{y}{b}+\frac{z}{c}-1+\lambda x=0. \tag{1}$$

It will be parallel to the second line, if it is parallel to the plane

$$\frac{x}{a}-\frac{z}{c}-1+\mu y=0 \tag{2}$$

through the second line.
From the condition of parallel planes

$$\frac{\lambda}{\frac{1}{a}}=\frac{\frac{1}{b}}{\mu}=\frac{\frac{1}{c}}{-\frac{1}{c}} \qquad \therefore \quad \lambda=-\frac{1}{a} \quad \text{and} \quad \mu=-\frac{1}{b}.$$

Putting this value of λ in (1)

$$-\frac{x}{a}+\frac{y}{b}+\frac{z}{c}-1=0, \quad \text{or} \quad \frac{x}{a}-\frac{y}{b}-\frac{z}{c}+1=0. \tag{3}$$

It is the equation of the plane.
$(a,0,0)$ is a point on the second line. The s.d. is the distance of the plane (3) from the point $(a,0,0)$.

$$\therefore \quad 2d=\frac{2}{\sqrt{\left(\frac{1}{a^2}+\frac{1}{b^2}+\frac{1}{c^2}\right)}}, \quad \text{or} \quad \frac{1}{d^2}=\frac{1}{a^2}+\frac{1}{b^2}+\frac{1}{c^2}.$$

Exercise

1. Find the equation of the line in symmetrical form

 (a) through the origin which has d. rs. $1,-1,4$;

 (b) through $(-1,-3,5)$ which has d. rs. $2,1,-3$;

 (c) through the points $(2,0,4)$ and $(1,-2,3)$;

 (d) through $(5,-7,8)$ and parallel to $\frac{x-2}{3}=\frac{y-2}{4}=\frac{z+1}{-5}$;

 (e) which is perpendicular to $\frac{x-5}{1}=\frac{y-2}{3}=\frac{z+1}{-3}$ and $\frac{x}{2}=\frac{y-3}{3}=\frac{z+1}{3}$ and passes through $(1,0,-1)$. [C.U. '90]

2. Find the equation of the line $8x+12y-13z-32=0=4x+4y-5z-12$ in symmetrical form.

3. (a) Find the points, where the line through the points $(5,-2,3)$ and $(3,0,1)$ pierces the co-ordinate planes. [C.U. '65]

(b) Find the point, where the line $x+3y-z=6, y-z=4$ meets the plane $2x+2y+z=0$. **[C.U. '75]**

4. Find the equations to the line passing through $(-1,-2,-3)$ and parallel to the line $2x+3y-3z+2=0=3x-4y+2z-4$.

5. (a) Find the equation to the plane through the line $3x-4y+5z=10, 2x+2y-3z=4$ and parallel to $x=2y=3z$.

(b) Find the equation of the plane which contains the line $\frac{x-2}{3}=\frac{y-3}{1}=\frac{z+1}{-2}$ and is perpendicular to the plane $x-y+2z=7$.

6. (a) Find the distance of the point of intersection of the line $\frac{x-2}{1}=\frac{y-3}{2}=\frac{z+1}{2}$ and the plane $2x+3y+4z+7=0$ from the point $(2,1,3)$.

(b) Find the distance of the point $(2,-1,1)$ from the plane $x+y+z=3$ measured parallel to the line whose d. rs. are $2,3,-4$. **[C.U. '89]**

7. (a) Prove that the following lines are coplanar and also find the equation of the plane.

(i) $\frac{x-1}{3}=\frac{y+2}{1}=\frac{z+3}{2}, \frac{x-5}{7}=\frac{y+8}{-5}=\frac{z-6}{11}$ **[C.U. '93]**

(ii) $\frac{x-1}{2}=\frac{y-2}{3}=\frac{z-3}{4}, 4x-3y+1=0=5x-3z+2$. **[C.U. '63]**

(iii) $x+2y-5z+9=0=3x-y+2z-5, 2x+3y-z-3=0=4x-5y+z+3$.

(b) Prove that the the lines $\frac{x+1}{3}=\frac{y+3}{5}=\frac{z+5}{7}; \frac{x-2}{1}=\frac{y-4}{3}=\frac{z-6}{5}$ intersect. Find their point of intersection and the equation of the plane in which they lie. **[C.U. '99]**

8. Show that the lines $\frac{x}{\alpha}=\frac{y}{\beta}=\frac{z}{\gamma}, \frac{x}{a\alpha}=\frac{y}{b\beta}=\frac{z}{c\gamma}, \frac{x}{l}=\frac{y}{m}=\frac{z}{n}$ will lie in the same plane if $\frac{l}{\alpha}(b-c)+\frac{m}{\beta}(c-a)+\frac{n}{\gamma}(a-b)=0$. **[C.U. '72, B.U. '87, '90]**

9. (a) Find the angle between the lines $3x+2y+z-5=0=x+y-2z-3$ and $2x-y-z=0=7x+10y-8z$.

(b) Find the angle between the line $\frac{x-1}{2}=\frac{y-2}{1}=\frac{z-3}{-2}$ and the plane $x+2y+z-3=0$.

10. Find the equation of the plane passing through the line $\frac{x+1}{-3}=\frac{y-3}{2}=\frac{z+2}{1}$ and the point $(0,7,-7)$. Show that the line $x=\frac{y-7}{-3}=\frac{z+7}{2}$ lies in this plane.

11. Find the value of k so that the lines $\frac{x+4}{k}=\frac{y+6}{5}=\frac{z-1}{-2}$ and $3x-2y+z+5=0=2x+3y+4z-4$ may intersect. What is the point of intersection?

12. Find the image of the point $(4,-2,3)$ in the plane $2x-3y+z=7$.

13. (a) Find the distance of the line $x+2y+3z+4=0=2x+3y+4z+5$ from the origin.

(b) Prove that the equation of the perpendicular from the point $(3,-1,11)$ to the line $\frac{x}{2}=\frac{y-2}{3}=\frac{z-3}{4}$ is $\frac{x-3}{1}=\frac{y+1}{-6}=\frac{z-11}{4}$ and that the co-ordinates of its foot are $(2,5,7)$.

14. Find the equations of the bisectors of the angles between the lines

$$\frac{x-1}{2}=\frac{y+1}{-3}=\frac{z+10}{8}, \frac{x-4}{1}=\frac{y+3}{-4}=\frac{z+1}{7}.$$

[HINTS. The lines intersect in the point $(5,-7,6)$. Now the bisectors are $\frac{x-5}{2\pm1}=\frac{y+7}{-3\mp4}=\frac{z-6}{8\pm7}$, i.e., $\frac{x-5}{3}=\frac{y+7}{-7}=\frac{z-6}{1}$ and $x-5=y+7=z-6$.]

15. Show that the lines $x + 7y - z - 16 = 0 = x - y + z - 4$ and $x + 11y - 2z = 0 = x - 5y + 2z - 4$ are parallel and find the distance between them. **[C.U. '74]**

16. Find the equations of the line which passes through the point $(3, -2, -4)$ is parallel to the plane $3x - 2y - 3z - 7 = 0$ and intersects the straight line $\frac{x-2}{3} = \frac{y+4}{-1} = \frac{z-1}{2}$. **[C.H. '78]**

17. Find the shortest distance and its equation between the lines

(a) $\frac{x-3}{2} = \frac{y+15}{-7} = \frac{z-9}{5}, \frac{x+1}{2} = \frac{y-1}{1} = \frac{z-9}{-3}$; **[C.U. '67, '89]**

(b) $3x - 9y + 5z = 0 = x + y - z, 6x + 8y + 3z - 13 = 0 = x + 2y + z - 3$;

(c) $x = y = z$ and $x + y = 2, z - x = 2$. **[C.U. '81]**

(d) $\frac{x-1}{5} = \frac{y-7}{-4} = \frac{z+3}{12}$ and y-axis. **[C.U. '92]**

18. (a) Find the s.d. between the lines $\frac{x-1}{2} = \frac{y-2}{4} = z - 3$ and $y - mx = z = 0$. Find what value of m will the two lines intersect? **[C.U. '98]**

(b) Show that the s.d. between the lines $x + a = 2y = -12z$ and $x = y + 2a = 6z - 6a$ is $2a$. **[B.U. '90, N.B. '89]**

19. (a) Find the equations of the line which can be drawn from the point $(2, -1, 3)$ to intersect the lines

$$\frac{x-1}{2} = \frac{y-2}{3} = \frac{z-3}{4}, \frac{x-4}{4} = \frac{y}{5} = \frac{z+3}{3}.$$

(b) A line with d. rs. $(2, 7, -5)$ intersects the lines

$$\frac{x-5}{3} = \frac{y-7}{-1} = \frac{z+2}{1} \quad \text{and} \quad \frac{x+3}{-3} = \frac{y-3}{2} = \frac{z-6}{4}.$$

Find the points of intersection and the length intercepted on it.

20. Find the equation to the straight line drawn from the origin to intersect the lines $3x + 2y + 4z - 5 = 0 = 2x - 3y + 4z + 1$, $2x - 4y + z + 6 = 0 = 3x - 4y + z - 3$. **[C.U. '80, B.U. '83]**

[HINTS. The general equation of the line intersecting the given lines can be written as $3x + 2y + 4z - 5 + \lambda(2x - 3y + 4z + 1) = 0 = 2x - 4y + z + 6 + \mu(3x - 4y + z - 3)$.

Since the line passes through $(0, 0, 0)$, $\lambda = 5, \mu = 2$.

$\therefore$ the line is

$$13x - 13y + 24z = 0 = 8x - 12y + 3z, \quad \text{or} \quad \frac{x}{249} = \frac{y}{153} = \frac{z}{-52}].$$

21. A variable plane makes intercepts on the co-ordinate axes, the sum of whose squares is constant and equal to k^2. Show that the locus of the foot of the perpendicular from the origin to the plane is

$$\left(\frac{1}{x^2} + \frac{1}{y^2} + \frac{1}{z^2}\right)(x^2 + y^2 + z^2)^2 = k^2.$$

[C.U. '92]

[HINTS. Let $\frac{x}{a}+\frac{y}{b}+\frac{z}{c}=1$ be the equation of the plane at one position. The intercepts made by the plane on the axes are a, b, c.

By the condition

$$a^2+b^2+c^2=k^2. \tag{1}$$

Let (α,β,γ) be the foot of the perpendicular from the origin to the plane. Then (α,β,γ) lies on the plane.

$$\therefore \quad \frac{\alpha}{a}+\frac{\beta}{b}+\frac{\gamma}{c}=1. \tag{2}$$

The d.rs. of the perpendicular are α,β,γ.

$$\therefore \quad \frac{\alpha}{\frac{1}{a}}=\frac{\beta}{\frac{1}{b}}=\frac{\gamma}{\frac{1}{c}}=\lambda \text{ (say)}. \tag{3}$$

From (3), $a=\frac{\lambda}{\alpha}, b=\frac{\lambda}{\beta}, c=\frac{\lambda}{\gamma}$.

Putting these values in (1) and (2),

$$\lambda^2\left(\frac{1}{\alpha^2}+\frac{1}{\beta^2}+\frac{1}{\gamma^2}\right)=k^2$$

$$\text{and} \quad \frac{1}{\lambda}(\alpha^2+\beta^2+\gamma^2)=1, \quad \text{or} \quad \alpha^2+\beta^2+\gamma^2=\lambda.$$

Eliminating λ, $(\alpha^2+\beta^2+\gamma^2)^2\left(\frac{1}{\alpha^2}+\frac{1}{\beta^2}+\frac{1}{\gamma^2}\right)=k^2$.

Hence the locus is $(x^2+y^2+z^2)\left(\frac{1}{x^2}+\frac{1}{y^2}+\frac{1}{z^2}\right)=k^2$.]

Answers

1. (a) $\frac{x}{1}=\frac{y}{-1}=\frac{z}{4}$; (b) $\frac{x+1}{2}=\frac{y+3}{1}=\frac{z-5}{-3}$; (c) $\frac{x-2}{1}=\frac{y}{2}=\frac{z-4}{1}$; (d) $\frac{x-5}{3}=\frac{y+7}{4}=\frac{z-8}{-5}$; (e) $\frac{x-1}{6}=\frac{y}{-3}=\frac{z+1}{-1}$; **2.** $\frac{x-1}{2}=\frac{y-2}{3}=\frac{z}{4}$; **3.** (a) $(2,1,0),(0,3,-2),(3,0,1)$; (b) $(2,0,-4)$.; **4.** $\frac{x+1}{6}=\frac{y+2}{13}=\frac{z+3}{17}$; **5.** (a) $x-20y+27z=14$; (b) $9x-2y-5z+4=0$; **6.** (a) $\sqrt{37}$; (b) $\sqrt{29}$; **7.** (a) (i) $21x-19y-22z-125=0$, (ii) $x-2y+z=0$, (iii) $83x-2y-7z=21$; (b) $\left(\frac{1}{2},-\frac{1}{2},-\frac{3}{2}\right), x-2y+z=0$; **9.** (a) $\pi/2$; (b) $\sin^{-1}\sqrt{\frac{2}{27}}$; **10.** $x+y+z=0$; **11.** $k=3,(2,4,-3)$; **12.** $\left(\frac{8}{7},\frac{16}{7},\frac{11}{7}\right)$; **13.** (a) $\sqrt{\frac{7}{3}}$; **15.** 3; **16.** $\frac{x-3}{5}=\frac{y+2}{-6}=\frac{z+4}{9}$; **17.** (a) $4\sqrt{3}, x=y=z$; (b) $\frac{11}{\sqrt{342}}, 10x-29y+16z=0=13x+82y+55z-109$; (c) $\sqrt{2}, z+x=3, 2y=3$; (d) $\frac{27}{13}, 5x+12z=0, 169y=1059$; **18.** $\frac{5m-10}{\sqrt{(5m^2-16m+17)}}, m=2$. **19.** (a) $12x+4y-9z+7=0=11x-10y+2z-38$; (b) $(2,8,-3),(0,1,2),\sqrt{78}$.

Chapter 5

Changes of Axes: Volume of a Tetrahedron

5.1 Transformation of One Set of Rectangular Axes to another without changing the position of Origin

Let OX, OY, OZ and OX', OY', OZ' be two sets of rectangular axes through the same origin O and $l_1, m_1, n_1; l_2, m_2, n_2; l_3, m_3, n_3$ be the d.cs. of OX', OY', OZ' respectively referred to OX, OY and OZ.

Let P be the point whose co-ordinates are (x, y, z) w.r.t. OX, OY, OZ and (x', y', z') w.r.t. OX', OY', OZ'.

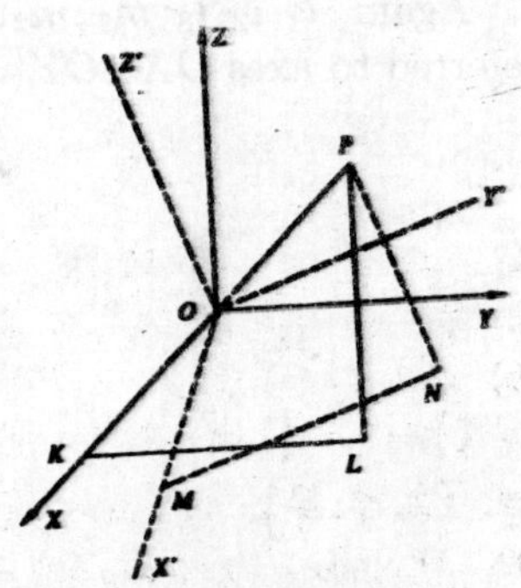

Fig. 5.1

In the fig. 5.1, $OK = x, KL = y, LP = z$; $OM = x', MN = y', NP = z'$.

Projection of OP on OX = proj. of OM on OX+ proj. of MN on OX+ proj. of NP on OX, or $x = l_1x' + l_2y' + l_3z'$.

Similarly $y = m_1x' + m_2y' + m_3z', z = n_1x' + n_2y' + n_3z'$.

Again projecting OP and OK, KL, LP on OX', OY' and OZ' in turn we have

$$x' = l_1x + m_1y + n_1z, y' = l_2x + m_2y + n_2z, z' = l_3x + m_3y + n_3z.$$

The above rules of transformation can be remembered in a compact form.

	x'	y'	z'
x	l_1	l_2	l_3
y	m_1	m_2	m_3
z	n_1	n_2	n_3

To obtain the value of x in terms of x', y', z' multiply each element of the row of x with the corresponding element of the first row and add, i.e., $x = l_1x' + l_2y' + l_3z'$. Similarly to obtain the value of x' in terms of x, y, z multiply each element of its column with the corresponding element of the first column and add, i.e., $x' = l_1x + m_1y + n_1z$.

5.2 Relations between the d. cs. of the Three Mutually Perpendicular Lines

Let $l_1, m_1, n_1; l_2, m_2, n_2; l_3, m_3, n_3$ be the d. cs. of the three mutually perpendicular lines OX', OY', OZ' respectively referred to a set of rectangular axes OX, OY and OZ.

Then

$$\left.\begin{aligned} l_1^2 + m_1^2 + n_1^2 = 1 \\ l_2^2 + m_2^2 + n_2^2 = 1 \\ l_3^2 + m_3^2 + n_3^2 = 1 \end{aligned}\right\} \tag{5.2.1}$$

$$\left.\begin{aligned} l_2l_3 + m_2m_3 + n_2n_3 = 0 \\ l_3l_1 + m_3m_1 + n_3n_1 = 0 \\ l_1l_2 + m_1m_2 + n_1n_2 = 0 \end{aligned}\right\}. \tag{5.2.2}$$

Again, $l_1, l_2, l_3; m_1, m_2, m_3; n_1, n_2, n_3$ are the d. cs. of OX, OY, OZ respectively referred to axes OX', OY' and OZ'.

$$\therefore \quad \left.\begin{aligned} l_1^2 + l_2^2 + l_3^2 = 1 \\ m_1^2 + m_2^2 + m_3^2 = 1 \\ n_1^2 + n_2^2 + n_3^2 = 1 \end{aligned}\right\} \tag{5.2.3}$$

$$\left.\begin{aligned} m_1n_1 + m_2n_2 + m_3n_3 = 0 \\ n_1l_1 + n_2l_2 + n_3l_3 = 0 \\ l_1m_1 + l_2m_2 + l_3m_3 = 0 \end{aligned}\right\} \tag{5.2.4}$$

The relations (5.2.1) and (5.2.2), (5.2.3) and (5.2.4) can be remembered in a compact form.

Considering the elements of horizontal rows and vertical columns as the d. cs. of two sets of three mutually perpendicular lines, the results of (5.2.1), (5.2.2), (5.2.3) and (5.2.4) are easily obtained.

l_1	m_1	n_1
l_2	m_2	n_2
l_3	m_3	n_3

(A) **Show that**

$$E = \begin{vmatrix} l_1 & m_1 & n_1 \\ l_2 & m_2 & n_2 \\ l_3 & m_3 & n_3 \end{vmatrix} = \pm 1.$$

$$E^2 = \begin{vmatrix} l_1 & m_1 & n_1 \\ l_2 & m_2 & n_2 \\ l_3 & m_3 & n_3 \end{vmatrix} \times \begin{vmatrix} l_1 & m_1 & n_1 \\ l_2 & m_2 & n_2 \\ l_3 & m_3 & n_3 \end{vmatrix}$$

$$= \begin{vmatrix} l_1^2+m_1^2+n_1^2 & l_1l_2+m_1m_2+n_1n_2 & l_3l_1+m_3m_1+n_3n_1 \\ l_1l_2+m_1m_2+n_1n_2 & l_2^2+m_2^2+n_2^2 & l_2l_3+m_2m_3+n_2n_3 \\ l_3l_1+m_3m_1+n_3n_1 & l_2l_3+m_2m_3+n_2n_3 & l_3^2+m_3^2+n_3^2 \end{vmatrix}$$

$$= \begin{vmatrix} 1 & 0 & 0 \\ 0 & 1 & 0 \\ 0 & 0 & 1 \end{vmatrix} = 1.$$

$\therefore \quad E = \pm 1.$

(B) **Show that**

$$l_1 = \pm(m_2n_3 - m_3n_2),\ m_1 = \pm(n_2l_3 - n_3l_2),\ n_1 = \pm(l_2m_3 - l_3m_2).$$

We have $l_1l_2 + m_1m_2 + n_1n_2 = 0$, $l_1l_3 + m_1m_3 + n_1n_3 = 0$.

$$\therefore \quad \frac{l_1}{m_2n_3 - m_3n_2} = \frac{m_1}{n_2l_3 - n_3l_2} = \frac{n_1}{l_2m_3 - l_3m_2}$$

$$= \pm\frac{\sqrt{(l_1^2+m_1^2+n_1^2)}}{\sqrt{[\Sigma(m_2n_3 - m_3n_2)^2]}} = \pm 1.$$

$[\because \sin\theta = \sqrt{[\Sigma(m_2n_3 - m_3n_2)^2]}$ and here $\theta = 90°.]$

$$\therefore \quad l_1 = \pm(m_2n_3 - m_3n_2),\ m_1 = \pm(n_2l_3 - n_3l_2),\ n_1 = \pm(l_2m_3 - l_3m_2).$$

5.3 If A_x, A_y and A_z be the projections of an area A on the co-ordinate planes YZ, ZX, XY respectively, then

$$A^2 = A_x^2 + A_y^2 + A_z^2.$$

Let l, m, n be the d. cs. of the normal to the plane of A.

Then l = cosine of the angle between the plane of A and YZ plane. Therefore $A_x = lA$. Similarly $A_y = mA, A_z = nA$.

Thus $(l^2 + m^2 + n^2)A^2 = A_x^2 + A_y^2 + A_z^2$, or $A^2 = A_x^2 + A_y^2 + A_z^2$.

(ii) *Area of a triangle whose vertices are given.*

Let $(x_1, y_1, z_1), (x_2, y_2, z_2), (x_3, y_3, z_3)$ be the co-ordinates of the vertices of a triangle of area A. The projections of these points on the YZ plane are $(0, y_1, z_1)$, $(0, y_2, z_2)$ and $(0, y_3, z_3)$.

These three points on the YZ plane form a triangle whose area is equal to the projection of A on the YZ plane. If A_x, A_y, A_z be the projections of A on the YZ, ZX and XY planes respectively, then

$$A_x = \frac{1}{2}\begin{vmatrix} y_1 & z_1 & 1 \\ y_2 & z_2 & 1 \\ y_3 & z_3 & 1 \end{vmatrix}, \quad A_y = \frac{1}{2}\begin{vmatrix} z_1 & x_1 & 1 \\ z_2 & x_2 & 1 \\ z_3 & x_3 & 1 \end{vmatrix}, \quad A_z = \frac{1}{2}\begin{vmatrix} x_1 & y_1 & 1 \\ x_2 & y_2 & 1 \\ x_3 & y_3 & 1 \end{vmatrix}.$$

Now $A^2 = A_x^2 + A_y^2 + A_z^2$. Hence A is obtained.

5.4 Volume of a Tetrahedron in terms of Co-ordinates of the Vertices

Let $ABCD$ be a tetrahedron and the co-ordinates of the vertices A, B, C, D be $(x_1, y_1, z_1), (x_2, y_2, z_2), (x_3, y_3, z_3)$ and (x_4, y_4, z_4) respectively.

If V, Δ and p be the volume of the tetrahedron, area of the triangle BCD and perpendicular distance from A to the opposite face BCD respectively, then

$$V = \frac{1}{3}\Delta p.$$

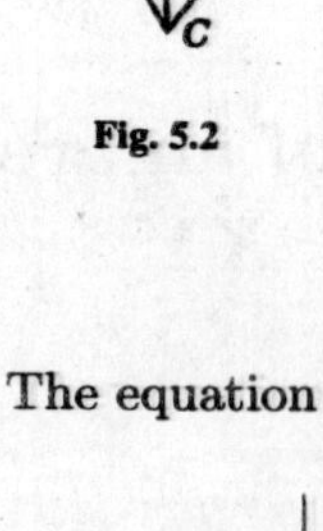

Fig. 5.2

If $\Delta_x, \Delta_y, \Delta_z$ are the projections of Δ on YZ, ZX and XY planes (co-ordinate planes) respectively, then

$$2\Delta_x = \begin{vmatrix} y_2 & z_2 & 1 \\ y_3 & z_3 & 1 \\ y_4 & z_4 & 1 \end{vmatrix}, \quad 2\Delta_y = \begin{vmatrix} z_2 & x_2 & 1 \\ z_3 & x_3 & 1 \\ z_4 & x_4 & 1 \end{vmatrix}, \quad 2\Delta_z = \begin{vmatrix} x_2 & y_2 & 1 \\ x_3 & y_3 & 1 \\ x_4 & y_4 & 1 \end{vmatrix}$$

and $\Delta^2 = \Delta_x^2 + \Delta_y^2 + \Delta_z^2$.

The equation of the plane BCD is

$$\begin{vmatrix} x & y & z & 1 \\ x_2 & y_2 & z_2 & 1 \\ x_3 & y_3 & z_3 & 1 \\ x_4 & y_4 & z_4 & 1 \end{vmatrix} = 0,$$

$$\text{or } x\begin{vmatrix} y_2 & z_2 & 1 \\ y_3 & z_3 & 1 \\ y_4 & z_4 & 1 \end{vmatrix} + y\begin{vmatrix} z_2 & x_2 & 1 \\ z_3 & x_3 & 1 \\ z_4 & x_4 & 1 \end{vmatrix} + z\begin{vmatrix} x_2 & y_2 & 1 \\ x_3 & y_3 & 1 \\ x_4 & y_4 & 1 \end{vmatrix} - \begin{vmatrix} x_2 & y_2 & z_2 \\ x_3 & y_3 & z_3 \\ x_4 & y_4 & z_4 \end{vmatrix} = 0.$$

$\therefore$ the coefficients of x, y, z are $2\Delta_x, 2\Delta_y$ and $2\Delta_z$ respectively.

$$p = \frac{x_1 \begin{vmatrix} y_2 & z_2 & 1 \\ y_3 & z_3 & 1 \\ y_4 & z_4 & 1 \end{vmatrix} + y_1 \begin{vmatrix} z_2 & x_2 & 1 \\ z_3 & x_3 & 1 \\ z_4 & x_4 & 1 \end{vmatrix} + z_1 \begin{vmatrix} x_2 & y_2 & 1 \\ x_3 & y_3 & 1 \\ x_4 & y_4 & 1 \end{vmatrix} - \begin{vmatrix} x_2 & y_2 & z_2 \\ x_3 & y_3 & z_3 \\ x_4 & y_4 & z_4 \end{vmatrix}}{2\sqrt{\Delta_x^2 + \Delta_y^2 + \Delta_x^2}}$$

$$= \frac{1}{2\Delta} \begin{vmatrix} x_1 & y_1 & z_1 & 1 \\ x_2 & y_2 & z_2 & 1 \\ x_3 & y_3 & z_3 & 1 \\ x_4 & y_4 & z_4 & 1 \end{vmatrix}.$$

$$\therefore \quad V = \frac{1}{3}\Delta p = \frac{1}{6} \begin{vmatrix} x_1 & y_1 & z_1 & 1 \\ x_2 & y_2 & z_2 & 1 \\ x_3 & y_3 & z_3 & 1 \\ x_4 & y_4 & z_4 & 1 \end{vmatrix} = \frac{1}{6} \begin{vmatrix} x_1 - x_2 & y_1 - y_2 & z_1 - z_2 \\ x_2 - x_3 & y_2 - y_3 & z_2 - z_3 \\ x_3 - x_4 & y_3 - y_4 & z_3 - z_4 \end{vmatrix}.$$

Corollary. If one vertex is the origin, i.e., $x_4 = y_4 = z_4 = 0$, then

$$V = \frac{1}{6} \begin{vmatrix} x_1 & y_1 & z_1 \\ x_2 & y_2 & z_2 \\ x_3 & y_3 & z_3 \end{vmatrix}.$$

Worked Out Examples

1. Find the form of the equation $x^2 + y^2 + z^2 + 12x - 6y + 8z - 3 = 0$ referred to new origin $(-6, 3, -4)$ without changing the direction of axes.

If (x, y, z) and (x', y', z') be the co-ordinates of the same points w.r.t. old system and new system, then

$$x = x' - 6,\ y = y' + 3,\ z = z' - 4.$$

For this transformation the equation changes to

$$(x' - 6)^2 + (y' + 3)^2 + (z' - 4)^2 + 12(x' - 6)$$
$$-6(y' + 3) + 8(z' - 4) - 3 = 0,$$
$$\text{or} \quad x'^2 + y'^2 + z'^2 = 64.$$

In the usual symbol of current co-ordinates this equation is

$$x^2 + y^2 + z^2 = 64.$$

2. Show that the equation $lx + my + nz = 0$ becomes $z = 0$ when referred to new axes through the origin with d. rs. $-m, l, 0; -ln, -mn, l^2 + m^2; l, m, n$. $[l^2 + m^2 + n^2 = 1]$

The d. cs. of the new axes are

$$\frac{-m}{\sqrt{l^2 + m^2}}, \frac{l}{\sqrt{l^2 + m^2}}, 0;\ \frac{-ln}{\sqrt{l^2 + m^2}}, \frac{-mn}{\sqrt{l^2 + m^2}}, \sqrt{l^2 + m^2};\ l, m, n.$$

By the rule of transformation

$$x = \frac{-m}{\sqrt{l^2+m^2}}x' - \frac{ln}{\sqrt{l^2+m^2}}y' + lz',$$

$$y = \frac{l}{\sqrt{l^2+m^2}}x' - \frac{mn}{\sqrt{l^2+m^2}}y' + mz'$$

and $z = \sqrt{l^2+m^2}y' + nz'$.

Now the equation changes to

$$l\left(-\frac{m}{\sqrt{l^2+m^2}}x' - \frac{ln}{\sqrt{l^2+m^2}}y' + lz'\right)$$

$$+m\left(\frac{l}{\sqrt{l^2+m^2}}x' - \frac{mn}{\sqrt{l^2+m^2}}y' + mz'\right) + n(\sqrt{l^2+m^2}y' + nz') = 0,$$

or $(l^2+m^2+n^2)z' = 0$, or $z' = 0$.

In the usual symbol of current co-ordinates this is $z = 0$.

3. Find the volume of the tetrahedron whose vertices are $(0,1,2)$, $(1,0,2)$, $(1,2,0)$ and $(1,2,1)$.

$$\text{Volume} = \frac{1}{6}\begin{vmatrix} 0 & 1 & 2 & 1 \\ 1 & 0 & 2 & 1 \\ 1 & 2 & 0 & 1 \\ 1 & 2 & 1 & 1 \end{vmatrix} = \frac{1}{6}\begin{vmatrix} 0 & 1 & 2 & 1 \\ 1 & -1 & 0 & 0 \\ 1 & 1 & -2 & 0 \\ 1 & 1 & -1 & 0 \end{vmatrix} \quad \left[\begin{array}{l}\text{subtracting the first} \\ \text{row from the rest}\end{array}\right]$$

$$= -\frac{1}{6}\begin{vmatrix} 1 & -1 & 0 \\ 1 & 1 & -2 \\ 1 & 1 & -1 \end{vmatrix} = -\frac{1}{6}\begin{vmatrix} 1 & 0 & 0 \\ 1 & 2 & -2 \\ 1 & 2 & -1 \end{vmatrix} \quad \left[\begin{array}{l}\text{adding the first} \\ \text{column with the} \\ \text{second}\end{array}\right]$$

$$= -\frac{1}{6}\begin{vmatrix} 2 & -2 \\ 2 & -1 \end{vmatrix} = -\frac{2}{6} = -\frac{1}{3}$$

$$\therefore \quad \text{Volume} = \frac{1}{3}.$$

4. Prove that the volume of the tetrahedron formed by the four planes $lx+my+nz = p$, $lx+my = 0$, $my+nz = 0$ and $nz+lx = 0$ is $\frac{2}{3}\frac{p^3}{lmn}$.

$$lx+my+nz = p, \tag{1}$$

$$lx+my = 0, \tag{2}$$

$$my+nz = 0, \tag{3}$$

$$nz+lx = 0. \tag{4}$$

The planes (1), (2) and (3) meet at $\left(\frac{p}{l}, -\frac{p}{m}, \frac{p}{n}\right)$.
The planes (1), (3) and (4) meet at $\left(\frac{p}{l}, \frac{p}{m}, -\frac{p}{n}\right)$.
The planes (1), (2) and (4) meet at $\left(-\frac{p}{l}, \frac{p}{m}, \frac{p}{n}\right)$.

The planes (2), (3) and (4) meet at $(0,0,0)$.
These are the vertices of the tetrahedron. Hence the volume

$$= \frac{1}{6}\begin{vmatrix} -\frac{p}{l} & \frac{p}{m} & \frac{p}{n} \\ \frac{p}{l} & -\frac{p}{m} & \frac{p}{n} \\ \frac{p}{l} & \frac{p}{m} & -\frac{p}{n} \end{vmatrix} = \frac{p^3}{6lmn}\begin{vmatrix} -1 & 1 & 1 \\ 1 & -1 & 1 \\ 1 & 1 & -1 \end{vmatrix} = \frac{2}{3}\frac{p^3}{lmn}.$$

Exercise

1. Find the form of the equation $x^2+y^2+z^2-4x-6y-8z-20=0$ referred to $(2,3,4)$ as new origin without changing the direction of axes.

2. Find the translation which will transform the equation

$$x^2+y^2-4z^2-2x+4y+24z-31=0 \text{ into } x'^2+y'^2-4z'^2=0.$$

[C.H. '82]

3. If OA, OB, OC are three mutually perpendicular lines through the origin with d. cs. $l_1, m_1, n_1; l_2, m_2, n_2; l_3, m_3, n_3$ and if $OA = OB = OC = a$, show that the equation of the plane ABC is $(l_1+l_2+l_3)x+(m_1+m_2+m_3)y+(n_1+n_2+n_3)z=0$.

4. Find the volume of the tetrahedron whose vertices are
(a) $(0,0,0), (\frac{d}{a},0,0), (0,\frac{d}{b},0), (0,0,\frac{d}{c})$;
(b) $(1,0,4), (1,1,1), (4,1,0), (0,2,3)$.

5. Find the volume of the tetrahedron formed by the planes

$$x+y=0,\ y+z=0,\ z+x=0 \text{ and } x+y+z=1.$$

6. Find the area of the triangle whose vertices are the points $(1,2,3), (-2,1,5)$ and $(3,4,2)$.

7. Show that the area of the triangle made by the plane $\frac{x}{a}+\frac{y}{b}+\frac{z}{c}=1$ with the co-ordinate axes is $\frac{1}{2}\sqrt{a^2b^2+b^2c^2+c^2a^2}$.

[HINTS. If the plane cuts the x-axis, y-axis and z-axis at the points A, B, C, then the co-ordinates of A, B, C are $(a,0,0), (0,b,0)$ and $(0,0,c)$. Let Δ be the area of the triangle ABC and $\Delta_x, \Delta_y, \Delta_z$ be the areas of the projections of the same on the co-ordinate planes. The vertices of the projection of the triangle on the yz-plane are $(0,0,0), (0,b,0)$ and $(0,0,c)$.

$$\therefore \quad \Delta_x = \frac{1}{2}bc.$$

Similarly

$$\Delta_y = \frac{1}{2}ca, \quad \Delta_z = \frac{1}{2}ab.$$

Now

$$\Delta^2 = \Delta_x^2+\Delta_y^2+\Delta_z^2 = \frac{1}{4}(b^2c^2+c^2a^2+a^2b^2)$$

$$\therefore \quad \Delta = \frac{1}{2}\sqrt{b^2c^2+c^2a^2+a^2b^2}.$$

8. If the co-ordinates of A, B, C be $(2, 1, -3), (0, -3, 2), (0, -2, 0)$, prove that the locus of the point P such that the volume of the tetrahedron $PABC$ is 2 is $3x - 4y - 2z = 20$.
[HINTS. Let the co-ordinates of P be (x, y, z).
Then the volume

$$V = \frac{1}{6}\begin{vmatrix} x & y & z & 1 \\ 2 & 1 & -3 & 1 \\ 0 & -3 & 2 & 1 \\ 0 & -2 & 0 & 1 \end{vmatrix}$$

$$= \frac{1}{6}\begin{vmatrix} x & y+2 & z & 0 \\ 2 & 3 & -3 & 0 \\ 0 & -1 & 2 & 0 \\ 0 & -2 & 0 & 1 \end{vmatrix} \quad \left[\begin{array}{l}\text{subtracting the 4th row}\\ \text{from the rest}\end{array}\right]$$

$$= \frac{1}{6}\begin{vmatrix} x & y+2 & z \\ 2 & 3 & -3 \\ 0 & -1 & 2 \end{vmatrix}$$

$$= \frac{1}{6}\{x(6-3) + 2(-z - 2y - 4)\}$$

$$= \frac{1}{6}(3x - 4y - 2z - 8).$$

By the condition $\frac{1}{6}(3x - 4y - 2z - 8) = 2$, or $3x - 4y - 2z = 20$.]

9. If the axes x, y and z are rectangular, prove that the substitutions $x = \frac{x'}{\sqrt{6}} + \frac{y'}{\sqrt{2}} + \frac{z'}{\sqrt{3}}$, $y = \frac{-2}{\sqrt{6}}x' + \frac{z'}{\sqrt{3}}$, $z = \frac{x'}{\sqrt{6}} - \frac{y'}{\sqrt{2}} + \frac{z'}{\sqrt{3}}$ give a transformation to another set of rectangular axes in which the plane $x + y + z = 0$ becomes the plane $z' = 0$ and hence prove that the intersection of the surface $yz + zx + xy + 1 = 0$ by the plane $x + y + z = 0$ is a circle of radius $\sqrt{2}$.

Answers

1. $x^2 + y^2 + z^2 = 49$; **2.** $(1, -2, 3)$; **4.** (a) $\frac{1}{6}\frac{d^3}{abc}$; (b) $\frac{7}{3}$; **5.** $\frac{2}{3}$; **6.** $\frac{\sqrt{13}}{2}$.

Chapter 6

Quadric Surfaces

6.1 Quadric Surface

A surface defined in space by an equation of the second degree in x, y, z is called a quadric surface. Here we discuss a few types of quadric surfaces.

[A]

Sphere

Definition 1 *A sphere is the locus of a point which moves in such a way that its distance from a fixed point is constant. The fixed point is called the centre and the distance of the moving point from the centre is known as radius.*

6.2 Equation of a Sphere

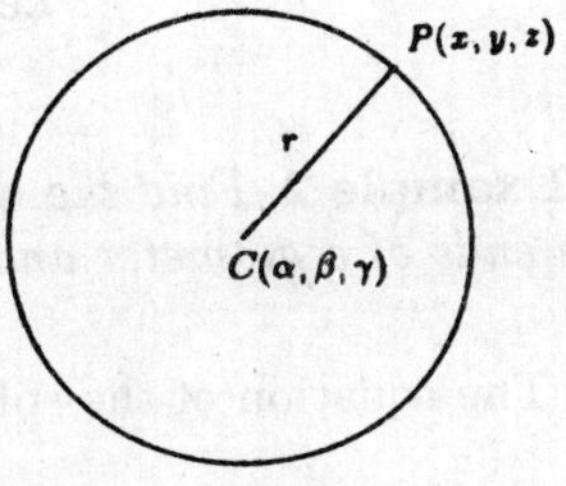

Fig. 6.1

If (α, β, γ) be the centre and r be the radius, then by definition the equation of this sphere is

$$(x-\alpha)^2 + (y-\beta)^2 + (z-\gamma)^2 = r^2.$$

This can be written as

$$x^2 + y^2 + z^2 - 2\alpha x - 2\beta y - 2\gamma z + \alpha^2 + \beta^2 + \gamma^2 - r^2 = 0.$$

Putting $\alpha = -u, \beta = -v, \gamma = -w$ and $\alpha^2 + \beta^2 + \gamma^2 - r^2 = d$, the equation reduces to $x^2 + y^2 + z^2 + 2ux + 2vy + 2wz + d = 0$.

It is the general equation of a sphere. The co-ordinates of the centre are $(-u, -v, -w)$ and radius $= \sqrt{u^2 + v^2 + w^2 - d}$.

Corollary. If the centre is at the origin, the equation of the sphere is $x^2 + y^2 + z^2 = r^2$. (Putting $\alpha = \beta = \gamma = 0$.)

► **Example 1** *Find the equation of the sphere whose centre is* $(2, 3, -4)$ *and radius is 5.*

The equation is

$$(x-2)^2 + (y-3)^2 + (z+4)^2 = 5^2,$$

$$\text{or} \quad x^2 + y^2 + z^2 - 4x - 6y + 8z + 4 = 0.$$

6.3 Diameter Form

To find the equation of the sphere having (x_1, y_1, z_1) *and* (x_2, y_2, z_2) *as the ends of a diameter.*

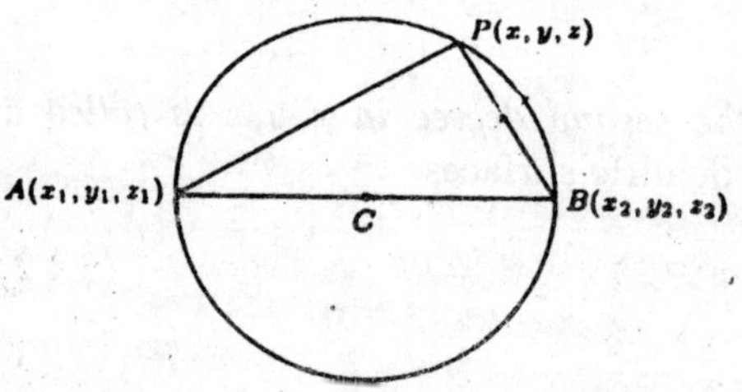

Fig. 6.2

Let the given points be A and B and $P(x, y, z)$ be a point on the sphere.

Since AB is a diamter, $\angle APB = 90°$, i.e., PA and PB are at right angle. The d. rs. of PA and PB are $x - x_1, y - y_1, z - z_1$ and $x - x_2, y - y_2, z - z_2$ respectively.

$$\because \ PA \perp PB, (x - x_1)(x - x_2) + (y - y_1)\cdot$$

$$(y - y_2) + (z - z_1)(z - z_2) = 0.$$

It is the required equation.

[The centre of the sphere is at the middle point of the line AB,

$$\text{i.e.,} \quad \left(\frac{x_1 + x_2}{2}, \frac{y_1 + y_2}{2}, \frac{z_1 + z_2}{2}\right).\Bigg]$$

► **Example 2** *Find the equation of the sphere which has* $(3, 4, -1)$ *and* $(-1, 2, 3)$ *as the ends of a diameter and find its centre and radius.* **[B.U. '66]**

The equation of the sphere is

$$(x-3)(x+1) + (y-4)(y-2) + (z+1)(z-3) = 0,$$

$$\text{or} \quad x^2 + y^2 + z^2 - 2x - 6y - 2z + 2 = 0.$$

The centre is $(1, 3, 1)$ and radius $= \sqrt{1 + 9 + 1 - 2} = 3$.

6.4 Through Four non-coplanar Points One and only One Sphere Passes

Let the points be $(x_i, y_i, z_i), i = 1, 2, 3, 4$.

Since the points are non-coplanar,

$$\begin{vmatrix} x_1 & y_1 & z_1 & 1 \\ x_2 & y_2 & z_2 & 1 \\ x_3 & y_3 & z_3 & 1 \\ x_4 & y_4 & z_4 & 1 \end{vmatrix} \neq 0.$$

This implies that the system of linear equations

$$\left.\begin{aligned} 2\alpha x_1 + 2\beta y_1 + 2\gamma z_1 + \delta &= -(x_1^2 + y_1^2 + z_1^2), \\ 2\alpha x_2 + 2\beta y_2 + 2\gamma z_2 + \delta &= -(x_2^2 + y_2^2 + z_2^2), \\ 2\alpha x_3 + 2\beta y_3 + 2\gamma z_3 + \delta &= -(x_3^2 + y_3^2 + z_3^2) \\ 2\alpha x_4 + 2\beta y_4 + 2\gamma z_4 + \delta &= -(x_4^2 + y_4^2 + z_4^2) \end{aligned}\right\} \tag{6.4.1}$$

has a unique solution.

Let the solution be $\alpha = u, \beta = v, \gamma = w, \delta = d$.

Thus the four points lie on the sphere

$$x^2 + y^2 + z^2 + 2ux + 2vy + 2wz + d = 0. \tag{6.4.2}$$

If

$$x^2 + y^2 + z^2 + 2u'x + 2v'y + 2w'z + d' = 0 \tag{6.4.3}$$

is another sphere which passes through the given points, then the equations (6.4.1) have solution (u', v', w', d'). But the system (6.4.1) has unique solution. So $u' = u, v' = v, w' = w, d' = d$.

Hence only one sphere passes through the given four points.

Note 1. If the four points are coplanar and any three of them are not collinear, many spheres can pass through them.

Note 2. If the four points are coplanar and three of them are collinear, no sphere passes through them.

► **Example 3** *Find the equation of the sphere through the four points* $(4, -1, 2)$, $(0, -2, 3), (1, -5, -1)$ *and* $(2, 0, 1)$.

Since

$$\begin{vmatrix} 4 & -1 & 2 & 1 \\ 0 & -2 & 3 & 1 \\ 1 & -5 & -1 & 1 \\ 2 & 0 & 1 & 1 \end{vmatrix} = 42 \neq 0,$$

the points are non-coplanar and the sphere passing through these four points is unique.

Let the equation of the sphere be $x^2 + y^2 + z^2 + 2ux + 2vy + 2wz + d = 0$.

Since the given points lie on it.

$$8u - 2v + 4w + d = -21 \tag{1}$$

$$-4v + 6w + d = -13 \tag{2}$$

$$2u - 10v - 2w + d = -27 \tag{3}$$

$$4u + 2w + d = -5. \tag{4}$$

Subtracting (2) from (1), (3) and (4),

$$4u + v - w = -4, \tag{5}$$

$$u - 3v - 4w = -7 \tag{6}$$

$$\text{and} \quad u + v - w = 2. \tag{7}$$

By (5) and (7), $u = -2$. Then $v - w = 4$ and $3v + 4w = 5$.
From these two $v = 3, w = -1$.
Putting these values of v and w in (2), $d = 5$.
Hence the equation is $x^2 + y^2 + z^2 - 4x + 6y - 2z + 5 = 0$.

6.5 Plane Section of a Sphere

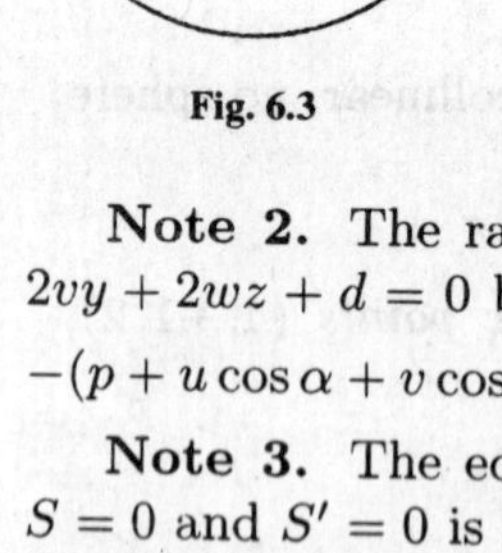

Fig. 6.3

Any section of a sphere by a plane is a circle.

Let PQ be a section of the sphere of centre C. CN is perpendicular to the plane of the section. N is the centre of the circle of this section and NP is the radius of it.

From the fig. 6.3, $CP^2 = CN^2 + NP^2$.

Note 1. Let $S = 0$ and $U = 0$ be the equations of a sphere and a plane respectively. If the plane intersects the sphere then the two equations taken together represent a circle. Again, $S + \lambda U = 0$ represents a sphere through the circle $S = 0, U = 0$.

Note 2. The radius of the circle of section of the sphere $x^2 + y^2 + z^2 + 2ux + 2vy + 2wz + d = 0$ by the plane $x\cos\alpha + y\cos\beta + z\cos\gamma = p$ is $[(u^2 + v^2 + w^2 - d) - (p + u\cos\alpha + v\cos\beta + w\cos\gamma)^2]^{1/2}$.

Note 3. The equation of the plane of the circle through the two given spheres $S = 0$ and $S' = 0$ is $S - S' = 0$. It is known as the *radical plane.* The equation of any sphere through the circle $S = 0, S' = 0$ is $S + \lambda S' = 0$, where λ is a parameter.

Worked Out Examples

1. Find the equation of the sphere through the four points $(0,0,0), (a,0,0), (0,b,0)$ and $(0,0,c)$. **[B.U. ' 87, '96 N.B. '91; C.U. '61, '71]**

Let the equation of the sphere be

$$x^2 + y^2 + z^2 + 2ux + 2vy + 2wz + d = 0.$$

As it passes through the points $(0,0,0), (a,0,0), (0,b,0)$ and $(0,0,c)$, $d = 0$,

$$a^2 + 2ua = 0, \quad \text{or} \quad 2u = -a \quad (\because\ a \neq 0),$$
$$b^2 + 2vb = 0, \quad \text{or} \quad 2v = -b \quad (\because\ b \neq 0),$$
$$c^2 + 2wc = 0, \quad \text{or} \quad 2w = -c \quad (\because\ c \neq 0).$$

$\therefore$ The equation of the sphere is $x^2 + y^2 + z^2 - ax - by - cz = 0$.

2. Prove that the sphere $x^2 + y^2 + z^2 + 2ux + 2vy + 2wz + d = 0$ cuts the sphere $x^2 + y^2 + z^2 + 2u'x + 2v'y + 2w'z + d' = 0$ in a great circle if $2(uu' + vv' + ww') = 2r'^2 + d + d'$, where r' is the radius of the second sphere.

The equation of the plane of the circle through the given spheres is $2(u-u')x + 2(v-v')y + 2(w-w')z + d - d' = 0$. If this circle is a great circle of the second sphere, then the centre $(-u', -v', -w')$ lies on it.

$$\therefore\ -2(u-u')u' - 2(v-v')v' - 2(w-w')w' + d - d' = 0,$$
$$\text{or} \quad 2(uu' + vv' + ww') = 2(u'^2 + v'^2 + w'^2 - d') + d + d',$$
$$\text{or} \quad 2(uu' + vv' + ww') = 2r'^2 + d + d'.$$

3. Find the greatest and least distance from the point $(2,-1,1)$ to the sphere

$$x^2 + y^2 + z^2 - 8x + 4y - 6z + 4 = 0.$$

The centre and radius of the sphere are $(4,-2,3)$ and 5.

The distance between the point $(2,-1,1)$ and the centre

$$= \sqrt{(4-2)^2 + (-2+1)^2 + (3-1)^2} = 3.$$

Since $3 < 5$, the point $(2,-1,1)$ is within the sphere.

The equation of the line through the point and the centre is

$$\frac{x-2}{2} = \frac{y+1}{-1} = \frac{z-1}{2} = r(\text{say}).$$

If the line meets the sphere at $(2r+2, -r-1, 2r+1)$ then $(2r+2)^2 + (r+1)^2 + (2r+1)^2 - 8(2r+2) - 4(r+1) - 6(2r+1) + 4 = 0$,

$$\text{or} \quad 9r^2 - 18r - 16 = 0, \quad \text{or} \quad r = \frac{8}{3}, -\frac{2}{3}.$$

$\therefore$ the line meets the sphere at points

$$\left(\frac{2}{3}, -\frac{1}{3}, -\frac{1}{3}\right) \quad \text{and} \quad \left(\frac{22}{3}, -\frac{11}{3}, \frac{19}{3}\right).$$

These points lie on the diameter through the given points.

Hence the least distance

$$= \sqrt{\left(2-\frac{2}{3}\right)^2 + \left(-1+\frac{1}{3}\right)^2 + \left(1+\frac{1}{3}\right)^2} = 2$$

and the greatest distance $= 10 - 2 = 8$.

4. Prove that the equation to a sphere circumscribing the tetrahedron whose faces are $\frac{y}{b}+\frac{z}{c}=0, \frac{z}{c}+\frac{x}{a}=0, \frac{x}{a}+\frac{y}{b}=0, \frac{x}{a}+\frac{y}{b}+\frac{z}{c}=1$ is

$$\frac{x^2+y^2+z^2}{a^2+b^2+c^2}-\frac{x}{a}-\frac{y}{b}-\frac{z}{c}=0.$$

$$\frac{y}{b}+\frac{z}{c}=0, \tag{1}$$

$$\frac{z}{c}+\frac{x}{a}=0, \tag{2}$$

$$\frac{x}{a}+\frac{y}{b}=0, \tag{3}$$

$$\frac{x}{a}+\frac{y}{b}+\frac{z}{c}=1. \tag{4}$$

If we solve these four equations taken three at a time, then we get the vertices of the tetrahedron. These are $(0,0,0), (-a,b,c), (a,-b,c)$ and $(a,b,-c)$.

Let the equation of the sphere through these four points be

$$x^2+y^2+z^2+2ux+2vy+2wz+d=0.$$

The vertices will be on the sphere, if $d=0$,

$$a^2+b^2+c^2-2ua+2vb+2wc=0, \tag{5}$$

$$a^2+b^2+c^2+2ua-2vb+2wc=0, \tag{6}$$

$$a^2+b^2+c^2+2ua+2vb-2wc=0. \tag{7}$$

Adding (5) and (6),

$$2w=-\frac{a^2+b^2+c^2}{c}.$$

Similarly

$$2v=-\frac{a^2+b^2+c^2}{b}, \quad 2u=-\frac{a^2+b^2+c^2}{a}.$$

$\therefore$ the required equation is

$$x^2+y^2+z^2-(a^2+b^2+c^2)\left(\frac{x}{a}+\frac{y}{b}+\frac{z}{c}\right)=0,$$

$$\text{or} \quad \frac{x^2+y^2+z^2}{a^2+b^2+c^2}-\frac{x}{a}-\frac{y}{b}-\frac{z}{c}=0.$$

5. Find the equation of the sphere which passes through the origin and touches the sphere $x^2+y^2+z^2=56$ at the point $(2,-4,6)$. **[C.U. '66, '96]**

$(2,-4,6)$ is a point on the given sphere and the origin is the centre of it. Since the required sphere passes through the origin, it touches the given sphere internally. Moreover it is described on the segment joining the points $(0,0,0)$ and $(2,-4,6)$ as a diameter. Hence the required equation is

$$x(x-2)+y(y+4)+z(z-6)=0,$$

$$\text{or} \quad x^2+y^2+z^2-2x+4y-6z=0.$$

6. Find the co-ordinates of the centre and the radius of the circle

$$x - 2y - 2z + 7 = 0, x^2 + y^2 + z^2 - 2x + 6y + 4z - 35 = 0.$$

The centre of the given sphere is $(1, -3, -2)$ and the radius is 7. The distance of the given plane from the centre

$$= \frac{1 + 6 + 4 + 7}{\sqrt{1^2 + 2^2 + 2^2}} = 6.$$

$\therefore$ radius of the circle $= \sqrt{7^2 - 6^2} = \sqrt{13}$.

The equation of the line perpendicular to the plane and passing through the centre $(1, -3, -2)$ is

$$\frac{x - 1}{1} = \frac{y + 3}{-2} = \frac{z + 2}{-2}.$$

Any point on the line is $(r + 1, -2r - 3, -2r - 2)$.

If this point satisfies the equation of the plane, then

$$r + 1 + 2(2r + 3) + 2(2r + 2) + 7 = 0, \quad \text{or} \quad 9r = -18, \quad \text{or} \quad r = -2.$$

$\therefore$ the co-ordinates of the centre are $(-1, 1, 2)$.

7. Obtain the equation of the circle lying on the sphere

$$x^2 + y^2 + z^2 - 2x + 2y - 4z + 3 = 0$$

and having its centre at the point $(2, 2, -3)$.

The equations of the circle are the given equation of the sphare and a plane passing through $(2, 2, -3)$ and normal to the line joining $(2, 2, -3)$ and the centre of the given sphere.

The centre of of the sphere is $(1, -1, 2)$. The d. rs. of the line joining $(2, 2, -3)$ and $(1, -1, 2)$ and $1, 3, -5$.

$\therefore$ the required plane is $(x - 2) + 3(y - 2) - 5(z + 3) = 0$, or $x + 3y - 5z - 23 = 0$.

Hence the equations of the circle are

$$x^2 + y + z^2 - 2x + 2y - 4z + 3 = 0$$

$$\text{and} \quad x + 3y - 5z - 23 = 0.$$

8. Find the equation of the sphere for which the circle $x^2 + y^2 + z^2 + 2x - 4y + 2z + 5 = 0, x - 2y + 3z + 1 = 0$ is a great circle. [C.U. '90]

[**Note.** The section of a sphere by a plane through the centre of it is called a great circle.]

The equation of a sphere through the given circle is

$$x^2 + y^2 + z^2 + 2x - 4y + 2z + 5 + \lambda(x - 2y + 3z + 1) = 0,$$

where λ is a parameter,

$$\text{or} \quad x^2 + y^2 + z^2 + 2\left(1 + \frac{\lambda}{2}\right)x - 2(2 + \lambda)y + 2\left(1 + \frac{3\lambda}{2}\right)z + 5 + \lambda = 0.$$

Its centre is

$$\left\{-\left(1 + \frac{\lambda}{2}\right), \quad 2 + \lambda, \quad -\left(1 + \frac{3\lambda}{2}\right)\right\}.$$

If the circle is a greate circle, this centre lies on the given plane.

$$\therefore\quad -\left(1+\frac{\lambda}{2}\right)-2(2+\lambda)-3\left(1+\frac{3\lambda}{2}\right)+1=0,$$

$$\text{or}\quad -7\lambda-7=0,\quad \text{or}\quad \lambda=-1.$$

Hence the required equation is

$$x^2+y^2+z^2+x-2y-z+4=0.$$

9. Obtain the equation of the sphere through the points $(1,1,2)$ and $(2,-2,3)$ and having its centre on the line $2x+3y=0=5x+y-z$. **[C.U. '98]**

Let the equation of the sphere be $x^2+y^2+z^2+2ux+2vy+2wz+d=0$.
The centre is $(-u,-v,-w)$.
This point lies on $2x+3y=0=5x+y-z$.

$$\therefore\quad 2u+3v=0\quad\text{and}\quad 5u+v-w=0.$$

From these two equations

$$v=-\frac{2}{3}u,\quad w=\frac{13}{3}u. \tag{1}$$

Since $(1,1,2)$ and $(2,-2,3)$ lie on the sphere

$$2u+2v+4w+d+6=0 \tag{2}$$

$$\text{and}\quad 4u-4v+6w+d+17=0. \tag{3}$$

By (1), (2) and (3) $u=-\frac{3}{4}, v=\frac{1}{2}, w=-\frac{13}{4}, d=\frac{15}{2}$.
Thus the equation of the sphere is

$$x^2+y^2+z^2-\frac{3}{2}x+y-\frac{13}{2}z+\frac{15}{2}=0,$$

$$\text{i.e.,}\quad 2(x^2+y^2+z^2)-3x+2y-13z+15=0.$$

10. A plane passes through a fixed point (p,q,r) and cuts the axes in A,B,C. Show that the locus of the centre of the sphere $OABC$ is

$$\frac{p}{x}+\frac{q}{y}+\frac{r}{z}=2.$$

Let the equation of the plane be

$$\frac{x}{a}+\frac{y}{b}+\frac{z}{c}=1. \tag{1}$$

Now the co-ordinates of A,B,C are $(a,0,0),(0,b,0)$ and $(0,0,c)$ respectively.
The equation of the sphere $OABC$ is

$$x^2+y^2+z^2-ax-by-cz=0.$$

If (α,β,γ) be the centre of this sphere, then

$$\alpha=\frac{a}{2},\quad \beta=\frac{b}{2}\quad\text{and}\quad \gamma=\frac{c}{2}. \tag{2}$$

Since the plane (1) passes through the point (p, q, r),

$$\frac{p}{a} + \frac{q}{b} + \frac{r}{c} = 1. \tag{3}$$

Putting the value of a, b and c from (2) in (3), we have

$$\frac{p}{\alpha} + \frac{q}{\beta} + \frac{r}{\gamma} = 2.$$

Hence the required locus is $\frac{p}{x} + \frac{q}{y} + \frac{r}{z} = 2$.

Exercise A

1. Find the centre and radius of the following sphere.

   (a) $x^2 + y^2 + z^2 - 8y + 10z - 10 = 0$.

   (b) $2(x^2 + y^2 + z^2) - 2x + 4y - 6z = 15$. **[B.U. '68, C.U. '89]**

2. Find the equation of the sphere passing through the following points.

   (a) $(0, 0, 0), (0, 1, -1), (-1, 2, 0), (1, 2, 3)$. **[C.U. '63]**

   (b) $(1, 1, 1), (-2, 1, 2), (3, -3, 1), (-1, 2, -1)$.

3. Find the equation of the sphere which has the line segment joining the points $(2, -2, 4)$ and $(-2, 2, 2)$ as diameter. **[B.U. '92]**

4. Discuss the position of the point $(2, -3, 0)$ w.r.t. the sphere $x^2 + y^2 + z^2 + 2x - 4y - 4z + 8 = 0$.

5. Find the equation of the sphere which passes through the origin and makes equal intercepts of unit length of the axes.

6. Find the equation of the sphere circumscribing the tetrahedron whose faces are $\frac{y}{3} + \frac{z}{4} = 0, \frac{z}{4} + \frac{x}{2} = 0, \frac{x}{2} + \frac{y}{3} = 0, \frac{x}{2} + \frac{y}{3} + \frac{z}{4} = 1$.

7. Show that the equation to the sphere passing through the circle $x^2 + y^2 + z^2 = 9, 2x + 3y + 4z = 5$ and the point $(1, 2, 3)$ is $3(x^2 + y^2 + z^2) - 2x - 3y - 4z - 22 = 0$.

8. Show that the equation of the sphere for which the circle $x^2 + y^2 + z^2 + 7y - 2z + 2 = 0, 2x + 3y + 4z - 8 = 0$ is a greate circle is $x^2 + y^2 + z^2 - 2x + 4y - 6z + 10 = 0$.

9. Find the co-ordinates of the centre and radius of the circle $x^2 + y^2 + z^2 - 2y - 4z - 11 = 0, x + 2y + 2z = 15$.

10. Show that the equations of that circle on the sphere $x^2 + y^2 + z^2 - 2x + 4y - 6z + 3 = 0$ whose centre is $(2, 3, -4)$ are $x^2 + y^2 + z^2 - 2x + 4y - 6z + 3 = 0$ and $x + 5y - 7z - 45 = 0$.

11. Find the equation of the sphere which passes through the circle $x^2 + y^2 = 4, z = 0$ and is cut by the plane $x + 2y + 2z = 0$ in a circle of radius 3. **[N.B. '82, '89]**

    [HINTS. Let the sphere through the circle $x^2 + y^2 = 4, z = 0$ be $x^2 + y^2 + z^2 - 4 + 2\lambda z = 0$. The centre is $(0, 0, -\lambda)$ and radius $= \sqrt{\lambda^2 + 4}$.

    The distance of the plane from the centre $= \frac{-2\lambda}{3}$.

    Here $(\lambda^2 + 4) - \frac{4\lambda^2}{9} = 9$, or $\lambda = \pm 3$.

    Hence there are two circles with the required property and the equations of them are $x^2 + y^2 + z^2 + 6z - 4 = 0$ and $x^2 + y^2 + z^2 - 6z - 4 = 0$.]

12. (a) A sphere of radius k passes through the origin and meets the axes in A, B, C. Prove that the locus of the centroid of the triangle ABC is the sphere $9(x^2+y^2+z^2) = 4k^2$. [B.U. '99]

(b) A sphere of constant radius r passes through the origin and cuts the axes in A, B, C. Prove that the locus of the foot of the perpendicular from O to the plane ABC is given by

$$(x^2+y^2+z^2)^2(x^{-2}+y^{-2}+z^{-2}) = 4r^2.$$ [B.H. '85]

13. Show that the greatest and least distance from the point $(1,-1,2)$ to the sphere $x^2+y^2+z^2-4x+6y-8z-71=0$ are 13 and 7 respectively.

14. Find the equation of the circle passing through the points $(2,0,1), (-2,1,0)$ and $(0,3,5)$.

[HINTS. It is the plane section of the sphere passing through the given points and the origin by the plane through the given points.]

15. Find the equation of the sphere passing through the points $(2,0,0), (0,2,0), (0,0,2)$ and having the least possible radius.

[HINTS. Let the sphere be $(x-a)^2+(y-b)^2+(z-c)^2=r^2$.

If it passes through the given points then

$$(2-a)^2+b^2+c^2=r^2, a^2+(2-b)^2+c^2=r^2, a^2+b^2+(2-c)^2=r^2.$$

From these three equations $a=b=c$.

$$(2-a)^2+a^2+a^2=r^2, \quad \text{or} \quad 3\left[\left(a-\frac{2}{3}\right)^2+\frac{8}{9}\right]=r^2.$$

Therefore the least value of r^2 is $\frac{8}{3}$ when $a=\frac{2}{3}$.

Thus the required equation is $3(x^2+y^2+z^2)-4(x+y+z)-4=0$.]

Answers

1. (i) $(0,4,-5), \sqrt{51}$; (ii) $\left(\frac{1}{2},-1,\frac{3}{2}\right), \sqrt{11}$; **2.** (i) $7(x^2+y^2+z^2)-15x-25y-11z=0$; (ii) $30(x^2+y^2+z^2)+275x-68y+39z-336=0$. **3.** $x^2+y^2+z^2-6z=0$; **4.** Outside; **5.** $x^2+y^2+z^2-x-y-z=0$; **6.** $x^2+y^2+z^2-29\left(\frac{x}{2}+\frac{y}{3}+\frac{z}{4}\right)=0$; **9.** $(1,3,4), \sqrt{7}$; **14.** $2(x^2+y^2+z^2)-3x-16y-4z=0, 7x+18y-10z-4=0$.

[B]
Cylinder

Definition 2 *A cylinder is a surface generated by one complete rotation of a straight line which is always parallel to a fixed straight line and intersects a given plane curve.*

The given curve is called the *guiding curve or directrix* and the revolving line is known as the *generator*. If the guiding curve is a circle and the fixed line is normal to the plane of the circle through the centre of it, then the cylinder is *right circular* and the fixed line is called the *axis* of this cylinder. The distance between the axis and any generator is known as the *radius of the right circular cylinder* and it is equal to the radius of the guiding circle.

6.6 Equation of the Cylinder whose Guiding Curve is $f(x, y) = 0$, $z = 0$ and Generators are parallel to $\frac{x}{l} = \frac{y}{m} = \frac{z}{n}$.

Let (α, β, γ) be a point on the cylinder. The equation of the generating line through (α, β, γ) is

$$\frac{x-\alpha}{l} = \frac{\gamma-\beta}{m} = \frac{z-\gamma}{n}.$$

On the guiding curve $z = 0$.

$\therefore$ this line meets the curve, where

$$\frac{x-\alpha}{l} = \frac{y-\beta}{m} = \frac{-\gamma}{n}, \quad \text{or} \quad x = \alpha - \frac{l\gamma}{n}, \quad y = \beta - \frac{m\gamma}{n}, \quad z = 0.$$

This point satisfies the equation of the guiding curve.

$$\therefore \quad f\left(\alpha - \frac{l\gamma}{n}, \quad \beta - \frac{m\gamma}{n}\right) = 0.$$

Hence the equation of the cylinder which is the locus of (α, β, γ) is

$$f\left(x - \frac{lz}{n}, \quad y - \frac{mz}{n}\right) = 0.$$

Corollary. $f(x, y) = 0$ represents a cylinder when the fixed line is the z-axis and the guiding curve is $f(x, y) = 0, z = 0$.

6.7 Equation of the Right Circular Cylinder whose Radius is a and Axis is given by

$$\frac{x-x_1}{l} = \frac{y-y_1}{m} = \frac{z-z_1}{n}.$$

Let (α, β, γ) be a point on the cylinder and the perpendicular from this point to the axis meet the axis at $(lr + x_1, mr + y_1, nr + z_1)$.

The d. rs. of the perpendicular are $lr + x_1 - \alpha, mr + y_1 - \beta, nr + z_1 - \gamma$.

From the condition of perpendicularity

$$l(lr + x_1 - \alpha) + m(mr + y_1 - \beta) + n(nr + z_1 - \beta) = 0,$$

$$\text{or} \quad r = \frac{l(\alpha - x_1) + m(\beta - y_1) + n(\gamma - z_1)}{l^2 + m^2 + n^2}. \tag{1}$$

Again this perpendicular distance is equal to the radius a.

$$\therefore \quad (lr + x_1 - \alpha)^2 + (mr + y_1 - \beta)^2 + (nr + z_1 - \gamma) = a^2,$$

$$\text{or} \quad (l^2 + m^2 + n^2)r^2 + 2r\{l(x_1 - \alpha) + m(y_1 - \beta)$$

$$+n(z_1 - \gamma)\} + (\alpha - x_1)^2 + (\beta - y_1)^2 + (\gamma - z_1)^2 = a^2,$$

$$\text{or} \quad (\alpha - x_1)^2 + (\beta - y_1)^2 + (\gamma - z_1)^2$$

$$-\frac{\{l(\alpha - x_1) + m(\beta - y_1) + n(\gamma - z_1)\}^2}{l^2 + m^2 + n^2} = a^2 \quad \text{[by (6.7.1)]}.$$

Hence the locus of (α, β, γ) or the equation of the cylinder is

$$(x - x_1)^2 + (y - y_1)^2 + (z - z_1)^2 - \frac{\{l(x - x_1) + m(y - y_1) + n(z - z_1)\}^2}{l^2 + m^2 + n^2} = a^2.$$

Corollary. $x^2 + y^2 = a^2$ represents a right circular cylinder with the z-axis as the axis of the cylinder.

Worked Out Examples

1. Find the equation of the quadric cylinder with generators parallel to z-axis and passing through the curve $ax^2 + by^2 + cz^2 = 1$, $lx + my + nz = p$.

$$ax^2 + by^2 + cz^2 = 1, \tag{1}$$

$$lx + my + nz = p. \tag{2}$$

Eliminating z between (1) and (2) we get the equation

$$ax^2 + by^2 + c\left(\frac{p - lx - my}{n}\right)^2 = 1. \tag{3}$$

The equation (3) represents a cylinder whose generators are parallel to z-axis and also it is satisfied by the common points of (1) and (2). Hence it is the required equation.

2. Find the equation of the cylinder whose directrix is $x^2 + y^2 = 9, z = 1$ and the fixed line is $\frac{x}{2} = \frac{y}{3} = \frac{z}{-1}$.

Let (α, β, γ) be a point on the cylinder. The generating line through this point is $\frac{x-\alpha}{2} = \frac{y-\beta}{3} = \frac{z-\gamma}{-1}$.

It meets the directrix, where $\frac{x-\alpha}{2} = \frac{y-\beta}{3} = \frac{1-\gamma}{-1}$, or $x = \alpha + 2\gamma - 2$, $y = \beta + 3\gamma - 3$, $z = 1$.

This point lies on the guiding curve.

$$\therefore \quad (\alpha + 2\gamma - 2)^2 + (\beta + 3\gamma - 3)^2 = 9.$$

Hence the locus of (α, β, γ), *i.e.*, the equation of the cylinder is $(x + 2z - 2)^2 + (y + 3z - 3)^2 = 9$.

3. Find the equation of the right circular cylinder of radius 2 whose axis is the straight line $\frac{x}{1} = \frac{y}{-2} = \frac{z}{2}$. **[B.U. '88, C.U. '56, '89]**

Let $P(\alpha, \beta, \gamma)$ be a point on the cyclinder. The distance of this point from the axis is 2.

The axis passes through the origin. The d.cs. of the line joining the origin $(0,0,0)$ and and the point (α, β, γ) are $\frac{\alpha}{OP}, \frac{\beta}{OP}, \frac{\gamma}{OP}$. The d. cs. of the axis are $\frac{1}{3}, \frac{-2}{3}, \frac{2}{3}$. The distance of (α, β, γ) from the axis $= OP \cdot \sin\theta$ (θ is the angle between OP and the axis)

$$= OP\sqrt{1-\cos^2\theta}$$

$$= OP\sqrt{\left\{1-\left(\frac{\alpha-2\beta+2\gamma}{3.OP}\right)^2\right\}}$$

$$= \frac{1}{3}\sqrt{\{9.OP^2-(\alpha-2\beta+2\gamma)^2\}}$$

$$= \frac{1}{3}\sqrt{\{9(\alpha^2+\beta^2+\gamma^2)-(\alpha-2\beta+2\gamma)^2\}}.$$

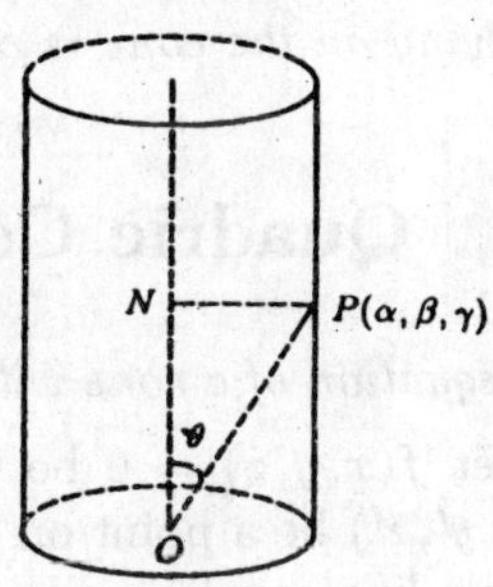

Fig. 6.4

$$\therefore \quad 2 = \frac{1}{3}\sqrt{\{9(\alpha^2+\beta^2+\gamma^2)-(\alpha-2\beta+2\gamma)^2\}},$$

$$\text{or} \quad 8\alpha^2+5\beta^2+5\gamma^2+4\alpha\beta+8\beta\gamma-4\gamma\alpha = 36.$$

Hence the required equation is

$$8x^2+5y^2+5z^2+4xy+8yz-4zx = 36.$$

Exercise B

1. Find the equation of the cylinder whose generators are parallel to x-axis and pass through the curve of intersection of $2x+3y-4z=5$ and $3x^2-4y^2+5z^2=4$.

2. Obtain the equation of the cylinder whose generators intersect the plane curve $ax^2+by^2=1, z=0$ and are parallel to the line $\frac{x}{l}=\frac{y}{m}=\frac{z}{n}$. **[C.U. '69]**

3. (a) Find the equation of the right circular cylinder of radius 5, whose axis passes through $(1,2,3)$ and is parallel to $\frac{x-4}{2}=\frac{y-3}{-1}=\frac{z-2}{2}$.

   (b) Find the equation of the right circular cylinder of radius 2, whose axis passes through the point $(1,2,3)$ and is parallel to the x-axis. **[N.B. '97]**

4. Show that the equation of the right circular cylinder whose guiding curve is the circle through the points $(1,0,0),(0,1,0),(0,0,1)$ is $x^2+y^2+z^2-yz-zx-xy=1$.

   [B.U. '66, '82, '97]

5. Show that $2x^2+5y^2+5z^2+4xy+2yz-4zx+16x+22y-10z-18=0$ is the equation of the cylinder which passes through the point $(3,-1,1)$ and has the axis $\frac{x-1}{2}=\frac{y+3}{-1}=\frac{z-2}{1}$.

Answers

1. $11y^2+68z^2-72yz-90y+120z+59=0$. **2.** $a(nx-lz)^2+b(ny-mz)^2=n^2$. **3.** (i) $5x^2+8y^2+5z^2+4yz-8zx+4xy+6x-48y-30z-135=0$. (ii) $(y-2)^2+(z-3)^2=4$.

[C]

Cone

Definition 3 *A cone is a surface generated by a straight line passing through a fixed point and intersecting a curve or touching a given surface. The fixed point is known as the vertex and the given curve is called the guiding curve or directrix or base. Any line lying on the cone is called its generator.*

6.8 Quadric Cone with the Vertex at the Origin

The equation of a cone with its vertex as origin is homogenous in x, y, z and conversely.

Let $f(x, y, z) = 0$ be the equation of a cone with the vertex at the origin O. If $P(x', y', z')$ be a point on the cone, the line OP will wholly lie on the cone. Any point on this line can be written as (rx', ry', rz'). Now (rx', ry', rz') satisfies the equation $f(x, y, z) = 0$ for all values of r. It is possible only when $f(x, y, z)$ is homogeneous in x, y, z.

Hence the proposition follows.

Conversely, if $f(x, y, z) = 0$ is an homogeneous equation in x, y, z, then (rx', ry', rz') will satisfy the equation when $P(x', y', z')$ satisfies the equation. Therefore the line OP wholly lies on $f(x, y, z) = 0$. Hence $f(x, y, z) = 0$ represents a cone with the vertex as origin.

Note. If $f(x, y, z)$ is factorizable into two linear factors, $f(x, y, z) = 0$ will represent a pair of planes. Otherwise it represents a cone.

Corollary. If $f(x, y, z) = 0$ be the homogeneous equation of a cone and $\frac{x}{l} = \frac{y}{m} = \frac{z}{n}$ be a generator then $f(l, m, n) = 0$.

6.9 Equation of a Cone with given Vertex and Directrix

(i) *To find the equation of the cone whose vertex is the origin and which passes through the curve of intersection of the plane $lx+my+nz = p$ and the surface $ax^2+by^2+cz^2 = 1$.*

Since the vertex is the origin, the equation of the cone is homogeneous. Thus making $ax^2 + by^2 + cz^2 = 1$ homogeneous by $lx + my + nz = p$, the required equation is obtained.

The equation is

$$ax^2 + by^2 + cz^2 = \left(\frac{lx + my + nz}{p}\right)^2,$$

$$\text{or} \quad p^2(ax^2 + by^2 + cz^2) = (lx + my + nz)^2.$$

(ii) *To find the equation of the cone whose vertex is the paint (α, β, γ) and the base is the conic $z = 0, f(x, y) = 0$.*

The equation of a line through (α, β, γ) is

$$\frac{x-\alpha}{l}=\frac{y-\beta}{m}=\frac{z-\gamma}{n}. \tag{6.9.1}$$

It meets the plane $z=0$, where $x=\alpha-\frac{l\gamma}{n}, y=\beta-\frac{m\gamma}{n}$.
If the line meets the curve $f(x,y)=0$, then

$$f\left(\alpha-\frac{l\gamma}{n},\quad \beta-\frac{m\gamma}{n}\right)=0. \tag{6.9.2}$$

Eliminating l, m, n from (6.9.2) by (6.9.1),

$$f\left(\alpha-\gamma\frac{x-\alpha}{z-\gamma}, \beta-\gamma\frac{y-\beta}{z-\gamma}\right)=0,$$

$$\text{or}\quad f\left(\frac{\alpha z-\gamma x}{z-\gamma}, \frac{\beta z-\gamma y}{z-\gamma}\right)=0.$$

It is the required equation.

6.10 Right Circular Cone

Definition 4 *A cone formed by a variable line passing through a fixed point (vertex) and making a constant acute angle with a given line through the vertex is known as a right circular cone. The given line and the acute angle are called the axis and the semi-vertical angle of the cone respectively. The section of a right circular cone by a plane perpendicular to the axis is circular.*

To find the equation of the right circular cone whose vertex is (α, β, γ), the semi-vertical angle is θ and the axis is

$$\frac{x-\alpha}{l}=\frac{y-\beta}{m}=\frac{z-\gamma}{n}.$$

Let A be the vertex, AN the axis and $P(x_1, y_1, z_1)$ a point on the cone.

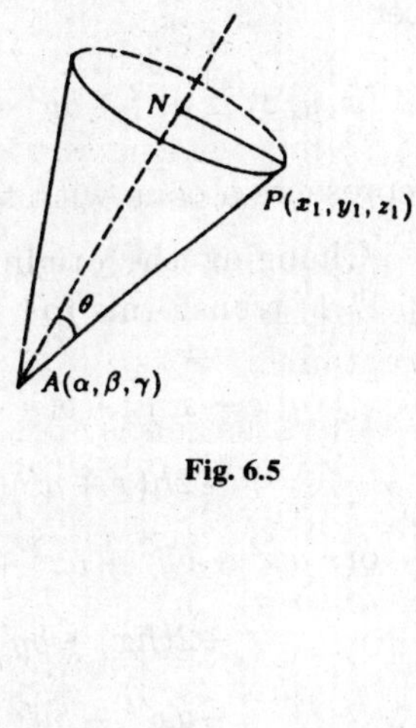

Fig. 6.5

The d. cs. of the line AP are

$$\frac{x_1-\alpha}{AP}, \quad \frac{y_1-\beta}{AP}, \quad \frac{z_1-\gamma}{AP}.$$

The d. cs. of the axis are

$$\frac{l}{\sqrt{l^2+m^2+n^2}}, \quad \frac{m}{\sqrt{l^2+m^2+n^2}}, \quad \frac{n}{\sqrt{l^2+m^2+n^2}}.$$

Since θ is the angle between these two lines,

$$\cos\theta=\frac{(x_1-\alpha)l+(y_1-\beta)m+(z_1-\gamma)n}{AP\cdot\sqrt{l^2+m^2+n^2}},$$

$$\text{or} \quad AP^2(l^2+m^2+n^2)\cos^2\theta$$
$$=\{l(x_1-\alpha)+m(y_1-\beta)+n(z_1-\gamma)\}^2,$$
$$\text{or} \quad \{(x_1-\alpha)^2+(y_1-\beta)^2+(z_1-\gamma)^2\}(l^2+m^2+n^2)\cos^2\theta$$
$$=\{l(x_1-\alpha)+m(y_1-\beta)+n(z_1-\gamma)\}^2.$$

Hence the locus of P, i.e., the equation of the cone is

$$\{(x-\alpha)^2+(y-\beta)^2+(z-\gamma)^2\}(l^2+m^2+n^2)\cos^2\theta$$
$$=\{l(x-\alpha)+m(y-\beta)+n(z-\gamma)\}^2.$$

Corollary 1. If the vertex is the origin, then $(\alpha,\beta,\gamma)\equiv(0,0,0)$. In this case the equation of the cone is

$$(x^2+y^2+z^2)(l^2+m^2+n^2)\cos^2\theta=(lx+my+nz)^2.$$

Note that it is a homogeneous equation of second degree.

Corollary 2. It the vertex is the origin and the z-axis is the axis of the cone, then $(\alpha,\beta,\gamma)\equiv(0,0,0)$ and $l=0=m, n=1$.

$\therefore$ the equation of the cone is

$$(x^2+y^2+z^2)\cos^2\theta=z^2, \quad \text{or} \quad x^2+y^2=z^2\tan^2\theta.$$

6.11 Condition for the General Equation of Second Degree to represent a Cone and coordinates of vertex

Let

$$F(x,y,z)\equiv ax^2+by^2+cz^2+2fyz+2gzx+2hxy+2ux+2vy+2wz+d=0 \quad (6.11.1)$$

represents a cone with the vertex at (x',y',z').

Changing the origin to (x',y',z') with the axes remaining parallel the equation (6.11.1) transforms to

$$a(x+x')^2+b(y+y')^2+c(z+z')^2+2f(y+y')(z+z')+2g(z+z')(x+x')$$
$$+2h(x+x')(y+y')+2u(x+x')+2v(y+y')+2w(z+z')+d=0,$$
$$\text{or} \quad ax^2+by^2+cz^2+2fyz+2gzx+2hxy+2(ax'+hy'+gz'+u)x$$
$$+2(hx'+by'+fz'+v)y+2(gx'+fy'+cz'+w)z$$
$$+ax'^2+by'^2+cz'^2+2fy'z'+2gz'x'+2hx'y'+2ux'+2vy'$$
$$+2wz'+d=0. \quad (6.11.2)$$

The equation of a cone with the vertex at the origin is a second degree homogeneous. Therefore the equation (6.11.2) should be homogeneous. For this the coefficients of x, y, z and the absolute term should be zero.

Thus

$$ax' + hy' + gz' + u = 0, \tag{6.11.3}$$

$$hx' + by' + fz' + v = 0, \tag{6.11.4}$$

$$gx' + fy' + cz' + w = 0, \tag{6.11.5}$$

and $ax'^2 + by'^2 + cz'^2 + 2fy'z' + 2gz'x' + 2hx'y' + 2ux' + 2vy' + 2wz' + d = 0,$

or $x'(ax' + by' + gz' + u) + y'(hx' + by' + fz' + v)$

$+z'(gx' + fy' + cz' + w) + (ux' + vy' + wz' + d) = 0,$

or $ux' + vy' + wz' + d = 0$ [by (6.11.3), (6.11.4), (6.11.5)]. (6.11.6)

Eliminating x', y', z' from (6.11.3), (6.11.4), (6.11.5) and (6.11.6), the required condition is

$$\begin{vmatrix} a & h & g & u \\ h & b & f & v \\ g & f & c & w \\ u & v & w & d \end{vmatrix} = 0.$$

The co-ordinates of the vertex are obtained by solving the equations (6.11.3), (6.11.4) and (6.11.5).

Note 1. If $F(x, y, z, t) \equiv ax^2 + by^2 + cz^2 + 2fyz + 2gzx + 2hxy + 2uxt + 2vyt + 2wzt + dt^2$, then

$$\frac{\delta F}{\delta x} = 2(ax + hy + gz + ut), \quad \frac{\delta F}{\delta y} = 2(hx + by + fz + vt),$$

$$\frac{\delta F}{\delta z} = 2(gx + fy + cz + wt), \quad \frac{\delta F}{\delta t} = 2(ux + vy + wz + dt).$$

Putting $t = 1$, the equations

$$\frac{\delta F}{\delta x} = 0, \quad \frac{\delta F}{\delta y} = 0, \quad \frac{\delta F}{\delta z} = 0, \quad \frac{\delta F}{\delta t} = 0$$

are satisfied by the vertex (x', y', z'). Elimination of (x', y', z') from these equations gives the condition for representing a cone.

Note 2. The equation $F(x, y, z) = 0$ represents a cone only when

$$\frac{\delta F}{\delta x} = 0, \quad \frac{\delta F}{\delta y} = 0, \quad \frac{\delta F}{\delta z} = 0, \quad \frac{\delta F}{\delta t} = 0$$

are consistent.

In the case of consistency, the vertex is obtained from any three of the above four equations.

► **Example 4** *Show that the equation* $x^2 + 2y^2 + z^2 - 4yz - 6zx - 2x + 8y - 2z + 9 = 0$ *represents a cone. Find its vertex.*

Making the equation homogeneous

$$F(x, y, z, t) \equiv x^2 + 2y^2 + z^2 - 4yz - 6zx - 2xt + 8yt - 2zt + 9t^2.$$

Putting $t = 1$ in

$$\frac{\delta F}{\delta x}, \quad \frac{\delta F}{\delta y}, \quad \frac{\delta F}{\delta z}, \quad \frac{\delta F}{\delta t}$$

and then equating to zero, the four equations are

$$x - 3z - 1 = 0, \tag{1}$$

$$y - z + 2 = 0, \tag{2}$$

$$3x + 2y - z + 1 = 0, \tag{3}$$

$$x - 4y + z - 9 = 0. \tag{4}$$

From (1), (2) and (3), $x = 1, y = -2, z = 0$.

These values satisfy the equation (4). Therefore the equations are consistent. Consequently the given equation represents a cone with the vertex as $(1, -2, 0)$.

6.12 Angle between Lines in which a Plane cuts a Cone

Let

$$ax^2 + by^2 + cz^2 + 2fyz + 2gzx + 2hxy = 0 \tag{6.12.1}$$

be a cone and

$$ux + vy + wz = 0 \tag{6.12.2}$$

be a plane which cuts the cone along two lines.

Let a line in which the plane cuts the cone be

$$\frac{x}{l} = \frac{y}{m} = \frac{z}{n}. \tag{6.12.3}$$

Since the line (6.12.3) lies on the plane (6.12.2),

$$ul + vm + wn = 0. \tag{6.12.4}$$

This line is a generator of the cone (6.12.1).

Therefore

$$al^2 + bm^2 + cn^2 + 2fmn + 2gnl + 2hlm = 0. \tag{6.12.5}$$

Eliminating n from (6.12.5) by (6.12.4)

$$al^2 + bm^2 + c\left(\frac{ul+vm}{w}\right)^2 - 2fm\frac{ul+vm}{w} - 2gl\frac{ul+vm}{w} + 2hlm = 0,$$

$$\text{or} \quad (cu^2 + aw^2 - 2gwu)l^2 + 2(hw^2 + cuv - fuw - guw)lm$$
$$+(bw^2 + cv^2 - 2fvw)m^2 = 0,$$

$$\text{or} \quad (cu^2 + aw^2 - 2gwu)\frac{l^2}{m^2} + 2(hw^2 + cuv - fuw - gvw)\frac{l}{m}$$
$$+(bw^2 + cv^2 - 2fvw) = 0. \tag{6.12.6}$$

If the d. rs. of the two lines of section are l_1, m_1, n_1 and l_2, m_2, n_2 then $\frac{l_1}{m_1}$ and $\frac{l_2}{m_2}$ are the roots of (6). Getting the values of $\frac{l_1}{m_1}, \frac{l_2}{m_2}$ from (6), the values of n_1 and n_2 in terms of l_1 and l_2 are obtained from (6.12.4).

If $m_1 = \lambda_1 l_1, n_1 = \mu_1 l_1$ and $m_2 = \lambda_2 l_2, n_2 = \mu_2 l_2$ then the lines of section are

$$\frac{x}{1} = \frac{y}{\lambda_1} = \frac{z}{\mu_1} \quad \text{and} \quad \frac{x}{1} = \frac{y}{\lambda_2} = \frac{z}{\mu_2}.$$

The angle between these lines is

$$\cos^{-1}\frac{1 + \lambda_1\lambda_2 + \mu_1\mu_2}{\sqrt{(1 + \lambda_1^2 + \mu_1^2)}\sqrt{(1 + \lambda_2^2 + \mu_2^2)}}.$$

Corollary 1. Condition for two perpendicular generators made by a plane.

From (6.12.6)

$$\frac{l_1 l_2}{m_1 m_2} = \frac{bw^2 + cv^2 - 2fvw}{cu^2 + aw^2 - 2gwu} \quad \text{(as product of the roots).}$$

From symmetry

$$\frac{l_1 l_2}{n_1 n_2} = \frac{bw^2 + cv^2 - 2fvw}{av^2 + bu^2 - 2huv}.$$

$$\therefore \quad \frac{l_1 l_2}{bw^2 + cv^2 - 2fvw} = \frac{m_1 m_2}{cu^2 + aw^2 - 2gwu} = \frac{n_1 n_2}{av^2 + bu^2 - 2huv} = k \text{ (say)}$$

$$\therefore \quad l_1 l_2 + m_1 m_2 + n_1 n_2 = k[a(v^2 + w^2) + b(w^2 + u^2) + c(u^2 + v^2)$$
$$-2fuv - 2gwu - 2huv]. \tag{6.12.7}$$

If the plane cuts the cone in two perpendicular generators then

$$l_1 l_2 + m_1 m_2 + n_1 n_2 = 0,$$

$$\text{i.e., } a(v^2 + w^2) + b(w^2 + u^2) + c(u^2 + v^2) - 2fuv - 2gwu - 2huv = 0. \tag{6.12.8}$$

Corollary 2. Necessary and sufficient condition for three mutually perpendicular generators.

Let the normal to the plane (6.12.2) through the origin, i.e., $\frac{x}{u}=\frac{y}{v}=\frac{z}{w}$ is a generator of the cone.

Then $au^2+bv^2+cw^2+2fvw+2gwu+2huv=0$.

In this case $l_1l_2+m_1m_2+n_1n_2=k(a+b+c)(u^2+v^2+w^2)$.

If the plane now cuts the cone in two perpendicular generators then

$$a+b+c=0. \quad [\because\ u^2+v^2+w^2\neq 0] \tag{6.12.9}$$

It is independent of u, v, w. Therefore $a+b+c=0$ is the necessary condition for three mutually perpendicular generators of the cone (6.12.1).

From the condition of $a+b+c=0$ it follows that the plane perpendicular to an arbitrary generator cuts the cone in two perpendicular generators. Therefore $a+b+c=0$ is the sufficient condition for three mutually perpendicular generators.

Different approach. If the cone (6.12.1) passes through the coordinate axes, then the d. cs. of the axes, i.e., 1, 0, 0; 0, 1, 0; 0, 0, 1 satisfy the equation (6.12.1).

Therefore $a=0=b=c$. $\therefore$ $fyz+gzx+hxy=0$ represents a cone passing through the coordinate axes.

If the equation (6.12.1) has three mutually perpendicular generators, let us transform the equation (6.12.1) into a new coordinate system whose axes are these three generators. The equation (6.12.1) will transform into the from $f'yz+g'zx+h'xy=0$.

By the property of invariant under the transformation of axes the sum of the coefficients of x^2, y^2 and z^2 remains unchanged.

$\therefore$ $a+b+c=0$ is the required condition.]

Corollary 3. If $a+b+c=0$, the cone (6.12.1) has an infinite number of triads of mutually perpendicular generators.

Worked Out Examples

1. Find the equation to the cone, whose vertex is the origin and base the circle $x=a, y^2+z^2=b^2$. **[C.U. '63]**

In the fig. 6.6, $OC=a, AC=b$ (radius of the base)

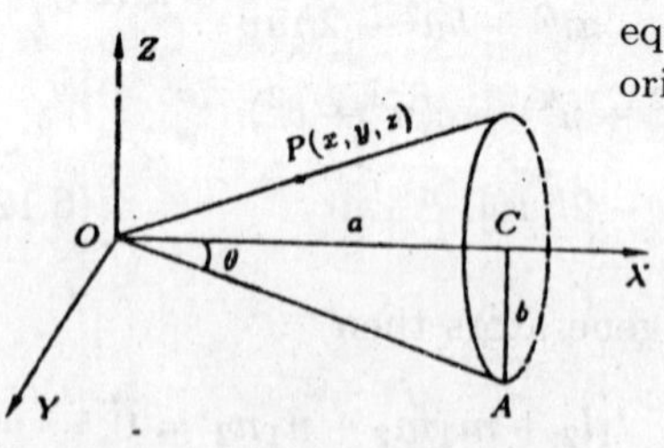

Fig. 6.6

If θ be the semi-vertical angle, then $\tan\theta=\frac{b}{a}$. The equation of the right circular cone whose vertex is the origin, the semi-vertical angle θ and the axis the x-axis is

$$y^2+z^2=x^2\tan^2\theta,$$

$$\text{or}\quad y^2+z^2=x^2\cdot\frac{b^2}{a^2},$$

$$\text{or}\quad a^2(y^2+z^2)=b^2x^2.$$

2. The axis of a right cone, vertex O, makes equal angles with the coordinate axes, and the cone passes through the line drawn from O, with d. cs. proportional to $(1,-2,2)$. Find the equation to the cone. **[C.U. '64]**

The d. cs. of the axis are $\frac{1}{\sqrt{3}}, \frac{1}{\sqrt{3}}, \frac{1}{\sqrt{3}}$ and the d. cs. of the given generating line are $\frac{1}{3}, \frac{-2}{3}, \frac{2}{3}$.

If θ be the semi-vertical angle, $\cos\theta = \frac{1-2+2}{3\sqrt{3}} = \frac{1}{3\sqrt{3}}$.

$\therefore$ the equation of the cone is

$$(x^2+y^2+z^2)\cdot\frac{1}{27} = \frac{1}{3}(x+y+z)^2, \quad \text{[by Cor. 1 of Section 6.12]}$$

$$\text{or}\quad 9(x+y+z)^2 = x^2+y^2+z^2.$$

3. Find the equations of the lines of intersection of the plane $3x+4y+z=0$ and the cone $15x^2-32y^2-7z^2=0$. **[C.U. '67; B.U. '69]**

Let the equation of the generating line in which the plane cuts the cone be $\frac{x}{l} = \frac{y}{m} = \frac{z}{n}$.

The line lies on the cone and also on the plane.

$$\therefore\quad 15l^2-32m^2-7n^2=0 \tag{1}$$

$$\text{and}\quad 3l+4m+n=0. \tag{2}$$

Eliminating n from (1) by (2)

$$15l^2-32m^2-7(3l+4m)^2 = 0, \quad\text{or}\quad 2l^2+7mn+6m^2=0,$$

$$\text{or}\quad (2l+3m)(l+2m) = 0, \quad\text{or}\quad m = -\frac{2}{3}l, -\frac{1}{2}l.$$

For $m=-\frac{2}{3}l, n=-(3l+4m) = -\left(3l-\frac{8}{3}l\right) = -\frac{l}{3}$.

For $m=-\frac{1}{2}l, n=-(3l+4m) = -(3l-2l) = -l$.

$\therefore$ the d. cs. are proportional to $1. -\frac{2}{3}, -\frac{1}{3}$ or $1, -\frac{1}{2}, -1$.

Hence the equations of the required generators are

$$\frac{x}{3} = \frac{y}{-2} = \frac{z}{-1} \quad\text{and}\quad \frac{x}{2} = \frac{y}{-1} = \frac{z}{-2}.$$

4. The plane $\frac{x}{a}+\frac{y}{b}+\frac{z}{c}=1$ meets the co-ordinate axes in A, B, C. Prove that the equation to the cone generated by lines drawn from O to meet the circle ABC is

$$yz\left(\frac{b}{c}+\frac{c}{a}\right)+zx\left(\frac{c}{a}+\frac{a}{c}\right)+xy\left(\frac{a}{b}+\frac{b}{a}\right)=0. \quad \textbf{[N.B. '69]}$$

The plane meets the axes at $(a,0,0), (0,b,0)$ and $(0,0,c)$. The equation of the sphere through $(0,0,0)$ and the above three points is $x^2+y^2+z^2-ax-by-cz=0$.

$\therefore$ the equation of the circle is

$$\frac{x}{a}+\frac{y}{b}+\frac{z}{c} = 1, \tag{1}$$

$$x^2+y^2+z^2-ax-by-cz = 0. \tag{2}$$

The equation of the cone whose vertex is the origin is homogeneous. Thus the equation to the cone is obtained by making (2) homoegeneous with the help of (1). Hence the required equations is

$$x^2+y^2+z^2-(ax+by+cz)\left(\frac{x}{a}+\frac{y}{b}+\frac{z}{c}\right) = 0,$$

$$\text{or}\quad yz\left(\frac{b}{c}+\frac{c}{b}\right)+zx\left(\frac{c}{a}+\frac{a}{c}\right)+xy\left(\frac{a}{b}+\frac{b}{a}\right) = 0.$$

5. Show that the condition that the plane $ax + by + cz = 0$ may cut the cone $yz + zx + xy = 0$ in perpendicular lines is

$$\frac{1}{a} + \frac{1}{b} + \frac{1}{c} = 0.$$ **[N.B. '84, '90, '98]**

Let $\frac{x}{l} = \frac{y}{m} = \frac{z}{n}$ be one of the generators of the cone made by the plane.

Then

$$al + bm + cn = 0 \tag{1}$$

$$\text{and} \quad mn + nl + lm = 0. \tag{2}$$

Eliminating n from (2) by (1)

$$-(l+m)\frac{al+bm}{c} + lm = 0,$$

$$\text{or} \quad al^2 + (a+b-c)lm + bm^2 = 0,$$

$$\text{or} \quad a\left(\frac{l}{m}\right)^2 + (a+b-c)\frac{l}{m} + b = 0.$$

Let the roots be

$$\frac{l_1}{m_1} \text{ and } \frac{l_2}{m_2}.$$

Then

$$\frac{l_1}{m_1} \cdot \frac{l_2}{m_2} = \frac{b}{a}.$$

From symmetry

$$\frac{l_1}{n_1} \cdot \frac{l_2}{n_2} = \frac{c}{a}$$

$$\therefore \quad al_1l_2 = bm_1m_2 = cn_1n_2.$$

If the generators are at right angle, then

$$l_1l_2 + m_1m_2 + n_1n_2 = 0, \quad \text{or} \quad \frac{1}{a} + \frac{1}{b} + \frac{1}{c} = 0.$$

6. Find the equation of the cone whose vertex is $(1, 0, -1)$ and which passes through the circle $x^2 + y^2 + z^2 = 4, x + y + z = 1$. **[B.H. '83]**

Let

$$\frac{x-1}{l} = \frac{y}{m} = \frac{z+1}{n}$$

be a generator. If it meets the base at the point $(lr + 1, mr, nr - 1)$ then

$$lr + 1 + mr + nr - 1 = 1,$$

$$\text{or} \quad (l + m + n)r = 1 \tag{1}$$

$$\text{and} \quad (lr+1)^2 + m^2r^2 + (nr-1)^2 = 4,$$

$$\text{or} \quad (l^2 + m^2 + n^2)r^2 + 2(l-n)r = 2. \tag{2}$$

By (1) and (2) $(l^2 + m^2 + n^2) + 2(l - n)(l + m + n) = 2(l + m + n)^2$.

Now eliminating l, m, n by the equation of the generator the equation of the cone is

$$(x-1)^2 + y^2 + (z+1)^2 + 2(x - z - 2)(x + y + z) = 2(x + y + z)^2.$$

7. Find the equation of the right circular cone with the vertex at $(1, -2, -1)$, semi-vertical angle $60°$ and the axis being $\frac{x-1}{3} = \frac{y+2}{-4} = \frac{z+1}{5}$. **[C.U. 2000]**

In the figure A is the vertex, AB is the axis and $P(x, y, z)$ is a point on the surface of the cone.

D. cs. of AP are $\frac{x-1}{AP}, \frac{y+2}{AP}, \frac{z+1}{AP}$.

D. cs. of AB are $\frac{3}{5\sqrt{2}}, \frac{-4}{5\sqrt{2}}, \frac{5}{5\sqrt{2}}$.

$60°$ is the angle between AP and AB.

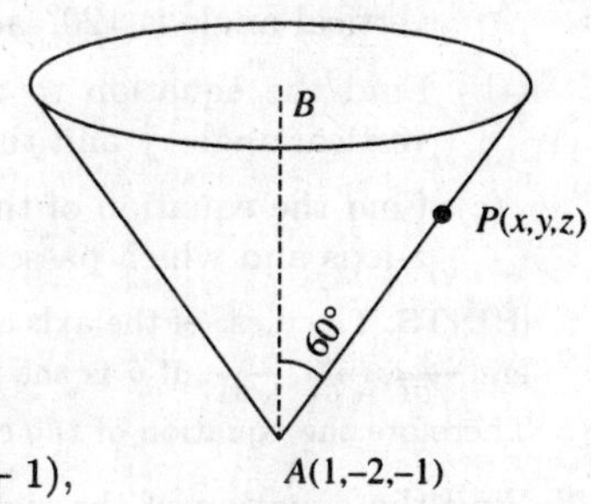

Fig 6.7

$$\therefore \cos 60° = \frac{3(x-1) - 4(y+2) + 5(z+1)}{AP \cdot 5\sqrt{2}}$$

$$\text{or} \quad AP \cdot 5\sqrt{2} \cdot \frac{1}{2} = 3(x-1) - 4(y+2) + 5(z+1),$$

$$\text{or} \quad 25AP^2 = 2(3x - 4y + 5z - 6)^2,$$

$$\text{or} \quad 25\left\{(x-1)^2 + (y+2)^2 + (z+1)^2\right\} = 2(3x - 4y + 5z - 6)^2,$$

$$\text{or} \quad 25(x^2 + y^2 + z^2 - 2x + 4y + 2z + 6) = 2(9x^2 + 16y^2 + 25z^2$$
$$+36 - 24xy - 30zx - 36x - 40yz + 48y - 60z),$$

$$\text{or} \quad 7x^2 - 7y^2 - 25z^2 + 48xy + 80yz - 60zx + 22x + 4y + 170z + 78 = 0.$$

It is the equation of the cone.

8. Lines are drawn through the origin to meet the circle in which the plane $x + y + z = 1$ cuts the sphere $x^2 + y^2 + z^2 - 4x - 6y - 8z + 4 = 0$. Show that they lie on the cone $x^2 - y^2 - 3z^2 - 6yz - 4zx - 2xy - = 0$ and meet the sphere again at points on the plane $y + 2z = 2$. **[C.H. '77]**

The circle is

$$x + y + z = 1, \qquad (1)$$

$$x^2 + y^2 + z^2 - 4x - 6y - 8z + 4 = 0. \qquad (2)$$

The equation of the cone is obtained by making (2) homogeneous with the help of (1). Hence the equation is

$$x^2 + y^2 + z^2 - 2(2x + 3y + 4z)(x + y + z) + 4(x + y + z)^2 = 0,$$

$$\text{or} \quad x^2 - y^2 - 3z^2 - 6yz - 4zx - 2xy = 0. \qquad (3)$$

Again by making (2) homogeneous with the help of $y + 2z = 2$, the equation is

$$x^2 + y^2 + z^2 - 2(2x + 3y + 4z)\frac{y + 2z}{2} + 4\left(\frac{y + 2z}{2}\right)^2 = 0,$$

$$\text{or} \quad x^2 - y^2 - 3z^2 - 6yz - 4zx - 2zy = 0. \qquad (4)$$

Since (3) and (4) are identical, the result follows.

Exercise C

1. (a) Find the equation to the right circular cone, whose vertex is at the origin, semi-vertical angle is 120° and the axis is along the z-axis. **[B.U. '85]**

   (b) Find the equation to the right circular cone, whose vertex is the origin, semi-vertical angle $\frac{\pi}{4}$ and the axis is $x = y = z$. **[B.U. '89]**

   (c) Find the equation of the right circular cone whose vertex is the origin, axis is the z-axis and which passes through the point $(3, -4, 6)$. **[B.H. '92]**

[HINTS. The d.cs. of the axis are $0, 0, 1$. The d.cs. of the generator through the point $(3, -4, 6)$ are $\frac{3}{\sqrt{61}}, \frac{-4}{\sqrt{61}}, \frac{6}{\sqrt{61}}$. If θ is the semi-vertical angle then $\cos\theta = \frac{6}{\sqrt{61}}$. Consequently $\tan\theta = \frac{5}{6}$. Therefore the equation of the cone is $x^2 + y^2 = z^2\tan^2\theta$, or $36(x^2 + y^2) = 25z^2$.]

2. Find the equation of the right circular cone which passes through the line $\frac{x}{2} = \frac{y}{3} = \frac{z}{4}$ and whose axis is $\frac{x}{1} = \frac{y}{-2} = \frac{z}{2}$.

3. Find the equation of a right circular cone whose vertex is $(1, 2, 3)$, the semi-vertical angle $\frac{\pi}{3}$ and the axis is parallel to $\frac{x}{2} = \frac{y}{5} = \frac{z}{7}$.

4. (a) Find the equation to the cone with vertex at the origin and which passes through the curve of intersection of $x^2 + 5y^2 - 7z^2 = 1$ and $2x - 3y + 4z = 1$.

   (b) Find the equation of the cone generated by lines drawn through the origin to meet the circle through the points $(3, 0, 0), (0, 2, 0), (0, 0, 1)$.

5. (a) Find the equation of the cone whose vertex is $(2, 3, 4)$ and the base is $x^2 + y^2 = 25$, $z = 0$.

   (b) Show that the equation of the cone with vertex $(5, 4, 3)$ and base $3x^2 + 2y^2 = 6$, $y + z = 0$ is $3(5y + 5z - 7x)^2 + 2(4x - 3y)^2 = 6(y + z - 7)^2$.

6. (a) Determine the angle between the lines of intersection of the plane $x - 3y + z = 0$ and a quadric cone $x^2 - 5y^2 + z^2 = 0$. **[C.U. '60]**

   (b) Find the equation of the lines in which the plane $2x - 6y - 5z = 0$ cuts the cone $xy + yz + zx = 0$. **[C.H. '81]**

7. If a right circular cone has three mutually perpendicular generators, the semi-vertical angle is $\tan^{-1}\sqrt{2}$.

8. (a) Find the equation of the cone whose vertex is (α, β, γ) and base is $ax^2 + by^2 = 1, z = 0$.

   (b) Show that the lines drawn from the point (α, β, γ) whose d. cs. satisfy $al^2 + bm^2 + cn^2 = 0$, generate the cone $a(x - \alpha)^2 + b(y - \beta)^2 + c(z - \gamma)^2 = 0$. **[B.U. '92]**

9. Find the equation of the cone obtained by revolving the straight line $x - 2y = 1$ about x-axis and find the coordinates of its vertex. **[C.U. '61]**

10. (a) Prove that the angle between the lines given by $x + y + z = 0$, $ayz + bzx + cxy = 0$ is $\frac{\pi}{2}$, if $a + b + c = 0$.

    (b) Prove that the plane $ux + vy + wz = 0$ may cut the cone $ax^2 + by^2 + cz^2 = 0$ in perpendicular generators if $(b + c)u^2 + (c + a)v^2 + (a + b)w^2 = 0$.

11. (a) Show that the equation

$$x^2 - 2y^2 + 3z^2 - 4xy + 5yz - 6zx + 8x - 19y - 2z - 20 = 0$$

represents a cone with vertex $(1, -2, 3)$.

(b) Show that the equation

$$7x^2 + 2y^2 + 2z^2 - 102x + 10xy + 26x - 2y + 2z - 17 = 0$$

represents a cone whose vertex is $(1, -2, 2)$.

12. Show that the locus of the points from which three mutually perpendicular lines can be drawn to intersect the conic $z = 0, ax^2 + by^2 = 1$ is $ax^2 + by^2 + (a + b)z^2 = 1$.

Answers

1. (a) $x^2 + y^2 = 3z^2$, (b) $x^2 + y^2 + z^2 - 4(yz + zx + xy) = 0$. **2.** $13x^2 + 100y^2 + 100z^2 - 232yz + 116zx - 116xy = 0$. **3.** $39\{(x-1)^2 + (y-2)^2 + z-3)^2\} = 2(2x + 5y + 7z - 33)^2$. **4.** (a) $3x^2 + 4y^2 + 23z^2 - 24yz + 16zx - 12xy = 0$. (b) $13xy + 15yz + 20zx = 0$. **5.** $16x^2 + 16y^2 - 12z^2 - 24yz - 16zx + 200z - 400 = 0$. **6.** (a) $\cos^{-1}\frac{5}{6}$; (b) $\frac{x}{2} = \frac{y}{-1} = \frac{z}{2}, \frac{x}{15} = \frac{y}{10} = \frac{z}{-6}$. **8.** $a(\alpha z - \gamma x)^2 + b(\beta z - \gamma y)^2 = (z - \gamma)^2$. **9.** $(x-1)^2 = 4(y + z^2), (1, 0, 0)$.

[D]
Surface of Revolution

Definition 5 *The surface produced by the rotation of a plane curve about a line in its plane is called a surface of revolution.*

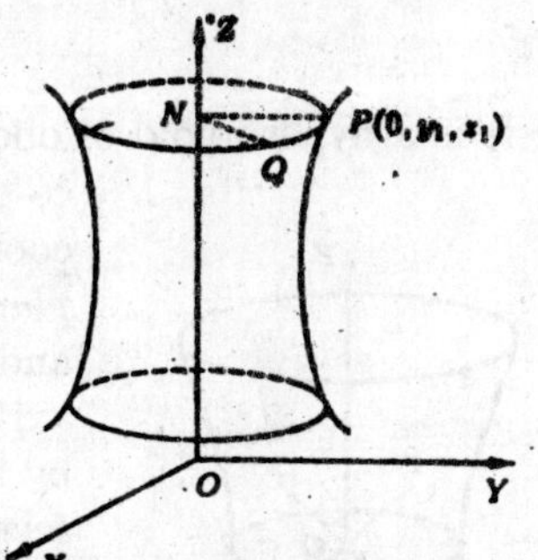

Fig. 6.8

Let $f(y, z) = 0$ be the equation of a plane curve in the YOZ plane and let it revolve about z-axis. Let $P(0, y_1, z_1)$ be a point on the curve.

As P moves about z-axis, its distance from the XOY plane remains unaltered and the distance from the z-axis is equal to y_1. If $Q(x, y, z)$ be the position of P on the other plane and QN is the distance of its from OZ, then $QN = \sqrt{x^2 + y^2}$.

Now $y_1 = \sqrt{x^2 + y^2}$ and $z_1 = z$.

We have $f(y_1, z_1) = 0$. Therefore the equation of the surface of revolution is $f(\sqrt{x^2 + y^2}, z) = 0$.

As for example, the sphere $x^2 + y^2 + z^2 = a^2$ is produced by the revolution of the circle $y^2 + z^2 = a^2, x = 0$ about OZ.

[E]
Conicoids

6.13 Ellipsoid

$$\frac{x^2}{a^2} + \frac{y^2}{b^2} + \frac{z^2}{c^2} = 1. \tag{6.13.1}$$

The equation (6.13.1) represents a closed surface which is known as ellipsoid. If (x, y, z) is a point on the surface of (6.13.1), then $(-x, -y, -z)$ is also a point on it. Thus the surface is symmetrical about the origin. Origin is the centre of it, so it is called a *central conicoid.* The surface is also symmetrical about the coordinate planes. These are called *principal planes* and the coordinate axes are called the *principal axes.* a, b, c denote lengths of *semi-axes.* The section by any plane parallel to a coordinate plane is an ellipse.

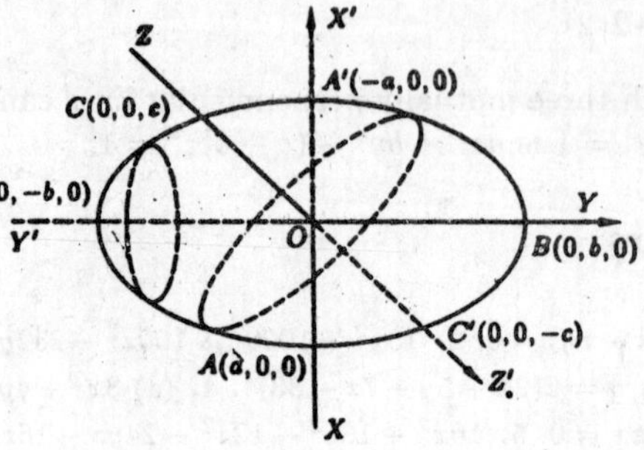

Fig. 6.9

If any two of a, b, c are equal, then the surface will be an ellipsoid of revolution. As by rotating $\frac{x^2}{a^2} + \frac{y^2}{b^2} = 1$, $z = 0$ about the y-axis, the generated surface is $\frac{x^2+z^2}{a^2} + \frac{y^2}{b^2} = 1$. The ellipsoid of revolution obtained by revolving an ellipse about its major axis is known as *prolate spheroid* and that obtained by revolving the ellipse about minor axis is known as *oblate spheroid.*

6.14 Hyperboloid of One Sheet

$$\frac{x^2}{a^2} + \frac{y^2}{b^2} - \frac{z^2}{c^2} = 1.$$

It is a hyperboloid of one sheet. It is symmetrical about the origin and also about co-ordinate planes. Thus it is a *central conicoid* and the coordinate planes and the coordinate axes are the *principal planes* and the *principal axes* respectively. Each of the x-axis and the y-axis meets the surface at two real points but the z-axis does not meet the surface at any real point. The section by any plane parallel to $Y\hat{O}Z$ and ZOX is a hyperbola. The section by any plane parallel to XOY plane is an ellipse. It is not a closed surface. If the hyperbola $\frac{x^2}{a^2} - \frac{z^2}{c^2} = 1, y = 0$ revolve about its conjugate axis, *i.e.*, z-axis, the surface whose equation is $\frac{x^2+y^2}{a^2} - \frac{z^2}{c^2} = 1$ is produced.

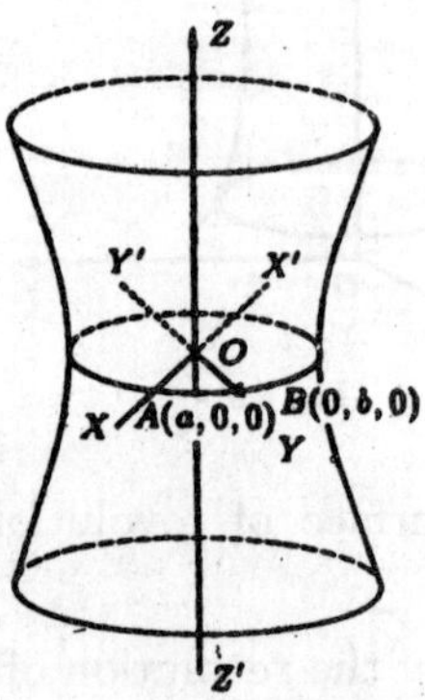

Fig. 6.10

It is a *hyperboloid of revolution of one sheet.*

$$\frac{x^2}{a^2} - \frac{y^2}{b^2} + \frac{z^2}{c^2} = 1 \quad \text{and} \quad -\frac{x^2}{a^2} + \frac{y^2}{b^2} + \frac{z^2}{c^2} = 1$$

are also the equations of hyperboloid of one sheet.

6.15 Hyperboloid of Two Sheets

$$-\frac{x^2}{a^2} - \frac{y^2}{b^2} + \frac{z^2}{c^2} = 1.$$

It is the equation of hyperboloid of two sheets. The surface is symmetrical about the origin and the coordinate planes. Thus it is a *central conicoid* and the coordinate planes and the coordinate axes are the *principal planes* and the *principal axes.* The z-axis meets the surface at two real points, namely $(0, 0, c)$ and $(0, 0, -c)$. The other two axes do not meet it at real points. Thus the surface is not closed. The section by a plane parallel to YOZ or ZOX plane is a hyperbola. The section by a plane parallel to XOY plane is an ellipse. If a hyperbola $\frac{z^2}{c^2} - \frac{x^2}{a^2} = 1, y = 0$ revolves about its transverse axis, *i.e.*, z-axis, the surface whose equation is $\frac{z^2}{c^2} - \frac{x^2+y^2}{a^2} = 1$ is produced. It is known as a *hyperboloid of revolution of two sheets.*

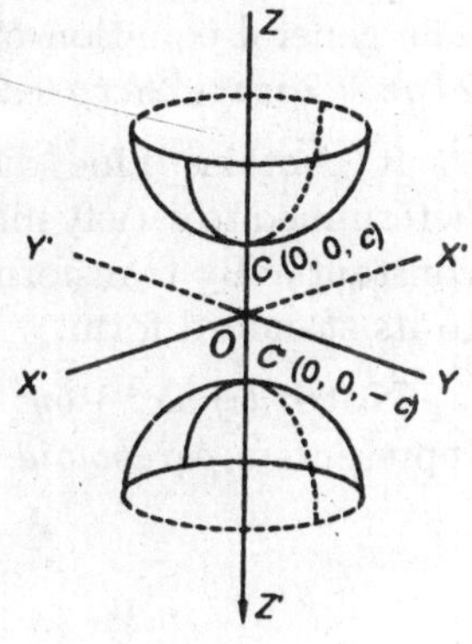

Fig. 6.11

$$\frac{x^2}{a^2} - \frac{y^2}{b^2} - \frac{z^2}{c^2} = 1, \quad \text{and} \quad -\frac{x^2}{a^2} + \frac{y^2}{b^2} - \frac{z^2}{c^2} = 1$$

also represent surfaces of hyperboloid to two sheets.

6.16 Paraboloid

(i) **Elliptic paraboloid:**

$$\frac{x^2}{a^2} + \frac{y^2}{b^2} = \frac{2z}{c}.$$

The surface represented by the above equation is generated by the variable ellipse.

$$z = k, \frac{x^2}{a^2} + \frac{y^2}{b^2} = \frac{2k}{c} \quad (k \text{ and } c \text{ are of the same sign}).$$

It the point (x, y, z) lies on the surface, then the point $(-x, -y, z)$ also lies on the surface. Thus the surface is symmetrical about $x = 0$ and $y = 0$ planes. These two planes are *principal planes.* The section by a plane parallel to XOY plane is an ellipse. The section by a plane parallel to YOZ plane is

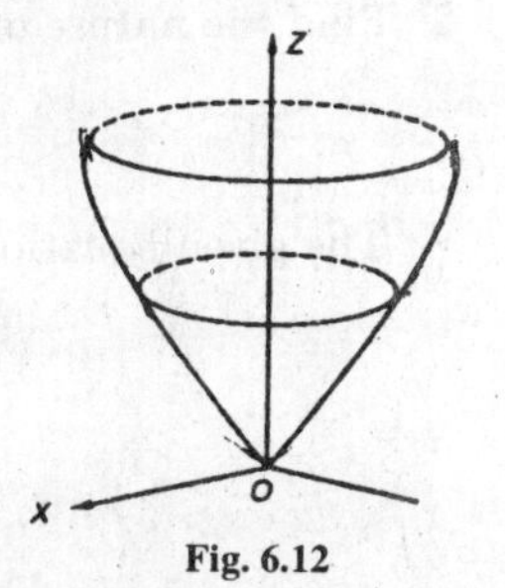

Fig. 6.12

$$\frac{y^2}{b^2} = \frac{2z}{c} - \frac{k^2}{a^2}, \; x = k, \quad \text{or} \quad y^2 = \frac{2b^2}{c}\left(z - \frac{ck^2}{2a^2}\right).$$

It is a parabola of latus rectum $\frac{2b^2}{c}$. Similarly the section by the plane parallel to ZOX is a parabola of latus rectum $\frac{2a^2}{c}$.

Revolving the parabola $y^2 = 2\lambda z, x = 0$ about its axis, , *i.e.*, z-axis, a surface of equation $x^2 + y^2 = 2\lambda z$ is obtained. It is a *paraboloid of revolution.*

(ii) **Hyperbolic paraboloid:**

$$\frac{x^2}{a^2} - \frac{y^2}{b^2} = \frac{2z}{c}.$$

The surface represented by the above equation is generated by the variable hyperbola, $\frac{x^2}{a^2} - \frac{y^2}{b^2} = \frac{2k}{c}, z = k$. The planes $x = 0$ and $y = 0$ are the principal planes. The sections by the planes parallel to YOZ or ZOX plane are parabolas. The section by a plane parallel to XOY plane is a hyperbola whose asymptotes are parallel to $\frac{x}{a} \pm \frac{y}{b} = 0, z = 0.$

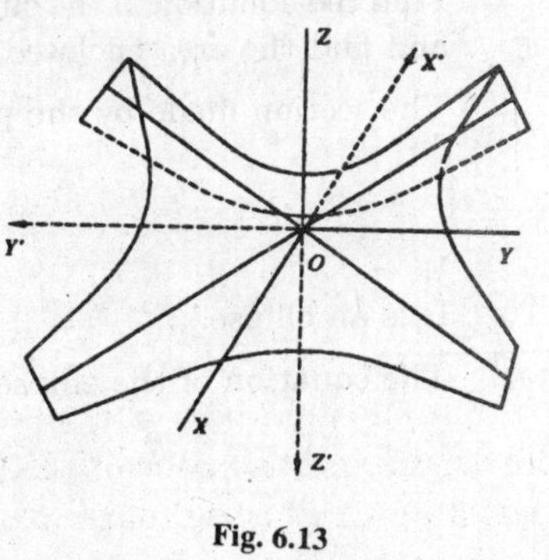

Fig. 6.13

6.17 General Equation

The general equation of the second degree in x, y, z is of the form $ax^2 + by^2 + cz^2 + 2fyz + 2gzx + 2hxy + 2ux + 2vy + 2wz + d = 0$.

It contains nine disposable constants. Thus a quadric or a conicoid can be determined to satisfy nine conditions each of which gives rise to one relation between the constants. By transformation of coordinate axes the general equation can be reduced to its standard form.

Note. (i) $ax^2 + by^2 + cz^2 = 1$ represents a *central conicoid* and (ii) $ax^2 + by^2 = 2cz$ represents a *paraboloid*.

Worked Out Examples

1. Find the nature of the conicoid, $3x^2 - 2y^2 - 12x - 12y - 6y = 0$. **[C.U. '66]**

The given equation can be written as

$$3(x-2)^2 - 2(y+3)^2 = 6(z+1), \quad \text{or} \quad \frac{(x-2)^2}{2} - \frac{(y+3)^2}{3} = z+1.$$

If the origin is shifted to $(2, -3, -1)$, the equation will reduce to $\frac{x^2}{2} - \frac{y^2}{3} = z$. It is a hyperbolic paraboloid with the principal planes $x = 0$ and $y = 0$. Hence the original equation is a hyperbolic paraboloid with the principal planes $x - 2 = 0$ and $y + 3 = 0$.

2. Find the nature of the quadric surface given by the equation

$$2x^2 + 5y^2 + 3z^2 - 4x + 20y - 6z - 5 = 0.$$

The given equation can be written as

$$2(x-1)^2 + 5(y+2)^2 + 3(z-1)^2 = 30,$$

$$\text{or} \quad \frac{(x-1)^2}{15} + \frac{(y+2)^2}{6} + \frac{(z-1)^2}{10} = 1.$$

Shifting the origin to $(1, -2, 1)$ the equation reduces to $\frac{x^2}{15} + \frac{y^2}{6} + \frac{z^2}{10} = 1$. It is an ellipsoid with the centre at the origin and the semi-axes are $\sqrt{15}, \sqrt{6}, \sqrt{10}$. Hence the original equation represents an ellipsoid whose centre is $(1, -2, 1)$ and the principal axes of it are parallel to coordinate axes. The principal planes are $x = 1, y = -2, z = 1$.

3. Find the equation of the curve in which the plane $z = h$ cuts the ellipsoid $\frac{x^2}{a^2} + \frac{y^2}{b^2} + \frac{z^2}{c^2} = 1$ and find the area enclosed by the curve. **[C.U. '61]**

The section made by the plane $z = h$ is

$$z = h, \quad \frac{x^2}{a^2} + \frac{y^2}{b^2} = 1 - \frac{h^2}{c^2}.$$

It is an ellipse.

The equation of the ellipse can be written as

$$\frac{x^2}{\frac{a^2(c^2-h^2)}{c^2}} + \frac{y^2}{\frac{b^2(c^2-h^2)}{c^2}} = 1.$$

Hence the area of this section

$$= \pi\sqrt{\frac{a^2(c^2-h^2)}{c^2}}\cdot\sqrt{\frac{b(c^2-h)}{c^2}} = \frac{\pi ab}{c^2}(c^2-h^2).$$

Exercise D

1. Find the nature of the following surfaces given by the equations:

   (a) $9x^2+36y^2+4z^2-36x+216y+32z+388=0$;

   (b) $x^2+4y^2-9z^2=36$;

   (c) $-3x^2+4y^2-5z^2=60$;

   (d) $4x^2+6y^2-16x+12y-24z-22=0$.

2. (a) Find the equation of the ellipsoid whose four vertices are $(\pm1,0,0)$, $(0,\pm2,0)$ and which passes through

$$\left(\frac{1}{2}, \frac{1}{\sqrt{2}}, \frac{3\sqrt{5}}{2\sqrt{2}}\right).$$

   (b) Obtain the equation of the surface generated by the rotation of $z^2=4x$, $y=0$ about z-axis.

3. Determine if the line $\frac{x-2}{2}=\frac{y-3}{-6}=\frac{z-1}{1}$ intersects the quadric $\frac{x^2}{16}+\frac{y^2}{9}-\frac{z^2}{4}=1$ and in case it does, find the points of intersection. **[C.H. '84]**

Answers

1. (a) Ellipsoid, centre $(2,-3,-4)$, semi-axes $2,1,3$, principal planes $x-2=0$, $y+3=0$, $z+4=0$; (b) Hyperboloid of one sheet, centre $(0,0,0)$, semi-axes $6,3,2$, principal plnes $x=0, y=0, z=0$; (c) Hyperboloid of two sheets, centre $(0,0,0)$, semi-axes $2\sqrt{5}, \sqrt{15}, 2\sqrt{3}$, principal planes $x=0, y=0, z=0$. (d) Elliptic paraboloid, vertex $(2,-1,0)$. **2.** (a) $\frac{x^2}{1}+\frac{y^2}{4}+\frac{z^2}{9}=1$; (b) $y^2+z^2=4x$. **3.** $(2,3,1), (4,-3,2)$.

Miscellaneous Examples

1. Find the equation of the cone whose vertex is at the origin and which passes through the curve given by $y^2+z^2=b^2$, $x=a$. Also identify the sections of the cone by the planes parallel to the xy plane.

   [HINTS. The equation of the cone is $y^2+z^2=b^2\cdot\frac{x^2}{a^2}$, or $a^2(y^2+z^2)=b^2x^2$. Section by $z=k$ is $a^2(y^2+k^2)=b^2x^2$, or $b^2x^2-a^2y^2=a^2k^2$. It is the equation of a hyperbola.]

2. Find the focus and length of a latus rectum of the conic obtained by the sections of $4x^2-9y^2=36+z$ by xz plane.

   [HINTS. Section by $y=0$ is $4x^2=36+z$, or $x^2=\frac{1}{4}(z+36)$. It is a parabola. Length of latus rectum $=\frac{1}{4}$ and the focus is $(0,0-35\frac{3}{4})$.]

3. Find the equation of the cylinder whose generators are parallel to z-axis and intersect the curve $ax^2 + by^2 = 2z$, $lx + my + nz = p$.

[HINTS. Let (α, β, γ) be a point on the cylinder. The equation of the generator through this point is $\frac{x-\alpha}{0} = \frac{y-\beta}{0} = \frac{z-\gamma}{1}$ (parallel to z-axis). If this generator meets curve at $(\alpha, \beta, \gamma + r)$, then r is given by $a\alpha^2 + b\beta^2 = 2(\gamma + r)$ and $l\alpha + m\beta + n(\gamma + r) = p$. Eliminating r, $2(l\alpha + m\beta) + n(a\alpha^2 + b\beta^2) = 2p$. Hence the equation of the cylinder is $n(ax^2 + by^2) + 2(lx + my) = 2p$.]

4. Obtain the equation of the sphere which passes through the circle $x^2 + y^2 = 4$, $x = 0$ and is cut by the plane $x + 2y + 2z = 0$ in a circle of radius 3.

[HINTS. Let the equation of the sphere be $x^2 + y^2 + z^2 - 4 + 2\lambda z = 0$. Centre is $(0, 0, -\lambda)$ and radius is $\sqrt{\lambda^2 + 4}$. Distance of the plane from the centre $= \frac{-2\lambda}{3}$. Considering the radius of the circle, $9 = \lambda^2 + 4 - \frac{4\lambda^2}{9}$, or $\lambda = \pm 3$. Hence the equations of spheres are $x^2 + y^2 + z^2 \pm 6z - 4 = 0$.]

5. Find the equations of circular cones which contain the three co-ordinate axes as generators.

[HINTS. The axes of the cone is equally inclined to the co-ordinate axes. If θ is the inclination then $\cos\theta = \frac{1}{\sqrt{3}}$.

If the axis of the cone in the first octant then its equations are $\frac{x}{1} = \frac{y}{1} = \frac{z}{1}$.

Equation of the cone is $(x + y + z)^2 = 3(x^2 + y^2 + z^2)\cdot\frac{1}{3}$, or $xy + yz + zx = 0$. Other possible equations for the axis of the cone are

$$\frac{x}{1} = \frac{y}{-1} = \frac{z}{-1} \quad \text{or} \quad \frac{x}{1} = \frac{y}{-1} = \frac{z}{-1} \quad \text{or} \quad \frac{x}{1} = \frac{y}{1} = \frac{z}{-1}.$$

The corresponding equations of cones are

$$(x - y - z)^2 = x^2 + y^2 + z^2 \quad \text{or} \quad (x - y + z)^2 = x^2 + y^2 + z^2$$

$$\text{or} \quad (x + y - z)^2 = x^2 + y^2 + z^2$$

$$\text{i.e.,} \quad xy - yz + zx = 0 \quad \text{or} \quad xy + yz - zx = 0 \quad \text{or} \quad xy - yz - zx = 0.]$$

Chapter 7

Tangent, Normal

7.1 To find the equation of the tangent plane at (x_1, y_1, z_1) to the quadric surface represented by

$$ax^2 + by^2 + cz^2 + 2fyz + 2gzx + 2hxy + 2ux + 2vy + 2wz + d = 0.$$

[In the subsequent discussion we shall denote

$$F(x, y, z) = ax^2 + by^2 + cz^2 + 2fyz + 2gzx + 2hxy + 2ux + 2vy + 2wz + d,$$

$$\phi(x, y, z) = ax^2 + by^2 + cz^2 + 2fyz + 2gzx + 2hxy,$$

$$\frac{\delta F}{\delta x_1}, \quad \frac{\delta F}{\delta y_1}, \quad \frac{\delta F}{\delta z_1}, \quad \frac{\delta \phi}{\delta x_1}, \quad \frac{\delta \phi}{\delta y_1}, \quad \frac{\delta \phi}{\delta z_1}$$

as partial derivatives of $F(x, y, z)$ and $\phi(x, y, z)$ w.r.t. x, y and z respectively at the point (x_1, y_1, z_1).]

The equations of the line having d. cs. l, m, n and passing through the point (x_1, y_1, z_1) are

$$\frac{x - x_1}{l} = \frac{y - y_1}{m} = \frac{z - z_1}{n} \quad (= r \text{ say}). \tag{7.1.1}$$

Any point on this line is $(lr + x_1, mr + y_1, nr + z_1)$. If the line meets the surface at this point, then

$$F(lr + x_1, mr + y_1, nr + z_1) = 0,$$

$$\text{or} \quad \phi(l, m, n)r^2 + \left(l\frac{\delta F}{\delta x_1} + m\frac{\delta F}{\delta y_1} + n\frac{\delta F}{\delta z_1}\right) r + F(x_1, y_1, z_1) = 0. \tag{7.1.2}$$

It is a quadratic equation in r. The two roots are the distances of the two points of intersection between the line (7.1.1) and the surface from (x_1, y_1, z_1). If the line be the tangent to the given surface at (x_1, y_1, z_1), both the roots will be zero. In this case the sum of the roots and product of the roots are zero. By the equation (7.1.2)

$$l\frac{\delta F}{\delta x_1} + m\frac{\delta F}{\delta y_1} + n\frac{\delta F}{\delta z_1} = 0,$$

$$\text{i.e., } (ax_1 + by_1 + gz_1 + u)l + (hx_1 + by_1 + fz_1 + v)m$$

$$+(gx_1 + fy_1 + cz_1 + w)n = 0 \quad (7.1.3)$$

$$\text{and } F(x_1, y_1, z_1) = 0. \quad (7.1.4)$$

The conditions (7.1.3) and (7.1.4) ensure that the line (7.1.2) is the tangent to the surface of $F(x, y, z) = 0$ at the point (x_1, y_1, z_1) on the surface. The tangent plane at (x_1, y_1, z_1) is the locus of the line (7.1.1) with the condition (7.1.3). Thus the equation of this tangent plane is obtained by eliminating l, m, n from (7.1.3) by (7.1.1) and its equation is

$$(ax_1 + hy_1 + gz_1 + u)(x - x_1) + (hx_1 + by_1 + fz_1 + v)(y - y_1)$$

$$+(gx_1 + fy_1 + cz_1 + w)(z - z_1) = 0,$$

$$\text{or } axx_1 + byy_1 + czz_1 + f(yz_1 + zy_1) + g(zx_1 + xz_1) + h(xy_1 + yx_1)$$

$$+u(x + x_1) + v(y + y_1) + w(z + z_1) + d = 0. \quad \text{[by (7.1.4)]} \quad (7.1.5)$$

Special Cases

Quadric surface	Equation of the tangent plane at (x_1, y_1, z_1)
(1) **Sphere** (i) $x^2 + y^2 + z^2 = a^2$ (ii) $x^2 + y^2 + z^2 + 2ux + 2vy$ $+2wz + d = 0$	 $xx_1 + yy_1 + zz_1 = a^2$ $xx_1 + yy_1 + zz_1 + u(x + x_1)$ $+v(y + y_1) + w(z + z_1) + d = 0$
(2) **Cone** $ax^2 + by^2 + cz^2 + 2fyz + 2gzx$ $+2hxy = 0$	 $axx_1 + byy_1 + czz_1 + f(yz_1 + zy_1)$ $+g(zx_1 + xz_1) + h(xy_1 + yx_1) = 0$
(3) **Central conicoid** $ax^2 + by^2 + cz^2 = 1$	 $axx_1 + byy_1 + czz_1 = 1.$
(4) **Paraboloid** $ax^2 + by^2 = 2cz$	 $axx_1 + byy_1 = c(z + z_1)$

Note.

(i) If the roots of (7.1.2) are real and unequal and they are r_1 and r_2 then the line (7.1.1) cuts the surface at two distinct points $(lr_1 + x_1, mr_1 + y_1, nr_1 + z_1)$ and $(lr_2 + x_1, mr_2 + y_1, nr_2 + z_1)$. In this case the line is called a *secant* of the surface.

(ii) If the roots of (7.1.2) are real and equal then the line is a tangent to the surface at the common point $(lr + x_1, mr + y_1, nr + z_1)$. This point is called the *point of contact.*

(iii) If the roots of (7.1.2) are complex, the line does not intersect the surface.

7.2 To find the condition that the plane $lx + my + nz = p$ is a tangent plane to (i) the sphere $x^2 + y^2 + z^2 + 2ux + 2vy + 2wz + d = 0$, (ii) the central conicoid $ax^2 + by^2 + cz^2 = 1$ and (iii) the paraboloid $ax^2 + by^2 = 2cz$

(i) (a) *The necessary condition.*

The plane will touch the sphere, if its distance from the centre of the sphere is equal to the radius of the sphere.

$\therefore$ the condition is

$$\frac{-lu - mv - nw - p}{\sqrt{l^2 + m^2 + n^2}} = \sqrt{(u^2 + v^2 + w^2 - d)},$$

$$\text{or} \quad (lu + mv + nw + p)^2 = (l^2 + m^2 + n^2)(u^2 + v^2 + w^2 - d). \tag{7.2.1}$$

This condition is necessary.

(b) *The sufficient condition.*

Let the condition (7.2.1) be satisfied.

Then

$$\left\{\frac{lu + mv + nw + p}{\sqrt{(l^2 + m^2 + n^2)}}\right\}^2 = u^2 + v^2 + w^2 - d,$$

$$\text{or} \quad \left(\frac{-lu - mv - nw - p}{\sqrt{l^2 + m^2 + n^2}}\right)^2 = u^2 + v^2 + w^2 - d,$$

$$\text{or} \quad \frac{-lu - mv - nw - p}{\sqrt{l^2 + m^2 + n^2}} = \sqrt{(u^2 + v^2 + w^2 - d)}.$$

It shows that the distance of the plane $lx+my+nz = p$ from the centre $(-u, -v, -w)$ is equal to the radius $\sqrt{(u^2 + v^2 + w^2 - d)}$. Therefore the plane touches the sphere. Hence the condition (7.2.1) is sufficient.

Note. *Point of contact.* Let the plane touch the sphere at (x_1, y_1, z_1). The equation of the tangent plane at this point is

$$(x_1 + u)x + (y_1 + v)y + (z_1 + w)z + ux_1 + vy_1 + wz_1 + d = 0.$$

It is identical with the given plane.

$$\therefore \quad \frac{x_1 + u}{l} = \frac{y_1 + v}{m} = \frac{z_1 + w}{n} = \frac{-(ux_1 + vy_1 + wz_1 + d)}{p}$$

$$= \frac{u(x_1 + u) + v(y_1 + v) + w(z_1 + w) - (ux_1 + vy_1 + wz_1 + d)}{lu + mv + nw + p}$$

$$= \frac{u^2 + v^2 + w^2 - d}{lu + mv + nw + p}.$$

Thus the point of contact is

$$\left\{\frac{l(u^2+v^2+w^2-d)}{lu+mv+nw+p}-u, \frac{m(u^2+v^2+w^2-d)}{lu+mv+nw+p}-v, \frac{n(u^2+v^2+w^2-d)}{lu+mv+nw+p}-w\right\}.$$

(ii) (a) *The necessary condition.*

Let the plane touch the central conicoid at (x_1, y_1, z_1).

The equation of the tangent plane at this point is

$$axx_1 + byy_1 + czz_1 = 1.$$

It is identical with the given plane

$$lx + my + nz = p.$$

Comparing the coefficients

$$\frac{ax_1}{l} = \frac{by_1}{m} = \frac{cz_1}{n} = \frac{1}{p}, \quad \text{or} \quad x_1 = \frac{l}{ap}, \quad y_1 = \frac{m}{bp}, \quad z_1 = \frac{n}{cp}.$$

Since (x_1, y_1, z_1) is a point on the conicoid

$$ax_1^2 + by_1^2 + cz_1^2 = 1,$$

$$\text{or} \quad a\frac{l^2}{a^2p^2} + b\frac{m^2}{b^2p^2} + c\frac{n^2}{c^2p^2} = 1,$$

$$\text{or} \quad \frac{l^2}{a} + \frac{m^2}{b} + \frac{n^2}{c} = p^2. \tag{7.2.2}$$

It is the necessary condition for the tangency of the plane.

(b) *The sufficient condition.*

Let the condition (7.2.2) hold.

Then from (7.2.2)

$$a\left(\frac{l}{ap}\right)^2 + b\left(\frac{m}{bp}\right)^2 + c\left(\frac{n}{ap}\right)^2 = 1.$$

Putting

$$\frac{l}{ap} = x_1, \quad \frac{m}{bp} = y_1, \quad \frac{n}{ap} = z_1,$$

we have $ax_1^2 + by_1^2 + cz_1^2 = 1$.

It shows that the point $\left(\frac{l}{ap}, \frac{m}{bp}, \frac{n}{cp}\right)$ lies on the conicoid.

Now the equation of the tangent plane at this point is

$$ax\frac{l}{ap} + by\frac{m}{bp} + cz\frac{n}{cp} = 1, \quad \text{or} \quad lx + my + nz = p.$$

Therefore the given plane is a tangent plane to the central conicoid. Hence the condition (7.2.2) is sufficient.

Note. The point of contact is $\left(\frac{l}{ap}, \frac{m}{bp}, \frac{n}{cp}\right)$.

(iii) (a) *The necessary condition.*

Let the plane touch the paraboloid at (x_1, y_1, z_1). The equation of the tangent plane at this point is $axx_1 + byy_1 = c(z+z_1)$. It is identical with $lx+my+nz = p$. Comparing the coefficients,

$$\frac{ax_1}{l} = \frac{by_1}{m} = \frac{-c}{n} = \frac{cz_1}{p}, \quad \text{or} \quad x_1 = -\frac{cl}{an}, \quad y_1 = \frac{-cm}{bn}, \quad z_1 = -\frac{p}{n}.$$

Since the point (x_1, y_1, z_1) is on the paraboloid

$$ax_1^2 + by_1^2 = 2cz_1,$$

$$\text{or} \quad a\frac{c^2l^2}{a^2n^2} + b\frac{c^2m^2}{b^2n^2} = -2c\frac{p}{n},$$

$$\text{or} \quad \frac{l^2}{a} + \frac{m^2}{b} + \frac{2np}{c} = 0. \tag{7.2.3}$$

It is the necessary condition.

(b) *The sufficient condition.*

Let the condition (7.2.3) hold.

From (7.2.3)

$$a\frac{c^2l^2}{a^2n^2} + b\frac{c^2m^2}{b^2n^2} = -2c\frac{p}{n}.$$

Putting

$$x_1 = -\frac{cl}{an}, \quad y_1 = -\frac{cm}{bn}, \quad z_1 = -\frac{p}{n},$$

we have $ax_1^2 + by_1^2 = 2cz_1$.

Therefore the point $\left(-\frac{cl}{an}, \frac{-cm}{bn}, -\frac{p}{n}\right)$ lies on the paraboloid.

Again the equation of the tangent plane at this point is

$$ax\left(-\frac{cl}{an}\right) + by\cdot\left(-\frac{cm}{bn}\right) = c\cdot\left(z - \frac{p}{n}\right), \quad \text{or} \quad lx + my + nz = p.$$

Thus the given plane is a tangent plane to the paraboloid. Hence the condition (7.2.3) is sufficient.

Note. *The point of contact is* $\left(-\frac{cl}{an}, -\frac{cm}{bn}, -\frac{p}{n}\right)$.

The plane

$$lx + my + nz = -\frac{c}{2n}\left(\frac{l^2}{a} + \frac{m^2}{b}\right)$$

touches the paraboloid.

7.3 To find the condition that the plane $lx + my + nz = 0$ is a tangent plane to the cone $ax^2 + by^2 + cz^2 + 2fyz + 2gzx + 2hxy = 0$.

(i) *The necessary condition.*

If (x_1, y_1, z_1) be the point of contact, the tangent plane is

$$(ax_1 + hy_1 + gz_1)x + (hx_1 + by_1 + fz_1)y + (gx_1 + fy_1 + cz_1)z = 0.$$

It is identical with $lx + my + nz = 0$.

Comparing the coefficients

$$\frac{ax_1 + hy_1 + gz_1}{l} = \frac{hx_1 + by_1 + fz_1}{m} = \frac{gx_1 + fy_1 + cz_1}{n} = k \text{ (say)}.$$

Then

$$ax_1 + hy_1 + gz_1 - kl = 0, \qquad (7.3.1)$$

$$hx_1 + by_1 + fz_1 - km = 0, \qquad (7.3.2)$$

$$gx_1 + fy_1 + cz_1 - kn = 0. \qquad (7.3.3)$$

Again

$$lx_1 + my_1 + nz_1 = 0. \qquad (7.3.4)$$

Eliminating x_1, y_1, z_1 from (7.3.1), (7.3.2), (7.3.3) and (7.3.4), we have

$$\begin{vmatrix} a & h & g & l \\ h & b & f & m \\ g & f & c & n \\ l & m & n & 0 \end{vmatrix} = 0 \qquad (7.3.5)$$

$$\text{or} \quad Al^2 + Bm^2 + Cn^2 + 2Fmn + 2Gnl + 2Hlm = 0. \qquad (7.3.6)$$

It is the necessary condition. Here $A, B, C, \ldots$ are the cofactors of $a, b, c, \ldots$ of the determinant.

$$\Delta = \begin{vmatrix} a & h & g \\ h & b & f \\ g & f & c \end{vmatrix}.$$

(ii) *The sufficient condition.*

Let the condition (7.3.5) hold.

Again

$$\begin{vmatrix} a & h & g \\ h & b & f \\ g & f & c \end{vmatrix} \neq 0.$$

For these two conditions the system of equations

$$ax + hy + gz + lw = 0,$$

$$hx + by + fz + mw = 0,$$

$$gx + fy + cz + mw = 0,$$

$$lx + my + nz = 0$$

has a non-trivial solution. If this solution be x_1, y_1, z_1, w_1, then

$$ax_1 + hy_1 + gz_1 = -lw_1,$$

$$hx_1 + by_1 + fz_1 = -mw_1,$$

$$gx_1 + fy_1 + cz_1 = -nw_1,$$

$$lx_1 + my_1 + nz_1 = 0.$$

Now the equation $lx + my + nz = 0$ can be written as

$$x(ax_1 + hy_1 + gz_1) + y(hx_1 + by_1 + fz_1 + z(gx_1 + fy_1 + cz_1) = 0.$$

It is the equation of the tangent to the cone at the point (x_1, y_1, z_i). Hence the condition (7.3.5) is sufficient for the tangency of the plane to the cone.

Corollary 1. *Point of contact.* Here $\Delta \neq 0$. Therefore the equations (7.3.1), (7.3.2), (7.3.3) and (7.3.4) in x_1, y_1 and z_1 are consistent. Thus x_1, y_1, z_1 are obtained from (7.3.1), (7.3.2) and (7.3.3).

Corollary 2. *Reciprocal cone.* The equation of the line through the origin perpendicular to the plane $lx + my + nz = 0$ is

$$\frac{x}{l} = \frac{y}{m} = \frac{z}{n}. \tag{7.3.7}$$

Eliminating l, m, n between (7.3.6) and (7.3.7),

$$Ax^2 + By^2 + Cz^2 + 2Fyz + 2Gzx + 2Hxy = 0. \tag{7.3.8}$$

Since

$$\Delta^2 = \begin{vmatrix} A & H & G \\ H & B & F \\ G & F & C \end{vmatrix} \quad \text{and} \quad \Delta \neq 0, \begin{vmatrix} A & H & G \\ H & B & F \\ G & F & C \end{vmatrix} \neq 0.$$

Therefore the equation (8) represents a cone with the vertex at the origin and generators are normals to the tangent planes to the cone

$$ax^2 + by^2 + cz^2 + 2fyz + 2gzx + 2hxy = 0 \tag{7.3.9}$$

at the origin. The cone represented by (7.3.8) is called the reciprocal to the cone (7.3.9).

The reciprocal to the cone (7.3.8) is

$$\bar{A}x^2 + \bar{B}y^2 + \bar{C}z^2 + 2\bar{F}yz + 2\bar{G}zx + 2\bar{H}xy = 0, \tag{7.3.10}$$

where

$$\bar{A} = BC - F^2 = a\Delta, \bar{B} = CA - G^2 = b\Delta, \bar{C} = AB - H^2 = c\Delta,$$

$$\bar{F} = GH - AF = f\Delta, \bar{G} = HF - BG = g\Delta, \bar{H} = FG - CH = h\Delta.$$

For these relations the eqation (7.3.10) is

$$ax^2 + by^2 + cz^2 + 2fyz + 2gzx + 2hxy = 0.$$

Hence the cones (7.3.9) and (7.3.8) are reciprocal to each other and they are called reciprocal cones.

Corollary 3. The condition for the cone (7.3.9) to possess three mutually perpendicular tangent planes is $A + B + C = 0$, i.e., $bc + ca + ab = f^2 + g^2 + h^2$. It implies that the cone (7.3.8) has three mutually perpendicular generators.

► **Example 1** *Prove that the tangent planes to the cone*

$$x^2 - y^2 + 2z^2 - 3yz + 4zx - 5xy = 0$$

are perpendicular to the generators of the cone

$$17x^2 + 8y^2 + 29z^2 + 28yz - 46zx - 16xy = 0.$$

Let us find the reciprocal cone to

$$x^2 - y^2 + 2z^2 - 3yz + 4zx - 5xy = 0.$$

Here $a = 1, b = -1, c = 2, f = -\frac{3}{2}, g = 2, h = -\frac{5}{2}$.

$$\therefore \quad A = bc - f^2 = -2 - \frac{9}{4} = -\frac{17}{4}, B = ca - g^2 = 2 - 4 = -2,$$

$$C = ab - h^2 = -1 - \frac{25}{4} = -\frac{29}{4}, F = gh - af = -5 + \frac{3}{2} = -\frac{7}{2},$$

$$G = hf - bg = \frac{15}{4} + 2 = \frac{23}{4}, H = fg - ch = -3 + 5 = 2.$$

$\therefore$ the reciprocal cone is

$$-\frac{17}{4}x^2 - 2y^2 - \frac{29}{4}z^2 - 7yz + \frac{23}{2}zx + 4xy = 0,$$

$$\text{or} \quad 17x^2 + 8y^2 + 29z^2 + 28yz - 46zx - 8xy = 0.$$

Hence the result follows.

► **Example 2** *Prove that the equation* $\sqrt{(fx)} \pm \sqrt{(gy)} \pm \sqrt{(hz)} = 0$ *represents a cone that touches the co-ordinate planes.*

The equation of the cone whose generators are the co-ordinate axes is of the form $fyz + gzx + hxy = 0$, or $2fyz + 2gzx + 2hxy = 0$.

Its reciprocal cone is

$$-f^2x^2 - g^2y^2 - h^2z^2 + 2ghyz + 2hfzx + 2fgxy = 0,$$

$$\text{or}\quad (fx + gy - hz)^2 = 4fgxy,$$

$$\text{or}\quad fx + gy - hz = \pm 2\sqrt{(fgxy)},$$

$$\text{or}\quad fx + gy \pm 2\sqrt{(fgxy)} = hz,$$

$$\text{or}\quad (\sqrt{fx} \pm \sqrt{gy})^2 = hz,$$

$$\text{or}\quad \sqrt{(fx)} \pm \sqrt{(gy)} \pm \sqrt{(hz)} = 0.$$

7.4 To find the equation of the normal at (x_1, y_1, z_1) to (i) the central conicoid $ax^2 + by^2 + cz^2 = 1$ and (ii) the paraboloid $ax^2 + by^2 = 2cz$.

(i) The equation of the tangent plane at (x_1, y_1, z_1) to the central conicoid is

$$axx_1 + byy_1 + czz_1 = 1.$$

The normal at (x_1, y_1, z_1) is a straight line passing through this point and perpendicular to the tangent plane. Therefore the d. rs. of this normal are ax_1, by_1, cz_1. Hence the equation of the normal at (x_1, y_1, z_1) is

$$\frac{x - x_1}{ax_1} = \frac{y - y_1}{by_1} = \frac{z - z_1}{cz_1}.$$

(ii) The equation of the tangent plane at (x_1, y_1, z_1) to the paraboloid is

$$axx_1 + byy_1 = c(z + z_1).$$

The normal is a straight line perpendicular to the tangent plane at the point of contact. Therefore the d. rs. of this normal are $ax_1, by_1, -c$. Hence the equation of normal is

$$\frac{x - x_1}{ax_1} = \frac{y - y_1}{by_1} = \frac{z - z_1}{-c}.$$

7.5 Number of normals through a given point

(a) **Central conicoid :** The equation of the normal at (x_1, y_1, z_1) to the central conicoid $ax^2 + by^2 + cz^2 = 1$ is

$$\frac{x - x_1}{ax_1} = \frac{y - y_1}{by_1} = \frac{z - z_1}{cz_1}.$$

If this normal passes through (α, β, γ), then

$$\frac{\alpha - x_1}{ax_1} = \frac{\beta - y_1}{by_1} = \frac{\gamma - z_1}{cz_1} = \lambda \text{ (say)}.$$

From these relations

$$x_1 = \frac{\alpha}{1 + a\lambda}, \quad y_1 = \frac{\beta}{1 + b\lambda}, \quad z_1 = \frac{\gamma}{1 + c\lambda}.$$

We have $ax_1^2 + by_1^2 + cz_1^2 = 1$.

$$\therefore \quad \frac{a\alpha^2}{(1 + a\lambda)^2} + \frac{b\beta^2}{(1 + b\lambda)^2} + \frac{c\gamma^2}{(1 + c\lambda)^2} = 1.$$

This equation gives six values of λ. Each value corresponds to a normal at a point on the conicoid. Hence six normals can be drawn in general from a point to a central conicoid.

(b) **Paraboloid :** The equation of the normal at (x_1, y_1, z_1) to the paraboloid $ax^2 + by^2 = 2cz$ is

$$\frac{x - x_1}{ax_1} = \frac{y - y_1}{by_1} = \frac{z - z_1}{-c}.$$

If this normal passes through (α, β, γ), then

$$\frac{\alpha - x_1}{ax_1} = \frac{\beta - y_1}{by_1} = \frac{\gamma - z_1}{-c} = \lambda \text{ (say)}.$$

From these relations

$$x_1 = \frac{\alpha}{1 + a\lambda}, \quad y_1 = \frac{\beta}{1 + b\lambda}, \quad z_1 = c\lambda + \gamma.$$

We have $ax_1^2 + by_1^2 = 2cz_1$.

$$\therefore \quad a\frac{\alpha^2}{(1 + a\lambda)^2} + b\frac{\beta^2}{(1 + b\lambda)^2} = 2c(c\lambda + \gamma).$$

This equation gives five values of λ. Each value corresponds to a normal to a point on the paraboloid. Hence five normals can be drawn in general to a paraboloid through a given point.

7.6 To find the locus of the point of intersection of three mutually perpendicular tangent planes to a conicoid.

(a) **Central conicoid :** Let $ax^2+by^2+cz^2=1$ be the equation of the central conicoid and the three mutually perpendicular tangent planes be

$$\left.\begin{aligned} l_1x+m_1y+n_1z &= \sqrt{\left(\frac{l_1^2}{a}+\frac{m_1^2}{b}+\frac{n_1^2}{c}\right)},\\ l_2x+m_2y+n_2z &= \sqrt{\left(\frac{l_2^2}{a}+\frac{m_2^2}{b}+\frac{n_2^2}{c}\right)},\\ \text{and}\quad l_3x+m_3y+n_3z &= \sqrt{\left(\frac{l_3^2}{a}+\frac{m_3^2}{b}+\frac{n_3^2}{c}\right)}.\end{aligned}\right\} \tag{7.6.1}$$

Here

$$\sum_{r=1}^{3} l_r^2 = 1 = \sum_{r=1}^{3} m_r^2 = \sum_{r=1}^{3} n_r^2$$

$$\text{and}\quad \sum_{r=1}^{3} l_rm_r = 0 = \sum_{r=1}^{3} m_rn_r = \sum_{r=1}^{3} n_rl_r. \tag{7.6.2}$$

Now squaring and adding the equations of (7.6.1) and using the conditions (7.6.2), the locus of the point of intersection of the planes of (7.6.1) is obtained. The locus is $x^2+y^2+z^2=\frac{1}{a}+\frac{1}{b}+\frac{1}{c}$. It is a sphere called the *director sphere.*

(b) **Paraboloid :** Let $ax^2+by^2=2cz$ be the equation of the paraboloid and the three mutually perpendicular tangents to it be

$$\left.\begin{aligned} l_1x+m_1y+n_1z+\frac{c}{2n_1}\left(\frac{l_1^2}{a}+\frac{m_1^2}{b}\right) &= 0,\\ l_2x+m_2y+n_2z+\frac{c}{2n_2}\left(\frac{l_2^2}{a}+\frac{m_2^2}{b}\right) &= 0,\\ \text{and}\quad l_3x+m_3y+n_3z+\frac{c}{2n_3}\left(\frac{l_3^2}{a}+\frac{m_3^2}{b}\right) &= 0.\end{aligned}\right\}. \tag{7.6.3}$$

Here

$$\sum_{r=1}^{3} l_r^2 = 1 = \sum_{r=1}^{3} m_r^2 = \sum_{r=1}^{3} n_r^2$$

$$\text{and}\quad \sum_{r=1}^{3} l_rm_r = 0 = \sum_{r=1}^{3} m_rn_r = \sum_{r=1}^{3} n_rl_r. \tag{7.6.4}$$

Now multiplying the equations of (7.6.3) by n_1, n_2, n_3 respectively and then adding and using the conditions (7.6.4) the required locus is obtained. This locus is $2z+c\left(\frac{1}{a}+\frac{1}{b}\right)=0$.

It is a plane perpendicular to the axis of the paraboloid.

Worked Out Examples

1. Find the equation of the tangent plane to $2x^2 - 5y^2 + 6z = 0$ at $(1, 2, 3)$.

   The point $(1, 2, 3)$ is on the conicoid. Therefore the equation of the tangent plane at this point is

$$2x \cdot 1 - 5y \cdot 2 + 3(z + 3) = 0, \quad \text{or} \quad 2x - 10y + 3z + 9 = 0.$$

2. Find the equation of the normal to the hyperboloid

$$3x^2 - 5y^2 + 7z^2 = 30 \text{ at } (2, 3, 3).$$

   The point $(2, 3, 3)$ is on the conicoid. Therefore the equation of the tangent plane at this point is

$$3x \cdot 2 - 5y \cdot 3 + 7z \cdot 3 = 30, \quad \text{or} \quad 2x - 5y + 7z = 10.$$

   The normal is a straight line perpendicular to this plane at $(2, 3, 3)$. Hence the required equation is

$$\frac{x-2}{2} = \frac{y-3}{-5} = \frac{z-3}{7}.$$

3. Find the equation of the sphere that passes through the points $(1, 0, 0)$, $(0, 1, 0)$, $(0, 0, 1)$ and touches the plane

$$2x + 2y - z = 15. \qquad \textbf{[C.U. '89]}$$

   Let the equation of the sphere be

$$x^2 + y^2 + z^2 + 2ux + 2vy + 2wz + d = 0.$$

   As it passes through the given points,

$$1 + 2u + d = 0, \tag{1}$$

$$1 + 2v + d = 0, \tag{2}$$

$$\text{and} \quad 1 + 2w + d = 0. \tag{3}$$

   Since the plane $2x + 2y - z = 15$ touches the sphere, the distance of the centre $(-u, -v, -w)$ from the plane is equal to the radius $\sqrt{u^2 + v^2 + w^2 - d}$.

$$\therefore \quad \frac{-2u - 2v + w - 15}{\sqrt{2^2 + 2^2 + 1^2}} = \sqrt{u^2 + v^2 + w^2 - d}. \tag{4}$$

   From (1), (2) and (3) $u = v = w$.
   Putting in (4)

$$\frac{-2u - 2u + u - 15}{3} = \sqrt{3u^2 - d}, \quad \text{or} \quad -(u + 5) = \sqrt{3u^2 - d},$$

$$\text{or} \quad 3u^2 - d = u^2 + 10u + 25, \quad \text{or} \quad 2u^2 - 10u - d - 25 = 0.$$

   Again $d = -(2u + 1)$.

$$\therefore \quad 2u^2 - 10u + 2u + 1 - 25 = 0,$$

$$\text{or} \quad 2u^2 - 8u - 24 = 0, \quad \text{or} \quad u^2 - 4u - 12 = 0,$$

$$\text{or} \quad (u - 6)(u + 2) = 0. \quad \therefore \quad u = 6, -2.$$

Corresponding to this values of $u, d = -13, 3$.

$\therefore$ the equation of the spheres are

$$x^2 + y^2 + z^2 + 12(x + y + z) - 13 = 0$$

$$\text{and} \quad x^2 + y^2 + z^2 - 4(x + y + z) + 3 = 0.$$

4. Find the equation of the tangent plane to the hyperbolic paraboloid $5x^2 - 2y^2 = 2z$, parallel to $10x - 6y - z = 7$ and also find the point of contact.

Let the equation of the tangent plane be $10x - 6y - z = k$.

Since it touches the conicoid

$$\frac{10^2}{5} + \frac{6^2}{-2} + \frac{2(-1) \cdot k}{1} = 0, \quad \text{or} \quad k = 1.$$

$\therefore$ the equation of the tangent plane is $10x - 6y - z = 1$.

The point of contact is

$$\left(-\frac{1.10}{5.-1}, \quad -\frac{1.-6}{-2.-1}, \quad -\frac{1}{-1}\right), \quad \text{or} \quad (2, 3, 1).$$

5. Find the equations of the tangent planes to $2x^2 - 6y^2 + 3z^2 = 5$, which pass through the line $x + 9y - 3z = 0 = 3x - 3y + 6z - 5$. **[C.U. '63]**

Let the equation of the tangent plane be

$$(3x - 3y + 6z - 5) + \lambda(x + 9y - 3z) = 0,$$

$$\text{or} \quad (3 + \lambda)x + (9\lambda - 3)y + (6 - 3\lambda)z - 5 = 0. \tag{1}$$

The equation of the conicoid can be written as

$$\frac{2x^2}{5} - \frac{6y^2}{5} + \frac{3z^2}{5} = 1.$$

From the condition of the tangency of a plane to a conicoid,

$$\frac{(3 + \lambda)^2}{\frac{2}{5}} + \frac{(9\lambda - 3)^2}{-\frac{6}{5}} + \frac{(6 - 3\lambda)^2}{\frac{3}{5}} = 25,$$

$$\text{or} \quad (3 + \lambda)^2 - 3(3\lambda - 1)^2 + 6(2 - \lambda)^2 = 10,$$

$$\text{or} \quad \lambda^2 = 1, \quad \text{or} \quad \lambda = \pm 1.$$

Putting these values of λ in (1), the tangent planes are

$$4x + 6y + 3z = 5 \quad \text{and} \quad 2x - 12y + 9z = 5.$$

6. Show that the plane $4x + 9y + 14z - 64 = 0$ touches the sphere

$$3(x^2 + y^2 + z^2) - 2x - 3y - 4z - 22 = 0.$$

Find also the point of contact.

The given sphere is

$$x^2 + y^2 + z^2 - \frac{2}{3}x - y - \frac{4}{3}z - \frac{22}{3} = 0.$$

Centre is $\left(\frac{1}{3}, \frac{1}{2}, \frac{2}{3}\right)$ and radius

$$= \sqrt{\left(\frac{1}{9} + \frac{1}{4} + \frac{4}{9} + \frac{22}{3}\right)} = \frac{\sqrt{293}}{6}.$$

Perpendicular distance from the centre to the plane

$$= \frac{\left|\frac{4}{3} + \frac{9}{2} + \frac{28}{3} - 64\right|}{\sqrt{16 + 81 + 196}} = \frac{\sqrt{293}}{6}.$$

Since radius = perpendicular distance, the plane touches the sphere.

The equations of the line passing through the centre $\left(\frac{1}{3}, \frac{1}{2}, \frac{2}{3}\right)$ and perpendicular to the plane $4x + 9y + 14z - 64 = 0$ are

$$\frac{x - \frac{1}{3}}{4} = \frac{y - \frac{1}{2}}{9} = \frac{z - \frac{2}{3}}{14} = r \text{ (say)}.$$

Any point on this line is

$$\left(4r + \frac{1}{3}, \quad 9r + \frac{1}{2}, \quad 14r + \frac{2}{3}\right).$$

This will lie on the given plane, if

$$4\left(4r + \frac{1}{3}\right) + 9\left(9r + \frac{1}{2}\right) + 14\left(14r + \frac{2}{3}\right) - 64 = 0, \quad \text{or} \quad r = \frac{1}{6}.$$

Hence the point of contact is

$$\left(4 \times \frac{1}{6} + \frac{1}{3}, \quad 9 \times \frac{1}{6} + \frac{1}{2}, \quad 14 \times \frac{1}{6} + \frac{2}{3}\right), \quad \text{i.e., } (1, 2, 3).$$

7. Find the equation of a sphere touching the three co-ordinate planes. Let the sphere be

$$x^2 + y^2 + z^2 + 2ux + 2vy + 2wz + d = 0. \tag{1}$$

The co-ordinates of the centre are $(-u, -v, -w)$ and the radius

$$= \sqrt{(u^2 + v^2 + w^2 - d)}.$$

If the sphere touches the yz plane, i.e., $x = 0$, then the distance of this plane from the centre is equal to the radius.

$$\therefore \quad -u = \sqrt{(u^2 + v^2 + w^2 - d)}, \quad \text{or} \quad v^2 + w^2 = d. \tag{2}$$

Similarly

$$w^2 + u^2 = d \tag{3}$$

$$\text{and} \quad u^2 + v^2 = d. \tag{4}$$

Adding (2), (3) and (4), $2(u^2 + v^2 + w^2) = 3d$.

$$\therefore \quad u^2 + v^2 + w^2 = \frac{3d}{2}. \tag{5}$$

By (2) and (5), $u^2 = \frac{d}{2} = a^2$ (say). $\therefore \quad u = \pm a$.

Similarly $v = \pm a$ and $w = \pm a$.

Hence the equation of the spheres are

$$x^2 + y^2 + z^2 \pm 2a(x + y + z) + 2a^2 = 0.$$

8. Show that the points $(12, -18, 8)$ and $(-6, 18, -10)$ are at the feet of the normals to the ellipsoid $x^2 + 2y^2 + 3z^2 = 984$ which lie in the plane $x + y + z = 2$.

[B.H. '91; C.H. '76, '84, '92]

The points $(12, -18, 8)$ and $(-6, 18, -10)$ are on the surface of the given quadric. Therefore the tangent planes at these points are

$$12x - 2.18y + 3.8z = 984, \quad \text{or} \quad x - 3y + 2z = 82 \tag{1}$$

$$\text{and} \quad -6x + 2.18y - 3.10z = 984, \quad \text{or} \quad x - 6y + 5z = -164. \tag{2}$$

Consequently the normals at $(12, -18, 8)$ and $(-6, 18, -10)$ are

$$\frac{x - 12}{1} = \frac{y + 18}{-3} = \frac{z - 8}{2} \tag{3}$$

$$\text{and} \quad \frac{x + 6}{1} = \frac{y - 18}{-6} = \frac{z + 10}{5}. \tag{4}$$

Since

$$\begin{vmatrix} -6 - 12 & 18 - (-18) & -10 - 8 \\ 1 & -3 & 2 \\ 1 & -6 & 5 \end{vmatrix} = \begin{vmatrix} -18 & 36 & -18 \\ 1 & -3 & 2 \\ 1 & -6 & 5 \end{vmatrix}$$

$$= -18(-15 + 12) + 1(108 - 180) + 1(72 - 54) = 54 - 72 + 18 = 0,$$

the lines (3) and (4) are coplanar.

The equation of the plane containing these normals is

$$\begin{vmatrix} x - 12 & y + 18 & z - 8 \\ 1 & -3 & 2 \\ 1 & -6 & 5 \end{vmatrix} = 0,$$

$$\text{or} \quad -3(x - 12) - 3(y + 18) - 3(z - 8) = 0, \quad \text{or} \quad x + y + z = 2.$$

9. If the normal at a point P on the ellipsoid

$$\frac{x^2}{a^2} + \frac{y^2}{b^2} + \frac{z^2}{c^2} = 1$$

meets the principal planes in G_1, G_2, G_3, then show that

$$PG_1 : PG_2 : PG_3 = a^2 : b^2 : c^2.$$

[C.U. '67]

Let the co-ordinates of P be (x_1, y_1, z_1). The equation of the narmal at this point is

$$\frac{x - x_1}{\frac{px_1}{a^2}} = \frac{y - y_1}{\frac{py_1}{b^2}} = \frac{z - z_1}{\frac{pz_1}{c^2}} = r \text{ (say)}.$$

Here p is the perpendicular distance from the origin to the tangent plane at (x_1, y_1, z_1).

$$\frac{px_1}{a^2}, \quad \frac{py_1}{b^2}, \quad \frac{pz_1}{c^2}$$

are the d. cs. of the normal. r is the distance of a point on the normal from (x_1, y_1, z_1). The co-ordinate planes are the principal planes. If G_1 is the point, where the normal meets the YOZ plane, then $x = 0$ and $r = PG_1$.

$\therefore \quad PG_1 = -\frac{a^2}{p}$. Similarly if the normal meets the ZOX and XOY plane at G_2 and G_3 then

$$PG_2 = -\frac{b^2}{p}, \quad PG_3 = -\frac{c^2}{p}.$$

Hence $PG_1 : PG_2 : PG_3 = a^2 : b^2 : c^2$.

10. Show that the locus of the line of intersection of tangent planes to cone $ax^2+by^2+cz^2 = 0$ which touch along the perpendicular generator is $a^2(b+c)x^2+b^2(c+a)y^2+c^2(a+b)z^2 = 0$.

[C.H. '83]

Let (x_1, y_1, z_1) and (x_2, y_2, z_2) be any two points on two perpendicular generators and the tangent planes along these generators meet at a line

$$\frac{x}{\lambda} = \frac{y}{\mu} = \frac{z}{\nu}. \tag{1}$$

The tangent plane at (x_1, y_1, z_1) is

$$axx_1 + byy_1 + czz_1 = 0. \tag{2}$$

Since the line (1) lies on the plane (2)

$$a\lambda x_1 + b\mu y_1 + c\nu z_1 = 0. \tag{3}$$

Similarly by considering the tangent plane at (x_2, y_2, z_2) it is obtained that

$$a\lambda x_2 + b\mu y_2 + c\nu z_2 = 0. \tag{4}$$

The conditions (3) and (4) suggest that the generators through (x_1, y_1, z_1) and (x_2, y_2, z_2) lie on the plane $a\lambda x + b\mu y + c\nu z = 0$.

If this plane cuts the cone at two perpendicular generators then

$$a(b^2\mu^2 + c^2\nu^2) + b(c^2\nu^2 + a^2\lambda^2) + c(a^2\lambda^2 + b^2\nu^2) = 0,$$

$$\text{or} \quad a^2(b+c)\lambda^2 + b^2(c+a)\mu^2 + c^2(a+b)\nu^2 = 0.$$

[See (6.12.8) of Chapter 6]

Eliminating λ, μ, ν by (1) the required locus is

$$a^2(b+c)x^2 + b^2(c+a)y^2 + c^2(a+b)z^2 = 0.$$

Exercise

1. Find the equation of the tangent plane to

(a) $2x^2 + 4y^2 + 5z^2 = 63$ at $(1, 2, 3)$;

(b) $3x^2 - 9y^2 + 11z = 0$ at $(1, 2, 3)$.

2. (a) Find the equation of the normal to the sphere $x^2 + y^2 + z^2 - 9x - 11y - 5z - 12 = 0$ at the point $(1, -2, 3)$.

(b) Find the equation of the normal to the hyperbolic paraboloid $2x^2 - 3y^2 = 10z$ at the point $(2, 4, -4)$. [B.U. '68]

3. Show that the tangent planes to the sphere $x^2 + y^2 + z^2 - 4x + 2y - 4 = 0$ at $(4, -2, 2)$ and $(0, 0, -2)$ are parallel.

4. Find the equation of the sphere with the centre at $(1, -1, 3)$ and touching the plane $2x + y - 3z = 5$. [B.U. '67]

5. Show that the plane $3x + 12y - 6z - 17 = 0$ touches the conicoid $3x^2 - 6y^2 + 9z^2 + 17 = 0$ and find the point of contact.

6. Obtain the equation of the sphere touching the three co-ordinate planes and passing through the point $(2, 1, 5)$. [C.U. '80]

7. Find the equation of the sphere passing through the points $(4, 1, 0)$, $(2, -3, 4)$, $(1, 0, 0)$ and touching the plane $2x + 2y - z = 11$.

8. Show that the equations to the spheres which pass through the circle $x^2 + y^2 + z^2 = 5, x + 2y + 3z = 3$ and touch the plane $4x + 3y - 15 = 0$ are $x^2 + y^2 + z^2 + 2x + 4y + 6z - 11 = 0$ and $5(x^2 + y^2 + z^2) - 4x - 8y - 12z - 13 = 0$.

9. Find the equations of the tangent planes to $\frac{x^2}{2} + \frac{y^2}{3} + \frac{z^2}{4} = 1$, parallel to $x + y + z = 0$.

10. Find the values of k for each of which $x + y + z = k$ is a tangent plane to the ellipsoid $x^2 + 2y^2 + 3z^2 = 66$. [C.U. '78]

11. Find the equations of the tangent planes to $3x^2 - 7y^2 + 5z^2 = 12$, perpendicular to $\frac{x}{2} = \frac{y}{3} = \frac{z}{4}$.

12. Find the equations of the pair of tangent planes to the paraboloid $2x^2 - y^2 = 4z$ which pass through the line $x + 1 = 0 = y + z - 1$ and find their points of contact. [C.U. '67]

13. (a) Find the equations of the two tangent planes to the quadric $\frac{x^2}{42} - \frac{y^2}{18} + \frac{z^2}{3} = 1$ which pass through the line $\frac{x-8}{-7} = \frac{y+4}{5} = \frac{z-1}{1}$. [C.H. '82]

[HINTS. The given line can be written as $5x + 7y - 12 = 0 = y - 5z + 9$. Any plane through this line is $5x + 7y - 12 + \lambda(y - 5z + 9) = 0$.
If it is the tangent plane to the given quadric then

$$5^2.42 - (7 + \lambda)^2.18 + 25\lambda^2.3 = (9\lambda - 12)^2,$$

$$\text{or} \quad 2\lambda^2 + 3\lambda - 2 = 0,$$

$$\text{or} \quad (\lambda + 2)(2\lambda - 1) = 0, \ \lambda = -2, \ \frac{1}{2}.$$

$\therefore$ tangent planes are $x + y + 2z - 6 = 0$ and $2x + 3y - z - 3 = 0$.]

(b) Find the equations of the tangent planes to the conicoid $2x^2 - 6y^2 + 3z^2 = 5$ which pass through the line $x + 3y - 1 = 0 = 2x + 3z - 3$. **[B.U. '98]**

(c) Find the equation of the tangent plane to the quadric $3x^2 + 2y^2 - 6z^2 = 6$ which passes through the point $(3, 4, -3)$ and is parallel to the line $x = y = -z$. **[C.H. '74, '86]**

14. Find the equation of the plane which touches the cone $x^2+4y^2-5z^2+4xy-6yz+2xz = 0$ along the generator with d. rs. 1, 1, 1. **[C. H. '82, '92]**

[HINTS. A plane along the generator $x = y = z$ is $x + (\lambda - 1)y - \lambda z = 0$. From the condition of the tangency of the plane $\lambda = \frac{7}{4}$ and the tangent plane is $4x + 3y - 7z = 0$.]

15. Show that the normals to the ellipsoid $x^2 + 2y^2 + 3z^2 = 50$ at the points $(6, 1, 2)$ and $(6, -1, 2)$ lie in the plane $x - z = 4$.

16. If the normal at a point P of the paraboloid $x^2 + y^2 = 4z$ meets the planes YOZ and ZOX at Q and R respectively, then show that $PQ = PR$. **[C.U. '66]**

17. Find the other two of the set of three mutually perpendicular generators of the cone $11yz + 6zx - 14xy = 0$ when one is $x = \frac{y}{2} = z$.

18. Show that the reciprocal cone of $ax^2 + by^2 + cz^2 = 0$ is $\frac{x^2}{a} + \frac{y^2}{b} + \frac{z^2}{c} = 0$.

Answers

1. (a) $2x + 8y + 15z = 63$; (b) $6x - 36y + 11z + 33 = 0$. **2.** (a) $\frac{x-1}{7} = \frac{-y+2}{15} = \frac{z-3}{-1}$, (b) $\frac{x-2}{4} = \frac{y-4}{-12} = \frac{z+4}{-5}$. **4.** $14(x^2+y^2+z^2) - 28x + 28y - 84z - 15 = 0$. **5.** $(-1, 2, \frac{2}{3})$. **6.** $x^2+y^2+z^2-10(x+y+z)+50 = 0$. **7.** $x^2 + y^2 + z^2 - 6x + 2y - 4z + 5 = 0$, $16(x^2 + y^2 + z^2) - 102x + 50y - 49z + 86 = 0$. **9.** $x + y + z = \pm 3$. **10.** ± 11. **11.** $2x + 3y + 4z = \pm\sqrt{\frac{1364}{35}}$. **12.** $y + z = 1, 2x + y + z + 1 = 0, (0, 2, -1), (-2, 2, 1)$. **13.** (c) $x + y + 2z = 1$. **17.** $\frac{x}{2} = \frac{y}{-3} = \frac{z}{4}$, $\frac{x}{-11} = \frac{y}{2} = \frac{z}{7}$.

Chapter 8

Generating Lines

8.1 Introduction

Definition 1 *The surfaces of cones and cylinders are generated by the motion of a straight line. The hyperboloid of one sheet and the hyperbolic paraboloid are also generated by the motion of a straight line. Such surfaces are called ruled surfaces.*

If through every point on a surface a straight line can be drawn in such a way that the straight line lies wholly on the surface, then this surface is called a ruled surface and the straight lines are named as generating lines (or shortly generators or rulings.)

8.2 To find the Condition that a given Straight Line should be a Generator of a given Conicoid

Let the equation to the conicoid and the line be

$$ax^2 + by^2 + cz^2 = 1 \tag{8.2.1}$$

$$\text{and} \quad \frac{x-\alpha}{l} = \frac{y-\beta}{m} = \frac{z-\gamma}{n} \quad (= r \text{ say}). \tag{8.2.2}$$

respectively.

Any point on the line (8.2.2) is $(lr+\alpha, mr+\beta, nr+\gamma)$.

If this point lies on the conicoid, then

$$(al^2 + bm^2 + cn^2)r^2 + 2(al\alpha + bm\beta + cn\gamma)r + (a\alpha^2 + b\beta^2 + c\gamma^2 - 1) = 0. \tag{8.2.3}$$

The line lies entirely on the surface, if the equation (8.2.3) is an identity.

For this

$$al^2 + bm^2 + cn^2 = 0, \tag{8.2.4}$$

$$al\alpha + bm\beta + cn\gamma = 0, \tag{8.2.5}$$

$$a\alpha^2 + b\beta^2 + c\gamma^2 = 1. \tag{8.2.6}$$

The condition (8.2.6) shows that (α, β, γ) lie on the conicoid; condition (8.2.5) shows that a generating line must lies on the tangent plane at (α, β, γ) to the conicoid and the condition (8.2.4) shows that the lines parallel to the generating lines and passing through the centre of the conicoid generate the asymptotic cone

$$ax^2 + by^2 + cz^2 = 0.$$

By (8.2.4) and (8.2.6),

$$(al^2 + bm^2)(a\alpha^2 + b\beta^2) + cn^2(1 - c\gamma^2) = 0,$$

$$\text{or} \quad (al^2 + bm^2)(a\alpha^2 + b\beta^2) + cn^2 - (a\alpha l + b\beta m)^2 = 0, \quad \text{[by (8.2.5)]}$$

$$\text{or} \quad ab(\beta l - \alpha m)^2 = -cn^2, \quad \text{or} \quad (\beta l - \alpha m)^2 = -\frac{c}{ab}n^2. \tag{8.2.7}$$

It shows that c and ab are of opposite signs. It is possible when two of a, b, c are positive and the other is negative. Therefore of all real central conicoids hyperboloid of one sheet is the only ruled surface.

From (8.2.7)

$$\beta l - \alpha m \pm \sqrt{-\frac{c}{ab}}n = 0. \tag{8.2.8}$$

By (8.2.5) and (8.2.8)

$$\frac{l}{c\gamma\alpha \pm b\beta\sqrt{-\frac{c}{ab}}} = \frac{m}{c\beta\gamma \mp a\alpha\sqrt{-\frac{c}{ab}}} = \frac{n}{-(a\alpha^2 + b\beta^2)}. \tag{8.2.9}$$

The relations of (8.2.9) exhibit that two and only two sets of values of l, m, n are admissible. Hence through a point on the hyperboloid of one sheet two and only two generators may be drawn. Thus it is a doubly ruled surface.

Note. By similar considerations it can be shown that of all paraboloids hyperbolic paraboloid is the only ruled surface and it is a doubly ruled surface.

8.3 Generating Lines of a Hyperboloid of One Sheet

Let the equation of the hyperboloid of one sheet be

$$\frac{x^2}{a^2} + \frac{y^2}{b^2} - \frac{z^2}{c^2} = 1.$$

We rewrite it as

$$\frac{x^2}{a^2} - \frac{z^2}{c^2} = 1 - \frac{y^2}{b^2}, \quad \text{or} \quad \left(\frac{x}{a} + \frac{z}{c}\right)\left(\frac{x}{a} - \frac{z}{c}\right) = \left(1 + \frac{y}{b}\right)\left(1 - \frac{y}{b}\right).$$

This can be put in either of the two forms

$$\frac{\frac{x}{a} - \frac{z}{c}}{1 - \frac{y}{b}} = \frac{1 + \frac{y}{b}}{\frac{x}{a} + \frac{z}{c}} = \lambda \text{ (say)} \quad \text{and} \quad \frac{\frac{x}{a} - \frac{z}{c}}{1 + \frac{y}{b}} = \frac{1 - \frac{y}{b}}{\frac{x}{a} + \frac{z}{c}} = \mu \text{ (say)},$$

$$\text{i.e.}\quad \frac{x}{a}-\frac{z}{c}=\lambda\left(1-\frac{y}{b}\right),\quad \frac{x}{a}+\frac{z}{c}=\frac{1}{\lambda}\left(1+\frac{y}{b}\right) \tag{8.3.1}$$

$$\text{and}\quad \frac{x}{a}-\frac{z}{c}=\mu\left(1+\frac{y}{b}\right),\quad \frac{x}{a}-\frac{z}{c}=\frac{1}{\mu}\left(1-\frac{y}{b}\right) \tag{8.3.2}$$

where λ and μ are variable parameters.

Each of (8.3.1) and (8.3.2) represents a line of intersection of the pair of planes. These are two system of generators. Let us call the generators of (8.3.1) as the first system and those of (8.3.2) as the second system.

Properties.

(i) *Every point on the lines of the first sysiem or the second system lies on the hyperboloid.*

Let (α,β,γ) be a point on the line (8.3.1). Then

$$\frac{\alpha}{a}-\frac{\gamma}{c}=\lambda\left(1-\frac{\beta}{b}\right)\quad\text{and}\quad \frac{\alpha}{a}+\frac{\gamma}{c}=\frac{1}{\lambda}\left(1+\frac{\beta}{b}\right).$$

Multiplying we have

$$\left(\frac{\alpha}{a}-\frac{\gamma}{c}\right)\left(\frac{\alpha}{a}+\frac{\gamma}{c}\right)=\left(1-\frac{\beta}{b}\right)\left(1+\frac{\beta}{b}\right),\quad\text{or}\quad \frac{\alpha^2}{a^2}+\frac{\beta^2}{b^2}-\frac{\gamma^2}{c^2}=1.$$

Hence (α,β,γ) is a point on the hyperboloid.

Similarly, any point on the line (8.3.2) lies on the hyperboloid.

(ii) *Through every point of the hyperboloid there passes one generator of each system.*

If (α,β,γ) be a point on

$$\frac{x^2}{a^2}+\frac{y^2}{b^2}-\frac{z^2}{c^2}=1,$$

then

$$\frac{\alpha^2}{a^2}+\frac{\beta^2}{b^2}-\frac{\gamma^2}{c^2}=1. \tag{8.3.3}$$

A generator of the first system will pass through (α,β,γ), if λ has a value equal to each of

$$\frac{\frac{\alpha}{a}-\frac{\gamma}{c}}{1-\frac{\beta}{b}}\quad\text{and}\quad \frac{1+\frac{\beta}{b}}{\frac{\alpha}{a}+\frac{\gamma}{c}}.$$

For the value of λ,

$$\frac{\frac{\alpha}{a}-\frac{\gamma}{c}}{1-\frac{\beta}{b}}=\frac{1+\frac{\beta}{b}}{\frac{\alpha}{a}+\frac{\gamma}{c}},\quad\text{or}\quad \frac{\alpha^2}{a^2}+\frac{\beta^2}{b^2}-\frac{\gamma^2}{c^2}=1.$$

It is true by virtue of (8.3.3).

Similarly, it can be shown that a generator of the second system will pass through (α,β,γ) corresponding to the value of

$$\mu=\frac{\frac{\alpha}{a}-\frac{\gamma}{c}}{1+\frac{\beta}{b}}=\frac{1-\frac{\beta}{b}}{\frac{\alpha}{a}+\frac{\gamma}{c}}.$$

Hence the result follows.

(iii) *No two generators of the same system intersect.*

Let the two generators of the first system be

$$\frac{x}{a}-\frac{z}{c}=\lambda_1\left(1-\frac{y}{b}\right),\quad \frac{x}{a}+\frac{z}{c}=\frac{1}{\lambda_1}\left(1+\frac{y}{b}\right) \tag{8.3.4}$$

and
$$\frac{x}{a}-\frac{z}{c}=\lambda_2\left(1-\frac{\lambda}{b}\right),\quad \frac{x}{a}+\frac{z}{c}=\frac{1}{\lambda_2}\left(1+\frac{y}{b}\right). \tag{8.3.5}$$

From (8.3.4) and (8.3.5)

$$(\lambda_1-\lambda_2)\left(1-\frac{y}{b}\right)=0,\quad \text{or}\quad y=b\quad (\because\quad \lambda_1\neq\lambda_2)$$

and
$$\left(\frac{1}{\lambda_1}-\frac{1}{\lambda_2}\right)\left(1+\frac{y}{b}\right)=0,\quad \text{or}\quad y=-b\quad (\because\quad \lambda_1\neq\lambda_2).$$

If (8.3.4) and (8.3.5) intersect, then on solving the equations of (8.3.4) and (8.3.5) we must have the same value of y. There is a contradiction since the values of y are different. Hence the two generators of the same system will not intersect.

(iv) *Any two generators belonging to two different systems intersect.*

Let two generators belonging to two systems be

$$\frac{x}{a}-\frac{z}{c}=\lambda\left(1-\frac{y}{b}\right),\ \frac{x}{a}+\frac{z}{c}=\frac{1}{\lambda}\left(1+\frac{y}{b}\right) \tag{8.3.6}$$

and
$$\frac{x}{a}-\frac{z}{c}=\mu\left(1+\frac{y}{b}\right),\ \frac{x}{a}+\frac{z}{c}=\frac{1}{\mu}\left(1-\frac{y}{b}\right). \tag{8.3.7}$$

From the first two equations of (8.3.6) and (8.3.7).

$$\lambda\left(1-\frac{y}{b}\right)=\mu\left(1+\frac{y}{b}\right),\quad \text{or}\quad y=b\frac{\lambda-\mu}{\lambda+\mu}.$$

Putting this value of y in the equations of (8.3.6)

$$\frac{x}{a}-\frac{z}{c}=\frac{2\lambda\mu}{\lambda+\mu},\quad \frac{x}{a}+\frac{z}{c}=\frac{2}{\lambda+\mu}.$$

From these

$$x=a\frac{1+\lambda\mu}{\lambda+\mu},\quad z=c\frac{1-\lambda\mu}{\lambda+\mu}.$$

These values of x, y, z satisfy the equations of (8.3.7). Thus the two generators of two systems intersest. *The point of intersection is*

$$\left(a\frac{1+\lambda\mu}{\lambda+\mu},\quad b\frac{\lambda-\mu}{\lambda+\mu},\quad c\frac{1-\lambda\mu}{\lambda+\mu}\right).$$

Note 1. The point of intersection satisfied the equation of the hyperboloid.

Note 2. $x = a\frac{1+\lambda u}{\lambda+\mu}, y = b\frac{\lambda-\mu}{\lambda+\mu}, z = c\frac{1-\lambda\mu}{\lambda+\mu}$ is the parametric equation of the hyperboloid.

(v) *Two intersecting generators belonging to two systems lie on a plane which is tangent to the hyperboloid at their point of intersection.*

Let us consider the planes

$$\frac{x}{a} - \frac{z}{c} - \lambda\left(1 - \frac{y}{b}\right) - k\left\{\left(\frac{x}{a} + \frac{z}{c}\right) - \frac{1}{\lambda}\left(1 - \frac{y}{b}\right)\right\} = 0$$

and $$\frac{x}{a} - \frac{z}{c} - \mu\left(1 + \frac{y}{b}\right) - k'\left\{\left(\frac{x}{a} + \frac{z}{c}\right) - \frac{1}{\mu}\left(1 - \frac{y}{b}\right)\right\} = 0.$$

If $k = k' = -\lambda\mu$, both of the equations reduce to the same form,

$$\frac{x}{a} \cdot \frac{1 + \lambda\mu}{\lambda + \mu} + \frac{y}{b}\frac{\lambda - \mu}{\lambda + \mu} - \frac{z}{c} \cdot \frac{1 - \lambda\mu}{\lambda + \mu} = 1.$$

Therefore, the two generators lie on this plane which is obviously the tangent plane to the conicoid at their common point.

8.4 Generating Lines of a Hyperbolic Paraboloid

Let the equation of the hyperbolic paraboloid be $\frac{x^2}{a^2} - \frac{y^2}{b^2} = 2z$.

It can be written as $\left(\frac{x}{a} - \frac{y}{b}\right)\left(\frac{x}{a} + \frac{y}{b}\right) = 2z$.

This can be put in either of the two forms

$$\frac{\frac{x}{a} - \frac{y}{b}}{z} = \frac{2}{\frac{x}{a} + \frac{y}{b}} = \lambda \text{ (say)} \quad \text{and} \quad \frac{\frac{x}{a} + \frac{y}{b}}{z} = \frac{2}{\frac{x}{a} - \frac{y}{b}} = \mu \text{ (say)}$$

$$\text{i.e.} \quad \frac{x}{a} - \frac{y}{b} = \lambda z, \quad \frac{x}{a} + \frac{y}{b} = \frac{2}{\lambda} \tag{8.4.1}$$

$$\text{and} \quad \frac{x}{a} + \frac{y}{b} = \mu z, \quad \frac{x}{a} - \frac{y}{b} = \frac{2}{\mu}. \tag{8.4.2}$$

Eliminating λ from the equations of (8.4.1) or eliminating μ from the equations of (8.4.2) we get the equation of the hyperbolic paraboloid. Hence any point on either the line (8.4.1) or (8.4.2) lies on the hyperbolic paraboloid. λ and μ are two variable parameters. For different values of λ and μ we get two families of straight lines such that each line of either family lies entirely on the conicoid. The lines of these two families are known as two systems of generators of the hyperbolic paraboloid.

Properties.

(i) *Through every point of the hyperbolic paraboloid there passes one generator of each system.*

(ii) *No two generators of the same system intersect.*

[Proofs of (i) and (ii) are similar to those of section 8.3.]

(iii) *Any two generators belonging to two different systems intersect.*

Let the two generators of the two systems be given by (8.4.1) and (8.4.2).

From the last two equations of (8.4.1) and (8.4.2)

$$x = a\frac{\lambda+\mu}{\lambda\mu}, \quad y = b\frac{\mu-\lambda}{\lambda\mu}.$$

Putting these values of x and y in the first equation of (8.4.1) and (8.4.2) $z = \frac{2}{\lambda\mu}$.

Hence the point of intersection is

$$\left(a\frac{\lambda+\mu}{\lambda\mu}, \quad b\frac{\mu-\lambda}{\lambda\mu}, \quad \frac{2}{\lambda\mu}\right).$$

For all values of λ and μ the above point satisfies the equation of the hyperbolic paraboloid. Thus *parametric equations* of it are

$$x = a\frac{\lambda+\mu}{\lambda\mu}, \quad y = b\frac{\mu-\lambda}{\lambda\mu}, \quad z = \frac{2}{\lambda\mu}.$$

(iv) *Two intersecting generators belonging to two systems lie on the tangent plane to the hyperbolic paraboloid at their point of intersection.*

Let us consider the planes

$$\frac{x}{a} - \frac{y}{b} - \lambda z + k\left\{\left(\frac{x}{a} + \frac{y}{b}\right) - \frac{2}{\lambda}\right\} = 0$$

$$\text{and} \quad \frac{x}{a} - \frac{y}{b} - \frac{2}{\mu} + k'\left\{\left(\frac{x}{a} + \frac{y}{b}\right) - \mu z\right\} = 0.$$

If $k = k' = \frac{\lambda}{\mu}$, both of the planes reduce to

$$\frac{x}{a}\frac{\lambda+\mu}{\lambda\mu} - \frac{y}{b}\frac{\mu-\lambda}{\lambda\mu} = z + \frac{2}{\lambda\mu}. \tag{8.4.3}$$

Therefore the generators (8.4.1) and (8.4.2) lie on the same plane whose equation is (8.4.3). Obviously it is the tangent plane to the hyperbolic paraboloid at the point of intersection of the generators.

8.5 Locus of the Point of Intersection of Perpendicular Generators

(a) Hyporboloid of one sheet:

Let

$$\frac{x}{a} - \frac{z}{c} = \lambda\left(1 - \frac{y}{b}\right), \quad \frac{x}{a} + \frac{z}{c} = \frac{1}{\lambda}\left(1 + \frac{y}{b}\right) \tag{8.5.1}$$

$$\text{and} \quad \frac{x}{a} - \frac{z}{c} = \mu\left(1 + \frac{y}{b}\right), \quad \frac{x}{a} + \frac{z}{c} = \frac{1}{\mu}\left(1 - \frac{y}{b}\right) \tag{8.5.2}$$

be the two generators of two systems of

$$\frac{x^2}{a^2}+\frac{y^2}{c^2}-\frac{z^2}{c^2}=1. \tag{8.5.3}$$

If l_1, m_1, n_1 are the d.cs. of the line (8.5.1), then

$$\frac{l_1}{a}+\frac{\lambda m_1}{b}-\frac{n_1}{c}=0 \quad \text{and} \quad \frac{l_1}{a}-\frac{m_1}{\lambda b}+\frac{n_1}{c}=0.$$

From these relations

$$\frac{\frac{l_1}{a}}{\lambda-\frac{1}{\lambda}}=\frac{\frac{m_1}{b}}{-1-1}=\frac{\frac{n_1}{c}}{-\frac{1}{\lambda}-\lambda}, \quad \text{or} \quad \frac{\frac{l_1}{a}}{\lambda^2-1}=\frac{\frac{m_1}{b}}{-2\lambda}=\frac{\frac{n_1}{c}}{-(\lambda^2+1)}. \tag{8.5.4}$$

Similarly if l_2, m_2, n_2 are the d. cs. of the line (8.5.2), then

$$\frac{\frac{l_2}{a}}{\mu^2-1}=\frac{\frac{m_2}{b}}{2\mu}=\frac{\frac{n_2}{c}}{-(\mu^2+1)}. \tag{8.5.5}$$

For perpendicularity of the generators

$$l_1l_2+m_1m_2+n_1n_2=0,$$

$$\text{or} \quad a^2(\lambda^2-1)(\mu^2-1)-4b^2\lambda\mu+c^2(\lambda^2+1)(\mu^2+1)=0.$$

It can be written as

$$a^2(1+\lambda\mu)^2+b^2(\lambda-\mu)^2+c^2(1-\lambda\mu)^2=(\lambda+\mu)^2(a^2+b^2-c^2),$$

$$\text{or} \quad a^2\left(\frac{1+\lambda\mu}{\lambda+\mu}\right)^2+b^2\left(\frac{\lambda-\mu}{\lambda+\mu}\right)^2+c^2\left(\frac{1-\lambda\mu}{\lambda+\mu}\right)^2=a^2+b^2-c^2.$$

This relation shows that the point of intersection of (8.5.1) and (8.5.2), namely

$$\left(a\frac{1+\lambda\mu}{\lambda+\mu},\quad b\frac{\lambda-\mu}{\lambda+\mu},\quad c\frac{1-\lambda\mu}{\lambda+\mu}\right),$$

lies on the director sphere $x^2+y^2+z^2=a^2+b^2-c^2$.

Hence the required locus is the curve of intersection of the hyperboloid and the director sphere.

(b) **Hyperbolic paraboloid:**

Let

$$\frac{x}{a}-\frac{y}{b}=\lambda z, \quad \frac{x}{a}+\frac{y}{b}=\frac{2}{\lambda} \tag{8.5.6}$$

$$\text{and} \quad \frac{x}{a}-\frac{y}{b}=\frac{2}{\mu}, \quad \frac{x}{a}+\frac{y}{b}=\mu z \tag{8.5.7}$$

be the two generators of two systems of

$$\frac{x^2}{a^2}-\frac{y^2}{b^2}=2z. \tag{8.5.8}$$

If l_1, m_1, n_1 be the d. cs. of the line (8.5.6), then

$$\frac{l_1}{a} - \frac{m_1}{b} - \lambda n_1 = 0, \quad \frac{l_1}{a} + \frac{m_1}{b} = 0, \quad \text{or} \quad \frac{l_1}{a\lambda} = \frac{m_1}{-b\lambda} = \frac{n_1}{2}.$$

Similarly, if l_2, m_2, n_2 are the d. cs. of the line (8.5.7), then

$$\frac{l_2}{a\mu} = \frac{m_2}{b\mu} = \frac{n_2}{2}.$$

For perpendicularity of the generators

$$l_1, l_2 + m_1 m_2 + n_1 n_2 = 0,$$

$$\text{or} \quad a^2\lambda\mu - b^2\lambda\mu + 4 = 0,$$

$$\text{or} \quad a^2 - b^2 + \frac{4}{\lambda\mu} = 0. \tag{8.5.9}$$

The point of intersection between (8.5.6) and (8.5.7) is

$$\left(a\frac{\lambda+\mu}{\lambda\mu}, \quad b\frac{\mu-\lambda}{\lambda\mu}, \quad \frac{2}{\lambda\mu}\right).$$

The relation (8.5.9) shows that this point of intersection lies on $a^2 - b^2 + 2z = 0$.

Hence the required locus is the curve of intersection of the hyperbolic paraboloid and the plane $a^2 - b^2 + 2z = 0$.

Worked Out Examples

1. Find the equations to the generating lines of the hyperboloid $\frac{x^2}{4} + \frac{y^2}{9} - \frac{z^2}{16} = 1$, which pass through the point $(2, 3, -4)$. **[C.U. '63, '68]**

Let l, m, n be the d. cs. of a generator through $(2, 3, -4)$. The equation of it is

$$\frac{x-2}{l} = \frac{y-3}{m} = \frac{z+4}{n} \quad (= r \text{ say}).$$

Any point on it is $(lr + 2, mr + 3, nr - 4)$. If it lies on the hyperboloid, then

$$\frac{(lr+2)^2}{4} + \frac{(mr+3)^2}{9} - \frac{(nr-4)^2}{16} = 1,$$

$$\text{or} \quad r^2\left(\frac{l^2}{4} + \frac{m^2}{9} - \frac{n^2}{16}\right) + 2r\left(\frac{2l}{4} + \frac{3m}{9} + \frac{4n}{16}\right) = 0.$$

Since the line is a generator it lies wholly on the hyperboloid.

$$\therefore \quad \frac{l^2}{4} + \frac{m^2}{9} - \frac{n^2}{16} = 0 \quad \text{and} \quad \frac{l}{2} + \frac{m}{3} + \frac{n}{4} = 0.$$

Eliminating n, $\frac{l^2}{4} + \frac{m^2}{9} - \left(\frac{l}{2} + \frac{m}{3}\right)^2 = 0$, or $lm = 0$.

When $l = 0$, then $\frac{m}{3} + \frac{n}{4} = 0$, or $\frac{m}{3} = \frac{n}{-4}$.

When $m = 0$, then $\frac{l}{2} + \frac{n}{4} = 0$, or $l = \frac{n}{-2}$.

Hence the two generators are

$$\frac{x-2}{0} = \frac{y-3}{3} = \frac{z+4}{-4} \quad \text{and} \quad \frac{x-2}{1} = \frac{y-3}{0} = \frac{z+4}{-2}.$$

2. Find the equations to the generating lines of the hyperboloid $yz + 2zx + 3xy + 6 = 0$ which pass through the point $(-1, 0, 3)$.

Let the equation of a generator be

$$\frac{x+1}{l} = \frac{y}{m} = \frac{z-3}{n} \quad (= r \text{ say}).$$

Any point on it is $(lr - 1, mr, nr + 3)$. If it lies on the hyperboloid, then

$$mr(nr+3) + 2(nr+3)(lr-1) + 3(lr-1).mr + 6 = 0.$$

$$\text{or} \quad r^2(mn + 2nl + 3lm) + 2r(3l - n) = 0.$$

Since the generator lies wholly on the hyperboloid,

$$mn + 2nl + 3lm = 0 \quad \text{and} \quad 3l - n = 0, \quad \text{or} \quad n = 3l.$$

Eliminating n we have $l^2 + lm = 0$, or $l(l+m) = 0$.

$$\therefore \quad l = 0, \quad \text{or} \quad l = -m.$$

When $l = 0$, then $n = 0$ and when $l = -m$, then $n = -3m$.

Therefore the generators are

$$x + 1 = 0 = z - 3$$

$$\text{and} \quad \frac{x+1}{1} = \frac{y}{-1} = \frac{z-3}{3}.$$

3. Find the equations to the generating lines of the paraboloid $(x+y+z)(2x+y-z) = 6z$, which pass through the point $(1, 1, 1)$. Hence find the angle between these generators.

[C.H. '85]

Let the equation of a generaor be

$$\frac{x-1}{l} = \frac{y-1}{m} = \frac{z-1}{n} \quad (= r \text{ say}).$$

Any point on the line is $(lr + 1, mr + 1, nr + 1)$.

If it lies on the paraboloid then

$$(lr + 1 + mr + 1 + nr + 1)(2lr + 2 + mr + 1 - nr - 1) = 6(nr + 1),$$

$$\text{or} \quad (l + m + n)(2l + m - n)r^2 + (8l + 5m - 7n)r = 0.$$

Since the generator lies wholly on the paraboloid,

$$(l + m + n)(2l + m - n) = 0 \quad \text{and} \quad 8l + 5m - 7n = 0.$$

For one generator $l + m + n = 0$, $8l + 5m - 7n = 0$.

From these $\frac{l}{4} = \frac{m}{-5} = \frac{n}{1}$.

For the other generator $2l + m - n = 0, 8l + 5m - 7n = 0$.

From these $\frac{l}{1} = \frac{m}{-3} = \frac{n}{-1}$.

Therefore, the generators are

$$\frac{x-1}{4}=\frac{y-1}{-5}=\frac{z-1}{1} \quad \text{and} \quad \frac{x-1}{1}=\frac{y-1}{-3}=\frac{z-1}{-1}.$$

The angle between these lines

$$=\cos^{-1}\frac{4.1+5.3-1.1}{\sqrt{(4^2+5^2+1^2)}.\sqrt{(1^2+3^2+1^2)}}=\cos^{-1}\frac{18}{\sqrt{462}}.$$

4. Obtain the equations to the generators of the hyperboloid $\frac{x^2}{a^2}+\frac{y^2}{b^2}-\frac{z^2}{c^2}=1$ through a point of the principal elliptic section.

Hence show that the projections of the generators of a hyperboloid on co-ordinate planes are tangents to the section of the hyperboloid by that plane.

[N.H. '84; C.H. '63, '64, '74, '76, '78, '84, '93, '99]

Any point on the principal elliptic section $\frac{x^2}{a^2}+\frac{y^2}{b^2}=1, z=0$ can be taken as $(a\cos\theta, b\sin\theta, 0)$.

Let

$$\frac{x-a\cos\theta}{l}=\frac{y-b\sin\theta}{m}=\frac{z-0}{n} \tag{1}$$

be a generator through this point. Any point on (1) is $(lr+a\cos\theta, mr+b\sin\theta, nr)$. It is a point on the hyperboloid.

$$\therefore \quad \frac{(lr+a\cos\theta)^2}{a^2}+\frac{(mr+b\sin\theta)^2}{b^2}-\frac{n^2r^2}{c^2}=1,$$

$$\text{or} \quad \left(\frac{l^2}{a^2}+\frac{m^2}{b^2}-\frac{n^2}{c^2}\right)r^2+2\left(\frac{l\cos\theta}{a}+\frac{m\sin\theta}{b}\right)r=0.$$

It is true for all values of r.

For this

$$\frac{l^2}{a^2}+\frac{m^2}{b^2}-\frac{n^2}{c^2}=0 \tag{2}$$

$$\text{and} \quad \frac{l\cos\theta}{a}+\frac{m\sin\theta}{b}=0. \tag{3}$$

By (2) and (3)

$$\frac{l}{a\sin\theta}=\frac{m}{-b\cos\theta}=\frac{\sqrt{\left(\frac{l^2}{a^2}+\frac{m^2}{b^2}\right)}}{\sqrt{(\sin^2\theta+\cos^2\theta)}}=\sqrt{\frac{n^2}{c^2}}=\frac{n}{\pm c}.$$

Hence the two generators through the point $(a\cos\theta, b\sin\theta, 0)$ are

$$\frac{x-a\cos\theta}{a\sin\theta}=\frac{y-b\sin\theta}{-b\cos\theta}=\frac{z}{\pm c}. \tag{4}$$

The section of the hyperboloid made by the co-ordinate plane $z=0$ is

$$\frac{x^2}{a^2}+\frac{y^2}{b^2}=1, \quad z=0. \tag{5}$$

The projection of one generator of (4) on the plane $z=0$ is

$$\frac{x-a\cos\theta}{a\sin\theta}=\frac{y-b\sin\theta}{-b\cos\theta},\quad z=0,$$

$$\text{or}\quad \frac{x}{a\sin\theta}+\frac{y}{b\cos\theta}=\frac{\cos\theta}{\sin\theta}+\frac{\sin\theta}{\cos\theta},\quad z=0,$$

$$\text{or}\quad \frac{x}{a}\cos\theta+\frac{y}{b}\sin\theta=1,\quad z=0. \tag{6}$$

It is clearly the tangent to the section (5) at $(a\cos\theta, b\sin\theta, 0)$.

Similarly, the projections of the generator on the planes $x=0$ and $y=0$ are tangents to the sections of the hyperboloid by these planes respectively.

Exercise

1. Find the equations to the generating lines of the hyperboloid $\frac{x^2}{4}+\frac{y^2}{9}-\frac{z^2}{16}=1$, which pass through the point $\left(2,-1,\frac{4}{3}\right)$.
2. Find the equations of the generators of the hyperbolic paraboloid $16x^2-9y^2=4z$, which pass through the point $(-2,2,7)$.
3. Show that the line $\frac{x+2}{2}=\frac{y}{3}=\frac{z-1}{-2}$ is a generator of the quadratic $\frac{x^2}{4}-\frac{y^2}{9}=z$. **[C.H. '80]**
4. Find the generators of the paraboloid $\frac{x^2}{a}-\frac{y^2}{b}=4z$ through $(\alpha,0,\gamma)$ and prove that the angle between them is $\cos^{-1}\frac{a-b+\gamma}{a+b+\gamma}$.
5. If $\frac{x-2}{3}=\frac{y+1}{6}=\frac{z-\frac{4}{3}}{10}$ is a generator of $\frac{x^2}{a^2}+\frac{y^2}{b^2}-\frac{y^2}{c^2}=1$, then show that $a=2, b=3, c=4$.
6. Show that the line $x-1=y-2=z+1$ lies entirely on the surface $z^2-xy+2x+y+2z-1=0$. **[C.U. '62]**
7. Show that the perpendiculars from the origin on the generators of the paraboloid $\frac{x^2}{a^2}-\frac{y^2}{b^2}=\frac{2z}{c}$ lie on the cones $\left(\frac{x}{a}\pm\frac{y}{b}\right)(ax\pm by)+2z^2=0$. **[C.H. '58]**

[HINTS. The generator of λ-system is

$$\frac{x}{a}-\frac{y}{a}=\frac{\lambda z}{c},\quad \frac{x}{a}+\frac{y}{b}=\frac{2}{\lambda},\quad \text{or}\quad \frac{x-\frac{a}{\lambda}}{a\lambda}=\frac{y-\frac{b}{\lambda}}{-b\lambda}=\frac{z-0}{2c}=r_1\ \text{(say)}. \tag{1}$$

A line through the origin is

$$\frac{x}{l}=\frac{y}{m}=\frac{z}{n}=r_2\ \text{(say)}. \tag{2}$$

If (2) is perpendicular to (1) then

$$a\lambda l-b\lambda m+2cn=0. \tag{3}$$

(1) and (2) intersect. At the common point

$$a\lambda r_1+\frac{a}{\lambda}=lr_2,\quad -b\lambda r_1+\frac{b}{\lambda}=mr_2,\quad 2cr_1=nr_2.$$

Eliminating r_1 and r_2, $\lambda=\frac{(bl-am)c}{nab}$.

From (3)

$$acl(bl-am)-bcm(bl-am)+2abcn^2 = 0,$$

$$\text{or}\quad l^2+m^2+2n^2-\frac{a^2+b^2}{ab}lm = 0.$$

$\therefore$ the required locus is

$$x^2+y^2+2z^2-\frac{a^2+b^2}{ab}xy = 0,$$

$$\text{or}\quad \left(\frac{x}{a}-\frac{y}{b}\right)(ax-by)+2z^2 = 0.]$$

8. Show that the perpendiculars from the origin to the generators of $\frac{x^2}{a^2}+\frac{y^2}{b^2}-\frac{z^2}{c^2}=1$ lie upon the cone

$$\frac{a^2(b^2+c^2)^2}{x^2}+\frac{b^2(c^2+a^2)^2}{y^2}=\frac{c^2(a^2-b^2)^2}{z^2}.$$

[C.H. '62]

9. Determine the two sets of generators of the paraboloid $(2x+3y+7z)^2-(x+y+z)^2 = 4x-y+3z$ and prove that they are parallel respectively to the two planes $(2x+3y+7z)+(x+y+z)=0$ and $(2x+3y+7z)-(x+y+z)=0$.

[C.H. '54]

10. Prove that the generators of the surface $yz+zx+xy+a^2=0$ through $(0, am, -\frac{a}{m})$ are $x(1\pm m)=am-y=\mp(mz+a)$. [C.H. '61]

[HINTS. The generators are $(y-\lambda z)+a(1+\lambda)=0$, $(\lambda+1)x+(y+\lambda a)=0$ and $\mu(x+y)-(y+a)=0, \mu(x+a)+z-a=0$.]

Answers

1. $\frac{x-2}{0}=\frac{y+1}{3}=\frac{z-\frac{4}{3}}{-4}, \frac{x-2}{3}=\frac{y+1}{6}=\frac{z-\frac{4}{3}}{10}$. 2. $\frac{x+2}{3}=\frac{y-2}{4}=\frac{z-7}{-84}, \frac{x-2}{3}=\frac{y-2}{-4}=\frac{z-7}{-12}$. 4. $\frac{x-\alpha}{2a}=\frac{y}{\pm 2\sqrt{ab}}=\frac{z-\gamma}{\alpha}$. 9. $(2x+3y+7z)-(x+y+z)=\lambda(4x-y+3z)$; $(2x+3y+7z)+(x+y+z)=1/\lambda$; $(2x+3y+7z)-(x+y+z)=1/\mu$; $(2x+3y+7z)+(x+y+z)=\mu(4x-y+3z)$.

Part Two

Short Questions

1. (a) Find the Co-ordinates of the projections of the points $(5,-1,7)$ and $(0,2,3)$ on the co-ordinate planes and axes.

   (b) Find the perpendicular distance of the point $P(x,y,z)$ from the x-axis. **[C.U. '96]**

2. Find the octants where the point (x,y,z) may lie for the following conditions: (a) $x-y=0$, (b) $yz<0$, (c) $xyz<0$.

3. What are the equations of the co-ordinate planes and co-ordinate axes?

4. A plane passing through the origin has no intercepts with the co-ordinate axes. Give arguments in support or otherwise.

5. Find the values of k, if the distance between the points $(5,-1,7)$ and $(k,5,1)$ is 11.

6. Find the ratio in which the yz plane divides the join of the points $(2,-6,3)$ and $(-3,4,5)$. Find the co-ordinates of this point of intersection.

7. The direction ratios of a line are $2,2,2$. Find the direction cosines of this line.

8. If $\cos\alpha$, $\cos\beta$, $\cos\gamma$ are the direction cosines of a line,show that $\sin^2\alpha+\sin^2\beta+\sin^2\gamma=2$. **[C.U. '93]**

9. The direction ratios of OA and OB are $2,3,5$ and $1,1,-1$ respectively. Show that $OA\perp OB$.

10. A line makes angles $45°$ with the positive direction of x and z axes. Find the direction cosines of it.

11. What is the equation of a plane whose intercepts with the cordinate axes are $\frac{1}{a},\frac{1}{b},\frac{1}{c}$?

12. The foot of the perpendicular from the origin to the plane is $(2,3,-1)$. Find the equation of the plane.

13. Write down the equations of the planes parallel to the co-ordinate planes.

14. Determine whether the point $(2,1,-2)$ lies on the plane passing through the points $(1,0,0),(0,1,0),(0,0,1)$. **[C.U. '82]**

15. Find the sum of the intercepts made by $12x+8y+6z=24$ on the axes.

16. What are the direction cosines of the co-ordinate axes?

17. Find out the equations of the planes perpendicular to the co-ordinate planes.

18. Find the equations of the straight line passing through the point $(1,2,3)$ and parallel to the line $\frac{x}{2}=\frac{y}{4}=\frac{z}{3}$. **[C.U. '82]**

19. (a) A plane makes equal intercepts on the co-ordinate axes and passes through $(1,2,4)$. Find the equation of the plane. **[C.U. '83]**

    (b) Does the point $(4,-6,0)$ lie on the plane which intersects the positive x,y and z-axes at distances $2,3,5$ units respectively. **[C.U. '84]**

20. What is the distance between the planes $x+y-2z-5=0$ and $x+y-2z-17=0$?

21. Find the equation of the plane which passes through the point $(3,2,1)$ and is perpendicular to the line joining the points $(-5,3,7)$ and $(2,-4,6)$.

22. Find the equation of the plane which passes through the point $(1,-1,2)$ and is parallel to the plane $2x+3y-z+1=0$. [C.U. '83]

23. Determine the value of h for which the planes $3x-2y+hz-1=0$ and $x+hy+5z+2=0$ may be perpendicular to each other. [C.U. '83, '97; N.B. '98]

24. Find the angle between the planes $2x-y+4z+2=0$ and $x+y+z+1=0$.

25. (a) Find the distance of the point $(0,1,-2)$ from the point where the line $\frac{x}{1}=\frac{y-1}{2}=\frac{z+2}{3}$ meets the plane $2x+3y+z=12$.

(b) Find the distance of the points $(1,1,1)$ from the line $\frac{x}{1}=\frac{y}{-1}=\frac{z}{0}$. [C.U. '89]

26. (a) Prove that the line $2x+3y-z+3=0=3x-2y+2z-6$ is not parallel to the z-axis.

(b) Show that the line $x=t-2, y=3-4t, z=5t+6$ is parallel to the plane $x-y-z=1$. [B.U. '91]

27. Show that the points $(3,3,3)$ and $(1,-1,3)$ do not lie on same side of the plane $5x+2y-7z+9=0$.

28. Find the equation of the sphere whose centre $(1,0,5)$ and radius 7.

29. Find the equation of the sphere passing through the circle $x^2+y^2+z^2=16$, $x+y=4$ and the origin.

30. Find the equation of the sphere passing through the circle of intersection of the spheres $x^2+y^2+z^2-2x+4y=11$ and $x^2+y^2+z^2-4x-6z=55$ and the origin.

31. Find the radius of the circle given by the equations $3x^2+3y^2+3z^2+x-5y-2=0, x+y=2$. [C.U. '81]

32. Find the equation of the sphere having $(2,0,5)$ and $(-1,2,3)$ as the ends of a diameter.

33. Show that the line $x-1=y-2=z+1$ lies entirely on the surface of $z^2-xy+2x+y+2z-1=0$.

34. Find the equations of the generator of the cone $x^2+y^2=z^2$ through the point $(3,4,5)$. [C.U. '82]

35. Find the equation of the right circular cyclinder whose radius is 1 and axis is the z-axis.

36. Find the equation of the right circular cone with vertex at the origin and semivertical angle α.

37. Find the equation of the tangent plane to the sphere $x^2+y^2+z^2=45$ at $(2,4,5)$.

38. For what value of p,the plane $lx+my+nz=p$ is a tangent plane to the conicoid $3x^2+4y^2+z^2=1$? Find also the point of contact.

39. What is the name of the conicoid represented by $\frac{x^2}{a^2}-\frac{y^2}{b^2}-\frac{z^2}{c^2}=1$? What conic will be the section of the conicoid by a plane $z=k$?

40. What conic will be the section of the conicoid by any plane parallel to $x = 0$ or $y = 0, ax^2 + by^2 - cz^2 = 1$?

41. Find the equation of the plane passing through the point $(4, -3, 5)$ and containing the y-axis. **[C.U. 2000]**

42. Find the angles (in degrees) that the line $\frac{x-1}{1} = \frac{y-1}{1} = \frac{z-1}{0}$ makes with the axes. **[N.B. '98]**

43. Which of the following statements are true? Justify your answers.
   (a) A plane is completely determined if the direction ratios of the normal to it are given.
   (b) A cylinder is a central conicoid.
   (c) All tangent planes to a circular cone pass through its vertex.
   (d) The equation of a sphere can be obtained if its centre and a point on it are known.
   (e) Two spheres can intersect in at most two points.

Answers

1. (a) Co-ordinates on the xy plane $(5, -1, 0), (0, 2, 0)$, Co-ordinates on the yz plane $(0, -1, 7)$, $(0, 2, 3)$, Co-ordinates on the zx plane $(5, 0, 7), (0, 0, 3)$, Co-ordinates on the x-axis $(5, 0, 0), (0, 0, 0)$, Co-ordinates on the y-axis $(0, -1, 0), (0, 2, 0)$, Co-ordinates on the z-axis $(0, 0, 7), (0, 0, 3)$ (b) $\sqrt{y^2 + z^2}$. **2.** (a) 1st,3rd,5th and 7th octants (b) 3rd,4th,5th and 6th octants (c) 2nd,4th,5th and 7th octants. **3.** xy plane: $z = 0$, yz plane: $x = 0$, zx plane: $y = 0$ x-axis: $y = 0 = z$, y-axis: $z = 0 = x$, z-axis: $x = 0 = y$. **4.** The intercepts cannot be determined. **5.** $12, -2$. **6.** $2 : 3, \left(0, -2, \frac{19}{5}\right)$. **7.** $\pm\frac{1}{\sqrt{3}}, \pm\frac{1}{\sqrt{3}}, \pm\frac{1}{\sqrt{3}}$. **10.** $\frac{1}{\sqrt{2}}, 0, \frac{1}{\sqrt{2}}$. **11.** $ax + by + cz = 1$. **12.** $2x + 3y - z = 14$. **13.** $z =$ const., $x =$ const., $y =$ const. **15.** 9. **16.** $1, 0, 0; 0, 1, 0; 0, 0, 1$. **17.** Plane perpendicular to x-axis. is x=constant. **18.** $\frac{x-1}{2} = \frac{y-2}{4} = \frac{z-3}{3}$. **19.** (a) $x + y + z = 7$. (b) No. **20.** $2\sqrt{6}$. **21.** $7x - 7y - z = 6$. **22.** $2x + 3y - z + 3 = 0$. **23.** -1. **24.** $\cos^{-1}\frac{5}{\sqrt{63}}$. **25** (a) $\sqrt{14}$, (b) $\sqrt{3}$. **28.** $x^2 + y^2 + z^2 - 2x - 10z - 23 = 0$. **29.** $x^2 + y^2 + z^2 - 4x - 4y = 0$. **30.** $2(x^2 + y^2 + z^2) - 3x + 5y + 3z = 0$. **31.** $\frac{1}{\sqrt{2}}$. **32.** $x^2 + y^2 + z^2 - x - 2y - 8z + 13 = 0$. **34.** $\frac{x}{3} = \frac{y}{4} = \frac{z}{5}$. **35.** $x^2 + y^2 = 1$. **36.** $(x^2 + y^2 + z^2)(l^2 + m^2 + n^2)\cos^2\alpha = (lx + my + nz)^2$. **37.** $2x + 4y + 5z = 45$. **38.** $p^2 = 4l^s + 3m^2 + 12n^2, \left(\frac{4l}{p}, \frac{3m}{p}, \frac{12n}{p}\right)$. **41.** $5x - 4z = 0$. **42.** $45°, 45°, 90°$. **43.** (a) False, (b) False, (c) True, (d) True, (e) False.

Part III

Vector Algebra

Chapter 1

Definitions and Elementary Operations

1.1 Definitions

Scalars. Quantities which have magnitude only are known as *scalars.* Mass, area, volume, temperature, work etc. are scalar quantities. A scalar can be represented by a number with the proper sign.

Vectors. Quantities which have magnitude as well as direction and which also obey certain law of addition are known as *vectors.* Force, displacement, velocity, moment, electric current etc. are vector quantities.

Representation of a vector. A vector can be represented by a directed line segment AB. The length of AB represents the magnitude of the vector to some convenient scale and the direction of it is indicated by an arrow as shown in the fig. 1.1.

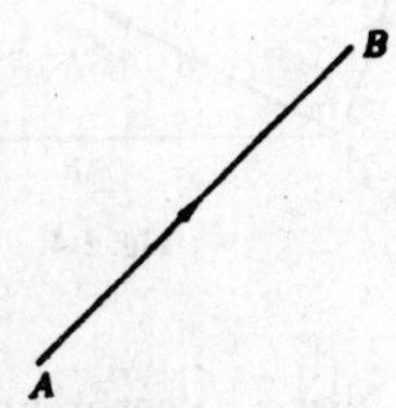

Fig. 1.1

A vector from the point A to the point B is generally denoted by $\overrightarrow{AB}$. Thus one can say that a vector is *directed a line segment.* For simple notation and complete distinction from others we denote the vectors by small letters with bars on them as $\overrightarrow{AB} = \bar{a}$. The magnitude or the *absolute value* or *module* or $\overrightarrow{AB}$ is written as $|\overrightarrow{AB}|$ or $|\bar{a}|$. It is always positive.

[Generally vectors are denoted in print by bold face letters as $\mathbf{a}, \mathbf{b}, \ldots$ It is difficult to write these in practice.]

Like Vectors. Vectors which have the same directions are said to be *like vectors.*

Position Vectors. Let O be an arbitrary point. It is called the *vector origin* or the *initial point.* The position of a point P w.r.t. O is represented by $\overrightarrow{OP}$. The vector $\overrightarrow{OP}$ is known as the position vector of P with respect to the initial point O.

Co-initial Vectors. Vectors having the same initial point are called *co-initial* vectors.

Unit Vector. A vector whose length is unity is called a *unit vector.*

Null Vector. (*zero vector*): It is a vector whose absolute value is zero. A null vector has no definite direction.

Fig. 1.2

Proper Vector. A vector whose length is not zero is called a *proper vector.*

Equal Vector. Two vectors are said to be equal, if they have the same magnitudes and directions. If $\bar{a}$ and $\bar{b}$ are equal, we will write $\bar{a} = \bar{b}$.

Collinear Vectors. Any set of vectors is said to be collinear if each member of the set is parallel to the same line.

If $\bar{a}$ is collinear with $\bar{b}$ then we can write $\bar{a} = x\bar{b}$, where x is a scalar.

Coplanar Vectors. A system of vectors is said to be coplanar if a plane can be drawn which is parallel to each of the vectors. Otherwise they are called non-coplanar.

1.2 Addition of Vectors (Triangle Law)

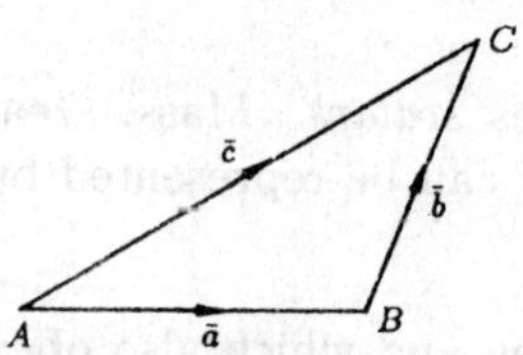

Fig. 1.3

If $\overrightarrow{AB} = \bar{a}, \overrightarrow{BC} = \bar{b}$ and $\overrightarrow{AC} = \bar{c}$, then $\bar{a} + \bar{b} = \bar{c}$, i.e., $\overrightarrow{AB} + \overrightarrow{BC} = \overrightarrow{AC}$.

It is the *triangle law* of addition.

Theorem 1 *Vector addition is commutative.* $(a + \bar{b} = \bar{b} + \bar{a})$.

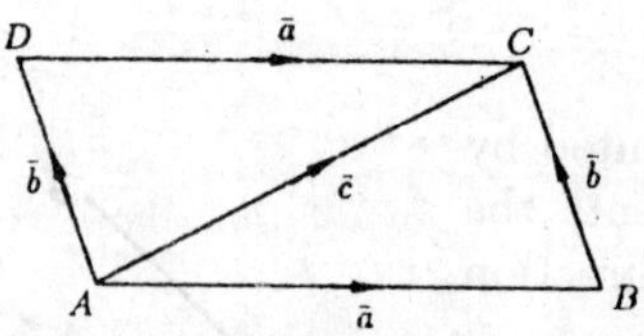

Fig. 1.4

Let $\overrightarrow{AB} = \bar{a}, \overrightarrow{BC} = \bar{b}$ and $\overrightarrow{AC} = \bar{c}$ in the fig. 1.4.

The parallelogram $ABCD$ is completed.

Now $\overrightarrow{AD} = \overrightarrow{BC} = \bar{b}, \overrightarrow{DC} = \overrightarrow{AB} = \bar{a}$.

From $\Delta ABC, \overrightarrow{AC} = \overrightarrow{AB} + \overrightarrow{BC} = \bar{a} + \bar{b}$.

From $\Delta ADC; \overrightarrow{AC} = \overrightarrow{AD} + \overrightarrow{DC} = \bar{b} + \bar{a}$.

$\therefore \quad \bar{a} + \bar{b} = \bar{b} + \bar{a} (= \bar{c})$.

Note. Vector addition can also be made by using parallelogram law (see fig. 1.4.) The addition of vectors represented by two adjacent sides at a parallelogram is represented by the diagonal of the parallelogram passing through the common point.

Theorem 2 *Vector addition is associative* $(\bar{a} + \bar{b}) + \bar{c} = \bar{a} + (\bar{b} + \bar{c})$.

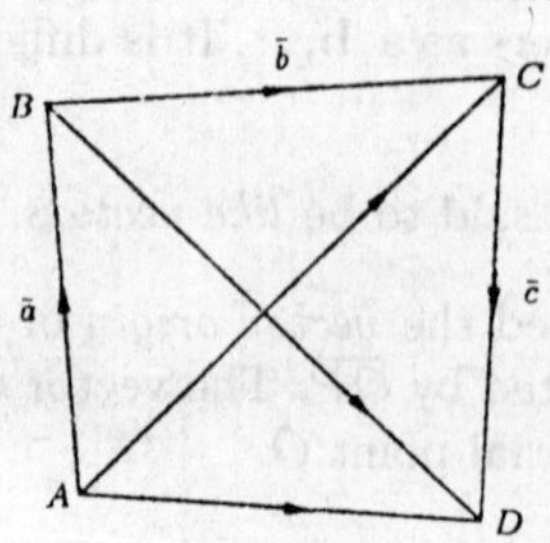

Fig. 1.5

Let $\overrightarrow{AB} = \bar{a}, \overrightarrow{BC} = \bar{b}, \overrightarrow{CD} = \bar{c}$ in the fig. 1.5.

From $\Delta ABC, \overrightarrow{AC} = \overrightarrow{AB} + \overrightarrow{BC} = \bar{a} + \bar{b}$.

From $\Delta ACD, \overrightarrow{AD} = \overrightarrow{AC} + \overrightarrow{CD} = (\bar{a} + \bar{b}) + \bar{c}$. (1.2.1)

From $\Delta BCD, \overrightarrow{BD} = \overrightarrow{BC} + \overrightarrow{CD} = \bar{b} + \bar{c}$.

From $\Delta ABD, \overrightarrow{AD} = \overrightarrow{AB} + \overrightarrow{BD} = \bar{a} + (\bar{b} + \bar{c})$. (1.2.2)

By (1.2.1) and (1.2.2) $(\bar{a} + \bar{b}) + \bar{c} = \bar{a} + (\bar{b} + \bar{c})$.

Hence the sum of the vectors is independent of the order in which they are added. Thus we write the sum as $\bar{a}+\bar{b}+\bar{c}$.

1.3 Subtraction of Vectors

The sum of two vectors is zero when they are equal in magnitudes but opposite in directions. $\overrightarrow{AB}+\overrightarrow{BA}=0$.

If $\overrightarrow{AB}=\bar{a}$ then $\overrightarrow{BA}=-\bar{a}$.

From the fig. 1.3, we have $\overrightarrow{AB}+\overrightarrow{BC}=\overrightarrow{AC}$.

It can be written as $\overrightarrow{AB}=\overrightarrow{AC}-\overrightarrow{BC}$, i.e., $\bar{a}=\bar{c}-\bar{b}$.

Subtraction can be conceived as addition of $\bar{c}$ with $-\bar{b}$.

1.4 Multiplication of a vector by a scalar

If $\bar{a}$ is a vector and m is a scalar, then their product is written as $m\bar{a}$ or $\bar{a}m$. It is also a vector in the same direction of $\bar{a}$ when m is positive and in the opposite direction of $\bar{a}$ when m is a negative but its magnitude is $|m|$ times that of $\bar{a}$.

Vectors obey the rules of ordinary algebra for addition, subtraction and multiplication by scalars.

As

(i) $(m+n)\bar{a}=m\bar{a}+n\bar{a}$,

(ii) $m(\bar{a}+\bar{b})=m\bar{a}+m\bar{b}$,

(iii) $m(n\bar{a})=n(m\bar{a})=mn\bar{a}$.

1.5 Point of division (Section ratio)

Let the point C divides AB in the ratio $m:n$. Here n is non-zero positive number and m is positive or negative according as C lies within or outside the segment AB.

Let the position vectors of A and B w.r.t. the vector origin O be $\bar{a}$ and $\bar{b}$, i.e., $\overrightarrow{OA}=\bar{a}$ and $\overrightarrow{OB}=\bar{b}$.

From $\Delta OAC, \overrightarrow{OA}+\overrightarrow{AC}=\overrightarrow{OC}$, or $\overrightarrow{AC}=\overrightarrow{OC}-\overrightarrow{OA}$.

From $\Delta OCB, \overrightarrow{OC}+\overrightarrow{CB}=\overrightarrow{OB}$, or $\overrightarrow{CB}=\overrightarrow{OB}-\overrightarrow{OC}$.

Since

$$\frac{AC}{CB}=\frac{m}{n},\quad \overrightarrow{AC}=\frac{m}{n}\overrightarrow{CB}.$$

$$\therefore\ n(\overrightarrow{OC}-\overrightarrow{OA})=m(\overrightarrow{OB}-\overrightarrow{OC}),$$

$$\text{or}\quad (m+n)\overrightarrow{OC}=m\overrightarrow{OB}+n\overrightarrow{OA},$$

$$\text{or}\quad \overrightarrow{OC}=\frac{m\overrightarrow{OB}+n\overrightarrow{OA}}{m+n}=\frac{m\bar{b}+n\bar{a}}{m+n}.$$

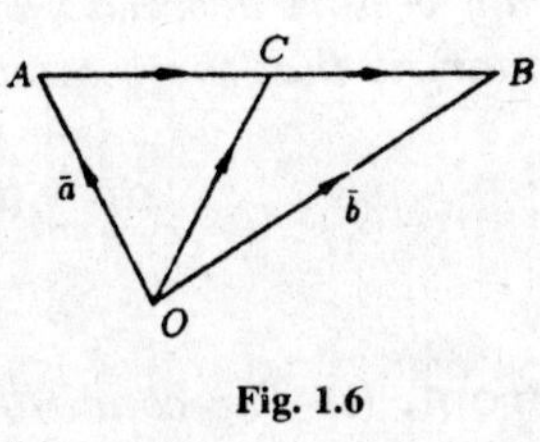

Fig. 1.6

It gives the position vector of C w.r.t. the same initial point O.

Corollary. If C is the mid-point of AB, then $m = n$. In this case $\overrightarrow{OC} = \frac{\bar{a}+\bar{b}}{2}$.

1.6 Collinearity of three points

Theorem. *The necessary and sufficient condition for the three distinct points A, B, C to lie on a straight line (i.e., collinear) is that there exist three scalars α, β, γ (not all zero) such that $\alpha\bar{a}+\beta\bar{b}+\gamma\bar{c} = \bar{0}$ and $\alpha+\beta+\gamma = 0$, where $\bar{a}, \bar{b}, \bar{c}$ are the position vectors of A, B, C respectively w.r.t. a certain vector origin.*

Proof. (i) *The condition is necessary.*

Let A, B, C lie on the same line. Then C must divide AB in some ratio $m : n$ which is not zero, ∞ or -1, so that $m, n, m+n$ are not zero.

Thus we write

$$\bar{c} = \frac{m\bar{b}+n\bar{a}}{m+n}, \quad \text{or} \quad n\bar{a}+m\bar{b}-(m+n)\bar{c} = \bar{0}.$$

Putting $n = \alpha, m = \beta$ and $-(m+n) = \gamma$, we have

$$\alpha\bar{a}+\beta\bar{b}+\gamma\bar{c} = \bar{0} \quad \text{and} \quad \alpha+\beta+\gamma = 0.$$

(ii) *The condition is sufficient.*

Let the relations $\alpha\bar{a}+\beta\bar{b}+\gamma\bar{c} = \bar{0}$ and $\alpha+\beta+\gamma = 0$ hold.

Then

$$\alpha+\beta = -\gamma \quad \text{and} \quad -\gamma\bar{c} = \alpha\bar{a}+\beta\bar{b}, \quad \text{or} \quad \bar{c} = \frac{\alpha\bar{a}+\beta\bar{b}}{-\gamma} = \frac{\alpha\bar{a}+\beta\bar{b}}{\alpha+\beta}.$$

This suggests that $\bar{c}$ lies on the line of join of $\bar{a}$ and $\bar{b}$ and also it divides AB in the ratio $\beta : \alpha$. Hence the points are collinear.

1.7 Coplanarity of four points

Theorem. *The necessary and sufficient condition for four points defined by the position vectors $\bar{a}, \bar{b}, \bar{c}, \bar{d}$ (no three of them are collinear) to lie on the same plane (i.e., coplanar) is that there exist four scalars $\alpha, \beta, \gamma, \delta$ (not all zero) such that*

$$\alpha\bar{a}+\beta\bar{b}+\gamma\bar{c}+\delta\bar{d} = \bar{0}, \quad \alpha+\beta+\gamma+\delta = 0.$$

Proof. (i) *The condition is necessary.*

Let A, B, C, D be the four points given by the position vectors $\bar{a}, \bar{b}, \bar{c}, \bar{d}$ lie on the same plane, where no three of them are collinear. In this case we can choose that AB

and CD meet at a point P (say). If $\bar{p}$ is the position vector of P w.r.t. the same vector origin and it divides AB and CD in the ratio $m : n$ and $r : s$ respectively, then

$$\bar{p} = \frac{n\bar{a} + m\bar{b}}{m+n} \quad \text{and} \quad \bar{p} = \frac{s\bar{c} + r\bar{d}}{r+s}.$$

$$\therefore \quad \frac{n\bar{a} + m\bar{b}}{m+n} = \frac{s\bar{c} + r\bar{d}}{r+s}, \quad \text{or} \quad \frac{n}{m+n}\bar{a} + \frac{m}{m+n}\bar{b} - \frac{s}{r+s}\bar{c} - \frac{r}{r+s}\bar{d} = \bar{0}.$$

Putting

$$\alpha = \frac{n}{m+n}, \quad \beta = \frac{m}{m+n}, \quad \gamma = -\frac{s}{r+s}, \quad \delta = -\frac{r}{r+s},$$

we have $\alpha\bar{a} + \beta\bar{b} + \gamma\bar{c} + \delta\bar{d} = \bar{0}$ and $\alpha + \beta + \gamma + \delta = 0$.

(ii) *The condition is sufficient.*

Let the relations $\alpha\bar{a} + \beta\bar{b} + \gamma\bar{c} + \delta\bar{d} = \bar{0}$ and $\alpha + \beta + \gamma + \delta = 0$ hold.

$$\therefore \quad \alpha\bar{a} + \beta\bar{b} = -(\gamma\bar{c} + \delta\bar{d}) \quad \text{and} \quad \alpha + \beta = -(\gamma + \delta).$$

Dividing one by the other

$$\frac{\alpha\bar{a} + \beta\bar{b}}{\alpha + \beta} = \frac{\gamma\bar{c} + \delta\bar{d}}{\gamma + \delta}.$$

It suggests that the line joining the points defined by the position vectors $\bar{a}$ and $\bar{b}$ is divided by a point in the ratio $\beta : \alpha$ and the line joining the points defined by the position vectors $\bar{c}$ and $\bar{d}$ is divided by the same point in the ratio $\delta : \gamma$. As these four points lie on two intersecting lines, they are coplanar.

1.8 Rectangular Components (Resolution of a Vector)

Let $\bar{i}, \bar{j}, \bar{k}$ be the unit vectors along the three rectangular axes OX, OY and OZ respectively and P be a point whose co-ordinates are (x, y, z) w.r.t. OX, OY and OZ.

In the fig. 1.7, PN is perpendicular to XOY plane and MN is parallel to OY.

Therefore

$$OM = x, \quad MN = y, \quad NP = z,$$

$$\text{and} \quad \overrightarrow{OM} = x\bar{i}, \quad \overrightarrow{MN} = y\bar{j}, \quad \overrightarrow{NP} = z\bar{k}.$$

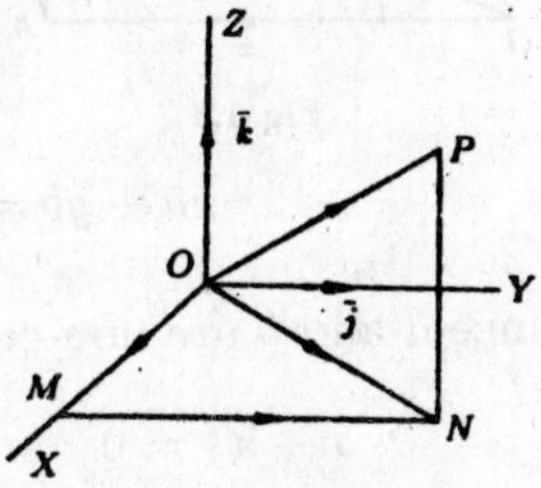

Fig. 1.7

From $\Delta OMN, \overrightarrow{ON} = \overrightarrow{OM} + \overrightarrow{MN} = x\bar{i} + y\bar{j}$.

From $\Delta ONP, \overrightarrow{OP} = \overrightarrow{ON} + \overrightarrow{NP} = x\bar{i} + y\bar{j} + z\bar{k}$.

$x\bar{i}, y\bar{j}, z\bar{k}$ are called the vector components of $\overrightarrow{OP}$ in the directions of OX, OY and OZ. Thus a vector can be resolved into three orthogonal components.

Here $|\overrightarrow{OP}| = \sqrt{x^2+y^2+z^2}$ and the d. cs. of OP are

$$\frac{x}{|\overrightarrow{OP}|}, \quad \frac{y}{|\overrightarrow{OP}|}, \quad \frac{z}{|\overrightarrow{OP}|}.$$

Sometimes the vector $\overrightarrow{OP}$ is denoted by (x, y, z) when the cartesian co-ordinates of P are (x, y, z).

Note 1. If $\overrightarrow{OP} = x\bar{i} + y\bar{j} + z\bar{k}$, (x, y, z) are the co-ordinates of P.

Note 2. If the co-ordinates of A and B are (x_1, y_1, z_1) and (x_2, y_2, z_2) then $\overrightarrow{AB} = \overrightarrow{OB} - \overrightarrow{OA} = (x_2 - x_1)\bar{i} + (y_2 - y_1)\bar{j} + (z_2 - z_1)\bar{k}$.

Note 3. It is customary to use the notations $\bar{i}, \bar{j}, \bar{k}$ for the unit vectors along the three rectangular axes.

1.9 Some Theorems

Theorem 1 *If $\bar{a}$ and $\bar{b}$ are two non-zero, non-collinear vectors and x, y are two scalars, then it is not possible to express $x\bar{a} + y\bar{b} = \bar{0}$ with $x \neq 0, y \neq 0$.*

Proof. Suppose $x \neq 0$. Then it can be written as $\bar{a} = -\frac{y}{x}\bar{b}$.

Since $-\frac{y}{x}$ is a scalar, $\bar{a}$ and $\bar{b}$ are collinear. It is a contradiction. Therefore $x = 0$. Similarly $y = 0$.

Theorem 2 *If $\bar{r}, \bar{a}, \bar{b}$ are coplanar vectors and $\bar{a}, \bar{b}$ are non-collinear then it can be uniquely expressed as $\bar{r} = x\bar{a} + y\bar{b}$, where x and y are scalars.*

Proof. Let $\bar{r} = \overrightarrow{OC}$ be in the plane of $\bar{a}$ and $\bar{b}$ or in the plane parallel to $\bar{a}$ and $\bar{b}$.

Through O and C lines are drawn parallel to $\bar{a}$ and $\bar{b}$ to form the parallelogram $OACB$. Since $\overrightarrow{OA}$ is parallel to $\bar{a}$, $\overrightarrow{OA} = x\bar{a}$, where x is a scalar. Similarly $\overrightarrow{OB} = y\bar{b}$, where y is a scalar.

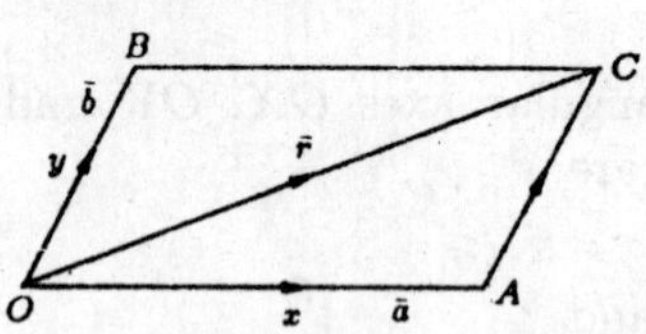

Fig. 1.8

Now $\overrightarrow{OC} = \overrightarrow{OA} + \overrightarrow{AC} = \overrightarrow{OA} + \overrightarrow{OB}$.

$$\therefore \quad \bar{r} = x\bar{a} + y\bar{b}.$$

The expression is unique.

Suppose $\bar{r} = x_1\bar{a} + y_1\bar{b}$, where x_1 and y_1 are scalars.

$$\therefore \quad x\bar{a} + y\bar{b} = x_1\bar{a} + y_1\bar{b}, \quad \text{or} \quad (x - x_1)\bar{a} + (y - y_1)\bar{b} = 0.$$

Since $\bar{a}$ and $\bar{b}$ are non-collinear.

$$x - x_1 = 0 \quad \text{and} \quad y - y_1 = 0, \quad \text{i.e.,} \quad x = x_1 \quad \text{and} \quad y = y_1.$$

Hence the theorem is proved.

Theorem 3 *If $\bar{a}, \bar{b}, \bar{c}$ are three non-coplanar vectors and x, y, z are three scalars then we can not express $x\bar{a} + y\bar{b} + z\bar{c} = \bar{0}$ with $x \neq 0, y \neq 0, z \neq 0$.*

Proof. Suppose $x \neq 0$. Then it can be written as $\bar{a} = -\frac{y}{x}\bar{b} - \frac{z}{x}\bar{c}$. Thus $\bar{a}$ is coplanar with $\bar{b}$ and $\bar{c}$. It leads to a contradiction. Hence $x = 0$. Similarly $y = 0$ and $z = 0$.

Theorem 4 *Any vector $\bar{r}$ in space can be uniquely expressed as the linear combination of three non-coplanar vectors $\bar{a}, \bar{b}$ and $\bar{c}$, i.e., $\bar{r} = x\bar{a} + y\bar{b} + z\bar{c}$, where x, y, z are scalars.*

Let $\bar{r} = \overrightarrow{OG}$ and OA, OC, OE be parallel to $\bar{a}, \bar{b}, \bar{c}$ respectively.

Since $\bar{a}$, $\bar{b}$, $\bar{c}$ are non-coplanar, OA, OC and OE are non-coplanar. Through G three planes parallel to the planes OAC, COE and EOA are drawn to meet OA, OC and OE at A, C, E respectively. Now the parallelopiped $OABCDEFG$ is obtained. OG is a diagonal of this parallelopiped. AG is joined. Again there exist three scalars x, y, z such that

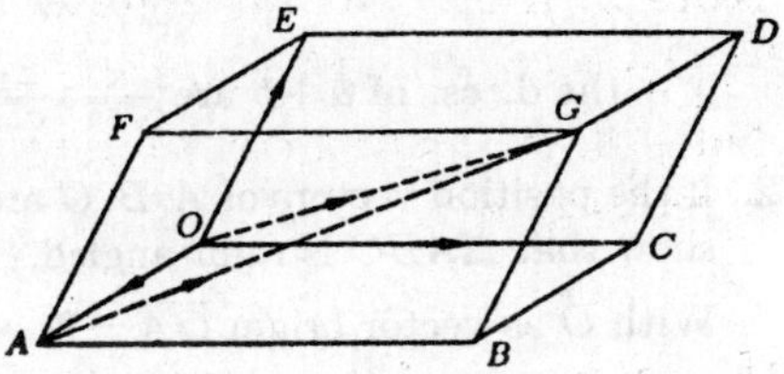

Fig. 1.9

$$\overrightarrow{OA} = x\bar{a}, \quad \overrightarrow{OC} = y\bar{b}, \quad \overrightarrow{OE} = z\bar{c}.$$

From the fig. 1.9, we have

$$\overrightarrow{OG} = \overrightarrow{OA} + \overrightarrow{AG} = \overrightarrow{OA} + \overrightarrow{AB} + \overrightarrow{BG} = \overrightarrow{OA} + \overrightarrow{OC} + \overrightarrow{OE}.$$

$$\therefore \quad \bar{r} = x\bar{a} + y\bar{b} + z\bar{c}.$$

We can now prove that the expression is unique.

Suppose $\bar{r} = x_1\bar{a} + y_1\bar{b} + z_1\bar{c}$.

Then

$$x\bar{a} + y\bar{b} + z\bar{c} = x_1\bar{a} + y_1\bar{b} + z_1\bar{c},$$

$$\text{or} \quad (x - x_1)\bar{a} + (y - y_1)\bar{b} + (z - z_1)\bar{c} = 0.$$

Since $\bar{a}, \bar{b}, \bar{c}$ are non-coplanar

$$x - x_1 = 0, y - y_1 = 0, z - z_1 = 0, \quad \text{i.e.} \quad x = x_1, y = y_1, z = z_1.$$

Thus the result follows.

Note 1. If $\bar{r} = x\bar{a} + y\bar{b}$ then $x\bar{a}$ and $y\bar{b}$ are called components of $\bar{r}$ and the scalars x, y are called the co-ordinates of $\bar{r}$ relative to the vectors $\bar{a}$ and $\bar{b}$.

Similarly in case of $\bar{r} = x\bar{a} + y\bar{b} + z\bar{c}, x\bar{a}, y\bar{b}, z\bar{c}$ are called components of $\bar{r}$ and the scalars x, y, z are called the co-ordinates of $\bar{r}$ relative to the vectors $\bar{a}, \bar{b}$ and $\bar{c}$.

Note 2. If $x\bar{a} + y\bar{b} = 0$ and x, y are not zero then $\bar{a}$ and $\bar{b}$ are linearly dependent and collinear.

Similarly if $x\bar{a} + y\bar{b} + z\bar{c} = 0$ and x, y, z are not all zero then the vectors $\bar{a}, \bar{b}, \bar{c}$ are linearly dependent and coplanar.

Worked Out Examples

1. If $\bar{a} = (1, 2, 3)$ and $\bar{b} = (2, -3, 4)$, find $\bar{a} + \bar{b}$ and $2\bar{b} - \bar{a}$. Calculate the module and d. cs. of $\bar{a} + \bar{b}$.

Here $\bar{a} = \bar{i} + 2\bar{j} + 3\bar{k}, \bar{b} = 2\bar{i} - 3\bar{j} + 4\bar{k}$, where $\bar{i}, \bar{j}, \bar{k}$ are the unit vectors along the three rectangular axes.

Now $\bar{a} + \bar{b} = (\bar{i} + 2\bar{j} + 3\bar{k}) + (2\bar{i} - 3\bar{j} + 4\bar{k}) = 3\bar{i} - \bar{j} + 7\bar{k} = (3, -1, 7)$, $2\bar{b} - \bar{a} = 2(2\bar{i} + 3\bar{j} + 4\bar{k}) - (\bar{i} + 2\bar{j} + 3\bar{k}) = 3\bar{i} - 8\bar{j} + 5\bar{k} = (3, -8, 5)$.

$$|\bar{a} + \bar{b}| = \sqrt{3^2 + 1^2 + 7^2} = \sqrt{59}.$$

$\therefore$ the d. cs. of $\bar{a} + \bar{b}$ are $\frac{3}{\sqrt{59}}, \frac{-1}{\sqrt{59}}, \frac{7}{\sqrt{59}}$.

2. If the position vectors of A, B, C are $2\bar{i} + 4\bar{j} - \bar{k}, 4\bar{i} + 5\bar{j} + \bar{k}$ and $3\bar{i} + 6\bar{j} - 3\bar{k}$ respectively, show that ΔABC is right-angled.

With O as vector origin $\overrightarrow{OA} = 2\bar{i} + 4\bar{j} - \bar{k}$,

$$\overrightarrow{OB} = 4\bar{i} + 5\bar{j} + \bar{k}, \overrightarrow{OC} = 3\bar{i} + 6\bar{j} - 3\bar{k}.$$

Now

$$\begin{aligned}
\overrightarrow{AB} &= \overrightarrow{OB} - \overrightarrow{OA} = (4\bar{i} + 5\bar{j} + \bar{k}) - (2\bar{i} + 4\bar{j} - \bar{k}) = 2\bar{i} + \bar{j} + 2\bar{k}, \\
\overrightarrow{BC} &= \overrightarrow{OC} - \overrightarrow{OB} = (3\bar{i} + 6\bar{j} - 3\bar{k}) - (4\bar{i} + 5\bar{j} + \bar{k}) = -\bar{i} + \bar{j} + 2\bar{k}, \\
\overrightarrow{CA} &= \overrightarrow{OA} - \overrightarrow{OC} = (2\bar{i} + 4\bar{j} - \bar{k}) - (3\bar{i} + 6\bar{j} - 3\bar{k}) = -\bar{i} - 2\bar{j} + 2\bar{k}. \\
|\overrightarrow{AB}| &= \sqrt{2^2 + 1^2 + 2^2} = \sqrt{9} = 3, \\
|\overrightarrow{BC}| &= \sqrt{1^2 + 1^2 + 4^2} = \sqrt{18} = 3\sqrt{2}, \\
|\overrightarrow{CA}| &= \sqrt{1^2 + 2^2 + 2^2} = \sqrt{9} = 3. \\
\therefore \quad AB^2 &+ CA^2 = 9 + 9 = 18 = BC^2.
\end{aligned}$$

Hence ΔABC is right-angled.

3. Show by vector method that the medians of a triangle are concurrent.

[C.H. '80, '84; C.U. '85, '87, '89, 2001; B.U. '82, '84, '86, '87, '92, '97; N.B. '81, '84, '98]

Let ABC be a triangle and $\bar{a}, \bar{b}, \bar{c}$ be the position vectors of A, B, C respectively.

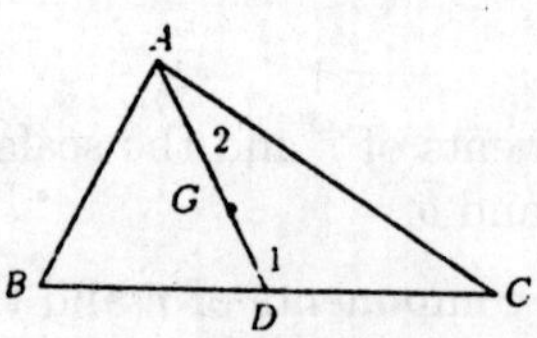

Fig. 1.10

If D be the mid-point of BC, then the position vector of D is $\frac{\bar{b}+\bar{c}}{2}$.

If the point G divides AD in the ratio $2 : 1$, then the position vector of G is

$$\frac{2\frac{\bar{b}+\bar{c}}{2} + 1\bar{a}}{2+1}, \quad \text{i.e.,} \quad \frac{\bar{a}+\bar{b}+\bar{c}}{3}.$$

The symmetry of this expression suggests that the other two medians will be divided in the ratio $2 : 1$ by the same point G.

Hence they are concurrent.

4. Show by the method of vectors that the join of the middle points of two sides of a triangle is parallel to the third side and is half of its length.

[N.B. '82, '98; B.U. '84, '89, '97, '99; C.U. '89, '97, '98]

Let ABC be a triangle and $\bar{a}, \bar{b}, \bar{c}$ be the position vectors of the vertices A, B, C respectively with respect to a certain vector origin.

If D and E are the mid-points of AB and AC then the position vectors of D and E are $\frac{\bar{a}+\bar{b}}{2}$ and $\frac{\bar{a}+\bar{c}}{2}$

Now $\overrightarrow{DE} = \frac{\bar{a}+\bar{b}}{2} - \frac{\bar{a}+\bar{c}}{2} = \frac{\bar{b}-\bar{c}}{2}$ and $\overrightarrow{BC} = \bar{c} - \bar{b}$. $\therefore \overrightarrow{DE} = \frac{\overrightarrow{BC}}{2}$.

$\therefore$ DE is parallel to BC and it is half of BC in length.

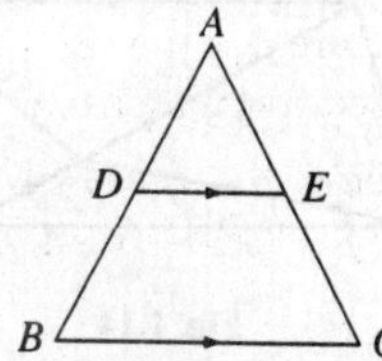

Fig 1.11

5. $ABCD$ is a parallelogram. P, Q are the mid-points of the sides AB and CD respectively. Show that DP and BQ trisect AC and are trisected by AC.

Let the position vectors A, B, C, D, P and Q be $\bar{a}, \bar{b}, \bar{c}, \bar{d}, \bar{p}$ and $\bar{q}$ respectively.

Since P and Q are the mid-points of AB and CD,

$$\bar{p} = \frac{\bar{a}+\bar{b}}{2}, \quad \bar{q} = \frac{\bar{c}+\bar{d}}{2}.$$

$ABCD$ is a parallelogram.

$$\therefore \quad \overrightarrow{AB} = \overrightarrow{DC}, \quad \text{or} \quad \bar{b} - \bar{a} = \bar{c} - \bar{d}. \quad (1)$$

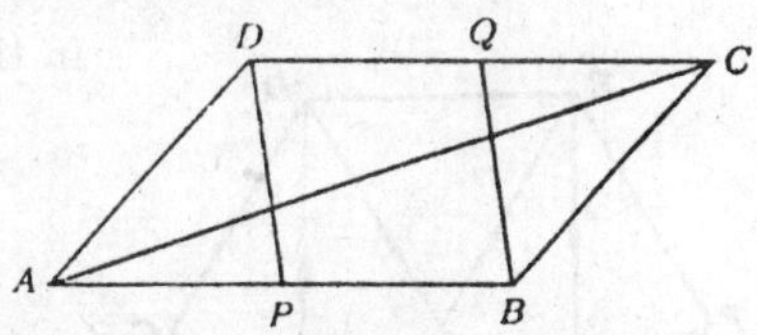

Fig. 1.12

Now

$$\bar{p} = \frac{\bar{a}+\bar{a}+\bar{c}-\bar{d}}{2} \quad \text{[by (1)]},$$

$$\text{or} \quad 2\bar{a}+\bar{c} = 2\bar{p}+\bar{d}, \quad \text{or} \quad \frac{2\bar{a}+\bar{c}}{3} = \frac{2\bar{p}+\bar{d}}{3}.$$

This result indicates that the common point of AC and PD intersects them in the ratio $1 : 2$.

Again

$$\bar{q} = \frac{\bar{c}+\bar{d}}{2} = \frac{\bar{c}+\bar{c}+\bar{a}-\bar{b}}{2} \quad \text{[by (1)]}, \quad \text{or} \quad \frac{2\bar{c}+\bar{a}}{3} = \frac{2\bar{q}+\bar{b}}{3}.$$

Therefore CA and QB are intersected by their common point in the ratio $1 : 2$.

Hence DP and BQ trisect AC and are trisected by AC.

6. Show that the lines joining the vertices to the centroids of the opposite faces of a tetrahedron are concurrent.

Let $ABCD$ be a tetrahedron and $\bar{a}, \bar{b}, \bar{c}, \bar{d}$ be the position vectors of its vertices.

If X is the centroid of ΔBCD, the position vector of X is $\frac{\bar{b}+\bar{c}+\bar{d}}{3}$.

Let the point G divide AX in the ratio $3 : 1$.

Then the position vector of G is $\frac{\bar{a}+\bar{b}+\bar{c}+\bar{d}}{4}$.

The symmetry of this expression suggests that the other three lines joining the other vertics to the centrodies of the opposite faces will be divided in the ratio $3 : 1$ by the same point G. Hence they are concurrent.

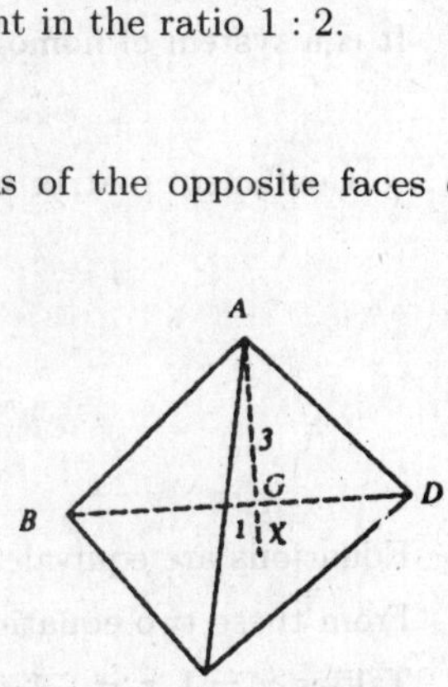

Fig. 1.13

7. P is a point on BC of ΔABC. If $\overrightarrow{PQ}$ is the resultant of $\overrightarrow{AP}, \overrightarrow{PB}, \overrightarrow{PC}$, show that $ABQC$ is a parallelogram and Q therefore is a fixed point.

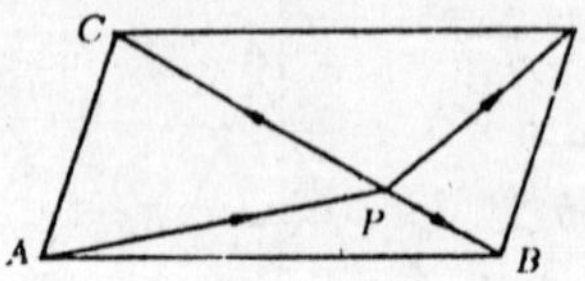

Fig. 1.14

Here $\overrightarrow{AP}+\overrightarrow{PB}+\overrightarrow{PC} = (\overrightarrow{AP}+\overrightarrow{PB})+\overrightarrow{PC} = \overrightarrow{AB}+\overrightarrow{PC}$.

CD is drawn parallel and equal to AB. Therefore $ABCD$ is a parallelogram.

Now $\overrightarrow{AB}+\overrightarrow{PC} = \overrightarrow{CD}+\overrightarrow{PC} = \overrightarrow{PC}+\overrightarrow{CD} = \overrightarrow{PD}$.

Since $\overrightarrow{PQ}$ is the resultant, D coincides with Q.

Therefore $ABQC$ is a parallelogram and Q is fixed w.r.t. A, B and C.

8. $\bar{a}$ and $\bar{b}$ are vectors representing consecutive sides of a hexagon. Find the vectors forming the other four sides.

Let $ABCDEF$ be a regular hexagon and $\overrightarrow{AB} = \bar{a}, \overrightarrow{BC} = \bar{b}$.

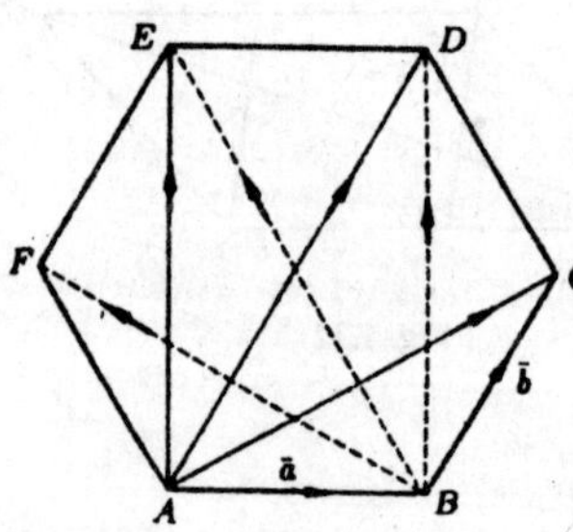

Fig. 1.15

Now $\overrightarrow{AC} = \overrightarrow{AB}+\overrightarrow{BC} = \bar{a}+\bar{b}$.

In the regular hexgon $AD \parallel BC$ and $AD = 2BC$.

$\therefore \quad \overrightarrow{AD} = 2\overrightarrow{BC} = 2\bar{b}$.

$AB \parallel ED$ and $AB = ED$.

$\therefore \quad \overrightarrow{DE} = -\overrightarrow{AB} = -\bar{a}$.

$\overrightarrow{CD} = \overrightarrow{AD} - \overrightarrow{AC} = 2\bar{b} - \bar{a} - \bar{b} = \bar{b} - \bar{a}$.

$\overrightarrow{EF} = -\overrightarrow{BC} = -\bar{b}$.

$\overrightarrow{FA} = -\overrightarrow{CD} = \bar{a} - \bar{b}$.

9. Show that the points $A(\bar{i} - 2\bar{j} + 3\bar{k})$, $B(2\bar{i} - 3\bar{j} + 4\bar{k})$ and $C(-2\bar{i} + \bar{j})$ are collinear. **[C.U. '67; B.U. '99]**

Let $x(\bar{i} - 2\bar{j} + 3\bar{k}) + y(2\bar{i} - 3\bar{j} + 4\bar{k}) + z(-2\bar{i} + \bar{j}) = \bar{0}$. ($x, y, z$ are scalars.)

It implies that $x + 2y - 2z = 0, -2x - 3y + z = 0, 3x + 4y = 0$.

It is a system of homogeneous equations in x, y, z.

$$\text{Coefficient matrix} = \begin{bmatrix} 1 & 2 & -2 \\ -2 & -3 & 1 \\ 3 & 4 & 0 \end{bmatrix} \sim \begin{bmatrix} 1 & 2 & -2 \\ 0 & 1 & -3 \\ 0 & -2 & 6 \end{bmatrix} \text{ by } R_2 + 2R_1 \text{ and } R_3 - 3R_1$$

$$\sim \begin{bmatrix} 1 & 2 & -2 \\ 0 & 1 & -3 \\ 0 & 0 & 0 \end{bmatrix} \text{ by } R_3 + 2R_2.$$

Equations are equivalent to $x + 2y - 2z = 0, y - 3z = 0$.

From these two equations $y = 3z, x = -4z$.

Taking $z = 1, x = -4, y = 3$.

Again $x + y + z = -4 + 3 + 1 = 0$.

$\therefore \quad$ the points are collinear.

10. Show by vector method that the points $P(1,5,-1)$, $Q(0,4,5)$, $R(-1,5,1)$ and $S(2,4,3)$ are coplanar. **[C.U. 2000]**

Let $x(1,5,-1)+y(0,4,5)+z(-1,5,1)+t(2,4,3)=(0,0,0)$, ($x,y,z,t$ are scalars.)

It implies that $x-z+2t=0$, $5x+4y+5z+4t=0$, $-x+5y+z+3t=0$. Coefficient matrix of this system of homogeneous equations is

$$\begin{bmatrix} 1 & 0 & -1 & 2 \\ 5 & 4 & 5 & 4 \\ -1 & 5 & 1 & 3 \end{bmatrix} \sim \begin{bmatrix} 1 & 0 & -1 & 2 \\ 0 & 4 & 10 & -6 \\ 0 & 5 & 0 & 5 \end{bmatrix} \sim \text{ by } R_2-5R_1 \text{ and } R_3+R_1$$

$$\sim \begin{bmatrix} 1 & 0 & -1 & 2 \\ 0 & 2 & 5 & -3 \\ 0 & 1 & 0 & 1 \end{bmatrix} \text{ by } \tfrac{1}{2}R_2 \text{ and } \tfrac{1}{5}R_3$$

$$\sim \begin{bmatrix} 1 & 0 & -1 & 2 \\ 0 & 1 & 0 & 1 \\ 0 & 2 & 5 & -3 \end{bmatrix} \text{ by interchanging } R_2 \text{ and } R_3$$

$$\sim \begin{bmatrix} 1 & 0 & -1 & 2 \\ 0 & 1 & 0 & 1 \\ 0 & 0 & 5 & -5 \end{bmatrix} \text{ by } R_3-2R_1.$$

Equations are equivalent to $x-z+2t=0$, $y+t=0$, $5z-5t=0$.

From these equations $z=t, y=-t, x=-t$.

Taking $t=1, x=-1, y=-1, z=1$.

Again $x+y+z+t=-1-1+1+1=0$.

$\therefore$ the points are coplanar.

11. Prove that the vectors $3\bar{\alpha}-7\bar{\beta}-4\bar{\gamma}$, $3\bar{\alpha}-2\bar{\beta}+\bar{\gamma}$, $\bar{\alpha}+\bar{\beta}+2\bar{\gamma}$ are coplanar. ($\bar{\alpha},\bar{\beta},\bar{\gamma}$ are any vectors.) **[B.U. '96]**

Let $\bar{r}_1=3\bar{\alpha}-7\beta-4\bar{\gamma}$, $\bar{r}_2=3\bar{\alpha}-2\bar{\beta}+\gamma$ and $\bar{r}_3=\bar{\alpha}+\bar{\beta}+2\bar{\gamma}$. If these vectors are coplanar, then one is expressible as the sum of the scalar multiplies of the other two.

Let $\bar{r}_3=x\bar{r}_1+y\bar{r}_2$ where x and y are scalars.

Then $\bar{\alpha}+\bar{\beta}+2\bar{\gamma}=x(3\bar{\alpha}-7\bar{\beta}-4\bar{\gamma})+y(3\bar{\alpha}-2\bar{\beta}+\bar{\gamma})$.

It implies that $3x+3y=1$, $7x+2y=-1$ and $4x-y=-2$. Solving the first two equations $x=-\frac{1}{3}, y=\frac{2}{3}$. These values satisfy the third equations.

Thus $\bar{r}_3=-\frac{1}{3}\bar{r}_1+\frac{2}{3}\bar{r}_2$. Hence the vectors are coplanar.

12. Show that the 4 points whose position vectors are $-3\bar{a}+2\bar{b}+\bar{c}$, $-\bar{a}+4\bar{b}-3\bar{c}$, $3\bar{a}+2\bar{b}-5\bar{c}$ and $-3\bar{a}+8\bar{b}-5\bar{c}$ are coplanar. **[C.U. '97]**

Let the given vectors be the position vectors of the points A,B,C,D with respect to the vector origin O.

Then

$$\overrightarrow{AB}=\overrightarrow{OB}-\overrightarrow{OA}=(-\bar{a}+4\bar{b}-3\bar{c})-(-3\bar{a}+2\bar{b}+\bar{c})=2\bar{a}+2\bar{b}-4\bar{c},$$

$$\overrightarrow{AC}=\overrightarrow{OC}-\overrightarrow{OA}=(3\bar{a}+2\bar{b}-5\bar{c})-(-3\bar{a}+2\bar{b}+\bar{c})=6\bar{a}-6\bar{c}$$

$$\overrightarrow{AD}=\overrightarrow{OD}-\overrightarrow{OA}=(-3\bar{a}+8\bar{b}-5\bar{c})-(-3\bar{a}+2\bar{b}+\bar{c})=6\bar{b}-6\bar{c}.$$

If the given points are coplanar then $\overrightarrow{AB}, \overrightarrow{AC}, \overrightarrow{AD}$ are coplanar. For this one is expressible as the sum of scalar multiplies of the other two.

Let $\overrightarrow{AD} = x\overrightarrow{AB} + y\overrightarrow{AC}$ where x and y are scalars,

$$\text{or} \quad 6\bar{b} - 6\bar{c} = x(2\bar{a} + 2\bar{b} - 4\bar{c}) + y(6\bar{a} - 6\bar{c}).$$

It implies that $2x + 6y = 0$, $2x = 6$, $-4x - 6y = -6$. From the first two $x = 3$, $y = -1$. These values satisfy the third equation. Therefore $\overrightarrow{AB}, \overrightarrow{AC}$ and $\overrightarrow{AD}$ are coplanar. Hence the given points are coplanar.

Exercise

1. If $\bar{a} = (2, 3, 4)$ and $\bar{b} = (-2, 1, 5)$, find $\bar{a} + \bar{b}$ and $\bar{a} - 2\bar{b}$.
   Calculate the module of $\bar{a} - 2\bar{b}$ and the d. cs. of it.
2. Determine the values of λ and μ for which the vectors $-3\bar{i} + 4\bar{j} + \lambda\bar{k}$ and $\mu\bar{i} + 8\bar{j} + 6\bar{k}$ are collinear. **[C.U. '81, 2001]**
3. If the vertices of a triangle are the points $\bar{i} - \bar{j} + 2\bar{k}$, $2\bar{i} + 2\bar{j} + 4\bar{k}$, $3\bar{i} + 3\bar{j} - 4\bar{k}$, what are the vectors determined by the sides? **[C.U. '62]**
4. If the position vectors of A, B, C are $3\bar{i} + 4\bar{j} - 6\bar{k}$, $4\bar{i} - 6\bar{j} + 3\bar{k}$ and $-6\bar{i} + 3\bar{j} + 4\bar{k}$, show that ABC is an equilateral triangle.
5. The vertices of a triangle are $(2, 4, -1)$, $(4, 5, 1)$ and $(3, 6, -3)$. Show that the triangle is an isosceles right-angled triangle. **[C.H. '84; C.U. '85; N.B. '83; B.H. '83]**
6. Prove by vect: method that if the diagonals of quadrilateral bisect each other, then the figure is a parallelogram. **[C.U. 2000]**
   [HINTS. Let the diagonals AC and BD of the quadrilateral $ABCD$ bisect at O. If $\overrightarrow{OA} = \bar{a}, \overrightarrow{OB} = \bar{b}$ then $\overrightarrow{OC} = -\bar{a}$ and $\overrightarrow{OD} = -\bar{b}$. From $\Delta OAB, \overrightarrow{AB} = \bar{b} - \bar{a}$ and from ΔOCD, $\overrightarrow{DC} = -\bar{a} + \bar{b} = \bar{b} - \bar{a}$.
   $\therefore$ AB and DC are equal in magnitude and parallel. Therefore $ABCD$ is a parallelogram.]
7. $ABCD$ is a parallelogram whose diagonals are AC and BD. Prove that $\overrightarrow{AC} + \overrightarrow{BD} = 2\overrightarrow{BC}, \overrightarrow{AC} - \overrightarrow{BD} = 2\overrightarrow{AB}$. Also prove that AC and BD bisect each other.
   [C.H. '83; C.U. '82, '88; B.U. '75, '83; N.B. '78]
8. If D, E, F are the mid-points of the sides BC, CA, AB of ΔABC, prove that $\overrightarrow{AD} + \overrightarrow{BE} + \overrightarrow{CF} = \bar{0}$. **[B.U. '91, '97; C.U. '84, '89, 2001; N.B. '70, '79]**
9. $ABCD$ is a quadrilateral and P is a variable point such that $\overrightarrow{PA} + \overrightarrow{PB} + \overrightarrow{PC} + \overrightarrow{PD}$ is constant. Prove that the locus of P is a circle.
10. Show that the lines joining the mid-points of the opposite sides of a tetrahedron are concurrent.
11. Prove by vector method that the line segment joining the mid-points of the oblique sides of a trapezium is parallel to the parallel sides and its length is half the sum of the lengths of the parallel sides. **[C.U. '88]**
12. Show that (a) the points $(1, 2, 1), (2, 3, 4)$ and $(4, 5, 10)$ are collinear; (b) the points $(\bar{a} - 2\bar{b} + 3\bar{c}), (-2\bar{a} + 3\bar{b} + 2\bar{c}), (-8\bar{a} + 13\bar{b})$ are linearly dependent.

13. Show that the points (a) $6\bar{a} - 4\bar{b} + 10\bar{c}$, $-5\bar{a} + 3\bar{b} - 10\bar{c}$, $4\bar{a} - 6\bar{b} - 10\bar{c}, 2\bar{b} + 10\bar{c}$ and (b) $(-1, 4, -3)$, $(3, 2, -5)$, $(-3, 8, -5)$, $(-3, 2, 1)$ are coplanar.

14. If $\bar{r}_1 = \bar{a} + \bar{b} + \bar{c}$, $\bar{r}_2 = 7\bar{a} + 6\bar{c}$, $\bar{r}_3 = 2\bar{a} - \bar{b} + \bar{c}$, $\bar{r}_4 = \bar{a} - \bar{b} - \bar{c}$, where $\bar{a}, \bar{b}, \bar{c}$ are non-zero, non-coplanar vectors, show that $2\bar{r}_1 = \bar{r}_2 - 3\bar{r}_3 + \bar{r}_4$.

15. Show that the following vectors are coplanar.

(a) $4\bar{a} + 5\bar{b} + \bar{c}$, $5\bar{a} + 9\bar{b} + 4\bar{c}$, $-\bar{b} - \bar{c}$.

(b) $\bar{a} - 3\bar{b} + 5\bar{c}$, $\bar{a} - 2\bar{b} + 3\bar{c}$, $-2\bar{a} + 3\bar{b} - 4\bar{c}$.

16. Show that the vector $\bar{i} - 3\bar{j} + 2\bar{k}, 2\bar{i} - 4\bar{j} - \bar{k}, 3\bar{i} + 2\bar{j} - \bar{k}$ are linearly independent.

17. $\bar{\alpha}$, $\bar{\beta}$, $\bar{\gamma}$, $\bar{\delta}$ are vectors from the origin to the points A, B, C, D respectively such that $\bar{\beta} - \bar{\alpha} = \bar{\gamma} - \bar{\delta}$, prove that $ABCD$ is a parallelogram. [C.U. '81]

18. If $\bar{i}$, $\bar{j}$ denote unit vectors having the directions of positive x, y axes respectively of a two-dimensional rectangular co-ordinate system and $\bar{\alpha} = 2\bar{i} + 3\bar{j}$, $\bar{\beta} = 4\bar{i} + \bar{j}, \bar{\gamma} = 5\bar{i} + d\bar{j}$ are three vectors having their initial points at the origin, find the value of d so that their extremities may be collinear. [C.H. '77]

Answers

1. $\bar{a} + \bar{b} = (0, 4, 9)$, $\bar{a} - 2\bar{b} = (6, 1, -6)$, $|\bar{a} - 2\bar{b}| = \sqrt{73}$, d.cs. $6/\sqrt{73}$, $1/\sqrt{73}$, $-6/\sqrt{73}$. **2.** $\lambda = 3$, $\mu = -6$. **3.** $\bar{i} + 4\bar{j} + 2\bar{k}$, $\bar{i} - 8\bar{k}$, $-2\bar{i} - 4\bar{j} + 6\bar{k}$. **18.** $d = 0$.

Chapter 2

Products of Vectors

2.1 Introduction

Multiplication between vectors and numbers obeys the rules of simple algebra. Now we define two operations between vectors, which are known as products. These have some properties in common with the products of numbers and some properties which are in striking disagreement with the usual products of numbers. The result of one of these products is scalar and of the other is vector so one will be called scalar product and the other will be vector product.

2.2 Scalar or Dot Product

Let $\bar{a}$ and $\bar{b}$ be two vectors and θ the smallest non-negative angle between them. The scalar product of these vectors is written as $\bar{a}\cdot\bar{b}$ and is defined as $|\bar{a}||\bar{b}|\cos\theta$.

$$\therefore \quad \bar{a}\cdot\bar{b} = |\bar{a}||\bar{b}|\cos\theta.$$

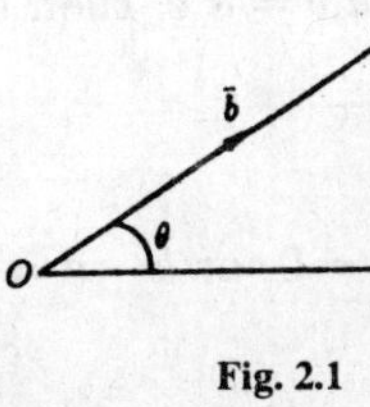

Fig. 2.1

It is a scalar quantity.

It is also denoted by $(\bar{a}, \bar{b})$.

If $\bar{a} = \overrightarrow{OA}, \bar{b} = \overrightarrow{OB}$ and BN is perpendicular to OA, then $\bar{a}\cdot\bar{b} = |\bar{a}||\bar{b}|\cos\theta = OA\cdot OB\cdot\cos\theta = OA\cdot ON$. Thus geometrically, the scalar or dot product of two vectors is equal to the product of the length of one and the projection of the length of the other to the former.

Properties.

(i) Scalar product is commutative. $(\bar{a}\cdot\bar{b} = \bar{b}\cdot\bar{a})$

Since

$$\bar{a}\cdot\bar{b} = |\bar{a}||\bar{b}|\cos\theta \quad \text{and} \quad \bar{b}\cdot\bar{a} = |\bar{b}||\bar{a}|\cos\theta = |\bar{a}||\bar{b}|\cos\theta, \bar{a}\cdot\bar{b} = \bar{b}\cdot\bar{a}.$$

(ii) $\bar{a}\cdot\bar{b}$ is positive, zero or negative accoridng as θ is acute, right or obtuse.

If $\bar{a}$ and $\bar{b}$ are proper vectors, i.e., non-zero vectors, then they will be at right angle when $\bar{a}\cdot\bar{b} = 0$.

(iii) $\bar{a}\cdot\bar{a} = |\bar{a}||\bar{a}|\cos\theta = |\bar{a}|^2 = \bar{a}^2$.

If $\bar{i}, \bar{j}, \bar{k}$ are the unit vectors along three rectangular axes, then

$$\bar{i}\cdot\bar{i} = \bar{i}^2 = 1 = \bar{j}^2 = \bar{k}^2.$$

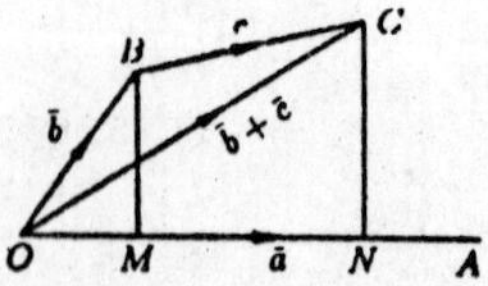

Fig. 2.2

Again

$$\bar{i}\cdot\bar{j} = 0 = \bar{j}\cdot\bar{k} = \bar{k}\cdot\bar{i}.$$

(iv) **Distributive law:** $\bar{a}\cdot(\bar{b}+\bar{c}) = \bar{a}\cdot\bar{b} + \bar{a}\cdot\bar{c}$.

Let $\overrightarrow{OA} = \bar{a}, \overrightarrow{OB} = \bar{b}$ and $\overrightarrow{BC} = \bar{c}$ then $\overrightarrow{OC} = \overrightarrow{OB} + \overrightarrow{BC} = \bar{b} + \bar{c}$.

Now $\bar{a}\cdot(\bar{b}+\bar{c}) = OA$. proj. of OC on $OA = OA.ON = OA.(OM+MN) = OA.OM + OA.MN = OA$. proj. of OB on $OA + OA$. proj. of BC on $OA = \bar{a}.\bar{b} + \bar{a}.\bar{c}$.

(v) If $\bar{a} = a_1\bar{i} + a_2\bar{j} + a_3\bar{k}$ and $\bar{b} = b_1\bar{i} + b_2\bar{j} + b_3\bar{k}$, then

$$\bar{a}\cdot\bar{b} = (a_1\bar{i} + a_2\bar{j} + a_3\bar{k}).(b_1\bar{i} + b_2\bar{j} + b_3\bar{k}) = a_1b_1 + a_2b_2 + a_3b_3.$$

Again $\bar{a}.\bar{b} = |\bar{a}||\bar{b}|\cos\theta$, where θ is the angle between $\bar{a}$ and $\bar{b}$.

$$\therefore \quad \cos\theta = \frac{\bar{a}\cdot\bar{b}}{|a||b|} = \frac{a_1b_1 + a_2b_2 + a_3b_3}{\sqrt{a_1^2+a_2^2+a_3^2}\sqrt{b_1^2+b_2^2+b_3^2}}.$$

(vi) If $\bar{a}.\bar{b} = \bar{a}.\bar{c}$, then $\bar{b}$ may not be equal to $\bar{c}$.

$$\bar{a}.b = \bar{a}.\bar{c}, \quad \text{or} \quad \bar{a}.(\bar{b}-\bar{c}) = 0.$$

Here $\bar{a} = 0$, or $\bar{b} - \bar{c} = 0$, i.e., $b = \bar{c}$, or $\bar{a}$ is perpendicular to $\bar{b} - \bar{c}$.

2.3 Vector or Cross-Product

Let $\bar{a}$ and $\bar{b}$ be two vectors and θ the smallest non-negative acute angle between them. The vector product between them is written as $\bar{a}\times\bar{b}$ and is defined as $|\bar{a}||\bar{b}|\sin\theta\bar{n}$, where $\bar{n}$ is a unit vector perpendicular to $\bar{a}$ and $\bar{b}$ and its direction coincides with the direction of the motion of a right-handed screw rotating from $\bar{a}$ to $\bar{b}$.

In the fig. 2.3, $\overrightarrow{OA} = \bar{a}, \overrightarrow{OB} = \bar{b}, \angle AOB = \theta$ and $\bar{n}$ is a unit vector perpendicular to the plane of OA and OB.

$$\therefore \quad \bar{a}\times\bar{b} = OA\cdot OB\sin\theta\bar{n}.$$

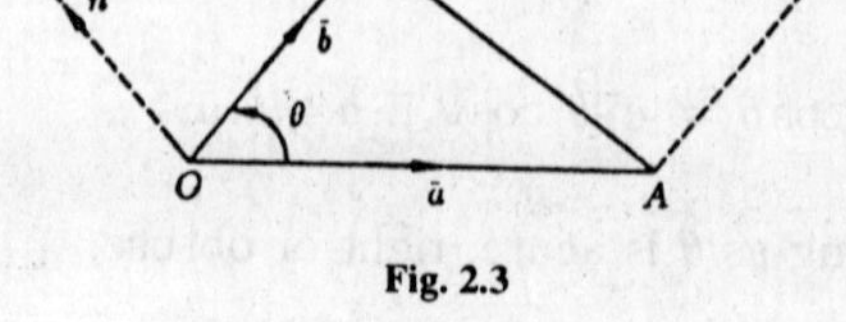

Fig. 2.3

It is a vector quantity.

It is also denoted by $[\bar{a}, \bar{b}]$.

Properties.

(i) Vector product is non-commutative, i.e., $a \times \bar{b} \neq \bar{b} \times \bar{a}$.

The directors of $\bar{a} \times \bar{b}$ and $\bar{b} \times \bar{a}$ are opposite to each other.

$$\therefore \quad \bar{a} \times \bar{b} = -\bar{b} \times \bar{a}.$$

(ii) $|\bar{a} \times \bar{b}| = |\bar{a}||\bar{b}| \sin\theta = OA.OB \sin\theta = 2\Delta OAB =$ area of the parallelogram whose adjacent sides are OA and OB.

Thus $\bar{a} \times \bar{b}$ is called a vector area of a parallelogram whose adjacent sides represent $\bar{a}$ and $\bar{b}$.

(iii) If $\bar{a}$ and $\bar{b}$ are both proper vectors and $\bar{a} \times \bar{b} = 0$, then $\bar{a}$ and $\bar{b}$ are parallel.

(iv) $\bar{a} \times \bar{a} = 0$, since $\theta = 0$.

If $\bar{i}, \bar{j}, \bar{k}$ are the unit vectors along three rectangular axes, then $\bar{i} \times \bar{i} = 0 = \bar{j} \times \bar{j} = \bar{k} \times \bar{k}$.

Again $\bar{i} \times \bar{j} = \bar{k}, \bar{j} \times \bar{k} = \bar{i}, \bar{k} \times \bar{i} = \bar{j}$ and $\bar{j} \times \bar{i} = -\bar{k}, \bar{k} \times \bar{j} = -\bar{i}, \bar{i} \times \bar{k} = -\bar{j}$.

(v) $(\bar{a} \times \bar{b})^2 = (\bar{a} \times \bar{b}) \cdot (\bar{a} \times \bar{b}) = |\bar{a} \times \bar{b}|^2 = |\bar{a}|^2|\bar{b}|^2 \sin^2\theta$

$$= |\bar{a}|^2|\bar{b}|^2(1 - \cos^2\theta) = |\bar{a}|^2|\bar{b}|^2 - |\bar{a}|^2|\bar{b}|^2 \cos^2\theta = \bar{a}^2\bar{b}^2 - (\bar{a} \cdot \bar{b})^2.$$

(vi) If $\bar{a} \times \bar{b} = \bar{c}$, then $\bar{c}$ is perpendicular to $\bar{a}$ and $\bar{b}$.

Consequently $\bar{a} \cdot \bar{c} = 0$ and $\bar{b} \cdot \bar{c} = 0$.

If $\bar{a} = a_1\bar{i} + a_2\bar{j} + a_3\bar{k}, \bar{b} = b_1\bar{i} + b_2\bar{j} + b_3\bar{k}$ then $\bar{c} = \begin{vmatrix} \bar{i} & \bar{j} & \bar{k} \\ a_1 & a_2 & a_3 \\ b_1 & b_2 & b_3 \end{vmatrix}$.

Let $\bar{c} = c_1\bar{i} + c_2\bar{j} + c_3\bar{k}$, then $a_1c_1 + a_2c_2 + a_3c_3 = 0, b_1c_1 + b_2c_2 + b_3c_3 = 0$.

By cross-multiplication,

$$\frac{c_1}{a_2b_3 - a_3b_2} = \frac{c_2}{a_3b_1 - a_1b_3} = \frac{c_3}{a_1b_2 - a_2b_1} \quad (= K \text{ say})$$

$$\therefore \quad c_1 = K(a_2b_3 - a_3b_2), c_2 = K(a_3b_1 - a_1b_3),$$

$$c_3 = K(a_1b_2 - a_2b_1).$$

Now

$$\begin{aligned}
\bar{c}^2 &= c_1^2 + c_2^2 + c_3^2 \\
&= K^2(a_2b_3 - a_3b_2)^2 + (a_3b_1 - a_1b_3)^2 + (a_1b_2 - a_2b_1)^2] \\
&= K^2[a_1^2(b_2^2 + b_3^2) + a_2^2(b_3^2 + b_1^2) + a_3^2(b_1^2 + b_2^2) \\
&\qquad -2(a_2b_2a_3b_3 + a_3b_3a_1b_1 + a_1b_1a_2b_2)] \\
&= K^2[(a_1^2 + a_2^2 + a_3^2)(b_1^2 + b_2^2 + b_3^2) - (a_1b_1 + a_2b_2 + a_3b_3)^2] \\
&= K^2[\bar{a}^2\bar{b}^2 - (\bar{a} \cdot \bar{b})^2] = K^2[\bar{a}^2\bar{b}^2 - \bar{a}^2\bar{b}^2 \cos^2\theta] \\
&= K^2\bar{a}^2\bar{b}^2(1 - \cos^2\theta) = K^2\bar{a}^2 \cdot \bar{b}^2 \sin^2\theta.
\end{aligned}$$

But $\bar{c}^2 = \bar{a}^2\bar{b}^2 \sin^2\theta$, $\quad\therefore\quad K^2 = 1$, or $K = \pm 1$.

If we take $\bar{a} = \bar{i}, \bar{b} = \bar{j}$, then $\bar{c} = \bar{i} \times \bar{j} = \bar{k}$.

In this case $a_1 = b_2 = c_3 = 1, a_2 = a_3 = b_1 = b_3 = c_1 = c_2 = 0$.

We have $c_3 = K(a_1b_2 - a_2b_1)$, or $1 = K.1$, or $K = 1$.

Thus taking

$$\begin{aligned} K &= 1, \bar{a} \times \bar{b} = c_1\bar{i} + c_2\bar{j} + c_3\bar{k} \\ &= (a_2b_3 - a_3b_2)\bar{i} + (a_3b_1 - a_1b_3)\bar{j} + (a_1b_2 - a_2b_1)\bar{k} \\ &= \begin{vmatrix} \bar{i} & \bar{j} & \bar{k} \\ a_1 & a_2 & a_3 \\ b_1 & b_2 & b_3 \end{vmatrix}. \end{aligned}$$

(vii) **Distributive law**: $\bar{a} \times (\bar{b} + \bar{c}) = \bar{a} \times \bar{b} + \bar{a} \times \bar{c}$.

Let $\bar{a} = a_1\bar{i} + a_2\bar{j} + a_3\bar{k}, \bar{b} = b_1\bar{i} + b_2\bar{j} + b_3\bar{k}, \bar{c} = c_1\bar{i} + c_2\bar{j} + c_2\bar{k}$.

Then $\bar{b} + \bar{c} = (b_1 + c_1)\bar{i} + (b_2 + c_2)\bar{j} + (b_3 + c_3)\bar{k}$.

By (vi) $\bar{a} \times (\bar{b} + \bar{c})$

$$\begin{aligned} &= \{a_2(b_3 + c_3) - a_3(b_2 + c_2)\}\bar{i} + \{a_3(b_1 + c_1) - a_1(b_3 + c_3)\}\bar{j} \\ &\qquad + \{a_1(b_2 + c_2) - a_2(b_1 + c_1)\}\bar{k} \\ &= \{(a_2b_3 - a_3b_2) + (a_2c_3 - a_3c_2)\}\bar{i} + \{(a_3b_1 - a_1b_3) + (a_3c_1 - a_1c_3)\}\bar{j} \\ &\qquad + \{(a_1b_2 - a_2b_1) + (a_1c_2 - a_2c_1)\}\bar{k} \\ &= \{(a_2b_3 - a_3b_2)\bar{i} + (a_3b_1 - a_1b_3)\bar{j} + (a_1b_2 - a_2b_1)\bar{k}\} + \{a_2c_3 - a_3c_2)\bar{i} \\ &\qquad + (a_3c_1 - a_1c_3)j + (a_1c_2 - a_2c_1)\bar{k}\} \\ &= \bar{a} \times \bar{b} + \bar{a} \times \bar{c}. \end{aligned}$$

Similarly $(\bar{b} + \bar{c}) \times \bar{a} = \bar{b} \times \bar{a} + \bar{c} \times \bar{a}$.

(viii) If $\bar{a} \times \bar{b} = \bar{a} \times \bar{c}, \bar{b}$ may not be equal to $\bar{c}$.

$$\bar{a} \times \bar{b} = \bar{a} \times \bar{c}, \quad \text{or} \quad \bar{a} \times (\bar{b} - \bar{c}) = 0.$$

It shows that either $\bar{a} = 0$, or $\bar{b} - \bar{c} = 0$, i.e., $\bar{b} = \bar{c}$, or $\bar{a}$ is parallel to $\bar{b} - \bar{c}$.

(ix) If $\bar{a} = (a_1, a_2, a_3), \bar{b} = (b_1, b_2, b_3)$ and θ be the angle between them, then

$$|\bar{a} \times \bar{b}| = |\bar{a}||\bar{b}| \sin\theta,$$

$$\text{or} \quad \sin\theta = \frac{\sqrt{\{(a_2b_3 - a_3b_2)^2 + (a_3b_1 - a_1b_3)^2 + (a_1b_2 - a_2b_1)^2\}}}{\sqrt{(a_1^2 + a_2^2 + a_3^2)} \cdot \sqrt{(b_1^2 + b_2^2 + b_3^2)}}.$$

2.4 Tripple Product

(a) Scalar triple product

Let $\bar{a}, \bar{b}$ and $\bar{c}$ be any three vectors. The expression $\bar{a} \cdot (\bar{b} \times \bar{c})$ is known as a scalor triple product of $\bar{a}, \bar{b}, \bar{c}$. It is a scalar and is numerically equal to the volume V of the parallelopiped whose coterminus edges represent $\bar{a}, \bar{b}$ and $\bar{c}$.

Let $\bar{a}, \bar{b}$ and $\bar{c}$ be the concurrent edges OA, OB, OC of the parallelopiped as shown in the fig. 2.4.

$\bar{b} \times \bar{c} = OB \cdot OC \cdot \sin BOC \;\bar{n}$, where $\bar{n}$ is the unit vector perpendicular to the plane $OBDC$ in the direction of OX.

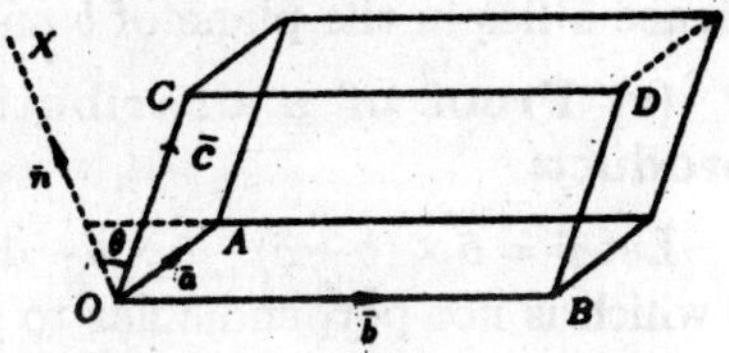

Fig. 2.4

Let θ be the angle between OX and OA.

Now

$$\begin{aligned}
\bar{a} \cdot (\bar{b} \times \bar{c}) &= (\bar{a} \cdot \bar{n}) \cdot (\text{area of parallelogram } OBDC) \\
&= (OA \cdot \cos\theta) \cdot (\text{area of } OBDC) \\
OA \cos\theta &= \text{projection of } OA \text{ on } OX \\
&= \text{distance of } A \text{ from the plane } OBDC \\
\therefore \quad \bar{a} \cdot (\bar{b} \times \bar{c}) &= (\text{area of } OBDC) \cdot (\text{distance of } A \text{ from the plane of } OBDC) \\
&= (\text{area of the base}) \cdot (\text{altitude of the parallelopiped}) \\
&= \text{volume of the parallelopiped.}
\end{aligned}$$

Properties.

(i) $\bar{a} \cdot (\bar{b} \times \bar{c})$ is denoted by $[\bar{a}\bar{b}\bar{c}]$ · (box notation)

$$[\bar{a}\bar{b}\bar{c}] = [\bar{b}\bar{c}\bar{a}] = [\bar{c}\bar{a}\bar{b}] = -[\bar{b}\bar{a}\bar{c}] = -[\bar{a}\bar{c}\bar{b}] = -[\bar{c}\bar{b}\bar{a}].$$

(ii) If $\bar{i}, \bar{j}, \bar{k}$ are the three unit vectors along the rectangular axes, then

$$[\bar{i}\bar{j}\bar{k}] = [\bar{j}\bar{k}\bar{i}] = [\bar{k}\bar{i}\bar{j}] = 1, \quad [\bar{i}\bar{k}\bar{j}] = [\bar{k}\bar{j}\bar{i}] = [\bar{j}\bar{i}\bar{k}] = -1.$$

(iii) If $\bar{a} = (a_1, a_2, a_3)$, $\bar{b} = (b_1, b_2, b_3)$ and $\bar{c} = (c_1, c_2, c_3)$, then

$$\bar{b} \times \bar{c} = (b_2c_3 - b_3c_2)\bar{i} + (b_3c_1 - b_1c_3)\bar{j} + (b_1c_2 - b_2c_1)\bar{k}.$$

Hence

$$\begin{aligned}
\bar{a} \cdot (\bar{b} \times \bar{c}) &= a_1(b_2c_3 - b_3c_2) + a_2(b_3c_1 - b_1c_3) + a_3(b_1c_2 - b_2c_1) \\
&= \begin{vmatrix} a_1 & a_2 & a_3 \\ b_1 & b_2 & b_3 \\ c_1 & c_2 & c_3 \end{vmatrix}.
\end{aligned}$$

(iv) $\bar{a} \cdot (\bar{b} \times \bar{c}) = 0$ is the necessary and sufficient condition for the coplanarity of the three proper vectors $\bar{a}, \bar{b}, \bar{c}$.

Proof. *The condition is necessary.*

Let the vectors $\bar{a}, \bar{b}, \bar{c}$ lie in the same plane.

Since $\bar{b} \times \bar{c}$ is a vector perpendicular to the plane of $\bar{b}$ and $\bar{c}$, it is perpendicular to $\bar{a}$. Hence $\bar{a}.(\bar{b} \times \bar{c}) = 0$.

The condition is sufficient.

Let $\bar{a}.(\bar{b} \times \bar{c}) = 0$. $\bar{a}$ is perpendicular to $\bar{b} \times \bar{c}$. $\bar{b} \times \bar{c}$ is a vector perpendicular to the plane of $\bar{b}$ and $\bar{c}$. Therefore $\bar{a}$ is perpendicular to the normal to the plane of $\bar{b}$ and $\bar{c}$. Hence $\bar{a}$ lies in the plane of $\bar{b}$ and $\bar{c}$, i.e., $\bar{a}, \bar{b}, \bar{c}$ lie in the same plane.

(v) **Proof of a distribution law of the cross-product by scalar triple product:**

Let $\bar{p} = \bar{a} \times (\bar{b} + \bar{c}) - \bar{a} \times \bar{b} - \bar{a} \times \bar{c}$. Considering the dot product with a proper vector $\bar{q}$ which is not perpendicular to $\bar{p}$,

$$\begin{aligned} \bar{q} \cdot \bar{p} &= \bar{q} \cdot \{\bar{a} \times (\bar{b} + \bar{c})\} - \bar{q} \cdot (\bar{a} \times \bar{b}) - \bar{q} \cdot (\bar{a} \times \bar{c}) \\ &= (\bar{q} \times \bar{a}) \cdot (\bar{b} + \bar{c}) - \bar{q} \cdot (\bar{a} \times \bar{b}) - \bar{q} \cdot (\bar{a} \times \bar{c}) \\ &= (\bar{q} \times \bar{a}) \cdot \bar{b} + (\bar{q} \times \bar{a}) \cdot \bar{c} - (\bar{q} \times \bar{a}) \cdot \bar{b} - (\bar{q} \times \bar{a}) \cdot \bar{c} = 0. \end{aligned}$$

$$\therefore \bar{p} = 0. \quad \therefore \bar{a} \times (\bar{b} + \bar{c}) = \bar{a} \times \bar{b} + \bar{a} \times \bar{c}.$$

(b) Vector triple product.

Let $\bar{a}, \bar{b}, \bar{c}$ be any three vectors. The expression $\bar{a} \times (\bar{b} \times \bar{c})$ is a vector and is called a vector triple product of $\bar{a}, \bar{b}$ and $\bar{c}$. This vector lies in the plane of $\bar{b}$ and $\bar{c}$.

Now we shall prove that $\bar{a} \times (\bar{b} \times \bar{c}) = (\bar{a} \cdot \bar{c})\bar{b} - (\bar{a} \cdot \bar{b})\bar{c}$.

Let $\bar{i}, \bar{j}, \bar{k}$ be the unit vectors along the three rectangular axes. Without any loss of generality we may choose $\bar{a}$ along $\bar{i}$, $\bar{b}$ in the plane of $\bar{i}$ and $\bar{j}$. Thus we can express $\bar{a}, \bar{b}$ and $\bar{c}$ in the following form:

$$\bar{a} = a_1\bar{i}, \bar{b} = b_1\bar{i} + b_2\bar{j}, \bar{c} = c_1\bar{i} + c_2\bar{j} + c_3\bar{k}.$$

Now

$$\bar{b} \times \bar{c} = \begin{vmatrix} \bar{i} & \bar{j} & \bar{k} \\ b_1 & b_2 & 0 \\ c_1 & c_2 & c_3 \end{vmatrix} = b_2c_3\bar{i} - b_1c_3\bar{j} + (b_1c_2 - b_2c_1)\bar{k},$$

$$\bar{a} \times (\bar{b} \times \bar{c}) = \begin{vmatrix} \bar{i} & \bar{j} & \bar{k} \\ a_1 & 0 & 0 \\ b_2c_3 & -b_1c_3 & b_1c_2 - b_2c_1 \end{vmatrix}$$

$$= -a_1(b_1c_2 - b_2c_1)\bar{j} - a_1b_1c_3\bar{k}$$

$$\bar{a}.\bar{c} = a_1\bar{i} \cdot (c_1\bar{i} + c_2\bar{j} + c_3\bar{k}) = a_1c_1,$$

$$(\bar{a}\cdot\bar{c})\bar{b} = a_1b_1c_1\bar{i} + a_1b_2c_1\bar{j}$$

$$\bar{a}\cdot\bar{b} = a_1\bar{i}\cdot(b_1\bar{i}+b_2\bar{j}) = a_1b_1.$$

$$(\bar{a}\cdot\bar{b})\bar{c} = a_1b_1c_1\bar{i} + a_1b_1c_2\bar{j} + a_1b_1c_3\bar{k}.$$

$$\therefore \quad (\bar{a}\cdot\bar{c})\bar{b} - (\bar{a}\cdot\bar{b})\bar{c} = -a_1(b_1c_2 - b_2c_1)\bar{j} - a_1b_1c_3\bar{k} = \bar{a}\times(\bar{b}\times\bar{c}).$$

Note 1. $\bar{a}\times(\bar{b}\times\bar{c}) = -(\bar{b}\times\bar{c})\times\bar{a}$.

$$\therefore \quad (\bar{b}\times\bar{c})\times a = (\bar{a}\cdot\bar{b})\bar{c} - (\bar{a}\cdot\bar{c})\bar{b}.$$

Note 2. If $\bar{a}\times(\bar{b}\times\bar{c}) = (\bar{a}\times\bar{b})\times\bar{c}$ then $(\bar{a}\cdot\bar{c})\bar{b} - (\bar{a}\cdot\bar{b})\bar{c} = (\bar{a}\cdot\bar{c})\bar{b} - (\bar{b}\cdot\bar{c})\bar{a}$.

$$\therefore \quad (\bar{b}\cdot\bar{c})\bar{a} = (\bar{a}\cdot\bar{b})\bar{c}$$

or $(\bar{c}\times\bar{a})\times\bar{b} = 0$. For this $\bar{c}$ is parallel to $\bar{a}$ or $\bar{b}$ is perpendicular to $\bar{a}$ and $\bar{c}$.

2.5 Products of four vectors

(a) Scalar product of four vectors: $(\bar{a}\times\bar{b})\cdot(\bar{c}\times\bar{d})$.

We shall prove that $(\bar{a}\times\bar{b})\cdot(\bar{c}\times\bar{d}) = \begin{vmatrix} \bar{a}.\bar{c} & \bar{a}.\bar{d} \\ \bar{b}.\bar{c} & \bar{b}.\bar{d} \end{vmatrix}$.

Let $\bar{a}\times\bar{b} = \bar{p}$.

$$\begin{aligned}(\bar{a}\times\bar{b})\cdot(\bar{c}\times\bar{d}) &= \bar{p}\cdot(\bar{c}\times\bar{d}) = (\bar{p}\times\bar{c})\cdot\bar{d} = \{(\bar{a}\times\bar{b})\times\bar{c}\}\cdot\bar{d} \\ &= \{(\bar{a}\cdot\bar{c})\bar{b} - (\bar{b}\cdot\bar{c})\bar{a}\}\cdot\bar{d} = (\bar{a}\cdot\bar{c})(\bar{b}\cdot\bar{d}) - (\bar{a}\cdot\bar{d})(\bar{b}\cdot\bar{c}) \\ &= \begin{vmatrix} \bar{a}\cdot\bar{c} & \bar{a}\cdot\bar{b} \\ \bar{b}\cdot\bar{c} & \bar{b}\cdot\bar{d} \end{vmatrix}.\end{aligned}$$

(b) Vector product of four vectors: $(\bar{a}\times\bar{b})\times(\bar{c}\times\bar{d})$

We shall prove that $(\bar{a}\times\bar{b})\times(\bar{c}\times\bar{d}) = [\bar{a}\bar{b}\bar{d}]\bar{c} - [\bar{a}\bar{b}\bar{c}]\bar{d} = [\bar{a}\bar{c}\bar{d}]\bar{b} - [\bar{b}\bar{c}\bar{d}]\bar{a}$.

Let $\bar{a}\times\bar{b} = \bar{p}$.

Now

$$\begin{aligned}(\bar{a}\times\bar{b})\times(\bar{c}\times\bar{d}) &= \bar{p}\times(\bar{c}\times\bar{d}) = (\bar{p}\cdot\bar{d})\bar{c} - (\bar{p}\cdot\bar{c})\bar{d} \\ &= \{(\bar{a}\times\bar{b})\cdot\bar{d}\}\bar{c} - \{(\bar{a}\times\bar{b})\cdot\bar{c}\}\bar{d} \\ &= [\bar{a}\bar{b}\bar{d}]\bar{c} - [\bar{a}\bar{b}\bar{c}]\bar{d}.\end{aligned}$$

Let $\bar{c}\times\bar{d} = \bar{q}$.

Then

$$(\bar{a}\times\bar{b})\times(\bar{c}\times\bar{d}) = (\bar{a}\times b)\times\bar{q} = (\bar{q}\cdot\bar{a})\bar{b}-(\bar{q}\cdot\bar{b})\bar{a}$$
$$= \{(\bar{c}\times\bar{d})\cdot\bar{a}\}\bar{b}-\{(\bar{c}\times\bar{d})\cdot\bar{b}\}\bar{a}$$
$$= [\bar{c}\bar{d}\bar{a}]\bar{b}-[\bar{c}\bar{d}\bar{b}]\bar{a} = [\bar{a}\bar{c}\bar{d}]\bar{b}-[\bar{b}\bar{c}\bar{d}]\bar{a}.$$

Hence the result follows.

Corollary 1. $[\bar{a}\bar{b}\bar{d}]\bar{c}-[\bar{a}\bar{b}\bar{c}]\bar{d} = [\bar{a}\bar{c}\bar{d}]\bar{b}-[\bar{b}\bar{c}\bar{d}]\bar{a}$,

$$\text{or}\quad [\bar{a}\bar{b}\bar{c}]\bar{d} = [\bar{b}\bar{c}\bar{d}]\bar{a}-[\bar{a}\bar{c}\bar{d}]\bar{b}+[\bar{a}\bar{b}\bar{d}]\bar{c} = [\bar{d}\bar{b}\bar{c}]\bar{a}+[\bar{d}\bar{c}\bar{a}]\bar{b}+[\bar{d}\bar{a}\bar{c}]\bar{c},$$

$$\text{or}\quad \bar{d} = \frac{[\bar{d}\bar{b}\bar{c}]}{[\bar{a}\bar{b}\bar{c}]}\bar{a}+\frac{[\bar{d}\bar{c}\bar{a}]}{[\bar{a}\bar{b}\bar{c}]}\bar{b}+\frac{[\bar{d}\bar{a}\bar{b}]}{[\bar{a}\bar{b}\bar{c}]}\bar{c}.$$

It shows that $\bar{d}$ can be expressed as a linear combination of three non-coplanar vectors $\bar{a},\bar{b},\bar{c}$.

Thus any vector $\bar{r}$ can be expressed as a linear combination of three non-coplanar vectors $\bar{a},\bar{b},\bar{c}$ in the form

$$\bar{r} = \frac{[\bar{r}\bar{b}\bar{c}]}{[\bar{a}\bar{b}\bar{c}]}\bar{a}+\frac{[\bar{r}\bar{c}\bar{a}]}{[\bar{a}\bar{b}\bar{c}]}\bar{b}+\frac{[\bar{r}\bar{a}\bar{b}]}{[\bar{a}\bar{b}\bar{c}]}\bar{c}.$$

Corollary 2. If $\bar{a},\bar{b},\bar{c},\bar{d}$ are coplanar then $[\bar{a}\bar{b}\bar{d}] = 0 = [\bar{a}\bar{b}\bar{c}]$.

$$\therefore\quad (\bar{a}\times\bar{b})\times(\bar{c}\times\bar{d}) = 0.$$

Worked Out Examples

1. If $\bar{\alpha}$, $\bar{\beta}$, $\bar{\gamma}$ be three vectors defined by $\bar{\alpha} = \lambda(2\bar{i}+3\bar{j}+6\bar{k})$, $\bar{\beta} = \lambda(3\bar{i}-6\bar{j}+2\bar{k})$ and $\bar{\gamma} = \lambda(6\bar{i}+2\bar{j}-3\bar{k})$, where λ is a scalar and $\bar{i},\bar{j},\bar{k}$ are the three orthogonal unit vectors, determine λ for which $\bar{\alpha},\bar{\beta}$ and $\bar{\gamma}$ may be each of length unity. Also prove that $\bar{\alpha},\bar{\beta}$ and $\bar{\gamma}$ are mutually perpendicular. **[B.U. '66]**

Here

$$|\bar{\alpha}| = \lambda\sqrt{2^2+3^3+6^2} = 7\lambda,$$
$$|\bar{\beta}| = \lambda\sqrt{3^2+6^2+2^2} = 7\lambda,$$
$$|\bar{\gamma}| = \lambda\sqrt{6^2+2^2+3^2} = 7\lambda.$$

$|\bar{\alpha}|,|\bar{\beta}|,|\bar{\gamma}|$ will be unity, if $\lambda = \frac{1}{7}$.

$$\bar{\alpha}\cdot\bar{\beta} = \lambda^2(2\bar{i}+3\bar{j}+6\bar{k})\cdot(3\bar{i}-6\bar{j}+2\bar{k}) = \lambda^2(6-18+12) = 0.$$
$$\bar{\beta}\cdot\bar{\gamma} = \lambda^2(3\bar{i}-6\bar{j}+2\bar{k})\cdot(6\bar{i}+2\bar{j}-3\bar{k}) = \lambda^2(18-12-6) = 0.$$
$$\bar{\gamma}\cdot\bar{\alpha} = \lambda^2(6\bar{i}+2\bar{j}-3\bar{k})\cdot(2i+3\bar{j}+6\bar{k}) = \lambda^2(12+6-18) = 0.$$

$\therefore\quad \bar{\alpha},\bar{\beta},\bar{\gamma}$ are mutually orthogonal.

2. The position vectors of two points P and Q are respectively $3\bar{i}+7\bar{j}-4\bar{k}$ and $6\bar{i}-2\bar{j}+12\bar{k}$, where $\bar{i},\bar{j},\bar{k}$ are unit vectors parallel to the axes of co-ordinates.

Calculate the angle between OP and OQ, where O is the origin. [C.U. '62]

Here $\overrightarrow{OP}=3\bar{i}+7\bar{j}-4\bar{k}, \overrightarrow{OQ}=6i-2\bar{j}+12\bar{k}$.

If θ be the angle between them

$$\overrightarrow{OP}\cdot\overrightarrow{OQ}=|\overrightarrow{OP}||\overrightarrow{OQ}|\cos\theta,$$

$$\text{or}\quad (3\bar{i}+7\bar{j}-4\bar{k})\cdot(6\bar{i}-2\bar{j}+12\bar{k})=|3\bar{i}+7\bar{j}-4\bar{k}||6\bar{i}-2\bar{j}+12\bar{k}|\cos\theta,$$

$$\text{or}\quad 18-14-48=\sqrt{(3^2+7^2+4^4)}\cdot\sqrt{(6^2+2^2+12^2)}\cos\theta,$$

$$\text{or}\quad -44=\sqrt{74}\cdot\sqrt{184}\cos\theta,$$

$$\text{or}\quad \cos\theta=\frac{-44}{\sqrt{74}\cdot\sqrt{184}}=\frac{-11}{\sqrt{37}\cdot\sqrt{23}}=\frac{-11}{\sqrt{851}}.$$

$$\therefore\quad \theta=\cos^{-1}\left(\frac{-11}{\sqrt{851}}\right).$$

3. Given $\bar{a}=\bar{i}-2\bar{j}, \bar{b}=\bar{j}+\bar{k}$, find the component of $\bar{a}$ along $\bar{b}$. Also find $\bar{a}\times\bar{b}$.

The component of $\bar{a}$ along $\bar{b}=|\bar{a}|\cos\theta\frac{\bar{b}}{|\bar{b}|}$, where θ is the angle between $\bar{a}$ and $\bar{b}$

$$=|\bar{a}|\frac{\bar{a}.\bar{b}}{|\bar{a}||\bar{b}|}\frac{\bar{b}}{|\bar{b}|}=\frac{\bar{a}.\bar{b}}{b^2}\bar{b}=\frac{(\bar{i}-2\bar{j})\cdot(\bar{j}+\bar{k})}{(\bar{j}+\bar{k})^2}(\bar{j}+\bar{k})$$

$$=\frac{-2}{2}(\bar{j}+\bar{k})=-(\bar{j}+\bar{k}).$$

$$\bar{a}\times\bar{b}=(\bar{i}-2\bar{j})\times(\bar{j}+\bar{k})=\begin{vmatrix}\bar{i}&\bar{j}&\bar{k}\\1&-2&0\\0&1&1\end{vmatrix}=-2\bar{i}-\bar{j}+\bar{k}.$$

4. Find the unit vector perpendicular to each of

$$\bar{a}=6\bar{i}+2\bar{j}+3\bar{k}\quad\text{and}\quad\bar{b}=3\bar{i}-6\bar{j}-2\bar{k}.$$

[C.U. '66, '85, '87]

$$\bar{a}\times\bar{b}=\begin{vmatrix}\bar{i}&\bar{j}&\bar{k}\\6&2&3\\3&-6&-2\end{vmatrix}=14\bar{i}+21\bar{j}-42\bar{k}.$$

It is a vector perpendicular to $\bar{a}$ and $\bar{b}$.

$\therefore$ the required unit vector

$$=\frac{14\bar{i}+21\bar{j}-42\bar{k}}{\sqrt{14^2+21^2+42^2}}=\frac{2\bar{i}+3\bar{j}-6\bar{k}}{7}.$$

5. Prove, by vector method, the trigonometrical formula

$$\cos C=\frac{a^2+b^2-c^2}{2ab}.$$

[B.U. '66; C.U. '62, '67, '70, '80, '99, 2001]

Let ABC be a triangle.

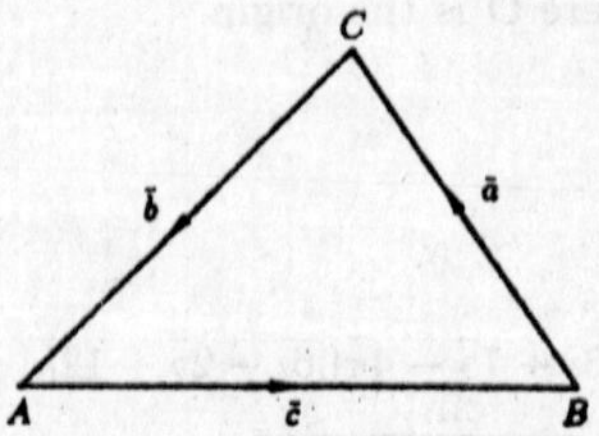

Fig. 2.5

Now

$$\overrightarrow{AB}+\overrightarrow{BC}=\overrightarrow{AC},$$

or $\overrightarrow{AB}=\overrightarrow{AC}-\overrightarrow{BC}=\overrightarrow{CB}-\overrightarrow{CA},$

or $\overrightarrow{AB}^2=\overrightarrow{CB}^2+\overrightarrow{CA}^2-2\overrightarrow{CA}\cdot\overrightarrow{CB},$

or $c^2=a^2+b^2-2ba\cos C,$

or $\cos C=\dfrac{a^2+b^2-c^2}{2ab}.$

6. **Prove the formula $c=a\cos B+b\cos A$, by vector method.** **[C.U. '66, '80, '93]**

In ΔABC (fig. 2.5),

$$\overrightarrow{AB}+\overrightarrow{BC}+\overrightarrow{CA}=0.$$

$\therefore\ \overrightarrow{AB}\cdot(\overrightarrow{AB}+\overrightarrow{BC}+\overrightarrow{CA})=0,$

or $\overrightarrow{AB}^2+\overrightarrow{AB}\cdot\overrightarrow{BC}+\overrightarrow{AB}\cdot\overrightarrow{CA}=0,$

or $c^2+ca\cos(\pi-B)+cb\cos(\pi-A)=0,$

or $c-a\cos B-b\cos A=0,$

or $c=a\cos B+b\cos A.$

7. **Prove by vector method $\frac{a}{\sin A}=\frac{b}{\sin B}=\frac{c}{\sin C}$ in a triangle ABC.**

[C.U. '82, '87, '89, '91, '96, 2000]

From ΔABC (fig. 2.5),

$$\overrightarrow{AB}+\overrightarrow{BC}+\overrightarrow{CA}=0.$$

$\therefore\ \overrightarrow{AB}\times(\overrightarrow{AB}+\overrightarrow{BC}+\overrightarrow{CA})=0,$

or $\overrightarrow{AB}\times\overrightarrow{BC}+\overrightarrow{AB}\times\overrightarrow{CA}=0,$

or $AB\cdot BC\sin(\pi-B)\bar{n}-AB\cdot CA\sin(\pi-A)\bar{n}=0,$

(where $\bar{n}$ is the unit vector perpendicular to the plane of ΔABC)

or $a\sin B-b\sin A=0$, or $\dfrac{a}{\sin A}=\dfrac{b}{\sin B}.$

Similarly considering cross-multiplication by $\overrightarrow{BC}$, we get

$$\frac{b}{\sin B}=\frac{c}{\sin C}.\quad\therefore\ \frac{a}{\sin A}=\frac{b}{\sin B}=\frac{c}{\sin C}.$$

8. **Show, by vector method, that the angle in a semi-circle is a right angle.**

[B.U. '90, C.U. '88; N.B. '76, '90, '96, 2000]

Let P be a point on the semi-circle APB, whose centre is O.

Considering O as vector origin,

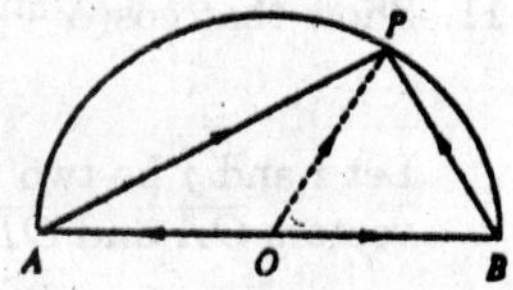

Fig. 2.6

$$\overrightarrow{AP}\cdot\overrightarrow{BP} = (\overrightarrow{OP}-\overrightarrow{OA})\cdot(\overrightarrow{OP}-\overrightarrow{OB})$$
$$= (\overrightarrow{OP}-\overrightarrow{OA})\cdot(\overrightarrow{OP}+\overrightarrow{OA}) \quad [\because \overrightarrow{OB} = -\overrightarrow{OA}]$$
$$= \overrightarrow{OP}^2 - \overrightarrow{OA}^2 = 0 \quad [\because OP = OA = \text{radius}]$$

$\therefore$ PA and PB are at right angle. Hence the result follows.

9. Show, by vector method, that the perpendicular from the vertices of a triangle to the opposite sides are concurrent.

[B.U. '89, '96; N.B. '80, '83, '85, '89, '99; C.U. '84; '88, '89,'92, '98]

Let ABC be a triangle and AD, BE perpendiculars to BC and CA from A and B respectively.
Let them meet at O. CO is joined and produced to meet AB at F.

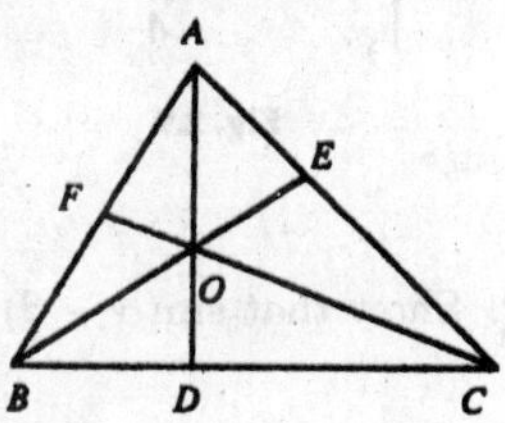

Fig. 2.7

With O as vector origin,

$$\overrightarrow{BC} = \overrightarrow{OC} - \overrightarrow{OB}, \quad \overrightarrow{CA} = \overrightarrow{OA} - \overrightarrow{OC}.$$

$\because$ AD is perpendicular to BC,

$$\overrightarrow{BC}\cdot\overrightarrow{OA} = 0, \quad \text{or} \quad (\overrightarrow{OC}-\overrightarrow{OB})\cdot\overrightarrow{OA} = 0. \tag{1}$$

Similarly

$$\overrightarrow{CA}\cdot\overrightarrow{OB} = 0, \quad \text{or} \quad (\overrightarrow{OA}-\overrightarrow{OC})\cdot\overrightarrow{OB} = 0. \tag{2}$$

From (1) and (2)

$$\overrightarrow{OC}\cdot\overrightarrow{OA} - \overrightarrow{OC}\cdot\overrightarrow{OB} = 0,$$
$$\text{or} \quad (\overrightarrow{OA}-\overrightarrow{OB})\cdot\overrightarrow{OC} = 0, \quad \text{or} \quad \overrightarrow{BA}\cdot\overrightarrow{OC} = 0.$$

$\therefore$ CF is perpendicular to AB.

Hence the result follows.

10. If the two medians of a triangle are equal, show that the triangle is isosceles.

[B.U. '98; C.U. '96]

Let BE and CF be two equal medians of ΔABC. Let $\overrightarrow{AB} = \bar{b}$ and $\overrightarrow{AC} = \bar{c}$ with respect to the vector origin A.

Now $\overrightarrow{CF} = \overrightarrow{AF} - \overrightarrow{AC} = \frac{\bar{b}}{2} - \bar{c}$ and $\overrightarrow{BE} = \overrightarrow{AE} - \overrightarrow{AB} = \frac{\bar{c}}{2} - \bar{b}$.

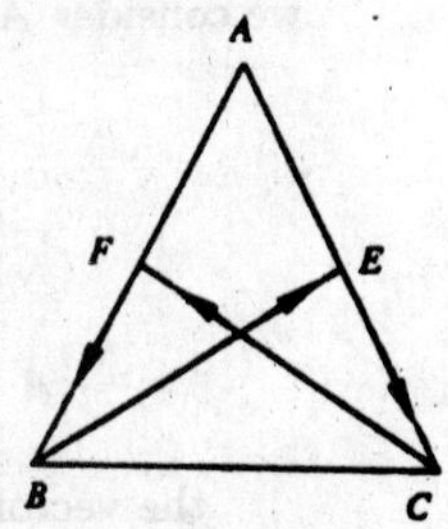

Fig. 2.8

$$\because \ |\overrightarrow{BE}| = |\overrightarrow{CF}|, \left(\frac{\bar{c}}{2} - \bar{b}\right)^2 = \left(\frac{\bar{b}}{2} - \bar{c}\right)^2,$$

$$\text{or} \quad \frac{\bar{c}^2}{4} + \bar{b}^2 - 2\left(\bar{b}\cdot\frac{\bar{c}}{2}\right) = \frac{\bar{b}^2}{4} + \bar{c}^2 - 2\left(\frac{\bar{b}}{2}\cdot\bar{c}\right), \quad \text{or} \quad \frac{3}{4}\bar{b}^2 = \frac{3}{4}\bar{c}^2,$$

$$\text{or} \quad \bar{b}^2 = \bar{c}^2, \quad \text{or} \quad AB^2 = AC^2, \quad \text{or} \quad AB = AC.$$

Hence the triangle is isosceles.

11. Show that $\cos(\alpha+\beta) = \cos\alpha\cos\beta - \sin\alpha\sin\beta$ by vector method.

[N.B. '91, '97]

Let $\bar{i}$ and $\bar{j}$ be two unit vectors in two rectangular directions OX and OY. Let the two vectors $\overrightarrow{OA}$ and $\overrightarrow{OB}$ make angles α and β with OX as shown in the fig. 2.9.

Taking $OA = r_1, OB = r_2$,

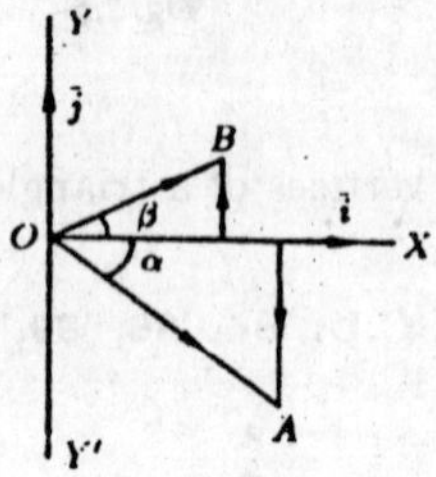

Fig. 2.9

$$\overrightarrow{OA} = \bar{i}r_1\cos\alpha - \bar{j}r_1\sin\alpha,$$

$$\overrightarrow{OB} = \bar{i}r_2\cos\beta + \bar{j}r_2\sin\beta.$$

The angle between OA and OB is $\alpha+\beta$.

$\therefore\quad \overrightarrow{OA}\cdot\overrightarrow{OB} = (\bar{i}r_1\cos\alpha - \bar{j}r_1\sin\alpha)\cdot(\bar{i}r_2\cos\beta + \bar{j}r_2\sin\beta),$

or $\quad r_1r_2\cos(\alpha+\beta) = r_1r_2(\cos\alpha\cos\beta - \sin\alpha\sin\beta),$

or $\quad \cos(\alpha+\beta) = \cos\alpha\cos\beta - \sin\alpha\sin\beta.$

12. Show that $\sin(\alpha-\beta) = \sin\alpha\cos\beta - \cos\alpha\sin\beta$ by vector method.

[C.U. '80, '92]

Let $\bar{i}$ and $\bar{j}$ be the unit vectors in the two rectangular axes OX and OY. Let $\overrightarrow{OA}$ and $\overrightarrow{OB}$ make angles α and β with OX as shown in the fig. 2.10.

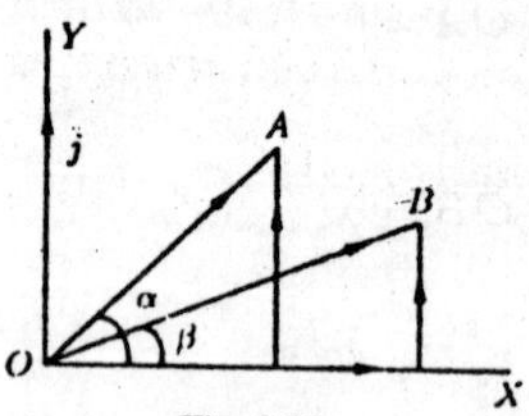

Fig. 2.10

If $OA = r_1$ and $OB = r_2$ then $\overrightarrow{OA} = \bar{i}r_1\cos\alpha + \bar{j}r_1\sin\alpha$ and $\overrightarrow{OB} = \bar{i}r_2\cos\beta + \bar{j}r_2\sin\beta$.

The angle between OA and OB is $\alpha-\beta$.

Now $|\overrightarrow{OB}\times\overrightarrow{OA}| = |r_1r_2(\bar{i}\cos\beta + \bar{j}\sin\beta)\times(\bar{i}\cos\alpha + \bar{j}\sin\alpha|$, or $r_1r_2\sin(\alpha-\beta) = r_1r_2|\bar{k}(\sin\alpha\cos\beta - \cos\alpha\sin\beta)|$, (where $\bar{k}$ is a unit vector perpendicular to $\bar{i}$ and $\bar{j}$) $= \pm r_1r_2(\sin\alpha\cos\beta - \cos\alpha\sin\beta)$, or $\sin(\alpha-\beta) = \pm(\sin\alpha\cos\beta - \cos\alpha\sin\beta)$.

To determine the exact sign, we choose $\beta = 0$. Then left hand side $= \sin\alpha$ and the right-hand side $= \pm\sin\alpha$.

Thus the $-$ve sign is discarded. Hence

$$\sin(\alpha-\beta) = \sin\alpha\cos\beta - \cos\alpha\sin\beta.$$

13. Show that the vectors $\bar{A} = 2\bar{i} - \bar{j} + \bar{k}, \bar{B} = \bar{i} - 3\bar{j} - 5\bar{k}, \bar{C} = 3\bar{i} - 4\bar{j} - 4\bar{k}$, form the sides of a right-angled triangle. [C.U. '90; N.B. '80, '91]

To form a triangle the vectors must be coplanar. To prove the coplanarity of the vectors we consider $\bar{A}\cdot(\bar{B}\times\bar{C})$.

$$\bar{B}\times\bar{C} = \begin{vmatrix} \bar{i} & \bar{j} & \bar{k} \\ 1 & -3 & -5 \\ 3 & -4 & -4 \end{vmatrix} = -8\bar{i} - 11\bar{j} + 5\bar{k}$$

$$\bar{A}\cdot(\bar{B}\times\bar{C}) = (2\bar{i} - \bar{j} + \bar{k})\cdot(-8\bar{i} - 11\bar{j} + \bar{k}) = -16 + 11 + 5 = 0.$$

$\therefore$ the vectors are coplanar.

Again $\bar{A}\cdot\bar{B} = 2 + 3 - 5 = 0$. Therefore $\bar{A}$ is perpendicular to $\bar{B}$. Hence the vectors form the sides of a right-angled triangle.

14. If $\bar{\alpha} = (2, -10, 2), \bar{\beta} = (3, 1, 2)$ and $\bar{\gamma} = (2, 1, 3)$, find the vector $\bar{\alpha} \times (\bar{\beta} \times \bar{\gamma})$ and interpret the result geometrically

$$\bar{\alpha} \times (\bar{\beta} \times \bar{\gamma}) = (\bar{\alpha} \cdot \bar{\gamma})\bar{\beta} - (\bar{\alpha} \cdot \bar{\beta})\bar{\gamma}.$$

[C.U. '64]

Here

$$\bar{\alpha} \cdot \bar{\gamma} = 2.2 - 10.1 + 2.3 = 0,$$

$$\bar{\alpha} \cdot \bar{\beta} = 2.3 - 10.1 + 2.2 = 0.$$

$$\therefore \quad \bar{\alpha} \times (\bar{\beta} \times \bar{\gamma}) = 0 \text{ and } \bar{\alpha} \text{ is perpendicular to } \bar{\beta} \text{ and } \bar{\gamma}.$$

15. Find the projection of the join of two points $P(x_1, y_1, z_1)$ and $Q(x_2, y_2, z_2)$ on a line whose direction cosines are l, m, n.

Let OS be the line with direction cosines l, m, n. If OS be of unit length then $\overrightarrow{OS} = l\bar{i} + m\bar{j} + n\bar{k}$.

Again

$$\overrightarrow{PQ} = \overrightarrow{OQ} - \overrightarrow{OP} = (\bar{i}x_2 + \bar{j}y_2 + \bar{k}z_2) - (\bar{i}x_1 + \bar{j}y_1 + \bar{k}z_1)$$

$$= \bar{i}(x_2 - x_1) + \bar{j}(y_2 - y_1) + \bar{k}(z_2 - z_2).$$

The required projection $= \overrightarrow{OS} \cdot \overrightarrow{PQ}$.

$$= l(x_2 - x_1) + m(y_2 - y_1) + n(z_2 - z_1).$$

16. Find a vector $\bar{\delta}$ which is perpendicular to both $\bar{\alpha} = 4\bar{i} + 5\bar{j} - \bar{k}$ and $\bar{\beta} = \bar{i} - 4\bar{j} + 5\bar{k}$ and $\bar{\delta} \cdot \bar{\gamma} = 21$, where $\bar{\gamma} = 3\bar{i} + \bar{j} - \bar{k}$. **[B.U. '90; C.H. '81; C.U. '81, 88, N.B. '89]**

Let $\bar{\delta} = x\bar{i} + y\bar{j} + z\bar{k}$. Here $\bar{\alpha} \cdot \bar{\delta} = 0$ and $\bar{\beta} \cdot \bar{\delta} = 0$.

$$\therefore \quad 4x + 5y - z = 0 \quad \text{and} \quad x - 4y + 5z = 0.$$

From these two equations $x = -y = -z = \lambda$ (say).

Again $\bar{\delta} \cdot \bar{\gamma} = 21$, or $3x + y - z = 21$, or $3\lambda = 21$, or $\lambda = 7$.

$$\therefore \quad \bar{\delta} = 7\bar{i} - 7\bar{j} - 7\bar{k}.$$

17. Solve $k\bar{r} + \bar{r} \times \bar{a} = \bar{b}$, where k is a non-zero scalar and $\bar{a}, \bar{b}$ are two given vectors.

$$k\bar{r} + \bar{r} \times \bar{a} = \bar{b}. \tag{1}$$

Considering dot product with $\bar{a}$,

$$k(\bar{a} \cdot \bar{r}) = (\bar{a} \cdot \bar{b}) \quad [\because \bar{a} \cdot (\bar{r} \times \bar{a}) = 0]. \tag{2}$$

Considering cross-product with $\bar{a}$,

$$k(\bar{a} \times \bar{r}) + \bar{a}^2\bar{r} - (\bar{a} \cdot \bar{r})\bar{a} = \bar{a} \times \bar{b},$$

$$\text{or} \quad k(k\bar{r} - \bar{b}) + \bar{a}^2\bar{r} - \frac{\bar{a} \cdot \bar{b}}{k}\bar{a} = \bar{a} \times \bar{b}, \quad \text{[by (1) and (2)]}$$

$$\text{or} \quad (k^2 + \bar{a}^2)\bar{r} = k\bar{b} + \frac{\bar{a} \cdot \bar{b}}{k}\bar{a} + \bar{a} \times \bar{b},$$

$$\text{or} \quad \bar{r} = \frac{1}{k^2 + \bar{a}^2}\left(k\bar{b} + \frac{\bar{a} \cdot \bar{b}}{k}\bar{a} + \bar{a} \times \bar{b}\right).$$

18. Show that $[\bar{\alpha}+\bar{\beta}, \bar{\beta}+\bar{\gamma}, \bar{\gamma}+\bar{\alpha}] = 2[\bar{\alpha}\bar{\beta}\bar{\gamma}]$, where $\bar{\alpha}, \bar{\beta}, \bar{\gamma}$ are any three vectors. **[N.B. '87; B.H. '83; C.U. '90, '97]**

$$\begin{aligned}[\bar{\alpha}+\bar{\beta}, \bar{\beta}+\bar{\gamma}, \bar{\gamma}+\bar{\alpha}] &= (\bar{\alpha}+\bar{\beta})\cdot\{(\bar{\beta}+\bar{\gamma})\times(\bar{\gamma}+\bar{\alpha})\}\\ &= (\bar{\alpha}+\bar{\beta})\cdot\{\bar{\beta}\times(\bar{\gamma}+\bar{\alpha})+\bar{\gamma}\times(\bar{\gamma}+\bar{\alpha})\}\\ &= (\bar{\alpha}+\bar{\beta})\cdot\{\bar{\beta}\times\bar{\gamma}-\bar{\alpha}\times\bar{\beta}+\bar{\gamma}\times\bar{\alpha}\}\\ &= \bar{\alpha}\cdot(\bar{\beta}\times\bar{\gamma})-\bar{\alpha}\cdot(\bar{\alpha}\times\bar{\beta})+\bar{\alpha}\cdot(\bar{\gamma}\times\bar{\alpha})\\ &\quad +\bar{\beta}\cdot(\bar{\beta}\times\bar{\gamma})-\bar{\beta}\cdot(\bar{\alpha}\times\bar{\beta})+\bar{\beta}\cdot(\bar{\gamma}\times\bar{\alpha})\\ &= [\bar{\alpha}\bar{\beta}\bar{\gamma}]-0+0+0-0+[\bar{\beta}\bar{\gamma}\bar{\alpha}]\\ &= [\bar{\alpha}\bar{\beta}\bar{\gamma}]+[\bar{\alpha}\bar{\beta}\bar{\gamma}] = 2[\bar{\alpha}\bar{\beta}\bar{\gamma}]\end{aligned}$$

Exercise

1. Find the dot product between $2\bar{i}-3\bar{j}+\bar{k}$ and $7\bar{i}+5\bar{j}+\bar{k}$ and hence find the angle between the vectors.

2. Show that the vectors $\bar{\alpha}=\bar{i}+2\bar{j}+\bar{k}, \bar{\beta}=\bar{i}+\bar{j}-3\bar{k}, \bar{\gamma}=7\bar{i}-4\bar{j}+\bar{k}$ are perpendicular to each other.

3. Find the value of $\bar{a}\times\bar{b}$, where $\bar{a}=2\bar{i}+3\bar{j}+4\bar{k}$ and $\bar{b}=4\bar{i}+3\bar{j}+\bar{k}$.

4. (a) Find the unit vector perpendicular to each of the vector $2\bar{i}+\bar{j}+\bar{k}$ and $\bar{i}-\bar{j}+2\bar{k}$. **[N.B. '70, '79]**

   (b) Prove tha the points $\bar{i}-2\bar{j}+3\bar{k}, 2\bar{j}+4\bar{k}$ and $5\bar{i}+3\bar{j}+5\bar{k}$ form the vertices of a right angled triangle (without using Pythagorian theorem). **[N.B. '84]**

   (c) Find a vector of magnitude 4 which is perpendicular to $6\bar{i}+2\bar{j}+3\bar{k}$ and $3\bar{i}-6\bar{j}-2\bar{k}$. **[N.B. '87]**

   (d) If the position vectors of the three points A, B, C are respectively $\bar{i}+\bar{j}+\bar{k}, 2\bar{i}+3\bar{j}+\bar{k}$ and $3\bar{i}-\bar{j}+4\bar{k}$, find a vector perpendicular to the plane ABC. **[B.U. '99]**

   [HINTS. $\overrightarrow{AB}=(2\bar{i}+3\bar{j}+\bar{k})-(\bar{i}+\bar{j}+\bar{k})=\bar{i}+2\bar{j}$ and $\overrightarrow{AC}=(3\bar{i}-\bar{j}+4\bar{k})-(\bar{i}+\bar{j}+\bar{k})=2\bar{i}-2\bar{j}+3\bar{k}$. $\overrightarrow{AB}\times\overrightarrow{AC}=(\bar{i}+2\bar{j})\times(2\bar{i}-2\bar{j}+3\bar{k})=3(2\bar{i}-\bar{j}-2\bar{k})$. $3(2\bar{i}-\bar{j}-2\bar{k})$ is perpendicular to AB and AC, so it is perpendicular to the plane ABC.]

5. Show, by vector method, that the perpendicular bisectors of the sides of a triangle are concurrent. **[N.B. '84, '92; C.U. '91]**

6. Show, by vector method, that the sum of the squares of the diagonals of a parallelogram is equal to the sum of the squares of its four sides. **[B.U. '78, '85; C.U. '65]**

7. (a) Show, by vector method, that a parallelogram will be rectangle, if the diagonals are equal. **[N.B. '80, '97]**

   (b) Show that the diagonals of a rhombus are at right angle. **[N.B. '77]**

8. (a) If $\bar{\alpha}, \bar{\beta}, \bar{\gamma}$ are three non-coplanar vectors and $\bar{\delta}\cdot\bar{\alpha}=\bar{\delta}\cdot\bar{\beta}=\bar{\delta}\cdot\bar{\gamma}=0$, show that $\bar{\delta}$ is a zero vector. **[C.U. '63]**

   [HINTS. $\because$ $\bar{\delta}$ is perpendicular to three non-coplanar vectors, $\bar{\delta}$ is zero.]

   (b) If $\bar{\alpha}\neq\bar{0}$ but $\bar{\alpha}\cdot\bar{\beta}=\bar{\alpha}\cdot\bar{\gamma}$ and $\bar{\alpha}\times\bar{\beta}=\bar{\alpha}\times\bar{\gamma}$ hold simultaneously, show that $\bar{\beta}=\bar{\gamma}$. **[C.U. '80]**

(c) If a straight line is equally inclined to three coplanar straight lines, show by vector methods, that it is perpendicular to their plane. [C.U. '64]

[HINTS. Let the coplanar lines be represented by unit vectors $\bar{\beta}, \bar{\gamma}, \bar{\delta}$ and the vector $\bar{\alpha}$ be equally inclined to them. Then $\bar{\alpha}\cdot\bar{\beta} = \bar{\alpha}\cdot\bar{\gamma} = \bar{\alpha}\cdot\bar{\delta}$.

$\therefore \quad \bar{\alpha}\cdot(\bar{\beta}-\bar{\gamma}) = 0$ and $\bar{\alpha}\cdot(\bar{\beta}-\bar{\delta}) = 0$. Thus $\bar{\alpha}$ is perpendicular to the coplanar lines $\bar{\beta} - \bar{\gamma}$ and $\bar{\beta} - \bar{\delta}$. Hence the result follows]

(d) If a line g makes angles α, β, γ with the co-ordinate axes, prove that the projection of a vector (X, Y, Z) on g is given by $X\cos\alpha + Y\cos\beta + Z\cos\gamma$. [C.U. '83]

9. $\bar{\alpha}, \bar{\beta}, \bar{\gamma}$ are three vectors satisfying the condition $\bar{\alpha} + \bar{\beta} + \bar{\gamma} = \bar{0}$. If $|\bar{\alpha}| = 3$, $|\bar{\beta}| = 4$ and $|\bar{\gamma}| = 5$, show that $\bar{\alpha}\cdot\bar{\beta} + \bar{\beta}\cdot\bar{\gamma} + \bar{\gamma}\cdot\bar{\alpha} = -25$. [C.H. '82]

10. (a) Use vector method to show that $\cos(\alpha-\beta) = \cos\alpha\cos\beta + \sin\alpha\sin\beta$. [N.B. '88]

(b) Use vector method to show that $\sin(\alpha+\beta) = \sin\alpha\cos\beta + \cos\alpha\sin\beta$. [N.B. '90]

11. (a) Find the area of the triangle ABC whose vertices are $A(1,1,2)$, $B(2,-1,1)$ and $C(3,2,-1)$. Also find the angles of ΔABC.

(b) Find the area of the triangle two of whose sides are $2\bar{i} + \bar{j} - 3\bar{k}$ and $3\bar{i} + 2\bar{k}$. [C.U. 2000]

12. If $\bar{e}_1$ and $\bar{e}_2$ are the two unit vectors inclined at angle θ then show that $2\sin\frac{\theta}{2} = |\bar{e}_1 - \bar{e}_2|$.

13. (a) Find the component of $\bar{a}(= 3\bar{i} - \bar{j})$ along $\bar{b}(= \bar{j} - 3\bar{k})$.

(b) Show that the component of $\bar{a}$ perpendicular to $\bar{b}$ is $\frac{\bar{b}\times(\bar{a}\times\bar{b})}{|\bar{b}|^2}$.

14. Find by the vector method, the area of the triangle whose vertices are $A(1,2,3)$, $B(2,5,-1)$, $C(-1,1,2)$.

[HINTS. Find $\frac{1}{2}|\overrightarrow{AB}\times\overrightarrow{AC}|$]

15. If $\bar{\alpha} = (-2,-2,4)$, $\bar{\beta} = (-2,4,-2)$ and $\bar{\gamma} = (4,-2,-2)$, calculate the value of $\bar{\alpha}\cdot(\bar{\beta}\times\gamma)$ in its simplest form and interpret your result geometrically. [C.U. '66]

16. If $\bar{\alpha} = 3\bar{i} - \bar{j} + 2\bar{k}$, $\bar{\beta} = 2\bar{i} + \bar{j} - \bar{k}$ and $\bar{\gamma} = \bar{i} - 2\bar{j} + 2\bar{k}$, show that $\bar{\alpha}\times(\bar{\beta}\times\bar{\gamma}) \neq (\bar{\alpha}\times\bar{\beta})\times\bar{\gamma}$. [C.U. '92, B.U. '84]

17. (a) Show that $\frac{1}{2}(\bar{b}\times\bar{c} + \bar{c}\times\bar{a} + \bar{a}\times\bar{b})$ is the vector area of the triangle ABC, where $\bar{a}, \bar{b}, \bar{c}$ are the position vector of A, B, C respectively. Interpret your result, if this is zero. [N.B. '92, '97; C.H. '82; C.U. '98]

[HINTS. $\overrightarrow{AB} = \bar{b} - \bar{a}$, $\overrightarrow{AC} = \bar{c} - \bar{a}$. Vector area of $\Delta ABC = \frac{1}{2}(\overrightarrow{AB}\times\overrightarrow{AC}) = \frac{1}{2}(\bar{b}\times\bar{c} + \bar{c}\times\bar{a} + \bar{a}\times\bar{b})$. If this area is zero then the points A, B, C are collinear.]

(b) Prove that $(\bar{a}-\bar{b})\times(\bar{a}+\bar{b}) = 2(\bar{a}\times\bar{b})$ and interpret it.

(c) Show that the three vectors $\bar{\alpha}\times(\bar{\beta}\times\gamma)$, $\bar{\beta}\times(\bar{\gamma}\times\bar{\alpha})$ and $\bar{\gamma}\times(\bar{\alpha}\times\bar{\beta})$ are coplanar. [C.U. '85]

(d) If $(\bar{a}\times\bar{b}) + (\bar{b}\times\bar{c}) + (\bar{c}\times\bar{a}) = \bar{0}$, show that the vectors $\bar{a}, \bar{b}, \bar{c}$ are coplanar. [N.B. '97]

18. Show that

(a) $\bar{a}\times(\bar{b}\times\bar{c}) + \bar{b}\times(\bar{c}\times\bar{a}) + \bar{c}\times(\bar{a}\times\bar{b}) = \bar{0}$, [C.H. '85; C.U. '84, '85, '89, '96, '97]

(b) $(\bar{b}\times\bar{c})\cdot(\bar{a}\times\bar{d}) + (\bar{c}\times\bar{a})\cdot(\bar{b}\times\bar{d}) + (\bar{a}\times\bar{d})\cdot(\bar{c}\times\bar{d}) = 0$,

(c) $[\bar{a}\times\bar{b}, \bar{b}\times\bar{c}, \bar{c}\times\bar{a}] = [\bar{a}\bar{b}\bar{c}]^2$. [N.B. '98; C.U. '81; C.H. '81, '83, '84]

(d) $|\bar{a} \times \bar{b}|^2 |\bar{a} \times \bar{c}|^2 - \{(\bar{a} \times \bar{b}) \cdot (\bar{a} \times \bar{c})\}^2 = |\bar{a}|^2 [\bar{a}\bar{b}\bar{c}]^2$. $\bar{a}, \bar{b}, \bar{c}, \bar{d}$ are any four vectors.

19. (a) Solve $\bar{r} \times \bar{b} = \bar{a}$, where $\bar{b}$ and $\bar{a}$ are given vectors and $\bar{a}$ is perpendicular to $\bar{b}$.

[HINTS. Let $\bar{r} = x\bar{a} + y\bar{b} + z(\bar{a} \times \bar{b})$ then $\bar{r} = y\bar{b} - \frac{1}{b^2} \bar{a} \times \bar{b}$]

(b) Solve $\bar{r} \times \bar{b} = \bar{a} \times \bar{b}$, where $\bar{a}, \bar{b}$ are two given vectors.

[HINTS. Here $(\bar{r} - \bar{a}) \times \bar{b} = 0$. $\therefore$ $\bar{r} - \bar{a}$ is parallel to $\bar{b}$.

Thus $\bar{r} - \bar{a} = t\bar{b}$, or $\bar{r} = \bar{a} + t\bar{b}$, where t is a scalar.]

20. Prove that $\sin(\alpha + \beta)\sin(\alpha - \beta) = \sin^2 \alpha - \sin^2 \beta$ and $\cos(\alpha + \beta)\cos(\alpha - \beta) = \cos^2 \alpha - \sin^2 \beta$ by vector method.

[HINTS. Use the identity $(\bar{b} \times \bar{c}) \cdot (\bar{a} \times \bar{d}) + (\bar{c} \times \bar{a}) \cdot (\bar{b} \times \bar{d}) + (\bar{a} \times \bar{b}) \cdot (\bar{c} \times \bar{d}) = 0$.]

Answers

1. 0, $\frac{\pi}{2}$, **3.** $-9\bar{i} + 14\bar{j} - 6\bar{k}$, **4.** $\frac{\bar{i} - \bar{j} - \bar{k}}{\sqrt{3}}$, **11.** (a) $\frac{5}{2}\sqrt{5}$; $\cos^{-1} \frac{3}{\sqrt{84}}, \cos^{-1} \frac{3}{\sqrt{84}}, \cos^{-1} \frac{11}{14}$, (b) $\frac{1}{2}\sqrt{182}$ sq. units, **13.** $-\frac{1}{\sqrt{10}} (\bar{j} - 3\bar{k})$, **14.** $\frac{\sqrt{155}}{2}$, **15.** 0, coplanar.

Chapter 3

Some Applications in Geometry and Mechanics

3.1 Straight Line

(i) **Equation of a line passing through a given point and parallel to a given vector** : (*Parametric form*)

Let A be a given point whose position vector is $\bar{a}$ w.r.t. the origin O and let $\bar{b}$ the given vector. Let P be a point on the line and its position vector be $\bar{r}$.

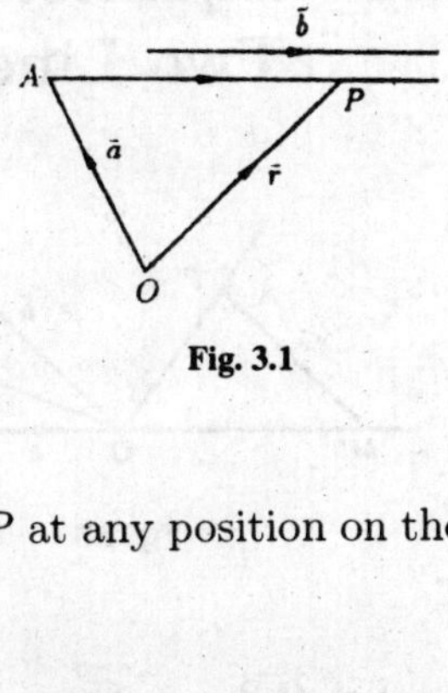

Fig. 3.1

Now $\overrightarrow{AP}$ is parallel to $\bar{b}$ and so we can write $\overrightarrow{AP} = t\bar{b}$, where t is a scalar.

From the fig. 3.1,

$$\bar{r} = \overrightarrow{OP} = \overrightarrow{OA} + \overrightarrow{AP} = \bar{a} + t\bar{b}. \qquad (3.1.1)$$

This relation gives the position vector of the moving point P at any position on the line for different values of t.

Hence it is the vector equation of the line.

Corollary. The equation of a line through the origin is

$$\bar{r} = t\bar{b}. \quad [\bar{a} = 0]$$

Note. The equation (3.1.1) can be written as $\bar{r} \times \bar{b} = \bar{a} \times \bar{b}$.

(ii) **Equation of a line passing through two points:**

Let A and B be two given points with the position vectors $\bar{a}$ and $\bar{b}$ referred to the vector origin O. Let P be a point on the line and $\bar{r}$ be its position vector.

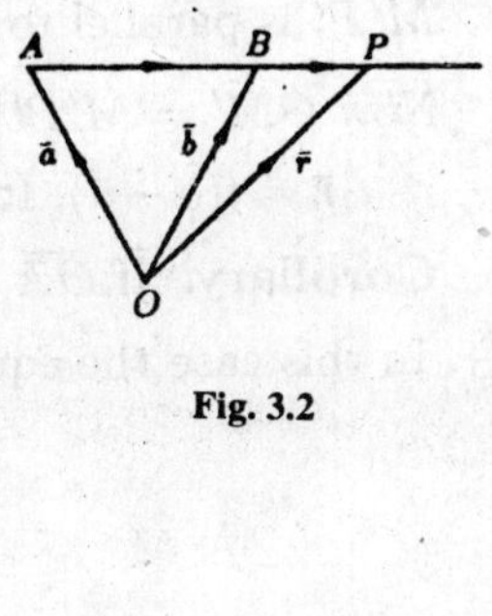

Fig. 3.2

Now $\overrightarrow{AB} = \overrightarrow{OB} - \overrightarrow{OA} = \bar{b} - \bar{a}$.

$\overrightarrow{AP}$ is collinear with $\overrightarrow{AB}$.

Therefore we can write $\overrightarrow{AB} = t(\bar{b} - \bar{a})$, where t is a scalar.

From $\Delta OAP, \overrightarrow{OP} = \overrightarrow{OA} + \overrightarrow{AP}$,

$$\text{or} \quad \bar{r} = \bar{a} + t(\bar{b} - \bar{a}) = (1-t)\bar{a} + t\bar{b}, \tag{3.1.2}$$

$$\text{or} \quad \bar{r} = s\bar{a} + (1-s)\bar{b}, \tag{3.1.3}$$

where $1 - t = s$.

Any one of (3.1.2) and (3.1.3) is the equation of the required line.

Note 1. The equation (3.1.3) can be written as $\bar{r} - s\bar{a} - (1-s)\bar{b} = 0$.

The sum of scalar co-efficients $= 1 - s - (1-s) = 0$.

The condition of collinearity of the three points A, B, P is satisfied.

Note 2. Let the co-ordinates of P, B, A be $(x, y, z), (x_1, y_1, z_1)$, and (x_2, y_2, z_2) w.r.t. three rectangular axes through the vector origin O.

From the equation (3.1.3), $(x, y, z) = s(x_2, y_2, z_2) + (1-s)(x_1 \cdot y_1, z_1)$.

It gives

$$\frac{x - x_1}{x_2 - x_1} = \frac{y - y_1}{y_2 - y_1} = \frac{z - z_1}{z_2 - z_1} = s.$$

It is the equation of a line through two given points in the cartesian system.

3.2 Equations of the Bisectors of Angles between Two Lines

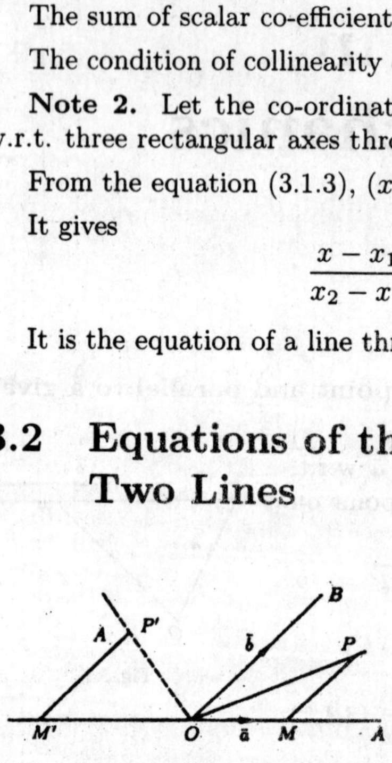

Fig. 3.3

Let OP be the internal bisector of the angle AOB of two lines OA and OB.

Let $\bar{a}$ and $\bar{b}$ be unit vectors along OA and OB.

We consider that O is the vector origin, P is any point on the bisector and MP is parallel to OB.

Here $OM = MP = t$ (say).

$$\therefore \quad \overrightarrow{OM} = t\bar{a}, \overrightarrow{MP} = t\bar{b}.$$

Let $\overrightarrow{OP} = \bar{r}$. From ΔOMP, $\overrightarrow{OP} = \overrightarrow{OM} + \overrightarrow{MP}$, or $\bar{r} = t(\bar{a} + \bar{b})$.

This relation gives the position vector of P at any position on the bisector for different values of t. Hence it is the equation of the internal bisector.

Let OP' be the external bisector and $\overrightarrow{OP'} = \bar{r}$.

$M'P'$ is parallel to OB and it meets AO (produced) at M'.

Now $OM' = M'P' = t$ say, then $\overrightarrow{OM'} = -t\bar{a}$ and $\overrightarrow{M'P'} = t\bar{b}$.

$\therefore \quad \bar{r} = t(\bar{b} - \bar{a})$. It is the equation of the external bisector.

Corollary. If $\overrightarrow{OA} = \bar{a}$, $\overrightarrow{OB} = \bar{b}$, then unit vectors along OA and OB are $\frac{\bar{a}}{|a|}$ and $\frac{\bar{b}}{|b|}$. In this case the equations are

$$\bar{r} = t\left(\frac{\bar{b}}{|\bar{b}|} \pm \frac{\bar{a}}{|\bar{a}|}\right).$$

3.3 Theorems on Bisectors

Theorem 1 *The internal bisector of an angle of a triangle divides the opposite side internally in the ratio of the other two sides.*

Let AD be the internal bisector of the angle A of ΔABC and $\overrightarrow{AB} = \bar{c}, \overrightarrow{AC} = \bar{b}$ with A as vector origin.

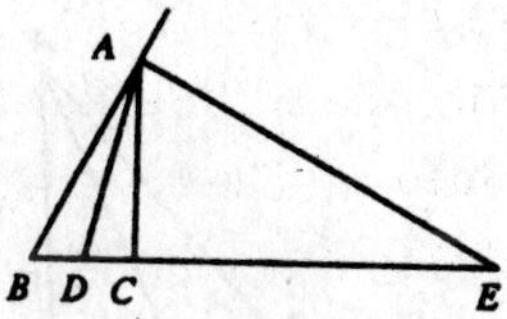

Fig. 3.4

Now the equation of AD is

$$\bar{r} = t\left(\frac{\bar{c}}{|\bar{c}|} + \frac{\bar{b}}{|\bar{b}|}\right) = t\frac{|\bar{b}|\bar{c} + |\bar{c}|\bar{b}}{|\bar{b}||\bar{c}|},$$

where t is a scalar.

Let us choose $t = \dfrac{|\bar{b}||\bar{c}|}{|\bar{b}| + |\bar{c}|}$. Then $\bar{r} = \dfrac{|\bar{b}|\bar{c} + |\bar{c}|\bar{b}}{|\bar{b}| + |\bar{c}|}$ is the position vector of a point on the line BC, which divides BC in the ratio $|\bar{c}| : |\bar{b}|$. So it is the common point of BC and AD. Hence the internal bisector AD divides BC in the ratio of $AB : AC$. This proves the theorem.

Theorem 2 *The external bisector of an angle of a triangle divides the opposite side externally in the ratio of the other two sides.*

Let AE be the external bisector of the angle A of ΔABC and $\overrightarrow{AB} = \bar{c}, \overrightarrow{AC} = \bar{b}$ w.r.t. the vector origin A (fig. 3.4). Now the equation of AE is

$$\bar{r} = t\left(\frac{\bar{b}}{|\bar{b}|} - \frac{\bar{c}}{|\bar{c}|}\right) = t\frac{|\bar{c}|\bar{b} - |\bar{b}|\bar{c}}{|\bar{b}||\bar{c}|},$$

where t is a scalar.

Let us choose $t = \dfrac{|\bar{b}||\bar{c}|}{|\bar{c}| - |\bar{b}|}$. Then $\bar{r} = \dfrac{|\bar{c}|\bar{b} - |\bar{b}|\bar{c}}{|\bar{c}| - |\bar{b}|}$. $\dfrac{|\bar{c}|\bar{b} - |\bar{b}|\bar{c}}{|\bar{c}| - |\bar{b}|}$ is the position vector of a point on BC, which divides BC externally in the ratio $|\bar{c}| : |\bar{b}|$. So it is the common point of BC and AE. Thus the external bisector divides the opposite side externally in the ratio of the other two sides.

Theorem 3 *The internal bisectors of the angles of a triangle are concurrent.*

Let $\bar{r}_1, \bar{r}_2, \bar{r}_3$ be the position vectors of the vertices A, B, C of ΔABC w.r.t. a certain vector origin.

Let $BC = a, CA = b, AB = c$.

If the internal bisector of the angle A meets the side BC at D, then D divides BC in the ratio $c : b$.

Therefore the position vector of D is

$$\frac{b\bar{r}_2 + c\bar{r}_3}{b + c}.$$

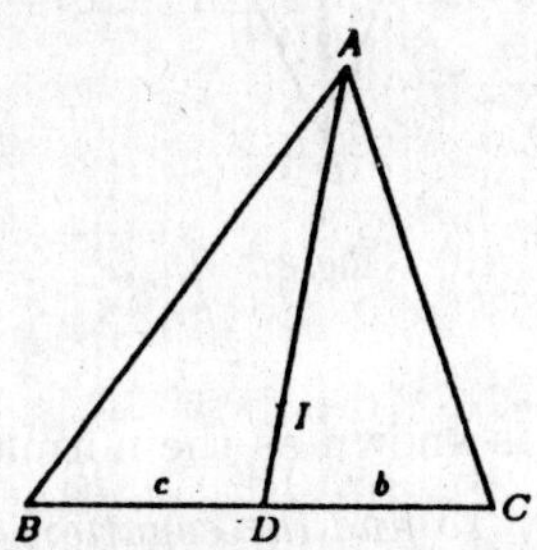

Fig. 3.5

Let the point I divide AD in the ratio $b + c : a$. Then the position vector of I is $\dfrac{a\bar{r}_1 + b\bar{r}_2 + c\bar{r}_3}{a + b + c}$. The symmetry of this expression suggests that this point must also lie on the other bisectors. Hence the bisectors are concurrent.

Theorem 4 *The internal bisector of one angle and the external bisectors of the other two angles of a triangle are concurrent.*

Let $\bar{r}_1, \bar{r}_2, \bar{r}_3$ be the position vectors of the vertices A, B, C of ΔABC w.r.t. a certain vector origin.

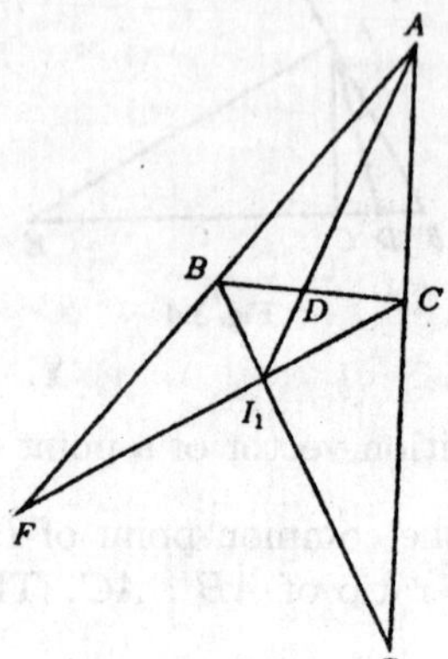

Fig. 3.6

If the external bisector of the angle B meets AC produced at E then the point E divides AC externally in the ratio $c : a$.

Therefore the position vector of E is $\frac{c\bar{r}_3 - a\bar{r}_1}{c-a}$.

Let the point I_1 divide BE in the ratio $c - a : b$. Then the position vector of I_1 is $\frac{b\bar{r}_2 + c\bar{r}_3 - a\bar{r}_1}{b+c-a}$.

If the external bisector of the angle C meets AB externally at F then the position vector of F is $\frac{b\bar{r}_2 - a\bar{r}_1}{b-a}$. Now CF is divided internally in the ratio $b - a : c$ by a point whose position is that of I_1. Similarly the internal bisector AD of the angle A is divided externally in the ratio $b + c : a$ by the point whose position vector is alike to that of I_1. Hence the result follows.

3.4 Equation of a Plane

(a) **Normal form** : *To find the equation of a plane perpendicular to the unit vector $\bar{n}$ and passing through a point whose position vector is $\bar{a}$.*

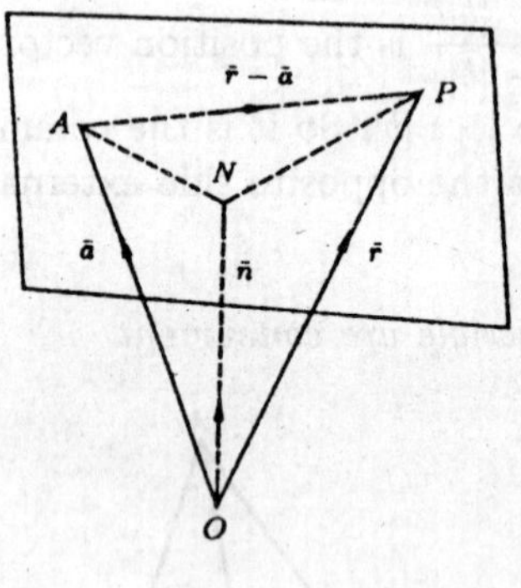

Fig. 3.7

Let O be the vector origin, ON perpendicular to the plane, A the given point on the plane and P any point on the plane. Here $\overrightarrow{OA} = \bar{a}, \overrightarrow{ON} = p\bar{n}$ and let

$$\overrightarrow{OP} = \bar{r}. \quad \left[|\overrightarrow{ON}| = p\right]$$

From the fig. 3.7, $\overrightarrow{AP} = \overrightarrow{OP} - \overrightarrow{OA} = \bar{r} - \bar{a}$.

AP lies in the plane. Therefore it is perpendicular to $\bar{n}$.

Thus $(\bar{r} - \bar{a}) \cdot \bar{n} = 0$. This relation holds for any position of P on the plane. Hence it is the equation of the plane.

Again $\bar{a} \cdot \bar{n}$ = projection of OA on ON = p.

$\therefore \quad \bar{r} \cdot \bar{n} = p.$

It is known as the normal form of the equation of plane.

(b) *To find the equation of a plane passing through a given point and parallel to two given vectors.*

Let the position vector of the given point be $\bar{a}$ and the given vectors be $\bar{b}$ and $\bar{c}$. Since $\bar{b}$ and $\bar{c}$ are parallel to the plane, $\bar{b} \times \bar{c}$ is perpendicular to the plane. Moreover the plane passes through the point whose position vector is $\bar{a}$. If $\bar{r}$ is a point on the plane, $\bar{r} - \bar{a}$ is perpendicular to $\bar{b} \times \bar{c}$.

Thus $(\bar{r}-\bar{a})\cdot(\bar{b}\times\bar{c})=0$, or $\bar{r}\cdot(\bar{b}\times\bar{c})=\bar{a}\cdot(\bar{b}\times\bar{c})$, or $[\bar{r}\bar{b}\bar{c}]=[\bar{a}\bar{b}\bar{c}]$.

It is the equation of the plane.

Note 1. The plane passing through two points with position vectors $\bar{a}$ and $\bar{b}$ and parallel to $\bar{c}$ is identical with the plane passsing through the point with position vector $\bar{a}$ and parallel to the vector $\bar{b}-\bar{a}$ and $\bar{c}$.

Note 2. The plane containing the line $\bar{r}=\bar{\alpha}+t\bar{\beta}$, where t is a scalar and perpendicular to the plane $\bar{r}\cdot\bar{\delta}=q$ is identical with the plane passing through the point having the position vector $\bar{\alpha}$ and parallel to the vectors $\bar{\beta}$ and $\bar{\delta}$.

(c) *To find the equation of the plane passing through three given points.*

Let $\bar{a},\bar{b},\bar{c}$ be the position vectors of the three given points. Then $\bar{b}-\bar{a}$ and $\bar{c}-\bar{a}$ lie in the same plane.

$\therefore$ $(\bar{b}-\bar{a})\times(\bar{c}-\bar{a})$ is perpendicular to the plane. If $\bar{r}$ is a point on the plane then $\bar{r}-\bar{a}$ is perpendicular to $(\bar{b}-\bar{a})\times(\bar{c}-\bar{a})$.

$$\therefore\ (\bar{r}-\bar{a})\cdot\{(\bar{b}-\bar{a})\times(\bar{c}-\bar{a})\}=0,$$

$$\text{or}\quad \bar{r}\cdot\{(\bar{b}-\bar{a})\times(\bar{c}-\bar{a})\}=\bar{a}\cdot\{(\bar{b}-\bar{a})\times(\bar{c}-\bar{a})\},$$

$$\text{or}\quad \bar{r}\cdot(\bar{b}\times\bar{c}+\bar{c}\times\bar{a}+\bar{a}\times\bar{b})=\bar{a}\cdot(\bar{b}\times\bar{c}+\bar{c}\times\bar{a}+\bar{a}\times\bar{b}),$$

$$\text{or}\quad \bar{r}\cdot\bar{m}=[\bar{a}\bar{b}\bar{c}].$$

[$\bar{m}=\bar{b}\times\bar{c}+\bar{c}\times\bar{a}+\bar{a}\times\bar{b}$ = twice the vector area of the triangle whose vertices are $\bar{a},\bar{b},\bar{c}$].

(d) *To find the equation of the plane through two given lines.*

Let the given lines be $\bar{r}=\bar{a}_1+t\bar{b}_1$ and $\bar{r}=\bar{a}_2+s\bar{b}_2$, where t and s are scalars.

Obviously the plane through these lines passes through the point $\bar{a}_1$ and it is parallel to $\bar{b}_1$ and $\bar{b}_2$. Therefore the equation of the plane is $[\bar{r}\bar{b}_1\bar{b}_2]=[\bar{a}_1\bar{b}_1\bar{b}_2]$. [by (b)]

Corollary. Condition of the coplanarity of two straight lines.

The plane passes through the point $\bar{a}_2$. Hence the condition of the coplanarity of the two given lines is $[\bar{a}_2\bar{b}_1\bar{b}_2]=[\bar{a}_1\bar{b}_1\bar{b}_2]$.

3.4.1 To find the distance of a point from a given plane.

Let the position vector of the given point A be $\bar{a}$ and the equation of the given plane be $\bar{r}\cdot\bar{n}=p$, where $\bar{n}$ is normal to the plane.

If AN is perpendicular to the plane then this line passes through the point $\bar{a}$ and it is parallel to $\bar{n}$. Therefore the equation of AN is $\bar{r}=\bar{a}+t\bar{n}$, where t is a scalar. At the point of intersection of the line AN and the plane,

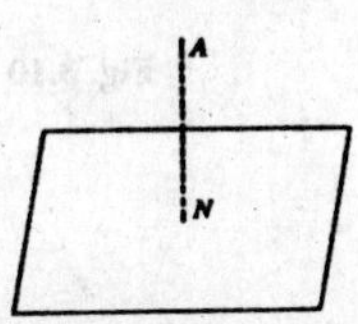

$$(\bar{a}+t\bar{n})\cdot\bar{n}=p,\quad\text{or}\quad \bar{a}\cdot\bar{n}+t\bar{n}^2=p,\quad\text{or}\quad t=\frac{p-\bar{a}\cdot\bar{n}}{\bar{n}^2}.$$

Fig. 3.8

$\therefore$ the position vector of the point N is

$$\bar{a}+t\bar{n}=\bar{a}+\frac{p-\bar{a}\cdot\bar{n}}{\bar{n}^2}\bar{n}.$$

Hence

$$\overrightarrow{AN} = \bar{a} + \frac{p - \bar{a}\cdot\bar{n}}{\bar{n}^2}\bar{n} - \bar{a} = \frac{p - \bar{a}\cdot\bar{n}}{\bar{n}^2}\bar{n}.$$

$\therefore$ the required distance

$$= \left|\overrightarrow{AN}\right| = \left|\frac{p - \bar{a}\cdot\bar{n}}{\bar{n}^2}\right| |\bar{n}| = \frac{|p - \bar{a}\cdot\bar{n}|}{|\bar{n}|}.$$

3.4.2 To find the distance of a point from a given line

Let the equation of the line BN be $\bar{r} = \bar{b} + t\bar{n}$, where t is a scalar, $\bar{b}$ the position vector of the point B and $\bar{n}$ the unit vector parallel to BN.

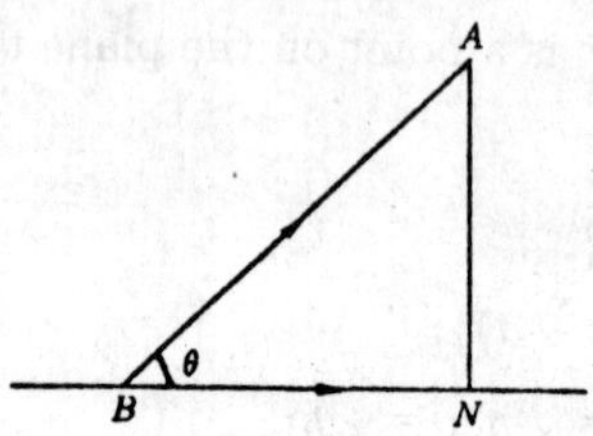

Fig. 3.9

Let $\bar{a}$ be the position vector of the point A.

Now $\overline{BA} = \bar{a} - \bar{b}$. $\quad\therefore\quad BA^2 = (\bar{a} - \bar{b})^2$.

AN is perpendicular to BN. BN is the projection of BA along the unit vector $\bar{n}$.

This projection

$$= BA\cos\theta = 1.BA\cos\theta = \bar{n}\cdot(\bar{a} - \bar{b}).$$

From the figure

$$AN^2 = BA^2 - BN^2 = (\bar{a} - \bar{b})^2 - \{\bar{n}\cdot(\bar{a} - \bar{b})\}^2$$

It gives the required distance.

Note 1. $\overrightarrow{NA} = \overrightarrow{BA} - \overrightarrow{BN} = (\bar{a} - \bar{b}) - \bar{n}\cdot(\bar{a} - \bar{b})\bar{n}$.

Note 2. If $\bar{n}$ is not the unit vector then $\bar{n}$ is replaced by $\bar{n}/|\bar{n}|$.

3.5 Shortest Distance between Two Skew Lines

Let LA and MC be the two skew lines and their equations be $\bar{r} = \bar{a} + t\bar{b}$ and $\bar{r} = \bar{c} + s\bar{d}$ respectively.

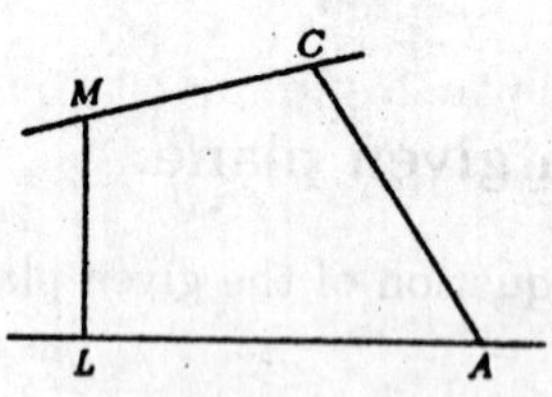

Fig. 3.10

Here t and s are scalars, $\bar{a}$ and $\bar{c}$ are the position vectors of A and C. LA and MC are parallel to $\bar{b}$ and $\bar{d}$.

Let LM be the shortest distance between the lines. Then it is perpendicular to both the lines, so it is parallel to $\bar{b} \times \bar{d}$. The shortest distance is the projection of AC on LM, i.e., the projection of $\bar{c} - \bar{a}$ on $(\bar{b} \times \bar{d})/|\bar{b} \times \bar{d}|$.

Hence

$$LM = (\bar{c} - \bar{a})\cdot(\bar{b} \times \bar{d})/|\bar{b} \times \bar{d}|. \tag{3.5.1}$$

The equation of the shortest distance : The shortest distance is the line of intersection of the two planes drawn through the two skew lines and the line of $s.d.$

The equation of the plane through $\bar{r} = \bar{a} + t\bar{b}$ and LM is

$$(\bar{r} - \bar{a})\cdot[\bar{b} \times (\bar{b} \times \bar{d})] = 0. \tag{3.5.2}$$

The equation of the plane through $\bar{r} = \bar{c} + s\bar{d}$ and LM is

$$(\bar{r} - \bar{c}) \cdot [\bar{d} \times (\bar{b} \times \bar{d})] = 0. \tag{3.5.3}$$

The planes (3.5.2) and (3.5.3) determine the equation of the shortest distance.

Note 1. Cartesian form of s.d.

$$(\bar{c} - \bar{a}) \cdot (\bar{b} \times \bar{d}) / |\bar{b} \times \bar{d}| = \frac{[\bar{c} - \bar{a}, \bar{b}, \bar{d}]}{|\bar{b} \times \bar{d}|}$$

$$= \begin{vmatrix} x_2 - x_1 & y_2 - y_1 & z_2 - z_1 \\ l_1 & m_1 & n_1 \\ l_2 & m_2 & n_2 \end{vmatrix} \div \sqrt{[\Sigma (m_1 n_2 - m_2 n_1)^2]}.$$

Here

$$\bar{a} = \bar{i}x_1 + \bar{j}y_1 + \bar{k}z_1, \bar{c} = \bar{i}x_2 + \bar{j}y_2 + \bar{k}z_2,$$
$$\bar{b} = \bar{i}l_1 + \bar{j}m_1 + \bar{k}n_1, \bar{d} = \bar{i}l_2 + \bar{j}m_2 + \bar{k}n_2.$$

Note 2. Condition for intersection of two lines.

Let $\bar{r} = \bar{a} + t\bar{b}$ and $\bar{r} = \bar{c} + s\bar{d}$ be the two lines.

If the lines intersect then s.d. between them is zero.

In this case $(\bar{c} - \bar{a}) \cdot (\bar{b} \times \bar{d}) = 0$, or $[\bar{c} - \bar{a}, \bar{b}, \bar{d}] = 0$.

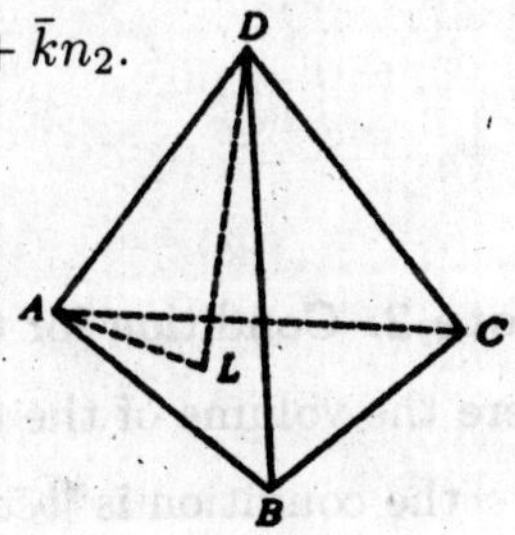

Fig. 3.11

3.6 Volume of a Tetrahedron

Let $ABCD$ be a tetrahedron and the position vectors of A, B, C, D be $\bar{a}, \bar{b}, \bar{c}, \bar{d}$ w.r.t. certain vector origin O.

DL is perpendicular to ΔABC.

Now $\overrightarrow{AB} \times \overrightarrow{AC} = (\bar{b} - \bar{a}) \times (\bar{c} - \bar{a})$.

It represents a vector parallel to LD and its magnitude $= 2\Delta ABC$.

Again LD is the projection of AD on LD.

$\therefore$ volume of the tetrahedron

$$= \frac{1}{3} \Delta ABC \cdot LD$$

$$= \frac{1}{6} (\overrightarrow{AB} \times \overrightarrow{AC}) \cdot \overrightarrow{AD} = \frac{1}{6} \{(\bar{b} - \bar{a}) \times (\bar{c} - \bar{a})\} \cdot (\bar{d} - \bar{a})$$

$$= \frac{1}{6} [\bar{b} - \bar{a}, \bar{c} - \bar{a}, \bar{d} - \bar{a}] = \frac{1}{6} \{[\bar{b}\bar{c}\bar{d}] + [\bar{c}\bar{a}\bar{d}] + [\bar{d}\bar{a}\bar{b}] - [\bar{a}\bar{b}\bar{c}]\}.$$

Note 1. Cartesian equivalence.

Let

$$\bar{a} = \bar{i}x_1 + \bar{j}y_1 + \bar{k}z_1, \bar{b} = \bar{i}x_2 + \bar{j}y_2 + \bar{k}z_2,$$
$$\bar{c} = \bar{i}x_3 + \bar{j}y_3 + \bar{k}z_3, \bar{d} = \bar{i}x_4 + \bar{j}y_4 + \bar{k}z_4.$$

Then

$$\bar{b} - \bar{a} = \bar{i}(x_2 - x_1) + \bar{j}(y_2 - y_1) + \bar{k}(z_2 - z_1),$$
$$\bar{c} - \bar{a} = \bar{i}(x_3 - x_1) + \bar{j}(y_3 - y_1) + \bar{k}(z_3 - z_1),$$
$$\bar{d} - \bar{a} = \bar{i}(x_4 - x_1) + \bar{j}(y_4 - y_1) + \bar{k}(z_4 - z_1).$$

$\therefore$ volume of the tetrahedron

$$= \frac{1}{6}[\bar{b} - \bar{a}, \bar{c} - \bar{a}, \bar{d} - \bar{a}]$$

$$= \frac{1}{6}\begin{vmatrix} x_2 - x_1 & y_2 - y_1 & z_2 - z_1 \\ x_3 - x_1 & y_3 - y_1 & z_3 - z_1 \\ x_4 - x_1 & y_4 - y_1 & z_4 - z_1 \end{vmatrix}.$$

Note 2. Condition for the coplanarity of four points.

Here the volume of the tetrahedron is zero.

$\therefore$ the condition is $[\bar{b}\bar{c}\bar{d}] + [\bar{c}\bar{a}\bar{d}] + [\bar{d}\bar{a}\bar{b}] - [\bar{a}\bar{b}\bar{c}] = 0$.

Note 3. Shortest distance between two opposite edges.

The s.d. between the opposite edges AC and BD is the numerical value of $(\bar{d} - \bar{a}).\{(\bar{c} - \bar{a}) \times (\bar{d} - \bar{b})\}/|(\bar{c} - \bar{a}) \times (\bar{d} - \bar{b})|$.

Note 4. If the vertex A is taken as vector origin, the volume of the tetratedron is $\frac{1}{6}[\bar{b}\bar{c}\bar{d}]$.

3.7 Centroids and Centre of Mass

Definition 1 Centroid. *If $\bar{a}, \bar{b}, \bar{c}, \ldots, \bar{l}$ be position vectors of n given points and $\bar{g} = \overrightarrow{OG} = \frac{1}{n}(\bar{a} + \bar{b} + \bar{c} + \ldots + \bar{l})$, then G is called the centroid (mean centre, centre of mean position) of the given points.* [O is the vector origin.]

If

$$\bar{g} = \overrightarrow{OG} = \frac{m_1\bar{a} + m_2\bar{b} + m_3\bar{c} + \ldots + m_n\bar{l}}{m_1 + m_2 + m_3 + \ldots + m_n},$$

then C is called the centroid of n weighted points.

(Here $m_1 + m_2 + m_3 + \ldots + m_n \neq 0$ and $m_1, m_2, \ldots, m_n$ are the weights of the points $\bar{a}, \bar{b}, \ldots, \bar{l}$ respectively.)

Definition 2 Centre of mass. *If $m_1, m_2, \ldots, m_n$ masses are situated at the points $\bar{a}_1, \bar{a}_2, \ldots, \bar{a}_n$ then the point*

$$\bar{a} = \frac{m_1\bar{a}_1 + m_2\bar{a}_2 + \ldots + m_n\bar{a}_n}{m_1 + m_2 + \ldots + m_n}$$

is called the centre of mass.

Definition 3 Centre of gravity. *It n parallel forces whose magnitudes are proportional to $m_1, m_2, \ldots, m_n$ act at the points $\bar{a}, \bar{a}_2, \ldots, \bar{a}_n$ then the point*

$$\bar{a} = \frac{m_1\bar{a}_1 + m_2\bar{a}_2 + \ldots + m_n\bar{a}_n}{m_1 + m_2 + \ldots + m_n}$$

is called the centre of gravity of these points. It is equivalent to the centre of mass.

3.8 Forces

Resultant of concurrent forces. If $\bar{F}_1, \bar{F}_2, \ldots, \bar{F}_n$ forces act at O, then resultant of these forces is $\bar{F} = \bar{F}_1 + \bar{F}_2 + \ldots + \bar{F}_n$. It acts at O.

If $\bar{F} = 0$, the forces are in equilibrium.

Theorem 5 Lami's Theorem. *If three coplanar forces acting at a point be in equilibrium then magnitude of each force is proportional to the sine of the angle between the other two.*

Let P, Q, R be the magnitudes of the forces and $\bar{a}, \bar{b}, \bar{c}$ be the unit vectors along them.

Since the forces are in equilibrium

$$P\bar{a} + Q\bar{b} + R\bar{c} = \bar{0}. \tag{3.8.1}$$

Taking vector product with $\bar{a}$,

$$P(\bar{a} \times \bar{a}) + Q(\bar{a} \times \bar{b}) + R(\bar{a} \times \bar{c}) = \bar{0},$$

$$\text{or} \quad Q(\bar{a} \times \bar{b}) = R(\bar{c} \times \bar{a}), \quad \text{or} \quad \frac{Q}{|\bar{c} \times \bar{a}|} = \frac{R}{|\bar{a} \times \bar{b}|}.$$

Taking vector product with $\bar{b}$,

$$P(\bar{b} \times \bar{a}) + Q(\bar{b} \times \bar{b}) + R(\bar{b} \times \bar{c}) = \bar{0}, \quad \text{or} \quad P(\bar{a} \times \bar{b}) = R(\bar{b} \times \bar{c}),$$

$$\text{or} \quad \frac{P}{|\bar{b} \times \bar{c}|} = \frac{R}{|\bar{a} \times \bar{b}|}, \quad \text{or} \quad \frac{P}{|\bar{b} \times \bar{c}|} = \frac{Q}{|\bar{c} \times \bar{a}|} = \frac{R}{|\bar{a} \times \bar{b}|},$$

$$\text{or} \quad \frac{P}{\sin \widehat{bc}} = \frac{Q}{\sin \widehat{ca}} = \frac{R}{\sin \widehat{ab}}.$$

Work done by a force. If a force $\bar{F}$ experiences a displacement $\bar{d}$ then the work done by the force is the product of the magnitudes of the displacement and the component of the force in the direction of the displacement.

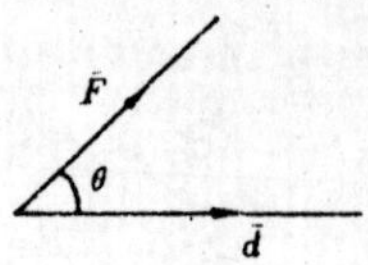

Fig. 3.12

$$\therefore \text{ work done} = |\bar{d}||\bar{F}|\cos\theta \quad (\theta \text{ is the angle between } \bar{F} \text{ and } \bar{d}.)$$
$$= \bar{F}\cdot\bar{d}.$$

Hence the dot product between a force and its displacement represents the work done by the force for this displacement.

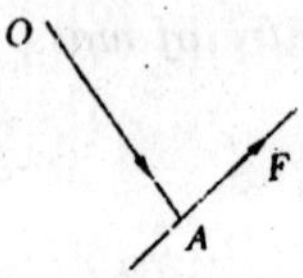

Fig. 3.13

Moment of a force about a point. Let a force $\bar{F}$ be applied on a body at A and O be any point.

Let $\overrightarrow{OA} = \bar{r}$.

$\bar{m} = \bar{r} \times \bar{F}$ is called the moment vector of $\bar{F}$ or torque of $\bar{F}$ about the point O. It is perpendicular to the plane through the point O and the line of $\bar{F}$.

Worked Out Examples

1. Find the equation of the plane through the points $A(-1, 1, 2), B(1, -2, 1)$ and $C(2, 2, 4)$.

Here $\overrightarrow{AB} = (2, -3, -1)$ and $\overrightarrow{AC} = (3, 1, 2)$.

Let $P(x, y, z)$ be any point on the plane.

$\overrightarrow{AB} \times \overrightarrow{AC} = (-5, -7, 11)$ is perpendicular to the plane.

$\overrightarrow{AP} = (x + 1, y - 1, z - 2)$. It lies in the plane. Therefore it is perpendicular to $\overrightarrow{AB} \times \overrightarrow{AC}$.

Thus $\overrightarrow{AP}\cdot(\overrightarrow{AB} \times \overrightarrow{AC}) = 0$,

$$\text{or} \quad -5(x+1) - 7(y-1) + 11(z-2) = 0,$$

$$\text{or} \quad 5x + 7y - 11z + 20 = 0.$$

If is the equation of the plane.

2. It $\bar{a}$ is the position vector of a given point A and $\bar{r}$ is the position vector of any point P in space, find the locus of P in each of the following cases.

(a) $(\bar{r} - \bar{a})\cdot\bar{a} = 0$; (b) $(\bar{r} - \bar{a})\cdot\bar{r} = 0$.

(a) Let $\bar{a}$ and $\bar{r}$ be the vector positions of points A and P w.r.t. the vector origin O. Then $\overrightarrow{AP} = \bar{r} - \bar{a}$.

Since $(\bar{r} - \bar{a})\cdot\bar{a} = 0$, $\overrightarrow{AP}\cdot\overrightarrow{OA} = 0$.

Therefore AP is perpendicular to OA. Hence the locus of P is a plane passing through the point A and perpendicular to OA.

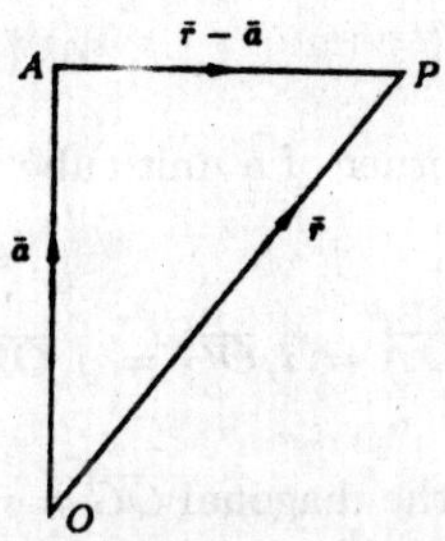

Fig. 3.14

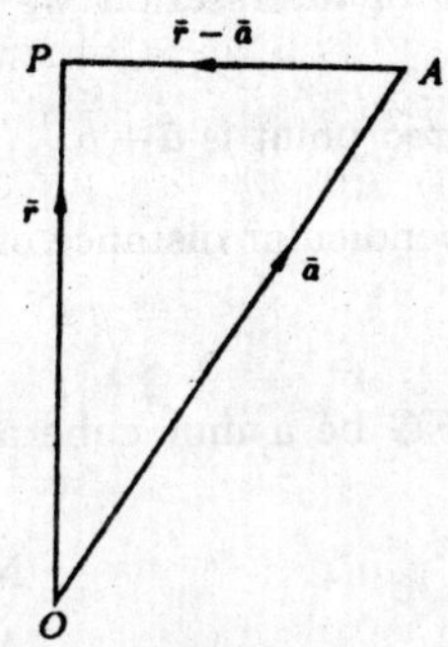

Fig. 3.15

(b) Here also $\bar{a}$ and $\bar{r}$ are position vectors of A and P w.r.t. the vector origin O.

Since $(\bar{r}-\bar{a})\cdot\bar{r}=0$, $\overrightarrow{AP}\cdot\overrightarrow{OP}=0$, i.e., AP is perpendicular to OP. Hence the locus of P is a sphere with OA as diameter.

3. If the vector equations of two lines are $\bar{r}=\bar{r}_1+t\bar{\alpha}$, $\bar{r}=\bar{r}_2+t\bar{\beta}$, where t is a scalar and $\bar{r}_1,\bar{\alpha},\bar{r}_2,\bar{\beta}$ are vectors with co-ordinates $(1,4,5),(2,1,2),(2,8,11)$ and $(-1,3,4)$ respectively, show that the lines are coplanar. **[C.H. '80]**

The lines will be coplanar, if $[\bar{r}_2\bar{\alpha}\bar{\beta}]=[\bar{r}_1\bar{\alpha}\bar{\beta}]$.

$$\bar{\alpha}\times\bar{\beta}=\begin{vmatrix}\bar{i} & \bar{j} & \bar{k}\\ 2 & 1 & 2\\ -1 & 3 & 4\end{vmatrix}=-2\bar{i}-10\bar{j}+7\bar{k},$$

where $\bar{i},\bar{j},\bar{k}$ are unit vectors parallel to x,y and z axes.

$$\bar{r}_2\cdot(\bar{\alpha}\times\bar{\beta}) = (2\bar{i}+8\bar{j}+11\bar{k})\cdot(-2\bar{i}-10\bar{j}+7\bar{k})=-4-80+77=-7.$$

$$\bar{r}_1\cdot(\bar{\alpha}\times\bar{\beta}) = (\bar{i}+4\bar{j}+5\bar{k})\cdot(-2\bar{i}-10\bar{j}+7\bar{k})=-2-40+35=-7.$$

Hence the lines are coplanar.

4. Prove that the lines $\bar{r}\times\bar{a}=\bar{b}\times\bar{a}$ and $\bar{r}\times\bar{b}=\bar{a}\times\bar{b}$ intersect, and find their point of intersection.

The first equation can be written as $(\bar{r}-\bar{b})\times\bar{a}=0$.

It implies that the line passes through the point of $\bar{b}$ and it is parallel to $\bar{a}$. Hence it can be written as

$$\bar{r}=\bar{b}+t\bar{a}, \tag{1}$$

where t is a scalar.

Similarly the second equation can be written as

$$\bar{r}=\bar{a}+s\bar{b}, \tag{2}$$

where s is a scalar.

If these lines intersect, they lie in the plane which is parallel to $\bar{b}-\bar{a}$, $\bar{a}$ and $\bar{b}$. For this the vector $\bar{b}-\bar{a},\bar{a}$ and $\bar{b}$ must be coplanar.

Since $(\bar{b}-\bar{a})\cdot(\bar{a}\times\bar{b})=\bar{b}\cdot(\bar{a}\times\bar{b})-\bar{a}\cdot(\bar{a}\times\bar{b})=0$, the vectors are coplanar. Hence the lines intersect.

For the point of intersection we should have identical values of $\bar{r}$. It happens for $t = s = 1$.

$\therefore$ the required point is $\bar{a} + \bar{b}$.

5. Find the perpendicular distance of a corner of a unit cube from a diagonal not passing through it.

Let $OABCDEFG$ be a unit cube and $\overrightarrow{OA} = \bar{i}, \overrightarrow{OC} = \bar{j}, \overrightarrow{OF} = \bar{k}, \bar{i}, \bar{j}, \bar{k}$ are three unit vectors.

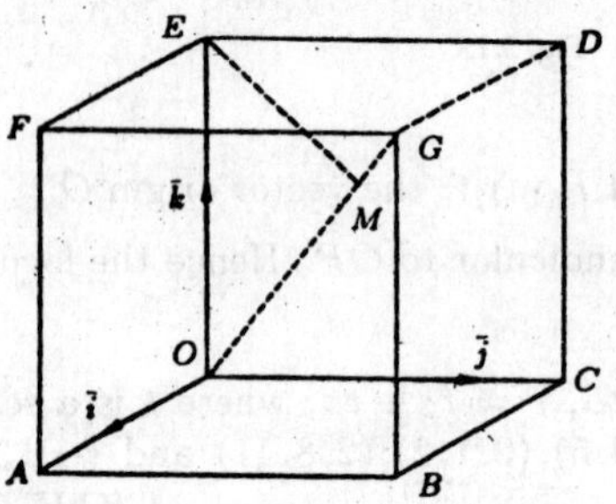

Fig. 3.16

Now the diagonal $\overrightarrow{OG} = \bar{i} + \bar{j} + \bar{k}$.

A unit vector along OG is $\frac{\bar{i}+\bar{j}+\bar{k}}{\sqrt{3}}$.

Let EM be perpendicular to OG.

Then OM is the projection of OE on OG.

$$\therefore \quad OM = \bar{k}\cdot \frac{\bar{i}+\bar{j}+\bar{k}}{\sqrt{3}} = \frac{1}{\sqrt{3}}.$$

$$\text{Again } OE = |\overrightarrow{OE}| = |\bar{k}| = 1.$$

$$\therefore \quad EM = \sqrt{(OE^2 - OM^2)} = \sqrt{\left(1 - \frac{1}{3}\right)} = \sqrt{\frac{2}{3}}.$$

6. Find the s.d. between two skew lines $\bar{r} = \bar{r}_1 + t\bar{\alpha}, \bar{r} = \bar{r}_2 + t\bar{\beta}$, where t is a scalar and $\bar{r}_1, \bar{\alpha}, \bar{r}_2, \bar{\beta}$ are vectors with co-ordinates $(1,-2,3), (2,1,1), (-2,2,-1)\cdot(-3,1,2)$ respectively. **[C.H. '81]**

Here s.d. $= \left|(\bar{r}_2 - \bar{r}_1)\cdot(\bar{\alpha}\times\bar{\beta})/|\bar{\alpha}\times\bar{\beta}|\right|$.

$$\bar{\alpha}\times\bar{\beta} = \begin{vmatrix} \bar{i} & \bar{j} & \bar{k} \\ 2 & 1 & 1 \\ -3 & 1 & 2 \end{vmatrix} = \bar{i} - 7\bar{j} + 5\bar{k}.$$

$$\bar{r}_2 - \bar{r}_1 = (-2\bar{i} + 2\bar{j} - \bar{k}) - (\bar{i} - 2\bar{j} + 3\bar{k}) = -3\bar{i} + 4\bar{j} - 4\bar{k}.$$

Now

$$\text{s.d.} = \left|\frac{(-3\bar{i} + 4\bar{j} - 4\bar{k})\cdot(\bar{i} - 7\bar{j} + 5\bar{k})}{|\bar{i} - 7\bar{j} + 5\bar{k}|}\right|$$

$$= \left|\frac{-3 - 28 - 20}{\sqrt{1 + 49 + 25}}\right| = \frac{51}{5\sqrt{3}} = \frac{17\sqrt{3}}{5}.$$

7. A particle, acted on by constant forces $4\bar{i} + \bar{j} - 3\bar{k}$ and $3\bar{i} + \bar{j} - \bar{k}$, is displaced from the point $\bar{i} + 2\bar{j} + 3\bar{k}$ to the point $5\bar{i} + 4\bar{j} + \bar{k}$. Find the work done by the force on the particle. **[C.U. '98; N.B. '90, '92]**

Let $\bar{F}_1 = 4\bar{i} + \bar{j} - 3\bar{k}, \bar{F}_2 = 3\bar{i} + \bar{j} - \bar{k}$ and $\bar{F}$ be the resultant of these forces. Then $\bar{F} = \bar{F}_1 + \bar{F}_2 = 7\bar{i} + 2\bar{j} - 4\bar{k}$.

The displacement vector $\bar{d} = (5\bar{i} + 4\bar{j} + \bar{k}) - (\bar{i} + 2\bar{j} + 3\bar{k}) = 4\bar{i} + 2\bar{j} - 2\bar{k}$.

Work done $= \bar{F}\cdot\bar{d} = (7\bar{i} + 2\bar{j} - 4\bar{k})\cdot(4\bar{i} + 2\bar{j} - 2\bar{k}) = 28 + 4 + 8 = 40$ units.

8. Find the torque about the point $A(1,2,3)$ of a force of magnitude 5 units acting through the point $(3,4,5)$ in the direction of the vector $2\bar{i}+3\bar{j}+4\bar{k}$. [C.U. '97]

Force

$$\bar{F}=5\frac{2\bar{i}+3\bar{j}+4\bar{k}}{\sqrt{2^2+3^2+4^2}}=\frac{5}{\sqrt{29}}(2\bar{i}+3\bar{j}+4\bar{k}).$$

Let the point on the force be B. Then $\overrightarrow{AB}=(3\bar{i}+4\bar{j}+5\bar{k})-(\bar{i}+2\bar{j}+3\bar{k})=2\bar{i}+2\bar{j}+2\bar{k}$.

Now the moment of the force about the point A

$$=\overrightarrow{AB}\times\bar{F}=(2\bar{i}+2\bar{j}+2\bar{k})\times\frac{5}{\sqrt{29}}(2\bar{i}+3\bar{j}+4\bar{k})$$

$$=\frac{10}{\sqrt{29}}(\bar{i}-2\bar{j}+\bar{k}).$$

9. Find the volume of the tetrahedron $ABCD$ where the position vectors of A,B,C,D are $(-1,1,1),(1,-1,1),(1,1,-1)$ and $(4,1,-3)$ respectively. [C.U. 2000]

Here

$$\overrightarrow{AB}=(\bar{i}-\bar{j}+\bar{k})-(-\bar{i}+\bar{j}+\bar{k})=2\bar{i}-2\bar{j},$$
$$\overrightarrow{AC}=(\bar{i}+\bar{j}-\bar{k})-(-\bar{i}+\bar{j}+\bar{k})=2\bar{i}-2\bar{k},$$
$$\overrightarrow{AD}=(4\bar{i}+\bar{j}-3\bar{k})-(-\bar{i}+\bar{j}+\bar{k})=5\bar{i}-4\bar{k}.$$

Volume of the tetrahedron

$$=\frac{1}{6}(\overrightarrow{AB}\times\overrightarrow{AC})\cdot\overrightarrow{AD}$$

$$=\frac{1}{6}\{(2\bar{i}-2\bar{j})\times(2\bar{i}-2\bar{k})\}\cdot(5\bar{i}-4\bar{k})=\frac{2}{3}\text{ units.}$$

Exercise

1. Find the equation of the line through the points $(2,3,4)$ and $(3,4,5)$.

2. Find the equation of a line through the point $(-2,1,0)$ and parallel to the vector $5\bar{i}-3\bar{j}+4\bar{k}$.

3. Find the equation of the plane passing through the point $2\bar{i}-3\bar{j}+\bar{k}$ and perpendicular to the line joining the points $3\bar{i}+4\bar{j}-\bar{k}$ and $2\bar{i}-\bar{j}+5\bar{k}$.

4. Find the equation of the plane through the points $(2,-1,4),(3,4,7)$ and $(-2,3,-1)$.

5. (a) Show that $\bar{r}=\bar{a}+t\bar{b}+s\bar{c}$ represents the vector equation of a plane through the point $\bar{a}$ and parallel to the vectors $\bar{b}$ and $\bar{c}$. t and s are scalars.

   (b) Find the vector equation of the plane passing through the origin and parallel to the vectors $\bar{i}+2\bar{j}+3\bar{k}$ and $2\bar{i}-\bar{j}-4\bar{k}$. [C.U. '92, '99]

6. Prove that the lines $\bar{r}=\bar{a}+t(\bar{b}\times\bar{c})$ and $\bar{r}=\bar{b}+s(\bar{c}\times\bar{a})$ will intersect if $\bar{a}\cdot\bar{c}=\bar{b}\cdot\bar{c}$.

7. Show that the lines $\bar{r}=\bar{r}_1+t\bar{\alpha},\bar{r}=\bar{r}_2+t\bar{\beta}$, where t is a scalar and $\bar{r}_1,\bar{\alpha},\bar{r}_2,\bar{\beta}$ are vectors with co-ordinates $(4,5,1),(-4,-6,-2),(3,9,4),(-7,-5,0)$ respectively, are coplanar.

8. The vector equations of two lines are $\bar{r} = \bar{r}_1 + t\bar{\alpha}, \bar{r} = \bar{r}_2 + t\bar{\beta}$, where t is a scalar and $\bar{r}_1, \bar{\alpha}, \bar{r}_2, \bar{\beta}$ are vectors with co-ordinates $(-14, 8, 6), (25, -4, -5), (3, 5, 5)$, $(3, 6, 3)$ respectively. Find the s.d. between the lines.

9. If $\bar{\alpha}$ be the position vector of a point P, find by vector method the distance of P from the line $\bar{r} = \bar{\beta} + t\bar{\gamma}$, where the vectors $\bar{\alpha}, \bar{\beta}, \bar{\gamma}$ have co-ordinates $(5, -6, 2), (1, -1, 2)$ and $(0, -4, -3)$ respectively. **[C.H. '82]**

10. $PQRS$ is a tetrahedron with $(-5, -4, 8), (2, 3, 1), (4, 1, 2), (6, 3, 7)$ as the co-ordinates of P, Q, R, S respectively. Find by vector method the distance of the point P from the plane determined by Q, R and S. **[C.H. '81]**

11. $P(1, 3, -1)$, $Q(0, 1, 6)$, $R(-1, 3, 1)$ are three points in space. Find the co-ordinates of a point S on the y-axis such that the volume of the tetrahedron $PQRS$ is 10. **[C.H. '80]**

12. Show that the work done by a particle acted on by a force $5\bar{i} + 10\bar{j} + 15\bar{k}$ for the displacement from the point $\bar{i} + 3\bar{k}$ to $3\bar{i} - \bar{j} - 6\bar{k}$ is 135 units.

13. If a force given by $\vec{F} = 2\bar{i} + 4\bar{j} - \bar{k}$ displaces a particle from the position A to B, where position vectors of A and B are given by $\bar{i} + \bar{j} + 2\bar{k}$ and $3\bar{i} - \bar{j} - \bar{k}$ respectively, find the work done by the force. **[B.U. '81, '97]**

14. Show that the moment of a force $4\bar{i} + 2\bar{j} + \bar{k}$ through the point $5\bar{i} + 2\bar{j} + 4\bar{k}$ about the point $3\bar{i} - \bar{j} + 3\bar{k}$ is $\bar{i} + 2\bar{j} - 8\bar{k}$.

15. A particle acted on by constant forces $2\bar{i} + 3\bar{j} - \bar{k}$ and $3\bar{i} - \bar{j} + 5\bar{k}$ is displaced from the point $A(1, 3, 2)$ to the point $B(4, 5, -1)$, find the work done by the forces. **[C.U. '93]**

Answers

1. $x - 2 = y - 3 = z - 4$; **2.** $\frac{x+2}{5} = \frac{y-1}{-3} = \frac{z}{4}$; **3.** $x + 5y - 6z + 19 = 0$; **4.** $37x + 7y - 24z + 29 = 0$; **5.** (b) $\bar{r} = t(\bar{i} + 2\bar{j} + 3\bar{k}) + s(2\bar{i} - \bar{j} - 4\bar{k})$; **8.** $\frac{29\sqrt{2}}{5}$; **9.** 5; **10.** $\frac{49}{\sqrt{17}}$; **11.** $(0, 8, 0), (0, -2, 0)$; **12.** 1; **15.** 7 units.

Part Three

Short Questions

1. What is geometric vector?

2. If $\bar{a} = 2\bar{i} - \bar{j} - \bar{k}$ and $\bar{b} = \bar{i} + 4\bar{j} + 4\bar{k}$, find $|\bar{a} + \bar{b}|$.

3. The position vectors of four points A, B, C, D are $\bar{i} + \bar{j} + \bar{k}$, $2\bar{i} + 3\bar{j}$, $4\bar{i} + 5\bar{j} - 3\bar{k}$, $\bar{i} + 3\bar{i} - 2\bar{k}$ respectively. Show that $\overrightarrow{AB}$ and $\overrightarrow{DC}$ are like vectors.

4. If $\bar{\alpha} + \bar{\beta} + \bar{\gamma} = \bar{0}$, prove that $\bar{\alpha} \times \bar{\beta} = \bar{\beta} \times \bar{\gamma} = \bar{\gamma} \times \bar{\alpha}$.

5. The co-ordinates of A and B are $(1, 1, 1)$ and $(2, 3, -2)$ respectively. Find the vector $\overrightarrow{AB}$.

6. If the vectors $\mu\bar{i} + 8\bar{j} + 6\bar{k}$ and $3\bar{i} + 4\bar{j} + \lambda\bar{k}$ are collinear, find the values of λ and μ.

7. Determine λ for which the vector $\bar{a} = \lambda(2\bar{i} - 2\bar{j} + 6\bar{k})$ is a unit vector.

8. The position vectors of A, B, C are $2\bar{i} + 4\bar{j} - \bar{k}$, $4\bar{i} - \bar{j} + 2\bar{k}$ and $-\bar{i} + 2\bar{j} + 4\bar{k}$. Show that ΔABC is equilateral.

9. If $\bar{\alpha} = 2\bar{i} - 3\bar{j} + \bar{k}$ and $\bar{\beta} = \bar{i} + \bar{j} - 3\bar{k}$, find $\bar{\alpha} \cdot \bar{\beta}$ and $\bar{\alpha} \times \bar{\beta}$.

10. If $|\bar{a} + \bar{\beta}| = |\bar{a} - \bar{\beta}|$, show that $\bar{a}$ and $\bar{\beta}$ are perpendicular to each other. **[N.B. '90, '92, C.U. '83, '88, '89, 2000]**

11. Find the angle between the vectors $3\bar{i} - \bar{j} + \bar{k}$ and $\bar{i} + \bar{j} - 3\bar{k}$.

12. Use geometric interpretation of scalar and vector products of two vectors $\bar{\alpha}$, $\bar{\beta}$ to show that $|\bar{\alpha} \times \bar{\beta}|^2 + (\bar{\alpha} \cdot \bar{\beta})^2 = |\bar{\alpha}|^2 |\bar{\beta}|^2$. **[C.U. '83]**

13. If $|\bar{\alpha}| = 10$, $|\bar{\beta}| = 1$ and $\bar{\alpha} \cdot \bar{\beta} = 0$, evaluate $|\bar{\alpha} \times \bar{\beta}|$. **[C.U. '83]**

14. If $\bar{\alpha}, \bar{\beta}, \bar{\gamma}, \bar{\delta}$ are coplanar vectors, show that $(\bar{\alpha} \times \bar{\beta}) \times (\bar{\gamma} \times \bar{\delta}) = \bar{0}$. **[C.U. '82]**

15. Find a vector of magnitude 2 which is orthogonal to two mutually orthogonal unit vectors $\bar{\alpha}$ and $\bar{\beta}$. **[C.U. '82]**

16. Find a vector $\bar{X}$ collinear with the vector $\bar{Y} = \{2, 1, 3\}$ satisfying $\bar{X} \cdot \bar{Y} = 16$.

17. What is the geometrical interpretation of $\bar{\alpha} \cdot (\bar{\beta} \times \bar{\gamma})$? What does $\bar{\alpha} \cdot (\bar{\beta} \times \bar{\gamma}) = 0$ imply?

18. Examine whether the vectors drawn from the origin to the points $(2, -5, 1)$, $(1, 4, 0)$, $(-3, 1, -1)$ are coplanar. **[C.U. '82]**

19. Find the simplest value of $\bar{\alpha} \times (\bar{\beta} \times \bar{\gamma})$, where

$$\bar{\alpha} = \bar{i} - 2\bar{j}, \bar{\beta} = \bar{j} - 2\bar{k}, \bar{\gamma} = \bar{k} - 2\bar{i}.$$

20. Find the volume of the parallelopied whose edges are

$$\overrightarrow{OA} = 2\bar{i} - 3\bar{j}, \overrightarrow{OB} = \bar{i} + \bar{j} - \bar{k}, \overrightarrow{OC} = 3\bar{i} - \bar{k}.$$

21. Find the unit vector perpendicular to each of the vectors

$$3\bar{i} - 6\bar{j} - 2\bar{k} \quad \text{and} \quad 6\bar{i} + 2\bar{j} + 3\bar{k}.$$

22. Show that the points with position vectors $60\bar{i}+3\bar{j}$, $40\bar{i}-8\bar{j}$, $a\bar{i}-52\bar{j}$ are collinear if $a=-40$.

23. If the vectors $2\bar{i}-\bar{k}$ and $\bar{j}+2\bar{k}$ are the adjacent sides of a triangle, find the area of the triangle.

24. Prove that $\frac{1}{\sqrt{6}}(2\bar{i}+\bar{j}+\bar{k})$ is a unit vector perpendicular to the plane determined by $P(1,-1,2)$, $Q(2,0,-1)$ and $R(0,2,1)$.

25. If the vectors $\bar{\alpha},\bar{\beta},\bar{\gamma}$ be of lengths 3, 4 and 5 respectively and $\bar{\alpha}$, $\bar{\beta}$, $\bar{\gamma}$ are perpendicular to $\bar{\beta}+\bar{\gamma}$, $\bar{\gamma}+\bar{\alpha}$, $\bar{\alpha}+\bar{\beta}$ respectively, then show that the length of the vector $\bar{\alpha}+\bar{\beta}+\bar{\gamma}$ is $5\sqrt{2}$.

26. If $\bar{\gamma}=\bar{\alpha}\times\bar{\beta}$ and $\bar{\beta}=\bar{\alpha}\times\bar{\gamma}$ and $\bar{\alpha}\neq\bar{0}$, show that $\bar{\beta}=\bar{0}$ and $\bar{\gamma}=\bar{0}$.

27. Find the value of $\bar{\alpha}\cdot(\bar{\beta}+\bar{\gamma})\times(\bar{\alpha}+\bar{\beta}+\bar{\gamma})$.

28. Show that the non-zero vectors $\bar{\alpha}$, $\bar{\beta}$, $\bar{\gamma}$ the relation $|(\bar{\alpha}\times\bar{\beta})\cdot\bar{\gamma}|=|\bar{\alpha}||\bar{\beta}||\bar{\gamma}|$ holds if and only if $\bar{\alpha}\cdot\bar{\beta}=\bar{\beta}\cdot\bar{\gamma}=\bar{\gamma}\cdot\bar{\alpha}=0$.

29. Interpret the vector equation $\bar{r}=t(\frac{\bar{a}}{a}\pm\frac{\bar{b}}{b})$.

30. If $\bar{f}$ be the position vector of any point in space and $\bar{\alpha}$ the position vector of a given point, find the locus of $\bar{f}$ if $(\bar{f}-\bar{\alpha})\cdot\bar{f}=0$.

31. Find the value of x so that the vectors $2\bar{i}-\bar{j}+\bar{k}$, $\bar{i}+2\bar{j}-\bar{k}$ and $x\bar{i}-4\bar{j}+5\bar{k}$ may be coplanar. **[C.U. '92]**

32. If $\bar{a}\cdot\bar{b}=0$ and $\bar{a}\times(\bar{b}\times\bar{c})=0$, find $\bar{a}\cdot\bar{c}$.

33. If $|\bar{a}|=4, |\bar{b}|=5$ and $\bar{a}\cdot\bar{b}=0$, find $\bar{a}\times\bar{b}$. **[C.U. 2000]**

34. Show that $(\bar{\beta}\times\bar{\gamma})\times(\bar{\gamma}\times\bar{\alpha})=[\bar{\alpha}\bar{\beta}\bar{\gamma}]\bar{\gamma}$. **[N.B. '97, '99]**

Answers

2. $3\sqrt{3}$; **5.** $\bar{i}+2\bar{j}-\bar{k}$; **6.** $\lambda=3,\mu=-6$; **7.** $\frac{1}{2\sqrt{11}}$; **9.** $8\bar{i}+7\bar{j}+5\bar{k}$; **11.** $\cos^{-1}\left(\frac{-1}{\sqrt{11}}\right)$; **13.** 10; **15.** $2(\bar{\alpha}\times\bar{\beta})$; **16.** $\left(\frac{16}{7},\frac{8}{7},\frac{24}{7}\right)$; **17.** Volume of a parallelopiped, coplanar; **19.** $-4\bar{i}-2\bar{j}+6\bar{k}$; **20.** 5; **21.** $\frac{-14\bar{i}-21\bar{j}+42\bar{k}}{49}$; **23.** $\frac{\sqrt{21}}{2}$; **27.** 0; **30.** sphere; **31.** 13; **32.** 0; **33.** $20\bar{n}$.

University Questions

Calcutta University

2000

[First Paper]

Group A

1. (a) Answer *any four* questions:

(i) Find the point on the conic $r = \frac{6}{1-\cos\theta}$ having the least radius vector.

[**Hints.** r is least when $1-\cos\theta$ is greatest. The greatest value of $1-\cos\theta = 2$ when $\theta = \pi$. $\quad\therefore\quad (3, \pi)$ is the point.]

(ii) Test whether the equation $x^2 + 6xy + 9y^2 - 5x - 15y + 6 = 0$ represents a pair of parallel straight lines.

[**Hints.** Here $\Delta = 0, D = 0$. Therefore, the equation represents a pair of parallel lines.]

(iii) Find the equation of the sphere through the circle

$$x^2 + y^2 + z^2 = 25,\ x + 2y - z + 2 = 0$$

and the point $(1, 1, 1)$.

[**Hints.** Let the sphere be $x^2 + y^2 + z^2 - 25 + \lambda(x + 2y - z + 2) = 0$. If the point $(1, 1, 1)$ lies on the sphere, then $-22 + 4\lambda = 0$, or, $\lambda = -\frac{11}{2}$. Thus the equation is $2(x^2 + y^2 + z^2) - 11x - 22y + 11z - 72 = 0$.]

(iv) If the expression $ax + by$ changes to $a'x' + b'y'$ by a rotation of the rectangular axes about the origin, prove that $a^2 + b^2 = a'^2 + b'^2$.

[**Hints.** See Ex.5 of page 13.]

(v) Find the direction cosines of the line, represented by the equation

$$x + y - z + 1 = 0 \quad \text{and} \quad 4x + y - 2z + 2 = 0.$$

[**Hints.** Let l, m, n be the d.cs. of the line. Then $l + m - n = 0$ and $4l + m - 2n = 0$. From these $\frac{l}{1} = \frac{m}{2} = \frac{n}{3} = \frac{1}{\sqrt{14}}$.]

(vi) Find the equation of the plane passing through the point $(4, -3, 5)$ and containing the y-axis.

[**Hints.** Let the plane be $ax + cz = 0$. It passes through the point $(4, -3, 5)$.

$\therefore\quad 4a + 5c = 0$, or, $a = -\frac{5c}{4}$. $\quad\therefore\quad$ the equation is $5x - 4z = 0$.]

(vii) Find the values of a and d for which the line $\frac{x-10}{-1} = \frac{y+8}{2} = \frac{z-3}{1}$ lies in the plane $ax + 2y - 3z + d = 0$.

[**Hints.**

$$10a - 16 - 9 + d = 0, \quad \text{or,} \quad 10a + d = 25 \tag{1}$$

$$\text{and} \quad -a + 4 - 3 = 0. \quad \therefore \quad a = 1, d = 15.]$$

Group D

[Answer *either* **Q. Nos. 10** and **11** *or* **Q. Nos. 12** and **13**.]

10. (a) Reduce the equation $x^2 - 2xy + 2y^2 - 4x - 6y + 3 = 0$ to the canonical form and state the nature of the conic.

[**Hints.** Here $\Delta = -26, D = 1$. The equation represents an ellipse. The canonical form is $(3 - \sqrt{5})x'^2 + (3 + \sqrt{5})y'^2 = 52$.]

(b) Find the equation of the plane which passes through the point $(1,2,3)$ and is perpendicular to the line $x+2y+3z-2=0=3x+2y+4z$.
[**Hints.** Let l,m,n be the d.rs. of the line.
$\therefore \quad l+2m+3n=0$ and $3l+2m+4n=0$. From these $\frac{l}{2}=\frac{m}{5}=\frac{n}{-4}$.
Let the plane be $ax+by+cz+d=0$. Here $\frac{a}{2}=\frac{b}{5}=\frac{c}{-4}=\lambda$ (say).
$\therefore \quad 2x+5y-4z+\frac{d}{\lambda}=0$. The point $(1,2,3)$ lies on this plane.
$\therefore \quad \frac{d}{\lambda}=0$. Thus the plane is $2x+5y-4z=0$.]

(c) Find the polar equation of the tangent at the point (r_1,θ_1) on the conic

$$\frac{l}{r}=1+e\cos\theta.$$

[**Hints.** See the section 8.8 of page 126.]

11. (a) If the three lines whose direction cosines are $l_1,m_1,n_1;l_2,m_2,n_2;l_3,m_3,n_3;$ are coplanar, prove that

$$\begin{vmatrix} l_1 & m_1 & n_1 \\ l_2 & m_2 & n_2 \\ l_3 & m_3 & n_3 \end{vmatrix}=0.$$

[**Hints.** See Ex.7 of page 163.]

(b) Find the equation of the plane containing the line $x+2y+3z-4=0=2x+y-z+5$ and perpendicular to the plane $5x+3y+6z+8=0$.
[**Hints.** Let the plane be

$$x+2y+3z-4+\lambda(2x+y-z+5)=0$$
$$\text{or,}\quad (1+2\lambda)x+(2+\lambda)y+(3-\lambda)z+5z-4=0.$$

It is perpendicular to the plane $5x+3y+6z+8=0$.
$\therefore \quad 5.(1+2\lambda)+3.(2+\lambda)+6.(3-\lambda)=0$, or, $\lambda=-\frac{29}{7}$.
$\therefore$ the equation of the plane is $51x+15y-50z+173=0$.]

(c) Find the distance of the point $(3,2,1)$ from the line $\frac{x-1}{3}=\frac{y}{4}=\frac{z-2}{1}$.
[**Hints.** See Ex.6 of page 190.]

12. (a) If PSP' and QSQ' be two mutually perpendicular focal chords of the conic

$$\frac{l}{r}=1-e\cos\theta,$$

show that $\frac{1}{PP'}+\frac{1}{QQ'}=\frac{2-e^2}{2l}$.
[**Hints.** See Ex.6 of page 133.]

(b) If the polar of a point (α,β) with respect to the parabola $y^2=4ax$ touches the parabola $x^2=4by$, show that the locus of (α,β) is $xy+2ab=0$.
[**Hints.** See Ex.16 (c) of page 83.]

(c) If the equation $ax^2+2hxy+by^2+2gx+2fy+c=0$ represents two parallel lines, show that the distance between them is

$$2\sqrt{\frac{g^2-ac}{a(a+b)}}.$$

[**Hints.** See Ex.14 of page 32.]

13. (a) Find the equation to the sphere whose centre lies on the line

$$\frac{x+1}{3}=\frac{y-3}{-4}=\frac{z+2}{5}$$

and which passes through the points $(3,4,5)$ and $(-3,0,1)$.
[**Hints.** Let the sphere be $x^2+y^2+z^2+2ux+2vy+2wz+d=0$. Since the points $(3,4,5)$ and $(-3,0,1)$ lie on the sphere

$$50+6u+8v+10w+d=0 \quad \textbf{(1)}$$

$$\text{and} \quad 10-6u+2w+d=0 \quad \textbf{(2)}$$

The centre $(-u,-v,-w)$ lies on the line.

$$\therefore \quad \frac{-u+1}{3}=\frac{-v-3}{4}=\frac{-w+2}{5}=\lambda \text{ (say)} \quad \textbf{(3)}$$

By (3) and (1), $52-36\lambda+d=0$.
By (3) and (2), $8+8\lambda+d=0$.
From these two $\lambda=1.d=-16$. $\therefore$ $u=-2, v=1, w=-3$.
Thus the equation of the sphere is $x^2+y^2+z^2-4x+2y-6z-16=0$.]

(b) Prove that the lines $\frac{x-1}{2}=\frac{y-2}{3}=\frac{z-3}{4}$ and $4x-3y+1=0=5x-3z+2$ are coplanar.
[**Hints.** Any plane through the second line is

$$4x-3y+1+\lambda(5x-3z+2)=0, \text{ or, } (4+5\lambda)x-3y-3\lambda z+1+2\lambda=0.$$

If the first line is parallel to this plane, then $2.(4+5\lambda)-3.3-4.3\lambda=0$, or, $\lambda=-\frac{1}{2}$.
Thus the equation of the plane is

$$x-2y+z=0 \quad \textbf{(1)}$$

Again, the point $(1,2,3)$ lies on (1). Hence, the lines are coplanar.]

(c) Find the equation of the right circular cone with the vertex $(1,-2,-1)$, semi-vertical angle $60°$ and the axis being $\frac{x-1}{3}=\frac{y+2}{-4}=\frac{z+1}{5}$.
[**Hints.** See Ex.7 of page 235.]

VECTOR

2000

[Third Paper]

Group A

[Answer **Q. No. 1** and **any one of the rest.**]

1. Answer *any two*:

(a) If $|\bar{a}|=4, |\bar{b}|=5$ and $\bar{a}\cdot\bar{b}=0$, find $\bar{a}\times\bar{b}$.
[**Hints.** Here angle between $\bar{a}$ and $\bar{b}$ is $90°$. Now $\bar{a}\times\bar{b}=|\bar{a}||\bar{b}|\sin 90°\bar{n}=20\bar{n}$ where $\bar{n}$ is a unit vector perpendicular to $\bar{a}$ and $\bar{b}$.]

(b) If the vectors $\lambda\bar{i}-4\bar{j}+3\bar{k}$ and $3\bar{i}+\lambda\bar{j}-\bar{k}$ where $\bar{i},\bar{j},\bar{k}$ are unit vectors parallel to the axes, be perpendicular to each other, find the value of λ.
[**Hints.** Here $(\lambda\bar{i}-4\bar{j}+3\bar{k})\cdot(3\bar{i}+\lambda\bar{j}-\bar{k})=0$, or, $3\lambda-4\lambda-3=0$, or, $\lambda=-3$.]

(c) Find the area of the triangle two of whose sides are $2\bar{i}+\bar{j}-3\bar{k}$ and $3\bar{i}+2\bar{k}$.
[**Hints.** Area $=\frac{1}{2}|(2\bar{i}+\bar{j}-3\bar{k})\times(3\bar{i}+2\bar{k})|=\frac{1}{2}\sqrt{182}$ sq. units.]

(d) If two vectors $\bar{a}$ and $\bar{b}$ are such that $|\bar{a}+\bar{b}|=|\bar{a}-\bar{b}|$, then find the angle between them.
[**Hints.**

$$|\bar{a}+\bar{b}|=|\bar{a}-\bar{b}|, \text{ or, } (\bar{a}+\bar{b})^2=(\bar{a}-\bar{b})^2,$$

$$\text{or, } \bar{a}^2+\bar{b}^2+2\bar{a}\cdot\bar{b}=\bar{a}^2+\bar{b}^2-2\bar{a}\cdot\bar{b},$$

$$\text{or, } 4\bar{a}\cdot\bar{b}=0, \text{ or, } \bar{a}\cdot\bar{b}=0$$

$\therefore$ the angle between the vectors is $90°$.]

2. (a) Prove by vector method that in any triangle ABC, $\frac{a}{\sin A} = \frac{b}{\sin B} = \frac{c}{\sin C}$ where the symbols have usual meaning.

[**Hints.** See Ex.7 of page 302.]

(b) Show by vector method that the four points $(1, 5, -1), (0, 4, 5), (-1, 5, 1)$ and $(2, 4, 3)$ are coplanar.

[**Hints.** See Ex.10 of page 289.]

(c) Find the moment of the force $\bar{F} = (4, 2, 1)$ passing through the point $(5, 2, 4)$ about the point $(3, -1, 3)$.

[**Hints.** If the points $(5, 2, 4)$ and $(3, -1, 3)$ are A and B, then moment $= \overrightarrow{BA} \times \bar{F}$ where $\bar{F}$ is the force $= (2\bar{i} + 3\bar{j} + \bar{k}) \times (4\bar{i} + 2\bar{j} + \bar{k}) = \bar{i} + 2\bar{j} - 8\bar{k}$.]

3. (a) Prove that $\bar{a} \times (\bar{b} \times \bar{c}) + \bar{b} \times (\bar{c} \times \bar{a}) + \bar{c} \times (\bar{a} \times \bar{b}) = \bar{0}$. Find also the condition for which $(\bar{a} \times \bar{b}) \times \bar{c} = \bar{a} \times (\bar{b} \times \bar{c})$.

[**Hints.**

$$\bar{a} \times (\bar{b} \times \bar{c}) + \bar{b} \times (\bar{c} \times \bar{a}) + \bar{c} \times (\bar{a} \times \bar{b})$$
$$= (\bar{a} \cdot \bar{c})\bar{b} - (\bar{a} \cdot \bar{b})\bar{c} + (\bar{b} \cdot \bar{a})\bar{c} - (\bar{b} \cdot \bar{c})\bar{a} + (\bar{c} \cdot \bar{b})\bar{a} - (\bar{c} \cdot \bar{a})\bar{b} = \bar{0}.$$

See **Note 2** of page 299.]

(b) Prove by vector method that the angle in a semicircle is a right angle.

[**Hints.** See Ex.8 of page 302.]

(c) Find the volume of the tetrahedron $ABCD$ where the position vectors of A, B, C, D are $(-1, 1, 1), (1, -1, 1), (1, 1, -1)$ and $(4, 1, -3)$ respectively.

[**Hints.** See Ex.9 of page 321.]

2001

[First Paper]

Group A

1. (a) Answer *any four* questions:

(i) Find the transformed equation of the line $y = \sqrt{3}x$, when the axes are rotated through an angle $\frac{\pi}{3}$ in the positive sense.

[**Hints.** See Ex.3 of page 5.]

(ii) Find the points on the conic $\frac{5}{r} = 1 + 2\cos\theta$ whose radius vector is 5.

[**Hints.** Here $\frac{5}{5} = 1 + 2\cos\theta$, or, $\cos\theta = 0$, $\therefore \theta = \frac{\pi}{2}$ or, $-\frac{\pi}{2}$. Thus the points are $(5, \frac{\pi}{2})$ and $(5, -\frac{\pi}{2})$.]

(iii) Find the nature of the conic $6x^2 - 5xy - 6y^2 + 14x + 5y + 4 = 0$.

[**Hints.** See Ex.6 of page 23.]

(iv) Find the distance between the planes

$$x + 2y - 3z + 5 = 0 \quad \text{and} \quad x + 2y - 3z - 7 = 0.$$

[**Hints.** Distances of the planes from the origin are $\frac{5}{\sqrt{14}}$ and $\frac{-7}{\sqrt{14}}$.

$\therefore$ required distance $= \frac{5}{\sqrt{14}} - \left(-\frac{7}{\sqrt{14}}\right) = \frac{12}{\sqrt{14}}$ units.]

(v) For what value of k the line $\frac{x-1}{2} = \frac{y-4}{1} = \frac{z-5}{2}$ intersects the line

$$\frac{x-2}{-1} = \frac{y-8}{k} = \frac{z-11}{4}?$$

[**Hints.** See Ex.3 of page 198.]

(vi) Find the centre and radius of the sphere

$$x^2 + y^2 + z^2 + 2x - 4y - 6z + 5 = 0.$$

[**Hints.** Centre is $(-1, 2, 3)$ and radius $= \sqrt{1^2 + 2^2 + 3^2 - 5} = 3$ units.]

(vii) Find the equation of the right circular cone whose vertex is the origin and the base is the circle $y = 2, x^2 + z^2 = 9$.
[**Hints.** The equation is $x^2 + z^2 = 9.\left(\frac{y}{2}\right)^2$, or, $4(x^2 + z^2) = 9y^2$.]

Group D

[Answer **any two** questions.]

10. (a) Find the equation to the pair of straight lines joining the origin to the intersection of the line $y = mx + c$ and the curve $x^2 + y^2 = a^2$. Prove also that if the lines are perpendicular to each other, then $2c^2 = a^2(1 + m^2)$.
[**Hints.** The equation $x^2 + y^2 = a^2$ is made homogeneous by $y = mx + c$.
The homogeneous equation is

$$x^2 + y^2 = a^2\left(\frac{y - mx}{c}\right)^2, \quad \text{or,} \quad c^2(x^2 + y^2) - a^2(y - mx)^2 = 0.$$

It is the equation of the required pair of lines. If these lines are at right angles, then coeff. of x^2 + coeff. of $y^2 = 0$, i.e., $(c^2 - a^2m^2) + (c^2 - a^2) = 0$, or, $2c^2 = a^2(1 + m^2)$.]

(b) Reduce the equation $7x^2 - 2xy + 7y^2 - 16x + 16y - 8 = 0$ to normal form and state the nature of the conic.
[**Hints.** Here $\Delta = 0$ and $D \neq 0$. Therefore, the equation represents a pair of intersecting lines.]

(c) Find the length of the focal chord of the parabola $\frac{3}{r} = 1 - \cos\theta$ which is inclined to the axis at angle $\frac{\pi}{3}$.
[**Hints.** The vectorial angles of the ends of the focal chord are $\frac{\pi}{3}$ and $\pi + \frac{\pi}{3}$. If r_1 and r_2 are distances from the pole, then

$$r_1 = \frac{3}{1 - \cos\frac{\pi}{3}} = 6 \quad \text{and} \quad r_2 = \frac{3}{1 - \cos\left(\pi + \frac{\pi}{3}\right)} = 2.$$

$\therefore$ the required length $= 6 + 2 = 8$ units.]

11. (a) Show that the lines $\frac{x+3}{2} = \frac{y+5}{3} = \frac{z-7}{-3}$ and $\frac{x+1}{4} = \frac{y+1}{5} = \frac{z+1}{-1}$ are coplanar, and find the equation of the plane containing them.
[**Hints.** Here

$$\begin{vmatrix} -1+3 & -1+5 & -1-7 \\ 2 & 3 & -3 \\ 4 & 5 & -1 \end{vmatrix} = 0.$$

Therefore, the lines are coplanar.
The equation of the plane

$$\begin{vmatrix} x+3 & y+5 & z-7 \\ 2 & 3 & -3 \\ 4 & 5 & -1 \end{vmatrix} = 0, \quad \text{or,} \quad 6x - 5y - z = 0.]$$

(b) Find the shortest distance and equation of the line of shortest distance between the lines $\frac{x-3}{-1} = \frac{y-4}{2} = \frac{z+2}{1}$ and $\frac{x-1}{1} = \frac{y+7}{3} = \frac{z+2}{2}$.
[**Hints.** Let the ends of s.d. be

$$(-r_1 + 3, 2r_1 + 4, r_1 - 2) \quad \text{and} \quad (r_2 + 1, 3r_2 - 7, 2r_2 - 2).$$

The d.rs. of s.d. are $r_2+r_1-2, 3r_2-2r_1-11, 2r_2-r_1$. It is perpendicular to both the lines. Therefore,

$$-1.(r_2+r_1-2)+2.(3r_2-2r_1-11)+1.(2r_2-r_1)=0,$$
$$\text{or, } 7r_2-6r_1-20=0 \text{ and } 1.(r_2+r_1-2)+3.(3r_2-2r_1-11)+2.(2r_2-r_1)=0,$$
$$\text{or, } 14r_2-7r_1-35=0.$$

From these $r_1=-1, r_2=2$. Therefore, ends are $(4,2,-3)$ and $(3,-1,2)$. Now length of s.d. $=\sqrt{1^2+3^2+5^2}=\sqrt{35}$ units and the equations of it are $\frac{x-4}{1}=\frac{y-2}{3}=\frac{z+3}{-5}$.]

(c) The foot of the perpendicular drawn from the origin to the plane is $(1,-5,2)$. Find the equation of the plane.
[**Hints.** The equation is $1.(x-1)-5.(y+5)+2.(z-2)=0$, or, $x-5y+2z-30=0$.]

12. (a) A point is such that the perpendicular from the centre on its polar with respect to the ellipse $\frac{x^2}{a^2}+\frac{y^2}{b^2}=1$ is constant and equal to c. Show that the locus of the point is $\frac{x^2}{a^4}+\frac{y^2}{b^4}=\frac{1}{c^2}$.
[**Hints.** Let (x_1,y_1) be the pole. The polar is $\frac{xx_1}{a^2}+\frac{yy_1}{b^2}=1$. Considering distance from the centre $(0,0)$, we have

$$\left|\frac{-1}{\sqrt{\frac{x_1^2}{a^4}+\frac{y_1^2}{b^4}}}\right|=c, \quad \text{or,} \quad \frac{x_1^2}{a^4}+\frac{y_1^2}{b^4}=\frac{1}{c^2}.$$

Thus the locus is $\frac{x^2}{a^4}+\frac{y^2}{b^4}=\frac{1}{c^2}$.]

(b) Find the condition that the line $\frac{l}{r}=A\cos\theta+B\sin\theta$ may touch the conic

$$\frac{l}{r}=1+e\cos\theta.$$

[**Hints.** See Ex.21 of page 139.]

(c) Find the equations of the two tangent planes to the sphere

$$x^2+y^2+z^2-2y-6z+5=0$$

which are parallel to the plane $2x+2y-z=0$.
[**Hints.** Let the tangent plane be $2x+2y-z+\lambda=0$. Its distance from the centre must be radius.

$$\therefore \quad \frac{2.1-3+\lambda}{\sqrt{4+4+1}}=\pm\sqrt{1^2+3^2-5}, \quad \text{or,} \quad \lambda=3\pm3\sqrt{5}.$$

Thus the planes are $2x+2y-z+3\pm3\sqrt{5}=0$.]

13. (a) Show that the lines $x+7y-z-16=0=x-y+z-4$ and

$$x+11y-2z=0=x-5y+2z-4 \text{ are parallel.}$$

[**Hints.** Let l_1, m_1, n_1 be the d.rs. of the first line. Then $l_1+7m_1-n_1=0$ and $l_1-m_1+n_1=0$. From these $\frac{l_1}{3}=\frac{m_1}{-1}=\frac{n_1}{-4}$. If l_2, m_2, n_2 are d.rs. of the second line, then $l_2+11m_2-2n_2=0$ and $l_2-5m_2+2n_2=0$. From these $\frac{l_2}{3}=\frac{m_2}{-1}=\frac{n_2}{-4}$. Hence, the lines are parallel.]

(b) Obtain the equation of the sphere having the circle

$$x^2+y^2+z^2+10y-4z-8=0,\ x+y+z=3$$

as the great circle.
[**Hints.** Let the sphere be

$$x^2+y^2+z^2+10y-4z-8+\lambda(x+y+z-3)=0,$$
$$\text{or, } x^2+y^2+z^2+\lambda x+(\lambda-10)y+(\lambda-4)z-3\lambda-8=0.$$

The centre is $\left(-\frac{\lambda}{2}, -\frac{\lambda-10}{2}, -\frac{\lambda-4}{2}\right)$. It lies on the given plane.

$$\therefore \quad -\frac{\lambda}{2} - \frac{\lambda-10}{2} - \frac{\lambda-4}{2} = 3, \quad \text{or,} \quad \lambda = \frac{8}{3}.$$

Thus the sphere is $3(x^2+y^2+z^2)+8x-22y-4z-48=0$.]

(c) Find the equation of the right circular cone whose vertex is the origin and whose axis is $\frac{x}{3} = \frac{y}{1} = \frac{z}{2}$ and semi-vertical angle is $\frac{\pi}{4}$.

[**Hints.** Let O be the vertex and $P(x,y,z)$ be a point on the cone.

The d.cs. of OP are $\frac{x}{OP}, \frac{y}{OP}, \frac{z}{OP}$.

The d.cs. of the axis are $\frac{3}{\sqrt{14}}, \frac{1}{\sqrt{14}}, \frac{2}{\sqrt{14}}$.

Now,

$$\frac{3x+y+2z}{OP.\sqrt{14}} = \cos\frac{\pi}{4}, \quad \text{or,} \quad (3x+y+2z)^2.2 = 14(x^2+y^2+z^2),$$

$$\text{or,} \quad 2x^2-6y^2-3z^2+6xy+4yz+12zx=0.$$

It is the equation of the cone.]

VECTOR

2001

[Third Paper]

Group A

[Answer **Q. No. 1** and **any one** of the rest.]

1. Answer *any two*:

(a) Find the length and the direction cosines of the vector $\bar{i}+2\bar{j}+2\bar{k}$; $\bar{i},\bar{j},\bar{k}$ having usual meanings.

[**Hints.** Length $=\sqrt{1^2+2^2+2^2}=3$. D.cs. are $\frac{1}{3}, \frac{2}{3}, \frac{2}{3}$.]

(b) ABC is a triangle; D,E,F are the mid-points of sides BC, CA and AB respectively. Show that $\overrightarrow{AD}+\overrightarrow{BE}+\overrightarrow{CF}=\bar{0}$.

[**Hints.** Let O be the vector origin and $\overrightarrow{OA}=\bar{a}, \overrightarrow{OB}=\bar{b}, \overrightarrow{OC}=\bar{c}$. Now

$$\overrightarrow{AD}+\overrightarrow{BE}+\overrightarrow{CF} = \overrightarrow{OD}-\overrightarrow{OA}+\overrightarrow{OE}-\overrightarrow{OB}+\overrightarrow{OF}-\overrightarrow{OC}$$

$$= \frac{\bar{b}+\bar{c}}{2}-\bar{a}+\frac{\bar{c}+\bar{a}}{2}-\bar{b}+\frac{\bar{a}+\bar{b}}{2}-\bar{c}=\bar{0}.]$$

(c) Two vectors $\bar{a}$ and $\bar{b}$ are such that $|\bar{a}+\bar{b}|^2 = |\bar{a}|^2+|\bar{b}|^2$. Prove that they are orthogonal.

[**Hints.** $|\bar{a}+\bar{b}|^2=|\bar{a}|^2+|\bar{b}|^2$, or, $|\bar{a}|^2+|\bar{b}|^2+2\bar{a}\cdot\bar{b}=|\bar{a}|^2+|\bar{b}|^2$, or, $\bar{a}\cdot\bar{b}=0$.

$\therefore$ $\bar{a}$ and $\bar{b}$ are orthogonal.]

(d) Determine the values of λ and μ for which the vectors

$$(-3\bar{i}+4\bar{j}+\lambda\bar{k}) \quad \text{and} \quad (\mu\bar{i}+8\bar{j}+6\bar{k})$$

are collinear; $\bar{i},\bar{j},\bar{k}$ having usual meanings.

[**Hints.** $\frac{-3}{\mu}=\frac{4}{8}=\frac{\lambda}{6}$. $\therefore$ $\lambda=3, \mu=-6$.]

2. (a) In any triangle, with usual notations, prove that, by vector method,

$$c^2=a^2+b^2-2ab\cos C.$$

[**Hints.** See Ex.5 of page 301.]

(b) The vertices of a quadrilateral are respectively given by

$$(3\bar{i} - 4\bar{j} + \bar{k}),\ (4\bar{i} + 3\bar{j} - \bar{k}),\ (2\bar{i} + 3\bar{j} - \bar{k}) \text{ and } (3\bar{i} + 2\bar{j} - 4\bar{k}).$$

Show that the diagonals are at right angles. ($\bar{i}, \bar{j}, \bar{k}$ have usual meanings.)
[**Hints.** Let $ABCD$ be the quadrilateral and

$$\overrightarrow{OA} = 3\bar{i} - 4\bar{j} + \bar{k}, \quad \overrightarrow{OB} = 4\bar{i} + 3\bar{j} - \bar{k},$$
$$\overrightarrow{OC} = 2\bar{i} + 3\bar{j} - \bar{k}, \quad \overrightarrow{OD} = 3\bar{i} + 2\bar{j} - 4\bar{k},$$

where O is the vector origin. AC and BD are diagonals.

$$\overrightarrow{AC} = -\bar{i} + 7\bar{j} - 2\bar{k}, \quad \overrightarrow{BD} = -\bar{i} - \bar{j} - 3\bar{k}.$$

Now, $\overrightarrow{AC} \cdot \overrightarrow{BD} = 1 - 7 + 6 = 0$. $\quad\therefore\quad$ diagonals are at right angles.]

(c) A particle acted on by constant forces $(3\bar{i} + 2\bar{j} + \bar{k})$ and $(2\bar{i} - 2\bar{j} + 8\bar{k})$ is displaced from the origin to the point $(2\bar{i} + \bar{j} + 2\bar{k})$. Find the total work done by the forces. ($\bar{i}, \bar{j}, \bar{k}$ have usual meanings.)
[**Hints.** If $\bar{F}$ is the resultant force, $\bar{F} = 5\bar{i} + 9\bar{k}$. The displacement vector
$\bar{d} = 2\bar{i} + \bar{j} + 2\bar{k}$. $\quad\therefore\quad$ work done $= \bar{F} \cdot \bar{d} = 10 + 18 = 28$ units.]

3. (a) Show that the three vectors $4\bar{i} + 2\bar{j} + \bar{k}$, $2\bar{i} - \bar{j} + 3\bar{k}$ and $8\bar{i} + 7\bar{k}$ are coplanar.
[**Hints.** Let $x(4\bar{i} + 2\bar{j} + \bar{k}) + y(2\bar{i} - \bar{j} + 3\bar{k}) + z(8\bar{i} + 7\bar{k}) = \bar{0}$.
It implies that $4x + 2y + 8z = 0$, $2x - y = 0$, $x + 3y + 7z = 0$.

$$\text{Coefficient matrix} = \begin{bmatrix} 1 & 3 & 7 \\ 2 & -1 & 0 \\ 4 & 2 & 8 \end{bmatrix} \sim \begin{bmatrix} 1 & 3 & 7 \\ 0 & -7 & -14 \\ 0 & -10 & -20 \end{bmatrix} \sim \begin{bmatrix} 1 & 3 & 7 \\ 0 & -7 & -14 \\ 0 & 0 & 0 \end{bmatrix}.$$

Rank $= 2$. Hence there exists nontrivial solution. $(1, 2, -1)$ is a nontrivial solution. Thus the vectors are coplanar.]

(b) Prove by vector method that the medians of a triangle are concurrent.
[**Hints.** See Ex.3 of page 286.]

(c) Show that the equation of the straight line through the points $(\bar{i} - 2\bar{j} + \bar{k})$ and $(3\bar{k} - 2\bar{j})$ is given by $\bar{r} = \bar{i} - 2\bar{j} + \bar{k} + t(2\bar{k} - \bar{i})$, where t is a scalar.
[**Hints.** The equation is

$$\bar{r} = t(3\bar{k} - 2\bar{j}) + (1 - t)(\bar{i} - 2\bar{j} + \bar{k}), \quad \text{or,} \quad \bar{r} = \bar{i} - 2\bar{j} + \bar{k} + t(2\bar{k} - \bar{i}).]$$

2002

[First Paper]

Group A

1. (a) Answer *any four* questions:

(i) Find the equation of the line $\frac{x}{a} + \frac{y}{b} = 1$ when the origin is shifted to the point (a, b).
[**Hints.** The transformed equation is $\frac{x'+a}{a} + \frac{y'+b}{b} = 1$, $\frac{x'}{a} + \frac{y'}{b} + 1 = 0$.]

(ii) Find the direction cosines of a line that makes equal angles with the coordinate axes.
[**Hints.** See Ex.2 of page 161.]

(iii) Find the pole of the straight line $3x + 2y = 4$ with respect to the circle

$$x^2 + y^2 = 8.$$

[**Hints.** Let (x_1, y_1) be the pole. The polar is $xx_1 + yy_1 = 8$. It is identical with

$$3x + 2y = 4. \qquad \therefore \quad \frac{x_1}{3} = \frac{y_1}{2} = \frac{8}{4}.$$

Thus the pole is $(6, 4)$.]

(iv) Find the equation of the plane passing through the point $(-2, 3, 10)$ and through the z-axis.
[**Hints.** Let the plane be $ax + by = 0$. Since it passes through the point

$$(-2, 3, 10), \quad -2a + 3b = 0, \quad \text{or,} \quad b = \frac{2a}{3}$$

$\therefore$ the equation of the plane is $3x + 2y = 0$.]

(v) Prove that the lines $\frac{x+1}{3} = \frac{y-2}{1} = \frac{z-3}{4}$ and $2x - 4y + z + 1 = 0 = 3x + y - 5z + 26$ are coplanar.
[**Hints.** Any plane through the second line

$$2x - 4y + z + 1 + \lambda(3x + y - 5z + 26) = 0,$$
$$\text{or,} \quad (2 + 3\lambda)x + (\lambda - 4)y + (1 - 5\lambda)z + (1 + 26\lambda) = 0.$$

If the first line is parallel to this plane, then

$$3.(2 + 3\lambda) + 1.(\lambda - 4) + 4.(1 - 5\lambda) = 0, \quad \text{or,} \quad \lambda = \frac{3}{5}.$$

$\therefore$ the plane is $19x - 17y - 10z + 83 = 0$. The point $(-1, 2, 3)$ lies on this plane. Therefore, the first line lies on this plane. Hence, the lines are coplanar.]

(vi) Find the equation of the sphere through the circle

$$x^2 + y^2 + z^2 = 9, \; x + y - 2z = 4 \text{ and origin.}$$

[**Hints.** Let the equation be $x^2 + y^2 + z^2 - 9 + \lambda(x + y - 2z - 4) = 0$. If it passes through the origin $-9 - 4\lambda = 0$, or, $\lambda = -\frac{9}{4}$.
$\therefore$ the equation is $4(x^2 + y^2 + z^2) - 9(x + y - 2z) = 0$.]

(vii) Find the equation of the right circular cone whose vertex is the origin and base is the circle $x = a, y^2 + z^2 = b^2$.
[**Hints.** The equation is $y^2 + z^2 = b^2 \cdot \left(\frac{x}{a}\right)^2$, or, $a^2(y^2 + z^2) = b^2x^2$.]

Group D

[Answer **any two** questions.]

10. (a) Reduce the equation $9x^2 + 24xy + 16y^2 - 126x + 82y - 59 = 0$ to its canonical form and find out the nature of the conic.
[**Hints.** Here $\Delta \neq 0, D = 0$. $\therefore$ the equation represents a parabola.
The equation can be written as

$$\begin{aligned}(3x + 4y)^2 &= 126x - 82y + 59\\ \text{or,} \quad (3x + 4y + \lambda)^2 &= (126 + 6\lambda)x + (8\lambda - 82)y + \lambda^2 + 59.\end{aligned}$$

λ is chosen in such a way that the lines

$$3x + 4y + \lambda = 0 \quad \text{and} \quad (126 + 6\lambda)x + (8\lambda - 82)y + \lambda^2 + 59 = 0$$

are at right angles. For this $3.(126 + 6\lambda) + 4.(8\lambda - 82) = 0$, or, $\lambda = -1$.

$$\therefore \quad (3x + 4y - 1)^2 = 30(4x - 3y + 2), \quad \text{or,} \quad \left(\frac{3x + 4y - 1}{5}\right)^2 = \frac{30}{5} \cdot \frac{4x - 3y + 2}{5}.$$

Putting $X = \frac{3x+4y-1}{5}, Y = \frac{4x-3y+2}{5}$, the canonical form is $X^2 = 6Y$.]

(b) In any conic, prove that the sum of the reciprocals of two perpendicular focal chords is constant.
[**Hints.** See Ex.6 of page 133.]

(c) A plane cuts the co-ordinate axes in A, B, C such that the centroid of the triangle ABC is the point (ξ, η, ζ). Show that the equation of the plane is $\frac{x}{\xi} + \frac{y}{\eta} + \frac{z}{\zeta} = 3$.
[**Hints.** See Ex.17 of page 181.]

11. (a) Show that the straight lines whose direction cosines are given by the equations $2l+2m-n=0$ and $mn+nl+lm=0$ are at right angles.

[**Hints.**

$$2l+2m-n=0 \quad (1)$$

$$mn+nl+lm=0 \quad (2)$$

From (1) $n=2l+2m$. Putting this value in (2), $2(l+m)^2+lm=0$, or, $2l^2+5lm+2m^2=0$, or, $(2l+m)(l+2m)=0$. $\therefore\ 2l+m=0$, or, $l+2m=0$.

$$2l+m=0,\ \text{or},\ m=-2l,\ n=-2l,\ \therefore\ \frac{l}{1}=\frac{m}{-2}=\frac{m}{-2}.$$

$$l+2m=0,\ \text{or},\ l=-2m,\ n=-2m,\ \therefore\ \frac{l}{2}=\frac{m}{-1}=\frac{n}{2}.$$

$\because\ 1.2+(-2).(-1)+(-2).2=0$, the lines are at right angles.]

(b) Show that the lines $\frac{x-3}{2}=\frac{y-5}{-3}=\frac{z+3}{-2}$ and $\frac{x-4}{-3}=\frac{y+1}{2}=\frac{z+4}{3}$ are coplanar and find the equation of the plane through them.

[**Hints.**

$$\because \begin{vmatrix} 4-3 & -1-5 & -4+3 \\ 2 & -3 & -2 \\ -3 & 2 & 3 \end{vmatrix}=0, \text{ the lines are coplanar.}$$

The equation of the plane is

$$\begin{vmatrix} x-3 & y-5 & z+3 \\ 2 & -3 & -2 \\ -3 & 2 & 3 \end{vmatrix}=0, \text{ or, } x+z=0.]$$

(c) Two planes $2x+3y-6z+14=0$ and $2x+3y-6z+42=0$ are tangents to a sphere whose centre lies on the line $2x+z=0, y=0$. Find the equation of the sphere.

[**Hints.** See Ex.19 of page 260.]

12. (a) Find the locus of the point of intersection of a pair of perpendicular tangents to the conic $\frac{l}{r}=1+e\cos\theta$.

[**Hints.** The tangents at α and β are

$$\frac{l}{r}=\cos(\theta-\alpha)+e\cos\theta \quad (1)$$

$$\text{and}\quad \frac{l}{r}=\cos(\theta-\beta)+e\cos\theta \quad (2)$$

If these tangents meet at (r_1,θ_1), then

$$\frac{l}{r_1}-e\cos\theta_1=\cos(\theta_1-\alpha)\ \text{and}\ \frac{l}{r_1}-e\cos\theta_1=\cos(\theta_1-\beta)$$

$$\therefore\ \cos(\theta_1-\alpha)=\cos(\theta_1-\beta).\ \because\ \alpha\neq\beta,\ \theta_1=\frac{1}{2}(\alpha+\beta)$$

$$\therefore\ \frac{l}{r_1}-e\cos\theta_1=\cos\frac{\alpha-\beta}{2} \quad (3)$$

(1) and (2) are at right angles.

$$\therefore\ (e+\cos\alpha)(e+\cos\beta)+\sin\alpha\sin\beta=0$$

$$\text{or, } 2\cos^2\frac{\alpha-\beta}{2}-1+2e\cos\frac{\alpha+\beta}{2}\cos\frac{\alpha-\beta}{2}+e^2=0$$

$$\therefore\ 2\left(\frac{l}{r_1}-e\cos\theta_1\right)^2-1+e^2+2e\cos\theta_1\left(\frac{l}{r_1}-e\cos\theta_1\right)=0$$

$$\text{or, } (e^2-1)r_1^2-2ler_1\cos\theta_1+2l^2=0$$

Hence, the locus is $(e^2-1)r^2-2ler\cos\theta+2l^2=0$.]

(b) If l_1, m_1, n_1 and l_2, m_2, n_2 are direction cosines of two perpendicular lines, show that the line whose direction ratios are $m_1n_2 - m_2n_1,\ n_1l_2 - n_2l_1,\ l_1m_2 - l_2m_1$ is perpendicular to both of them.
[**Hints.** See Ex.4 of page 161.]

(c) Find the length and equations of the line of shortest distance between the lines $\frac{x+3}{-4} = \frac{y-6}{3} = \frac{z}{2}$ and $\frac{x+2}{-4} = \frac{y}{1} = \frac{z-7}{1}$.
[**Hints.** Let the ends of s.d. be $(-4r_1 - 3, 3r_1 + 6, 2r_1)$ and $(-4r_2 - 2, r_2, r_2 + 7)$. D.rs. of s.d. are $-4r_2 + 4r_1 + 1, r_2 - 3r_1 - 6, r_2 - 2r_1 + 7$, s.d. is perpendicular to both the lines.

$$\therefore \quad -4.(-4r_2 + 4r_1 + 1) + 3.(r_2 - 3r_1 - 6) + 2.(r_2 - 2r_1 + 7) = 0$$
$$\text{and} \quad -4.(-4r_2 + 4r_1 + 1) + 1.(r_2 - 3r_1 - 6) + 1.(r_2 - 2r_1 + 7) = 0$$

From these $r_1 = -1, r_2 = -1$. The ends are $(1, 3, -2)$ and $(2, -1, 6)$.
$\therefore$ length of s.d. $= 9$ units and the equations are $\frac{x-1}{1} = \frac{y-3}{-4} = \frac{z+2}{8}$.]

13. (a) Find the equation of the plane which passes through $(2, 1, 4)$ and is perpendicular to each of the plane $9x - 7y + 6z + 48 = 0, x + y - z = 0$.
[**Hints.** Let the equation of the plane be $a(x-2) + b(y-1) + c(z-4) = 0$. It is perpendicular to the given planes.

$$\therefore \quad 9a - 7b + 6c = 0 \quad \text{and} \quad a + b - c = 0.$$

From these $\frac{a}{1} = \frac{b}{15} = \frac{c}{16}$. Thus the equation is $x + 15y + 16z - 81 = 0$.]

(b) Find the equation of the sphere for which the circle

$$x + y^2 + z^2 + 2x - 4y + 5 = 0,\ x - 2y + 3z + 1 = 0$$

is a great circle.
[**Hints.** Let the sphere be

$$x^2 + y^2 + z^2 + 2x - 4y + 5 + \lambda(x - 2y + 3z + 1) = 0,$$
$$\text{or,} \quad x^2 + y^2 + z^2 + (2+\lambda)x - (4 + 2\lambda)y + 3\lambda z + 5 + \lambda = 0.$$

Here the centre $\left(-\frac{2+\lambda}{2}, 2 + \lambda, \frac{3\lambda}{2}\right)$ lies on $x - 2y + 3z + 1 = 0$.

$$\therefore \quad -\frac{2+\lambda}{2} - 4 - 2\lambda + \frac{9\lambda}{2} + 1 = 0, \quad \text{or,} \quad \lambda = 2.$$

Thus the equation is $x^2 + y^2 + z^2 + 4x - 8y + 6z + 7 = 0$.]

(c) Find the equation of a right circular cone which passes through the line $2x = 3y = 5z$ and has the line $x = y = z$ as its axis.
[**Hints.**

$$2x = 3y = 5z \quad \text{or,} \quad \frac{x}{15} = \frac{y}{10} = \frac{z}{6} \qquad (1)$$
$$x = y = z \qquad (2)$$

The semivertical angle of the cone $= \cos^{-1}\frac{15+10+6}{\sqrt{3}\cdot 19} = \cos^{-1}\frac{31}{19\sqrt{3}}$.
Here the origin O is the vertex. If $P(x, y, z)$ is a point on the cone, then the angle between OP and the axis is $\cos^{-1}\frac{31}{19\sqrt{3}}$.

$$\therefore \quad \frac{x.1 + y.1 + z.1}{\sqrt{3}.\sqrt{x^2 + y^2 + z^2}} = \frac{31}{19\sqrt{3}}.$$

Thus the equation is

$$361(x + y + z)^2 = 961(x^2 + y^2 + z^2)$$
$$\text{or,} \quad 300(x^2 + y^2 + z^2) - 361(xy + yz + zx) = 0.]$$

VECTOR

2002

[Third Paper]

Group A

[Answer **Q. No. 1** and **any one** of the rest.]

1. Answer *any two* questions:

(a) Show that in a triangle ABC, $\overrightarrow{AB}+\overrightarrow{AC}=2\overrightarrow{AD}$, where D is the midpoint of BC.
[**Hints.** $\overrightarrow{AB}+\overrightarrow{BD}=\overrightarrow{AD}$ and $\overrightarrow{AC}+\overrightarrow{CD}=\overrightarrow{AD}$.
$\therefore$ $\overrightarrow{AB}+\overrightarrow{AC}=2\overrightarrow{AD}$, since $\overrightarrow{BD}$ and $\overrightarrow{CD}$ are equal in magnitude but opposite in sign.]

(b) Define scalar product of two vectors. Show that $\bar{a}.\bar{a}=|\bar{a}|^2$.
[**Hints.** See the section 2.2 of page 203. $\bar{a}.\bar{a}=|\bar{a}||\bar{a}|\cos 0=|\bar{a}|^2$.]

(c) Show that $2\bar{i}+2\bar{j}+3\bar{k}, 3\bar{i}+\bar{j}-\bar{k}$ and $\bar{i}-\bar{j}-4\bar{k}$ form the sides of a triangle.
[**Hints.** Let $\bar{a}=2\bar{i}+2\bar{j}+3\bar{k}, \bar{b}=3\bar{i}+\bar{j}-\bar{k}, \bar{c}=\bar{i}-\bar{j}-4\bar{k}$.

$$x\bar{a}+y\bar{b}+z\bar{c}=\bar{0} \quad\Rightarrow\quad 2x+3y+z=0, 2x+y-z=0, 3x-y-4z=0.$$

Equations are satisfied by $x=1, y=-1, z=1$. Therefore, the vectors are coplanar.
No. two of $\bar{a},\bar{b},\bar{c}$ are parallel. Hence they form a triangle.]

(d) Find the unit vector perpendicular to both $\bar{i}$ and $\bar{j}+\bar{k}$.
[**Hints.** Let $x\bar{i}+y\bar{j}+z\bar{k}$ be perpendicular to $\bar{i}$ and $\bar{j}+\bar{k}$. Then $x=0$ and $y+z=0$.
$\therefore$ the vector is $y(\bar{j}-\bar{k})$. The unit vector is $\frac{\bar{j}-\bar{k}}{\sqrt{2}}$.]

2. (a) Using vectorial methods prove that the angle in a semicircle is a right angle.
[**Hints.** See Ex.8 of page 302.]

(b) Show that the points with position vectors $\bar{i}-2\bar{j}+3\bar{k}, 2\bar{i}+3\bar{j}-4\bar{k}$ and $-7\bar{j}+10\bar{k}$ are collinear.
[**Hints.** Let $x(\bar{i}-2\bar{j}+3\bar{k})+y(2\bar{i}+3j-4\bar{k})+z(-7\bar{j}+10\bar{k})=\bar{0}$ where x,y,z are scalars.
It implies that $x+2y=0, -2x+3y-7z=0, 3x-4y+10z=0$.
We take another equation $x+y+z=0$. Here

$$\text{coefficient matrix} = \begin{bmatrix} 1 & 2 & 0 \\ -2 & 3 & -7 \\ 3 & -4 & 10 \\ 1 & 1 & 1 \end{bmatrix} \sim \begin{bmatrix} 1 & 2 & 0 \\ 0 & 7 & -7 \\ 0 & -10 & 10 \\ 0 & -1 & 1 \end{bmatrix} \sim \begin{bmatrix} 1 & 2 & 0 \\ 0 & 1 & -1 \\ 0 & 0 & 0 \\ 0 & 0 & 0 \end{bmatrix}.$$

Rank = 2. $\therefore$ the system has nontrivial solution. Hence, the points are collinear.]

(c) Find the moment of the force $3\bar{i}+\bar{k}$ acting at the point $(2\bar{i}-\bar{j}+3\bar{k})$ about the $\bar{i}+2\bar{j}-\bar{k}$.
[**Hints.**

$$\begin{aligned}\text{Moment} &= (2\bar{i}-\bar{j}+3\bar{k}-\bar{i}-2\bar{j}+k)\times(3\bar{i}+\bar{k})\\ &= (\bar{i}-3\bar{j}+4\bar{k})\times(3\bar{i}+\bar{k}) = -3\bar{i}+11\bar{j}+9\bar{k}.]\end{aligned}$$

3. (a) Show that $[\bar{\alpha}+\bar{\beta}\quad \bar{\beta}+\bar{\gamma}\quad \bar{\gamma}+\bar{\alpha}]=2[\bar{\alpha}\bar{\beta}\bar{\gamma}]$.
[**Hints.** See Ex.18 of page 306.]

(b) Show that $[(2\bar{i}+\bar{j}+3\bar{k})\times(-\bar{i}+2\bar{j}+\bar{k})]\times(3\bar{i}+\bar{j}+2\bar{k})=-15\bar{i}+25\bar{j}+10\bar{k}$.
[**Hints.** L.H.S. $=(-5\bar{i}-5\bar{j}+5\bar{k})\times(3\bar{i}+\bar{j}+2\bar{k})=-15\bar{i}+25\bar{j}+10\bar{k}=$ R.H.S.]

(c) Find the vector equation of a plane passing through the origin and parallel to the vectors $\bar{i}+2\bar{j}+3\bar{k}$ and $2\bar{i}-\bar{j}-4\bar{k}$.
[**Hints.** If the given vectors are $\bar{b}$ and $\bar{c}$ and $\bar{r}$ is a point on the plane, then the equation of the plane is $\bar{r}.(\bar{b}\times\bar{c})=0$.]

2003

[First Paper]

Group A

2. (a) Answer *any three* questions:

(i) Determine the nature and the length of the latus rectum of the conic $r = \frac{8}{4-5\cos\theta}$.
[**Hints.** The equation is

$$\frac{8}{r} = 4 - 5\cos\theta, \quad \text{or,} \quad \frac{2}{r} = 1 - \frac{5}{4}\cos\theta.$$

Eccentricity $= \frac{5}{4} > 1$. $\quad\therefore\quad$ the conic is a hyperbola.
Length of latus rectum = 4 units.]

(ii) Show that the equation $4x^2 + 9y^2 + 12xy - 16x - 24y + 16 = 0$ represents a pair of coincident straight lines.
[**Hints.** $4x^2 + 9y^2 + 12xy - 16x - 24y + 16 = (2x + 3y - 4)^2$.
$\therefore$ the equation represents a pair of coincident straight lines.]

(iii) Find the value of m for which the line $\frac{x+1}{3} = \frac{y-2}{m} = \frac{z+3}{-2}$ is parallel to the plane $x - 3y + 6z + 7 = 0$.
[**Hints.** Here $1.3 - 3.m + 6.(-2) = 0$, or, $m = -3$.]

(iv) Find the equation of the plane passing through the point $(4, -3, 5)$ and containing the y-axis.
[**Hints.** Let the equation of the plane be $ax + cz = 0$.
Here $4a + 5c = 0$, or, $c = -\frac{4a}{5}$. Thus the equation of the plane be $5x - 4z = 0$.]

(v) Find the transformed equation of the line $y = \sqrt{3}x$ when the axes are rotated through angle $\frac{\pi}{3}$ in the positive direction.
[**Hints.** The transformed equation is

$$x' \sin\frac{\pi}{3} + y' \cos\frac{\pi}{3} = \sqrt{3}\left(x' \cos\frac{\pi}{3} - y' \sin\frac{\pi}{3}\right)$$

$$\text{or,} \quad x' \cdot \frac{\sqrt{3}}{2} + y' \cdot \frac{1}{2} = \sqrt{3}\left(x' \cdot \frac{1}{2} - y' \cdot \frac{\sqrt{3}}{2}\right), \quad \text{or,} \quad y' = 0.]$$

(vi) Find the centre and radius of the sphere $x^2 + y^2 + z^2 + 2x - 4y - 6z + 5 = 0$.
[**Hints.** Centre is $(-1, 2, 3)$ and radius $= \sqrt{1^2 + 2^2 + 3^2 - 5} = 3$.]

(b) Answer *any two* questions:

(i) If two vectors $\bar{a}$ and $\bar{b}$ are perpendicular to each other, then prove that

$$|(\bar{a} + \bar{b}) \times (\bar{a} - \bar{b})| = 2|\bar{a}||\bar{b}|.$$

[**Hints.** $|(\bar{a} + \bar{b}) \times (\bar{a} - \bar{b})| = |-\bar{a} \times \bar{b} + \bar{b} \times \bar{a}| = |-2(\bar{a} \times \bar{b})| = 2|\bar{a}||\bar{b}|$.]

(ii) Show that the three points whose position vectors are $-\bar{i} - 2\bar{j} + 7\bar{k}$, $3\bar{i} + \bar{k}$ and $5\bar{i} + \bar{j} - 2\bar{k}$ are collinear.
[**Hints.** Let $x(-\bar{i} - 2\bar{j} + 7\bar{k}) + y(3\bar{i} + \bar{k}) + z(5\bar{i} + \bar{j} - 2\bar{k}) = \bar{0}$.
It implies that $-x + 3y + 5z = 0, -2x + z = 0, 7x + y - 2z = 0$.

$$\text{Coefficient matrix} = \begin{bmatrix} -1 & 3 & 5 \\ -2 & 0 & 1 \\ 7 & 1 & -2 \end{bmatrix} \sim \begin{bmatrix} -1 & 3 & 5 \\ 0 & -6 & -9 \\ 0 & 22 & 33 \end{bmatrix} \sim \begin{bmatrix} -1 & 3 & 5 \\ 0 & 2 & 3 \\ 0 & 0 & 0 \end{bmatrix}.$$

$\therefore$ Equations are equivalent to $-x + 3y + 5z = 0, 2y + 3z = 0$.
Taking $z =$ $\quad y = -3, x = 1$. Again, $x + y + z = 0$. Therefore, the points are collinear.]

(iii) Under what condition $\bar{a} \times (\bar{b} \times \bar{c}) = (\bar{a} \times \bar{b}) \times \bar{c}$ is satisfied?
[Hints. See Note 2 of page 299.]

(iv) Find the area of the parallelogram formed by two vectors $3\bar{i} + 2\bar{j}$ and $2\bar{j} - 4\bar{k}$.
[Hints.

$$\text{Area} = |(3\bar{i} + 2\bar{j}) \times (2\bar{j} - 4\bar{k})| = |6\bar{k} + 12\bar{j} - 8\bar{i}|$$
$$= \sqrt{6^2 + 12^2 + 8^2} = \sqrt{244} \text{ sq. units.}]$$

Group C

[Answer any two questions.]

8. (a) Reduce the equation $5x^2 - 2xy + 5y^2 - 6x + 6y + 2 = 0$ to canonical form and state the nature of the conic represented by it.
[Hints. Here $\Delta = -24, D = 24$.
$\therefore$ the equation represents an ellipse.
Its canonical form is $(5 - \sqrt{2})x'^2 + (5 + \sqrt{2})y'^2 = 1$.]

(b) Determine the angle through which the axes must be turned so that the equation $lx + my + n = 0 (l \neq 0)$ may reduce to the form $ax + b = 0$.
[Hints. Let the axes be rotated through an angle θ. The equation $lx + my + n = 0$ transforms to $l(x' \cos\theta - y' \sin\theta) + m(x' \sin\theta + y' \cos\theta) + n = 0$.
Here coefficient of $y' = 0$, i.e., $m\cos\theta - l\sin\theta = 0$, or, $\tan\theta = \frac{m}{l}$.]

(c) The gradient of one of the straight lines represented by $ax^2 + 2hxy + by^2 = 0$ is twice that of the other. Show that $8h^2 = 9ab$.
[Hints. Let

$$y^2 + \frac{2h}{b}xy + \frac{a}{b}x^2 = (y - mx)(y - 2mx).$$

Equating the coefficients,

$$3m = -\frac{2h}{b}, 2m^2 = \frac{a}{b}. \quad \therefore \quad 2 \cdot \frac{4h^2}{9b^2} = \frac{a}{b}, \quad \text{or,} \quad 8h^2 = 9ab.]$$

9. (a) Find the condition that straight line $r\cos(\theta - \alpha) = p$ touches the conic

$$\frac{l}{r} = 1 + e\cos\theta.$$

[Hints. Let $r\cos\theta\cos\alpha + r\sin\theta\sin\alpha = p$ touch the conic at $\theta = \beta$. The equation of the tangent at $\theta = \beta$ is $\frac{l}{r} = e\cos\theta + \cos(\theta - \beta)$. It is identical with the given line.

$$\therefore \quad \frac{e + \cos\beta}{\cos\alpha} = \frac{\sin\beta}{\sin\alpha} = \frac{l}{p}$$

From these relations

$$\cos\beta = \frac{l\cos\alpha}{p} - e, \; \sin\beta = \frac{l}{p}\sin\alpha.$$

$$\therefore \quad 1 = \left(\frac{l\cos\alpha}{p} - e\right)^2 + \left(\frac{l\sin\alpha}{p}\right)^2, \quad \text{or,} \quad l^2 - 2lep\cos\alpha + e^2p^2 = p^2.]$$

(b) Find the angle between the planes $x - y + 2z = 9$ and $2x + y + z = 7$.
[Hints.

$$\cos\theta = \frac{1.2 - 1.1 + 2.1}{\sqrt{6}.\sqrt{6}} = \frac{1}{2} \quad \therefore \quad \theta = \frac{\pi}{3}.]$$

(c) A variable plane is always at a constant distance p from the origin and meets the axes in A, B, C. Prove that the locus of the points of intersection of the three planes through A, B, C parallel to the co-ordinate planes is given by $\frac{1}{x^2} + \frac{1}{y^2} + \frac{1}{z^2} = \frac{1}{p^2}$.
[Hints. See Q. No. 6.(a)(ii) of B.U. 2002]

10. (a) Find the equation of the plane which bisects the acute angle between the planes $x + 2y + 2z = 9$ and $4x - 3y + 12z + 13 = 0$.

[**Hints.** See Ex.9 of page 174.]

(b) Find the equation of the plane through the point $(2, -3, 1)$ normal to the line joining the points $(3, 4, -1)$ and $(2, -1, 5)$.

[**Hints.** The d.rs. of the line segment are $1, 5, -6$.

Let the equation of the plane be $a(x-2) + b(y+3) + c(z-1) = 0$.

Here $\frac{a}{1} = \frac{b}{5} = \frac{c}{-6}$.

$\therefore$ the equation of the plane is

$$x - 2 + 5(y+3) - 6(z-1) = 0, \quad \text{or,} \quad x + 5y - 6z + 19 = 0.]$$

(c) For what value of h the planes $3x - 2y + hz - 1 = 0$ and $x + hy + 5z + 2 = 0$ may be perpendicular to each other.

[**Hints.** Here $3.1 - 2.h + h.5 = 0$, or, $h = -1$.]

11. (a) Find the equation to the sphere whose centre lies on the line $\frac{x+1}{3} = \frac{y-3}{-4} = \frac{z+2}{5}$ and which passes through the points $(3, 4, 5)$ and $(-3, 0, 1)$. Find the co-ordinates of the centre of the sphere.

[**Hints.** Let (u, v, w) be the centre. Then $\frac{u+1}{3} = \frac{v-3}{-4} = \frac{w+2}{5} = \lambda$ (say).

Let the equation of the sphere be $(x-u)^2 + (y-v)^2 + (z-w)^2 = r^2$.

Here

$$\begin{aligned}
&(3-u)^2 + (4-v)^2 + (5-w)^2 = r^2 = (-3-u)^2 + v^2 + (1-w)^2, \\
&\text{or, } 3u + 2v + 2w = 10, \quad \text{or,} \quad 3(3\lambda - 1) + 2(-4\lambda + 3) + 2(5\lambda - 2) = 10, \\
&\text{or, } \lambda = \tfrac{1}{11}.
\end{aligned}$$

Thus the centre is $(-\frac{8}{11}, -\frac{29}{11}, -\frac{17}{11})$.]

(b) Show that the lines $\frac{x+3}{2} = \frac{y+5}{3} = \frac{z-7}{-3}$ and $\frac{x+1}{4} = \frac{y+1}{5} = \frac{z+1}{-1}$ are coplanar and find the equation of the plane containing them.

[**Hints.** Here

$$\begin{vmatrix} -1-(-3) & -1-(-5) & -1-7 \\ 4 & 5 & -1 \\ 2 & 3 & -3 \end{vmatrix} = \begin{vmatrix} 2 & 4 & -8 \\ 2 & 2 & 2 \\ 2 & 3 & -3 \end{vmatrix} = \begin{vmatrix} 2 & 4 & -8 \\ 0 & -2 & 10 \\ 0 & -1 & 5 \end{vmatrix} = 0.$$

$\therefore$ lines are coplanar. The equation of the plane is

$$\begin{vmatrix} x+3 & y+5 & z-7 \\ 4 & 5 & -1 \\ 2 & 3 & -3 \end{vmatrix} = 0, \quad \text{or,} \quad 6x - 5y - z = 0.]$$

Group D

[Answer **any one** question.]

12. **(a)** **Prove by vector method that in any triangle ABC, $\cos C = \frac{a^2+b^2-c^2}{2ab}$.**

[Hints. See Ex.5 of page 301.]

(b) **If $\bar{a}.\bar{b} = \bar{a}.\bar{c}$ and $\bar{a} \times \bar{b} = \bar{a} \times \bar{c} (\bar{a} \neq \bar{0})$ are satisfied simultaneously, then prove that $\bar{b} = \bar{c}$.**

[Hints. $\bar{a}.\bar{b} = \bar{a}.\bar{c}$, or, $\bar{a}.(\bar{b} - \bar{c}) = 0$. $\bar{a}$ is perpendicular to $\bar{b} - \bar{c}$ or, $\bar{b} = \bar{c}$.

$$\bar{a} \times \bar{b} = \bar{a} \times \bar{c}, \quad \text{or,} \quad \bar{a} \times (\bar{b} - \bar{c}) = 0.$$

$\bar{a}$ is parallel to $\bar{b} - \bar{c}$, or, $\bar{b} = \bar{c}$. $\bar{a}$ cannot be simultaneously parallel and perpendicular to $\bar{b} - \bar{c}$. $\therefore$ $\bar{b} = \bar{c}$.]

(c) **Prove that the three vectors $\bar{a} = 2\bar{i} - \bar{j} + \bar{k}$, $\bar{b} = \bar{i} - 3\bar{j} - 5\bar{k}$ and $\bar{c} = 3\bar{i} - 4\bar{j} - 4\bar{k}$ form the sides of a right-angled triangle.**

[Hints. See Ex.13 of page 304.]

13. **(a)** **Find the torque about the point $\bar{i} + 2\bar{j} - \bar{k}$ of a force represented by $3\bar{i} + \bar{j} + \bar{k}$ acting at a point $2\bar{i} - \bar{j} + 3\bar{k}$.**

[Hints. Let the point be A and the point on the force be B. $\bar{F}$ is the force.

Torque $= \overrightarrow{AB} \times \bar{F} = (\bar{i} - 3\bar{j} + 4\bar{k}) \times (3\bar{i} + \bar{j} + \bar{k}) = -7\bar{i} + 11\bar{j} + 10\bar{k}$.]

(b) **State and prove the necessary and sufficient condition for collinearity of three distinct points A, B, C whose position vectors are $\bar{\alpha}, \bar{\beta}$ and $\bar{\gamma}$.**

[Hints. See the section 1.6 of page 282.]

(c) **Find the vector equation of the plane passing through the origin and parallel to the vectors $\bar{i} + 2\bar{j} + 3\bar{k}$ and $2\bar{i} - \bar{j} - 4\bar{k}$.**

[Hints. $\bar{r} \cdot \{(\bar{i} + 2\bar{j} + 3\bar{k}) \times (2\bar{i} - \bar{j} - 4\bar{k})\} = 0$.]

Burdwan University

2000

[First Paper]

Group A

1. Answer the following:

(viii) Find the polar coordinates of the centre of the circle $r = a\cos\theta + b\sin\theta$.
[**Hints.** If $a = 2d\cos\alpha$ and $b = 2d\sin\alpha$, then $r = 2d\cos(\theta-\alpha)$.
$\therefore$ coordinates of the centre are (d, α) where d is the radius of the circle.]

(ix) Find the vertex and latus rectum of the curve $3x^2 + 12x - 8y = 0$.
[**Hints.**

$$3x^2 + 12x - 8y = 0,$$
$$\text{or, } 3(x+2)^2 = 8y + 12,$$
$$\text{or, } (x+2)^2 = \tfrac{8}{3}\left(y + \tfrac{3}{2}\right).$$

It is a parabola. Vertex is $(2, \frac{3}{2})$ and latus rectum $= \frac{8}{3}$ units.]

(x) Find the equation to the curve $9x^2 + 4y^2 + 18x - 16y = 11$ referred to parallel axes through the point $(-1, 2)$.
[**Hints.** Transformed equation is

$$9(x-1)^2 + 4(y+2)^2 + 18(x-1) - 16(y+2) = 11, \quad \text{or,} \quad 9x^2 + 4y^2 = 36.]$$

(xi) If a straight line makes angles A, B, C respectively with the axes of coordinates, show that $\sin^2 A + \sin^2 B + \sin^2 C = 2$.
[**Hints.** See **Corollary 2** of Page 158.]

(xii) Obtain in the symmetrical form the equation of the straight line

$$x - 2y + 3z = 4, 2x - 3y + 4z = 5.$$

[**Hints.** See Ex.1 of Page 185.]

(xiii) If the point P moves such that its distance from z-axis is same as its distance from xy-plane, find the locus of P.
[**Hints.** If (x_1, y_1, z_1) are coordinates of P at a certain position, then

$$\sqrt{x_1^2 + y_1^2} = z_1. \quad \therefore \text{ locus is } x^2 + y^2 = z^2.]$$

(xiv) Find the value of n if $\bar{a} = 4\bar{i} + n\bar{j} - 3\bar{k}$ and $|\bar{a}| = 13$.
[**Hints.** Here $4n^2 + n^2 + 3^2 = 13^2$, or, $n^2 = 144$, or, $n = \pm 12$.]

(xv) If $\bar{a} = 2\bar{i} + 3\bar{j}$, $\bar{b} = \bar{i} + \bar{j} - \bar{k}$, $\bar{c} = 3\bar{i} - \bar{k}$, find $[\bar{a}\bar{b}\bar{c}]$.
[**Hints.** $\bar{b} \times \bar{c} = \bar{i} - 2\bar{j} - 3\bar{k}$. $[\bar{a}\bar{b}\bar{c}] = \bar{a}.(\bar{b} \times \bar{c}) = -2 - 6 = -8$.]

Group C

5. Answer either (a) or (b):

(a) (i) Transform the polar equation $\sqrt{r}\cos\frac{\theta}{2} = \sqrt{a}$ to a Cartesian equation.
[**Hints.**

$$\sqrt{r}\cos\frac{\theta}{2} = \sqrt{a}, \quad \text{or,} \quad r\cos^2\frac{\theta}{2} = a, \quad \text{or,} \quad r(1+\cos\theta) = 2a.$$

$$\therefore \quad \sqrt{x^2+y^2} + x = 2a, \quad \text{or,} \quad x^2 + y^2 = (2a - x)^2, \quad \text{or,} \quad y^2 = 4a(a-x).]$$

(ii) If by rotation of rectangular axes about the origin, the expression $(ax^2 + 2hxy + by^2)$ changes to $(c'x'^2 + 2hx'y' + b'y'^2)$, then prove that $a + b = a' + b'$.
[**Hints.** See the Section 1.7 (iv) of page 9.]

(b) (i) Find the nature of the following conic, $\frac{2}{r} = 3 - 3\cos\theta$.

[**Hints.** $\frac{2}{r} = 3 - 3\cos\theta$, or, $\frac{2/3}{r} = 1 - \cos\theta$. It is the equation of a parabola with semi-latus rectum $\frac{2}{3}$.]

(ii) Show that the equation $4x^2 - 4xy + y^2 + 2x - 26y + 9 = 0$ represents a parabola whose latus rectum is $2\sqrt{5}$ units.

[**Hints.** The equation can be written as

$$(2x - y)^2 = -2x + 26y - 9,$$
$$\text{or, } (2x - y + \lambda)^2 = (4\lambda - 2)x + (26 - 2\lambda)y + \lambda^2 - 9.$$

λ is chosen in such a way that $2x - y + \lambda = 0$ and

$$(4\lambda - 2)x + (26 - 2\lambda)y + \lambda^2 - 9 = 0$$

are at right angles.

For this $2 \cdot (4\lambda - 2) - 1 \cdot (26 - 2\lambda) = 0$, or, $\lambda = 3$.

$$\therefore \quad (2x - y + 3)^2 = 10x + 20y = 10(x + 2y),$$
$$\text{or, } \left(\frac{2x-y+3}{\sqrt{5}}\right)^2 . 5 = 10 \cdot \frac{x+2y}{\sqrt{5}} \cdot \sqrt{5}.$$

Putting $\frac{2x-y+3}{\sqrt{5}} = Y$, $\frac{x+2y}{\sqrt{5}} = X$, $Y^2 = 2\sqrt{5}X$. It is the equation of a parabola with latus rectum $2\sqrt{5}$ units. Hence, the result follows.]

6. Answer *any three* questions:

(a) (i) Find the distance of the point $(1, 2, 0)$ from the plane $4x + 3y + 12z + 16 = 0$.

[**Hints.**

$$\text{Distance} = \left|\frac{4.1 + 3.2 + 0 + 16}{\sqrt{4^2 + 3^2 + 12^2}}\right| \text{ units} = 2 \text{ units.]}$$

(ii) A variable plane which is at a constant distance $3p$ from the origin O cuts the axes in A, B, C. Show that the locus of the centroid of the triangle ABC is $\frac{1}{x^2} + \frac{1}{y^2} + \frac{1}{z^2} = \frac{1}{p^2}$.

[**Hints.** See Ex. 18 of page 181.]

(b) (i) Show that the equations of the straight line through the point $(3, 4, 5)$ which is equally inclined to the axes are $x - 3 = y - 4 = z - 5$.

[**Hints.** If the line makes angle α with each coordinate axis, then the equations of the line are

$$\frac{x-3}{\cos\alpha} = \frac{y-4}{\cos\alpha} = \frac{z-5}{\cos\alpha}, \quad \text{or,} \quad x - 3 = y - 4 = z - 5.]$$

(ii) Find the image of the point $(1, -2, 3)$ in the plane $2x - 3y + 2z + 3 = 0$.

[**Hints.** The equations of the line passing through the point $(1, -2, 3)$ and perpendicular to the plane are

$$\frac{x-1}{2} = \frac{y+2}{-3} = \frac{z-3}{2}.$$

Let the image point be $(2r + 1, -3r - 2, 2r + 3)$.

The midpoint of the line segment joining the given point and the image point is

$$\left(r + 1, \quad \frac{-3r - 4}{2}, \quad r + 3\right).$$

This point lies on the plane.

$$\therefore \quad 2(r + 1) + 3 \cdot \frac{3r + 4}{2} + 2(r + 3) + 3 = 0, \quad \text{or,} \quad r = -2.$$

$\therefore$ the image point is $(-3, 4, -1)$.]

(c) (i) Show that the equation of the plane making intercepts a, b, c on the coordinate axes is $\frac{x}{a} + \frac{y}{b} + \frac{z}{c} = 1$.

[**Hints.** See the Section 3.3 of page 168.]

(ii) Find the equation of the plane passing through the line of intersection of the planes $x+2y+2z-4=0$, $2x+y-z+5=0$ and perpendicular to the plane $4x+5y-3z=8$.

[**Hints.** Let the plane be

$$x+2y+2z-4+\lambda(2x+y-z+5)=0,$$

$$\text{or,}\quad (1+2\lambda)x+(2+\lambda)y+(2-\lambda)z+5\lambda-4=0.$$

It is perpendicular to the plane $4x+5y-3z=8$.

$$\therefore\quad 4(1+2\lambda+5(2+\lambda)-3(2-\lambda)=0,\quad \text{or,}\quad \lambda=-\frac{1}{2}.$$

$\therefore$ the equation of the plane is $3y+5z-13=0$.]

(d) (i) Find the distance between the parallel planes $x+2y-2z+1=0$ and $2x+4y-4z+5=0$.

[**Hints.** Distances of the planes from the origin $\frac{1}{3}$ and $\frac{5}{6}$.

$\therefore$ required distance $=\frac{5}{6}-\frac{1}{3}=\frac{1}{2}$ unit.]

(ii) Find the shortest distance between the straight lines $\frac{x-2}{1}=\frac{y-1}{-2}=\frac{z-6}{1}$ and $\frac{x+3}{7}=\frac{y+3}{-6}=\frac{z+3}{1}$.

[**Hints.** $A(2,1,6)$ and $B(-3,-3,-3)$ are points on the lines.

The d.cs. of AB are $\frac{5}{AB},\frac{4}{AB},\frac{9}{AB}$. If l,m,n are d.cs. of s.d., then $l-2m+n=0$ and $7l-6m+n=0$. D.cs. of s.d. are $\frac{2}{\sqrt{29}},\frac{3}{\sqrt{29}},\frac{4}{\sqrt{29}}$.

If θ is the angle between AB and s.d.. then s.d. $=AB\cos\theta=2\sqrt{29}$ units.]

(e) (i) Find the equation of the tangent plane to the sphere $x^2+y^2+z^2=5$ at the point $(2,0,1)$.

[**Hints.** The equation of the tangent plane is $2x+z=5$.]

(ii) Find the equation of the cone whose vertex is $(1,2,3)$ and guiding curve is the circle $x^2+y^2+z^2=4, x+y+z=1$.

[**Hints.** Let $\frac{x-1}{l}=\frac{y-2}{m}=\frac{z-3}{n}$ be a generator.

If it meets the base at $(lr+1, mr+2, nr+3)$ then

$$lr+1+mr+2+nr+3=1,\quad\text{or,}\quad (l+m+n)r+5=0 \tag{1}$$

$$\text{and}\quad (lr+1)^2+(mr+2)^2+(nr+3)^2=4$$

$$\text{or,}\quad (l^2+m^2+n^2)r^2+(2l+4m+6n)r+10=0. \tag{2}$$

By (1) and (2),

$$25(l^2+m^2+n^2)-5(2l+4m+6n)(l+m+n)+10(l+m+n))^2=0,$$

$$\text{or,}\quad 5(l^2+m^2+n^2)-2(l+2m+3n))(l+m+n)+2(l+m+n)^2=0.$$

Eliminating l,m,n by the equation of the generator the equation of the cone is

$$5\{(x-1)^2+(y-2)^2+(z-3)^2\}-2(x+2y+3z-14)$$
$$(x+y+z-6)+2(x+y+z-6)^2=0.]$$

7. Answer either (a) or (b):

(a) (i) Prove that $\bar{a}\times(\bar{b}+\bar{c})+\bar{b}\times(\bar{c}+\bar{a})+\bar{c}\times(\bar{a}+\bar{b})=0$.

[**Hints.**

$$\bar{a}\times(\bar{b}+\bar{c})+\bar{b}\times(\bar{c}+\bar{a})+\bar{c}\times(\bar{a}+\bar{b})$$
$$=a\times\bar{b}+\bar{a}\times\bar{c}+\bar{b}\times\bar{c}+\bar{b}\times\bar{a}+\bar{c}\times\bar{a}+\bar{c}\times\bar{b}$$
$$=\bar{a}\times\bar{b}-\bar{c}\times\bar{a}+\bar{b}\times\bar{c}-\bar{a}\times\bar{b}+\bar{c}\times\bar{a}-\bar{b}\times\bar{c}=\bar{0}.]$$

(ii) Prove that the three points whose position vectors are $\bar{i}+2\bar{j}+3\bar{k}$, $-\bar{i}-2\bar{j}+8\bar{k}$ and $-4\bar{i}+4\bar{j}+6\bar{k}$ form an equilateral triangle.

[**Hints.** Let the position vectors of the points A,B,C be the given vectors. Then

$$\overrightarrow{AB}=-2\bar{i}-4\bar{j}+5\bar{k},\ \overrightarrow{BC}=-3\bar{i}+6\bar{j}-2\bar{k},\ \overrightarrow{CA}=5\bar{i}-2\bar{j}-3\bar{k}.$$
$$|\overrightarrow{AB}|=\sqrt{45},\ |\overrightarrow{BC}|=7,\ |\overrightarrow{CA}|=\sqrt{38}.$$

Sides are not equal, so the triangle is not equilateral.]

(b) (i) Prove by vector method that for any triangle ABC, $\cos A = \frac{b^2+c^2-a^2}{2bc}$ with usual notations.

[**Hints.** See Ex.5 of page 301.]

(ii) Prove that $\bar{i} \times (\bar{a} \times \bar{i}) + \bar{j} \times (\bar{a} \times \bar{j}) + \bar{k} \times (\bar{a} \times \bar{k}) = 2\bar{a}$.

[**Hints.**

$$\begin{aligned} &\bar{i} \times (\bar{a} \times \bar{i}) + \bar{j} \times (\bar{a} \times \bar{j}) + \bar{k} \times (\bar{a} \times \bar{k}) \\ &= (\bar{i}.\bar{i})\bar{a} - (\bar{i}.\bar{a})\bar{i} + (\bar{j}.\bar{j})\bar{a} - (\bar{j}.\bar{a})\bar{j} + (\bar{k}.\bar{k})\bar{a} - (\bar{k}.\bar{a})\bar{k} \\ &= 3\bar{a} - |\bar{a}|(\bar{i}\cos\alpha + \bar{j}\cos\beta + \bar{k}\cos\gamma) = 3\bar{a} - \bar{a} = 2\bar{a}. \end{aligned}$$

$\bar{a}$ makes angles α, β, γ with the axes.]

2001

Part I

1. Answer the following:

(h) Find the nature of the curve whose polar equation is given by $r = \frac{12}{2-\cos\theta}$.

[**Hints.** The equation is $\frac{6}{r} = 1 - \frac{1}{2}\cos\theta$. It represents an ellipse with eccentricity $\frac{1}{2}$ and semi-latus rectum = 6 units.]

(i) By shifting origin to the point (α, β) the equations $x + y = 1$ and $2x - y = 1$ are transformed to $a'x' + b'y' = 0$ type. Find (α, β).

[**Hints.** The equations transform to

$$x' + y' + \alpha + \beta = 1 \quad \text{and} \quad 2x' - y' + 2\alpha - \beta = 1.$$

$$\therefore \quad \alpha + \beta = 1 \quad \text{and} \quad 2\alpha - \beta = 1.$$

$$\therefore \quad \alpha = \frac{2}{3}, \quad \beta = \frac{1}{3}.]$$

(j) Find the polar equation of a circle when the pole is on circumference and the initial line is along a diameter.

[**Hints.** See **Corollary 3** of the Section 8.6 of page 122.]

(k) Justify with proper logic if it is possible to have a line making angles 45°, 135° and 60° with x-axis, y-axis and z-axis respectively.

[**Hints.**

$$\cos^2 45^\circ + \cos^2 135^\circ + \cos^2 60^\circ = \frac{1}{2} + \frac{1}{2} + \frac{1}{4} \neq 1.$$

$\therefore$ a line cannot make the given angles with the axes.]

(l) Find the coordinates of the point where the line

$$x - 4y + z - 7 = 0 = x + 2y - z + 8$$

meets xz plane.

[**Hints.** On the xz plane $y = 0$.

$$\therefore \quad x + z = 7 \quad \text{and} \quad x - z = -8.$$

From these $x = -\frac{1}{2}, z = \frac{15}{12}$. Thus the point is $(-\frac{1}{2}, 0, \frac{15}{2})$.]

(m) Find the equation of the sphere through (0, 0, 0) and (2, −2, 4) as extrimity of the diameter.

[**Hints.** The equation is

$$x(x-2) + y(y+2) + z(z-4) = 0, \quad \text{or,} \quad x^2 + y^2 + z^2 - 2x + 2y - 4z = 0.]$$

(n) Find the vector parallel to the vector $\bar{\alpha} = \bar{i} + 2\bar{j} + 3\bar{k}$ having length 5 units.

[**Hints.** Let the vector be $\lambda\bar{\alpha}$. Then $\lambda|\bar{\alpha}| = 5$, or, $\lambda\sqrt{14} = 5$, or, $\lambda = \frac{5}{\sqrt{14}}$.

Thus the required vector is $\frac{5}{\sqrt{14}}(\bar{i} + 2\bar{j} + 3\bar{k})$).]

(o) **Find sine of the angle between the vectors $2\bar{i} - \bar{j} + \bar{k}$ and $3\bar{i} + 4\bar{k} - \bar{j}$.**
[**Hints.** If θ is the angle between the vectors, then

$$|(2\bar{i} - \bar{j} + \bar{k}) \times (3\bar{i} - \bar{j} + 4\bar{k})| = |2\bar{i} - \bar{j} + \bar{k}||3\bar{i} - \bar{j} + 4\bar{k}) \sin\theta,$$

$$\text{or,} \quad \sin\theta = \frac{\sqrt{35}}{\sqrt{6}\sqrt{26}}.]$$

Part II

Group B

5. **Answer either (a) or (b):**

(a) (i) **If r_1 and r_2 be two mutually perpendicular radius vectors of the ellipse $r^2 = \frac{b^2}{1-e^2\cos^2\theta}$, then show that $\frac{1}{r_1^2} + \frac{1}{r_2^2} = \frac{1}{a^2} + \frac{1}{b^2}$ where $e^2 = 1 - \frac{b^2}{a^2}$.**
[**Hints.**

$$\frac{1}{r^2} = \frac{1 - e^2\cos^2\theta}{b^2} \quad \therefore \quad \frac{1}{r_1^2} = \frac{1 - e^2\cos^2\alpha}{b^2}$$

$$\text{and} \quad \frac{1}{r_2^2} = \frac{1 - e^2\cos^2\left(\frac{\pi}{2} + \alpha\right)}{b^2} = \frac{1 - e^2\sin^2\alpha}{b^2}$$

where α is the vectorial angle corresponding to the radius vector r_1.
Now

$$\frac{1}{r_1^2} + \frac{1}{r_2^2} = \frac{2 - e^2}{b^2} = \frac{1 + \frac{b^2}{a^2}}{b^2} = \frac{1}{a^2} + \frac{1}{b^2}.]$$

(ii) **Find the angle through which the axes are to be rotated so that the equation $17x^2 + 18xy - 7y^2 = 1$ is reduced to the form $Ax^2 + By^2 = 1$.**
[**Hints.** Let the axes be rotated through an angle θ. The equation transforms to

$$17(x\cos\theta - y\sin\theta)^2 + 18(x\cos\theta - y\sin\theta)(x\sin\theta + y\cos\theta)$$
$$-7(x\sin\theta + y\cos\theta)^2 = 1,$$
$$\text{or,} \quad (17\cos^2\theta + 18\sin\theta\cos\theta - 7\sin^2\theta)x^2$$
$$+\{18(\cos^2\theta - \sin^2\theta) - 48\sin\theta\cos\theta\}xy$$
$$+(17\sin^2\theta - 18\sin\theta\cos\theta - 7\cos^2\theta)y^2 = 1.$$

Since the form is $Ax^2 + By^2 = 1$, the coefficient of $xy = 0$, i.e.,

$$18\cos 2\theta - 24\sin 2\theta = 0, \quad \text{or,} \quad \tan 2\theta = \frac{3}{4}.$$

Now,

$$A = \frac{17}{2}(1 + \cos 2\theta) + 9\sin 2\theta - \frac{7}{2}(1 - \cos 2\theta)$$
$$= \frac{17}{2}\left(1 + \frac{4}{5}\right) + 9\cdot\frac{3}{5} - \frac{7}{2}\left(1 - \frac{4}{5}\right) = 20.$$

$$B = 10. \quad \theta = \frac{1}{2}\tan^{-1}\frac{3}{4}.]$$

(b) (i) **If the latus rectum of a conic is 6 units and its eccentricity is $\frac{1}{2}$, find the length of the focal chord making an angle 45° with the major axis.**
[**Hints.** The equation of the conic is $\frac{3}{r} = 1 - \frac{1}{2}\cos\theta$. The vectorial angles of the ends of the focal chord are 45° and 180° + 45°.

$$\frac{3}{r_1} = 1 - \frac{1}{2}\cos 45^\circ = 1 - \frac{1}{2\sqrt{2}}, \quad \text{or,} \quad r_1 = \frac{6\sqrt{2}}{2\sqrt{2} - 1}$$

$$\frac{3}{r_2} = 1 - \frac{1}{2}\cos(180^\circ + 45^\circ) = 1 + \frac{1}{2\sqrt{2}}, \quad \text{or,} \quad r_2 = \frac{6\sqrt{2}}{2\sqrt{2} + 1}.$$

$$\therefore \quad r_1 + r_2 = \frac{48}{7} \quad \text{units.}]$$

(ii) If $ax+by$ annd $cx+dy$ are changed to $a'x'+b'y'$ and $c'x'+d'y'$ respectively by rotation of axes, show that $ad-bc=a'd'-b'c'$.
[**Hints.** Let the axes be rotated through an angle θ.
Then $ax+by$ transforms to $(a\cos\theta+b\sin\theta)x'+(b\cos\theta-a\sin\theta)y'$.

$$\therefore \quad a'=a\cos\theta+b\sin\theta, \quad b'=b\cos\theta-a\sin\theta.$$

Similarly, $c'=c\cos\theta+d\sin\theta, d'=d\cos\theta-c\sin\theta$.
Now

$$\begin{aligned} a'd'-b'c' &= (a\cos\theta+b\sin\theta)(d\cos\theta-c\sin\theta) \\ &\quad -(b\cos\theta-a\sin\theta)(c\cos\theta+d\sin\theta) \\ &= ad-bc.] \end{aligned}$$

6. **Answer** *any three* **of the following:**

(a) (i) Find the equation of the plane which bisects the line joining $(1,2,3)$ and $(3,4,5)$ at right angles and show that it makes equal intercepts on axes.
[**Hints.** See Ex.4 of page 177.]

(ii) Find the coordinates of the points on x-axis and y-axis which are equidistant from the planes $x+2y+2z=9$ and $x+2y+2z=27$.
[**Hints.** Let the points be $(x_1,0,0)$ and $(0,y_2,0)$.
Here, $x_1-27=-x_1+9$, or, $x_1=18$ and $2y_2-27=-2y_2+9$, or, $y_2=9$.]

(b) (i) Find the equation of image of the line $\frac{x}{2}=\frac{y-2}{3}=\frac{z-3}{4}$ on the plane

$$x+2y+z=1.$$

[**Hints.** Let the line meet the plane at the point $(2r, 3r+2, 4r+3)$.
Then $2r+6r+4+4r+3=1$, or, $12r=-6$, or, $r=-\frac{1}{2}$. Therefore, the point is $(-1,\frac{1}{2},1)$. If the image of the point $(0,2,3)$) with respect to the plane is (x_1,y_1,z_1), then $\left(\frac{x_1}{2},\frac{y_1+2}{2},\frac{z_1+3}{2}\right)$ lies on the plane. Moreover, the line joining the points is perpendicular to the plane.

$$\therefore \quad \frac{x_1}{1}=\frac{y_1-2}{2}=\frac{z_1-3}{1}=\lambda \text{ (say)}.$$

$$\frac{x_1}{2}+y_1+2+\frac{z_1+3}{2}=1,$$

$$\text{or,} \quad \frac{\lambda}{2}+2\lambda+2+2+\frac{\lambda+3+3}{2}=1, \quad \text{or,} \quad \lambda=-2.$$

$\therefore$ the image point is $(-2,-2,1)$.
$\therefore$ the required line passes through the points $(1,\frac{1}{2},1)$ and $(-2,-2,2)$.
The equatiions of it are

$$\frac{x+2}{1}=\frac{2(y+2)}{5}=\frac{z-1}{0}, \quad \text{or,} \quad \frac{x+2}{2}=\frac{y+2}{5}, \ z=1.]$$

(ii) Find the plane containing the lines $\frac{x-5}{3}=\frac{y-7}{-1}=\frac{z+2}{2}$ and $\frac{x+3}{-3}=\frac{y-3}{2}=\frac{z-6}{-4}$. Also find its distance from the point $(2,7,-5)$.
[**Hints.** Let the plane be $ax+by+cz+d=0$.
Since the lines lie on the plane $3a-b+2c=0$ and $-3a+2b-4c=0$.
$\therefore$ $a=0$, $b=2c$. $(5,7,-2)$ lies on the plane.
$\therefore$ $5a+7b-2c+d=0$, or, $d=-12c$.
$\therefore$ the plane is $2y+z-12=0$. Distance from the point

$$(2,7,-5)=\left|\frac{14-5-12}{\sqrt{5}}\right|=\frac{3}{\sqrt{5}} \text{ units.}]$$

(c) (i) Find the distance of the point $(-4, 5, 5)$ from the line $\frac{x+7}{8} = \frac{y-5}{-4} = \frac{z-9}{1}$.
[**Hints.** Let the point A be $(-7, 5, 9)$ and the point P be $(-4, 5, 5)$. D.cs. of AP are $\frac{3}{AP}, 0, \frac{-4}{AP}$. D.cs. of the given line are $\frac{8}{9}, \frac{-4}{9}, \frac{1}{9}$. If θ is the angle between AP and the line, then the distance

$$= AP \sin\theta = AP\sqrt{1-\cos^2\theta} = AP\sqrt{1-\left(\frac{20}{9AP}\right)^2}$$

$$= \sqrt{AP^2 - \frac{400}{81}} = \sqrt{25 - \frac{400}{81}} = \frac{5}{9}\sqrt{65} \text{ units.}]$$

(ii) Determine the value of k so that the lines $\frac{x-1}{2} = \frac{y-4}{1} = \frac{z-5}{2}$ and $\frac{x-2}{-1} = \frac{y-8}{k} = \frac{z-11}{4}$ may intersect.
[**Hints.** See Ex.3 of page 198.]

(d) (i) If $2c$ is the shortest distance between the lines $\frac{x}{l} - \frac{z}{n} = 1, y = 0$ and $\frac{y}{m} + \frac{z}{n} = 1, x = 0$, show that $\frac{1}{c^2} = \frac{1}{l^2} + \frac{1}{m^2} + \frac{1}{n^2}$.
[**Hints.** See Ex.9 of page 201.]

(ii) Find the equation of the sphere which passes through the circle $x^2 + y^2 = 4$, $z = 0$ and is cut by the plane $x + 2y + 2z = 0$ in a circle of radius 3 units.
[**Hints.** See Ex.11 of page 221.]

(e) (i) Find the greatest and least distance from the point $(1, -1, 2)$ to the sphere

$$x^2 + y^2 + z^2 - 4x + 6y - 8z - 71 = 0.$$

[**Hints.** The centre is $(2, -3, 4)$ and radius $= \sqrt{2^2 + (-3)^2 + 4^2 + 71} = 10$.
Distance between $(2, -3, 4)$) and $(1, -1, 2) = \sqrt{1^2 + (-2)^2 + 2^2} = 3 < 10$.
$\therefore$ the point $(1, -1, 2)$ is inside the sphere.
$\therefore$ greatest distance $= 10 + 3 = 13$ units and least distance $= 10 - 3 = 7$ units.]

(ii) Determine the angle between the lines in which the plane $x - 3y + z = 0$ cuts the quadric cone $x^2 - 5y^2 + z^2 = 0$.
[**Hints.** Let the plane cut the cone by the plane $\frac{x}{l} = \frac{y}{m} = \frac{z}{n}$.
Then $l - 3m + n = 0$ and $l^2 - 5m^2 + n^2 = 0$.
Eiminating n,

$$l^2 - 5m^2 + (l - 3m)^2 = 0,$$

$$\text{or,}\quad l^2 - 3lm + 2m^2 = 0, \quad \text{or,} \quad (l-m)(l-2m) = 0.$$

$$\therefore\ l - m = 0, \quad \text{or,} \quad l - 2m = 0.$$

If $l = m$, then $n = 2m$. $\frac{l}{1} = \frac{m}{1} = \frac{n}{2}$.
If $l = 2m$, then $n = m$.

$$\therefore\ \frac{l}{2} = \frac{m}{1} = \frac{n}{1}. \quad \therefore \text{ the required angle} = \cos^{-1}\frac{5}{6}.]$$

7. Answer either (a) or (b):

(a) (i) If $\bar{a} = 2\bar{i} + 3\bar{j} + 6\bar{k}$ and $\bar{b} = 3\bar{i} - 6\bar{j} + 2\bar{k}$ and $\bar{c} = 6\bar{i} + 2\bar{j} - 3\bar{k}$, show that

$$\bar{a} \times (\bar{b} \times \bar{c}) = (\bar{a} \times \bar{b}) \times \bar{c}.$$

[**Hints.** $\bar{b} \times \bar{c} = 7(2\bar{i} + 3\bar{j} + 6\bar{k})$.
It is parallel to $\bar{a}$. $\therefore$ $\bar{a} \times (\bar{b} \times \bar{c}) = \bar{0}$. $\bar{a} \times \bar{b} = 7(6\bar{i} + 2\bar{j} - 3\bar{k})$.
It is parallel to $\bar{c}$. $\therefore$ $(\bar{a} \times \bar{b}) \times \bar{c} = \bar{0}$]

(ii) A line makes angles $\alpha, \beta, \gamma, \delta$ with the diagonals of a cubic. Prove by using vector method that $\cos^2\alpha + \cos^2\beta + \cos^2\gamma + \cos^2\delta = \frac{4}{3}$.
[**Hints.** See the figure of Ex.8 of page 163. The diagonal vectors

$$\overrightarrow{OG} = \bar{i} + \bar{j} + \bar{k}, \quad \overrightarrow{AD} = -\bar{i} + \bar{j} + \bar{k},$$

$$\overrightarrow{BE} = -\bar{i} - \bar{j} + \bar{k}, \quad \overrightarrow{CF} = \bar{i} - \bar{j} + \bar{k}.$$

Let the vector $\bar{i}x + \bar{j}y + \bar{k}z$ make angles $\alpha, \beta, \gamma, \delta$ with $\overrightarrow{OG}, \overrightarrow{AD}, \overrightarrow{BE}$ and $\overrightarrow{CF}$ respectively. Then $\cos^2\alpha + \cos^2\beta + \cos^2\gamma + \cos^2\delta$

$$= \frac{(x+y+z)^2 + (-x+y+z)^2 + (-x-y+z)^2 + (x-y+z)^2}{3(x^2+y^2+z^2)} = \frac{4}{3}.]$$

(b) (i) If $\bar{a}, \bar{b}, \bar{c}$ are three non-coplanar vectors, then find after suitable simplification the value of $(\bar{a}-\bar{b}).\{(\bar{b}-\bar{c}) \times (\bar{c}-\bar{a})\}$.
[**Hints.**

$$(\bar{a}-\bar{b}) \cdot \{(\bar{b}-\bar{c}) \times (\bar{c}-\bar{a})\} = (\bar{a}-\bar{b}) \cdot (\bar{b}\times\bar{c} - \bar{b}\times\bar{a} + \bar{c}\times\bar{a})$$
$$= [\bar{a}\bar{b}\bar{c}] + [\bar{b}\bar{c}\bar{a}] = 2[\bar{a}\bar{b}\bar{c}].]$$

(ii) The position vectors of A, B, C are $\bar{i}+2\bar{k}, \bar{i}+2\bar{j}$ and $2\bar{i}+3\bar{j}+4\bar{k}$ respectively. Find by vector method, the area of the triangle ABC.
[**Hints.**

$$\text{Area} = \frac{1}{2}|\overrightarrow{AB} \times \overrightarrow{AC}| = \frac{1}{2}|(2\bar{j}-2\bar{k}) \times (\bar{i}+3\bar{j}+2\bar{k})| = 3\sqrt{3} \text{ square units.}]$$

2002

First Paper

Part I

1. Answer the following:

(x) The axes are rotated through an angle of 60° without change of origin. The coordinates of a point are $(4, \sqrt{3})$ in the new system. Find the coordinates in the old system.
[**Hints.**

$$x = 4\cos 60° + \sqrt{3}\sin 60° = 2 + \frac{3}{2} = \frac{7}{2},$$
$$y = -4\sin 60° + \sqrt{3}\cos 60° = -\frac{3\sqrt{3}}{2}.]$$

(xi) Find the equation of the plane through the points $(1,-2,4)$ and $(3,-4,5)$ and parallel to x-axis.
[**Hints.** Let the equation of the plane be $by + cz + d = 0$. It passes through the points $(1,-2,4)$ and $(3,-4,5)$.
Therefore, $-2b+4c+d=0$ and $-4b+5c+d=0$. From these $b = -\frac{d}{6}, c = -\frac{d}{3}$.
∴ the equation of the plane is $y + 2z - 6 = 0$.]

(xii) If the straight line makes A, B, C respectively with the axes of coordinates, then show that $\sin^2 A + \sin^2 B + \sin^2 C = 2$.
[**Hints.** See **Corollary 2** of page 158.]

(xiii) What is the equation of a right circular cone whose vertex is at the origin and z-axis is the axis.
[**Hints.** See **Corollary 2** of page 228.]

(xiv) Find a unit vector parallel to the resultant of vectors

$$\bar{\alpha} = 2\bar{i} + 4\bar{j} - 5\bar{k} \quad \text{and} \quad \bar{\beta} = \bar{i} + 2\bar{j} + 3\bar{k}.$$

[**Hints.** Resultant vector $= \bar{\alpha} + \bar{\beta} = 3\bar{i} + 6\bar{j} - 2\bar{k}$
∴ the unit vector $= \frac{3\bar{i}+6\bar{j}-2\bar{k}}{7}$.]

(xv) Find the projection of the vector $\bar{\alpha} = \bar{i} - 2\bar{j} + \bar{k}$ on the vector $\bar{\beta} = 4\bar{i} - 4\bar{j} + 7\bar{k}$.
[**Hints.** Projection $= (\bar{\alpha}.\bar{\beta})/|\bar{\beta}| = \frac{19}{9}$.]

Part II
Group B

5. Answer either (a) or (b):

(a) (i) Find the points on the conic $\frac{15}{r} = 1 - 4\cos\theta$ whose radius vector is 5.
[**Hints.**

$$\frac{15}{5} = 1 - 4\cos\theta, \quad \text{or,} \quad 3 = 1 - 4\cos\theta,$$

$$\text{or,} \quad \cos\theta = -\frac{1}{2}, \quad \theta = \frac{2\pi}{3}, \quad \text{or,} \quad \frac{4\pi}{3}.$$

The points are $(5, \frac{2\pi}{3})$ and $(5, \frac{4\pi}{3})$.]

(ii) Show that the equation $14x^2 - 4xy + 11y^2 - 44x - 58y + 71 = 0$ represents an ellipse whose centre is at $(2, 3)$.
[**Hints.** $\Delta \neq 0, D > 0$. The equation represents an ellipse.
To find the centre $28x - 4y - 44 = 0$ and $-4x + 22y - 58 = 0$.
From these two equations $x = 2, y = 3$. Thus the centre is $(2, 3)$.]

(b) (i) A moving point is always at a distance 2 units from a fixed point $(3, \frac{\pi}{3})$. Show that the locus of the point is $r^2 - 6r\cos(\theta - \frac{\pi}{3}) + 5 = 0$.
[**Hints.** Let (r, θ) be the coordinates of the moving point at a certain position. Now

$$r^2 + 3^2 - 2.r.3\cos\left(\theta - \frac{\pi}{3}\right) = 4, \text{ or, } r^2 - 6r\cos\left(\theta - \frac{\pi}{3}\right) + 5 = 0.]$$

(ii) By the rotation of axes after transferring the origin to the point $(2, 3)$ and equation $3x^2 + 2xy + 3y^2 - 18x - 22y + 50 = 0$ is transformed to $4x^2 + 2y^2 = 1$. Find the inclination of the latter axes to the former.
[**Hints.** See Ex.1 of page 11.]

6. Answer *any three* of the following:

(a) (i) Find the equation of a plane which passes through the origin and contains two straight lines with direction ratios $(1, 0, 2)$ and $(-1, 5, 0)$.
[**Hints.** Let the equation of the plane be $ax + by + cz = 0$. Here $a + 2c = 0$ and $-a + 5b = 0$. $\therefore \quad c = -\frac{a}{2}$ and $b = \frac{a}{5}$.
$\therefore$ the equation is $10x + 2y - 5z = 0$.]

(ii) A variable plane which is at a constant distance p from the origin O cuts the axes in A, B, C. Show that the locus of point of intersection of the planes through A, B, C parallel to the coordinate planes is $\frac{1}{x^2} + \frac{1}{y^2} + \frac{1}{z^2} = \frac{1}{p^2}$.
[**Hints.** Let the plane be $\frac{x}{a} + \frac{y}{b} + \frac{z}{c} = 1$. Here

$$p = \frac{1}{\sqrt{\frac{1}{a^2} + \frac{1}{b^2} + \frac{1}{c^2}}}, \quad \text{or,} \quad \frac{1}{a^2} + \frac{1}{b^2} + \frac{1}{c^2} = \frac{1}{p^2}.$$

The coordinates of A, B, C are $(a, 0, 0), (0, b, 0)$ and $(0, 0, c)$.
The planes are $x = a, y = b, z = c$. These planes meet at (a, b, c).

$$\therefore \quad \text{the locus is } \frac{1}{x^2} + \frac{1}{y^2} + \frac{1}{z^2} = \frac{1}{p^2}.]$$

(b) (i) Find the coordinates of the point in which the straight line $\frac{x-2}{3} = \frac{y+1}{4} = \frac{z-2}{12}$ meets the plane $x - y + z = 5$.
[**Hints.** $(3r + 2, 4r - 1, 12r + 2)$ is a point on the line. If this point lies on the plane, then $3r + 2 - 4r + 1 + 12r + 2 = 5$, or, $r = 0$.
$\therefore$ the line meets the plane at the point $(2, -1, 2)$.]

(ii) Find the equation of the plane containing the straight line $\frac{x-1}{3} = \frac{y+6}{4} = \frac{z+1}{2}$ and parallel to the straight line $\frac{x-2}{2} = \frac{y-1}{-3} = \frac{z+4}{5}$.
[**Hints.** Let the equation of the plane be $a(x-1) + b(y+6) + c(z+1) = 0$ where

$$3a + 4b + 2c = 0 \quad \text{and} \quad 2a - 3b + 5c = 0.$$

From the two conditions

$$\frac{a}{26} = \frac{b}{-11} = \frac{c}{-17}.$$

$\therefore$ the equation is $26(x-1) - 11(y+6) - 17(z+1) = 0$.]

(c) (i) Find a point on the x-axis such that it is equidistant from the planes

$$4x - 12y + 6z + 2 = 0, 2x - y + 2z - 3 = 0.$$

[**Hints.** Let the point be $(x_1, 0, 0)$. By the condition

$$\frac{4x_1 + 2}{14} = \pm\frac{2x_1 - 3}{3}, \; x_1 = 3, \; \text{or,} \; \frac{9}{10}.$$

$\therefore$ the points are $(3,0,0)$ and $(\frac{9}{10}, 0, 0)$.]

(ii) Find the distance of the point $(2,3,1)$ from the straight line

$$y + z - 1 = 0 = 2x - 3y - 2z + 4.$$

[**Hints.** The equations of the line in symmetric form $\frac{x+\frac{1}{2}}{1} = \frac{y-1}{2} = \frac{z}{-2}$.
The point $A(\frac{-1}{2}, 1, 0)$ lies on the line. Let the point P be $(2,3,1)$. D.cs. of AP are $\frac{5}{2AP}, \frac{2}{AP}, \frac{1}{AP}$ and d.cs. of the line are $\frac{1}{3}, \frac{2}{3}, \frac{-2}{3}$. If θ is the angle betweeen the line and AP, then $\cos\theta = \frac{3}{2AP}$.

$$\text{Distance} = AP\sin\theta = AP\sqrt{1-\cos^2\theta} = \sqrt{AP^2 - \frac{9}{4}}$$

$$= \sqrt{\frac{45}{4} - \frac{9}{4}} = 3 \text{ units.}]$$

(d) (i) Find the shortest distance between the lines

$$\frac{x-3}{3} = \frac{y-8}{-1} = \frac{z-3}{1} \quad \text{and} \quad \frac{x+3}{-3} = \frac{y+7}{2} = \frac{z-6}{4}$$

and the equation of s.d.
[**Hints.** See Ex.7 of page 192.]

(ii) Find the equation of the sphere which has the line segment joining the points $(2,3,4)$ and $(0,-1,2)$ as diameter.
[**Hints.** The equation is

$$(x-2).x + (y-3)(y+1) + (z-4)(z-2) = 0,$$

$$\text{or,} \quad x^2 + y^2 + z^2 - 2x - 2y - 6z + 5 = 0.]$$

(e) (i) Find the equation of the normal to the hyperboloid $3x^2 - 5y^2 + 7z^2 = 30$ at $(2,3,3)$.
[**Hints.** $(2,3,3)$ is a point on the hyperboloid. The equation of the tangent plane at this point is $2x - 5y + 7z = 10$.
$\therefore$ the equation of the normal at $(2,3,3)$ is $\frac{x-2}{2} = \frac{y-3}{-5} = \frac{z-3}{7}$.]

(ii) Find the equation of the cylinder whose guiding curve is $4x^2 + y^2 = 1, z = 0$ and axis is $\frac{x}{2} = \frac{y}{-1} = \frac{z}{3}$.
[**Hints.** Let (α, β, γ) be a point on the cylinder. The generating line through this point is

$$\frac{x-\alpha}{2} = \frac{y-\beta}{-1} = \frac{z-\gamma}{3}.$$

If this line meets the guiding curve at $(x_1, y_1, 0)$, then

$$\frac{x_1-\alpha}{2}=\frac{y_1-\beta}{-1}=-\frac{\gamma}{3}. \quad \therefore \quad x_1=\alpha-\frac{2}{3}\gamma, y_1=\beta+\frac{\gamma}{3}.$$

We have

$$4x_1^2+y_1^2=1, \quad \text{or,} \quad 4\left(\alpha-\frac{2}{3}\gamma\right)^2+\left(\beta+\frac{\gamma}{3}\right)^2=1.$$

The locus of (α, β, γ), i.e., the equation of the cylinder is

$$4\left(x-\frac{2}{3}z\right)^2+\left(y+\frac{z}{3}\right)^2=1.]$$

7. Answer either (a) or (b):

(a) (i) Show that the unit vector perpendicular to both the vectors $3\bar{i}+\bar{j}+2\bar{k}$ and $2\bar{i}-2\bar{j}+4\bar{k}$ is $\frac{1}{\sqrt{3}}(\bar{i}-\bar{j}-\bar{k})$ and the angle between them is $\sin^{-1}\frac{2}{\sqrt{7}}$.

[**Hints.**

$$(3\bar{i}+\bar{j}+2\bar{k})\times(2\bar{i}-2\bar{j}+4\bar{k})=8(\bar{i}-\bar{j}-\bar{k}).$$

It is perpendicular to both the vectors. Thus the unit vector is $\frac{1}{\sqrt{3}}(\bar{i}-\bar{j}-\bar{k})$.

$$|(3\bar{i}+\bar{j}+2\bar{k})\times(2\bar{i}-2\bar{j}+4\bar{k})|=|3\bar{i}+\bar{j}+2\bar{k}||2\bar{i}-2\bar{j}+4\bar{k}|\sin\theta$$

where θ is the angle between the vectors.

$$\therefore \quad |8(\bar{i}-\bar{j}-\bar{k})|=\sqrt{14}.\sqrt{24}\sin\theta,$$

$$\text{or,} \quad \sin\theta=\frac{8\sqrt{3}}{\sqrt{14}\sqrt{24}}=\frac{2}{\sqrt{7}}, \quad \text{or,} \quad \theta=\sin^{-1}\frac{2}{\sqrt{7}}.]$$

(ii) Prove by vector method that for any triangle ABC, $c^2=a^2+b^2-2ab\cos C$ with usual notation.

[**Hints.** See Ex.5 of page 301.]

(b) (i) If $\bar{a}$ and $\bar{b}$ are non-collinear vectors and $\bar{A}=(x+4y)\bar{a}+(2x+y+1)\bar{b}$ and $\bar{B}=(y-2x+2)\bar{a}+(2x-3y-1)\bar{b}$. Find x and y such that $3\bar{A}=2\bar{B}$.

[**Hints.**

$$\begin{aligned}3\bar{A}&=2\bar{B}, \quad \text{or,} \quad 3(x+4y)\bar{a}+3(2x+y+1)\bar{b}\\&=2(y-2x+2)\bar{a}+2(2x-3y-1)\bar{b}.\end{aligned}$$

It implies that $3(x+4y)=2(y-2x+2)$ and $3(2x+y+1)=2(2x-3y-1)$.

From these two equations $x=2, y=-1$.]

(ii) Find the work done by the force $\bar{F}=(-2,3,4))$ whose point of application is given a displacement from the point $A(2,-1,-2)$ to the point $B(-1,2,3)$.

[**Hints.** Displacement vector $=(-3,3,5)$. Work done $=(-2,3,4).(-3,3,5)=35$ units.]

North Bengal University

2000

Group A

[Answer **Q. No. 1** and **any two** from the rest.]

1. Answer *any five*:

(a) Determine the nature of the conic $r = \frac{12}{2-\cos\theta}$.
[**Hints.**

$$r = \frac{12}{2-\cos\theta}, \quad \text{or,} \quad \frac{12}{r} = 2 - \cos\theta, \quad \text{or,} \quad \frac{6}{r} = 1 - \frac{1}{2}\cos\theta.$$

Here eccentricity $= \frac{1}{2}$.
$\therefore$ the conic is an ellipse with the semi-latus rectum of 6 units.]

(b) Which curve is represented by the equation $7x^2 - 2xy + 7y^2 - 16x + 16y - 8 = 0$?
[**Hints.** Here $\Delta \neq 0$, $D > 0$. $\therefore$ the equation represents an ellipse.]

(c) Find the equation to the cone whose vertex is the origin and the base, the circle $x = a, y^2 + z^2 = b^2$.
[**Hints.** The equation is

$$y^2 + z^2 = b^2 \cdot \left(\frac{x}{a}\right)^2, \quad \text{or,} \quad a^2(y^2 + z^2) = b^2x^2.]$$

(d) Show that the circles $x^2 + y^2 + 2x + 4y - 20 = 0$ and $x^2 + y^2 + 6x -$ $+ 10 = 0$ cut orthogonally.
[**Hints.** Here $2.1.3 + 2.2.(-4) = -10$ and $-20 + 10 = -10$.
$\therefore$ the circles cut orthogonally.]

(e) Do the points $(4, 1, 2)$ and $(2, 3, -1)$ lie on the same or opposite sides of the plane

$$5x - 7y - 6z + 3 = 0?$$

[**Hints.** Distances of the points from the plane are

$$\frac{5.4 - 7.1 - 6.2 + 3}{\sqrt{5^2 + 7^2 + 6^2}} = \frac{4}{\sqrt{110}} \quad \text{and} \quad \frac{5.2 - 7.3 + 6.1 + 3}{\sqrt{110}} = \frac{-2}{\sqrt{110}}.$$

Since these are of opposite signs the points lie on the opposite sides of the plane.]

(f) Find the equation of the sphere which has $(1, 0, 0)$ and $(-1, 0, 0)$ for extremities of a diameter.
[**Hints.** The equation is $(x-1)(x+1) + y^2 + z^2 = 0$, or, $x^2 + y^2 + z^2 = 1$.]

(g) Find the direction cosines of the straight line $x = 2y = 3z$.
[**Hints.** The equation can be written as $\frac{x}{6} = \frac{y}{3} = \frac{z}{2}$. Thus d.cs. are $\frac{6}{7}, \frac{3}{7}, \frac{2}{7}$.]

2. (a) Show that the product of the perpendicular from (x', y') on the lines

$$ax^2 + 2hxy + by^2 = 0 \text{ is } \frac{ax'^2 + 2hx'y' + by'^2}{\sqrt{(a-b)^2 + 4h^2}}.$$

[**Hints.** See Ex.7 of page 28.]

(b) Show that the area of the triangle formed by the tangents to the circle

$$x^2 + y^2 = a^2$$

from (α, β) and the chord of contact is $\frac{a(\alpha^2 + \beta^2 - a^2)^{3/2}}{\alpha^2 + \beta^2}$.

[**Hints.** See Ex.22 of page 80.]

3. (a) Reduce the equation $9x^2 + 24xy + 16y^2 - 126x + 82y - 59 = 0$ to normal form.
[**Hints.** The equation is

$$(3x+4y)^2 = 126x - 82y + 59,$$
$$\text{or,}\quad (3x+4y+\lambda)^2 = (126+6\lambda)x - (82-8\lambda)y + 59 + \lambda^2.$$

λ is chosen in such a way that

$$3x+4y+\lambda = 0 \quad\text{and}\quad (126+6\lambda)x - (82-8\lambda)y + 59 + \lambda^2 = 0$$

are at right angles. For this $3(126+6\lambda) - 4(82-8\lambda) = 0$, or, $\lambda = -1$.
Now the equation is

$$(3x+4y-1)^2 = 120x - 90y + 60,$$
$$\text{or,}\quad \left(\frac{3x+4y-1}{\sqrt{3^2+4^2}}\right)^2 \times 25 = 30\frac{4x-3y+2}{\sqrt{4^2+3^2}} \times 5.$$

Putting

$$\frac{3x+4y-1}{\sqrt{3^2+4^2}} = Y \quad\text{and}\quad \frac{4x-3y+2}{\sqrt{4^2+3^2}} = X, Y^2 = 6X.$$

It is the normal form. The equation represents a parabola.]

(b) Prove that the sum of the reciprocals of two perpendicular focal chords of a conic $\frac{l}{r} = 1 + e\cos\theta$ is constant.
[**Hints.** See Ex.6 of page 133.]

4. (a) Find the length and equation of the line of shortest distance between the lines $\frac{x+3}{-4} = \frac{y-6}{3} = \frac{z}{2}$ and $\frac{x+2}{-4} = \frac{y}{1} = \frac{z-7}{1}$.
[**Ans.** 9 units. $32x + 34y + 13z - 108 = 0 = 4x + 11y + 5z - 27$.]

(b) Find the equation of the tangent plane to the cone

$$ax^2 + by^2 + cz^2 + 2fyz + 2gzx + 2hxy = 0$$

at the point $P(x', y', z')$.
[**Ans.**

$$axx' + byy' + czz' + f(yz' + zy') + g(zx' + xz') + h(xy' + yx') = 0,$$
$$\text{or,}\quad (ax' + hy' + gz')x + (hx' + by' + fz')y + (gx' + fy' + cz')z = 0.]$$

5. (a) The plane $bcx + cay + abz = abc$ cuts the axes in A, B, C. Find the equation of the circle ABC.
[**Hints.** The plane meets the axes at $(a,0,0), (0,b,0)$ and $(0,0,c)$.
The equation of the sphere passing through the points $(0,0,0), (a,0,0), (0,b,0)$ and $(0,0,c)$ is $x^2 + y^2 + z^2 - ax - by - cz = 0$.
The equation of the circle ABC is $x^2 + y^2 + z^2 - ax - by - cz = 0, bcx + cay + abz = abc$.]

(b) Find the equation of the cylinder whose generators are parallel to the line $2x = y = 3z$ and which passes through the circle $y = 0, x^2 + z^2 = 8$.
[**Ans.** $(x - \frac{y}{2})^2 + (z - \frac{y}{3})^2 = 8$.]

Group C

1. Answer *any four*:

(a) Prove, by vector method, that the medians of a triangle are concurrent.
[**Hints.** See Ex.3 of page 286.]

(b) If $|\bar{a} - \bar{b}| = |\bar{a} + \bar{b}|$, show that $\bar{a}, \bar{b}$ are perpendicular to each other.
[**Hints.** $(\bar{a} - \bar{b})^2 = (\bar{a} + \bar{b})^2$, or, $4\bar{a}\cdot\bar{b} = 0$. $\therefore$ $\bar{a}$ is perpendicular to $\bar{b}$.]

(c) $\bar{\alpha}, \bar{\beta}, \bar{\gamma}$ are three vectors such that $\bar{\alpha} + \bar{\beta} + \bar{\gamma} = \bar{0}$ and $|\bar{\alpha}| = 3, |\bar{\beta}| = 5, |\bar{\gamma}| = 7$, find the angle between $\bar{\alpha}$ and $\bar{\beta}$.

[**Hints.**

$$\bar{\alpha} + \bar{\beta} = -\bar{\gamma}. \quad \therefore \quad (\bar{\alpha} + \bar{\beta})^2 = \bar{\gamma}^2,$$

$$\text{or,} \quad \bar{\alpha}^2 + \bar{\beta}^2 + 2\bar{\alpha}\cdot\bar{\beta} = \bar{\gamma}^2, \quad \text{or,} \quad 2\bar{\alpha}\cdot\bar{\beta} = 49 - 9 - 25 = 15,$$

$$\text{or,} \quad 2|\bar{\alpha}||\bar{\beta}|\cos\theta = 15, \cos\theta = \frac{1}{2}, \quad \text{or,} \quad \theta = \frac{\pi}{3}.]$$

(d) Show that $[(\bar{\alpha} + \bar{\beta})(\bar{\beta} + \bar{\gamma})(\bar{\gamma} + \bar{\alpha})] = 2[\bar{\alpha}\bar{\beta}\bar{\gamma}]$.

[**Hints.** See Ex.18 of page 306.]

(e) The position vectors of the vertices of a triangle are

$$\bar{i} + \bar{j} + 2\bar{k}, \ 2\bar{i} + 2\bar{j} + 3\bar{k}, \ 3\bar{i} - \bar{j} - \bar{k}.$$

Find the area of the triangle.

[**Hints.** Let ABC be the triangle and position vectors of A, B, C be

$$\bar{i} + \bar{j} + 2\bar{k}, 2\bar{i} + 2\bar{j} + 3\bar{k}, 3\bar{i} - \bar{j} - \bar{k}.$$

Area $= \frac{1}{2}|\overrightarrow{AB} \times \overrightarrow{AC}| = \frac{1}{2}|(\bar{i} + \bar{j} + \bar{k}) \times (2\bar{i} - 2\bar{j} - 3\bar{k})| = 21\sqrt{2}$ sq. units.]

(f) Two constant forces $3\bar{i} + \bar{j} - 4\bar{k}$ and $\bar{i} - 3\bar{j} + \bar{k}$ act on a particle which is displaced from the point $3\bar{i} + \bar{j} + 2\bar{k}$ to the point $\bar{i} + 4\bar{j} - 5\bar{k}$. Find the work done.

[**Hints.** Resultant of the forces $= (3\bar{i} + \bar{j} - 4\bar{k}) + (\bar{i} - 3\bar{j} + \bar{k}) = 4\bar{i} - 2\bar{j} - 3\bar{k}$.
The displacement vector $= (\bar{i} + 4\bar{j} - 5\bar{k}) - (3\bar{i} + \bar{j} + 2\bar{k}) = -2\bar{i} + 3\bar{j} - 7\bar{k}$.

$$\therefore \quad \text{work done} = (4\bar{i} - 2\bar{j} - 3\bar{k}).(-2\bar{i} + 3\bar{j} - 7\bar{k})$$
$$= -8 - 6 + 21 = 7 \text{ units.}]$$

2001

Group A

[Answer Q. No. 1 and any two from the rest.]

1. Answer *any five*:

(a) Transform the equation $r^2\cos 2\theta = a^2$ to the cartesian form.

[**Hints.** $r^2\cos 2\theta = a^2$ or, $r^2(\cos^2\theta - \sin^2\theta) = a^2$, or, $x^2 - y^2 = a^2$.]

(b) Determine the nature of the conic $x^2 - 2xy + 2y^2 - 4x - 6y + 3 = 0$.

[**Hints.** Here, $\Delta \neq 0, D > 0$. $\therefore$ the equation represents an ellipse.]

(c) Find the radical axis of the circles

$$x^2 + y^2 + 2x + 4y + 7 = 0 \quad \text{and} \quad x^2 + y^2 + 5x + 5y + 4 = 0.$$

[**Ans.** $3x + y - 3 = 0$.]

(d) For what values of k, the following equation $2x^2 - y^2 + xy - 2x - 5y + k = 0$ will represent a pair of lines?

[**Ans.** $k = -4$.]

(e) Can the numbers $\frac{1}{\sqrt{2}}, \frac{1}{\sqrt{2}}, \frac{1}{\sqrt{2}}$ be the direction cosines of any directed line? Give reason.

[**Hints.** See Ex.1 of page 161.]

(f) Show that the planes $2x - y + 3z + 5 = 0$ and $4x + 2y - 2z + 3 = 0$ are perpendicular.

[**Hints.** Since $2.4 - 1.2 + 3.(-2) = 0$, the planes are perpendicular to each other.]

(g) Find the equation of the sphere through the circle $x^2+y^2+z^2=9, 2x+3y+4z=5$ and the point $(1,2,3)$.
[**Hints.** Let the equation be $x^2+y^2+z^2-9+\lambda(2x+3y+4z-5)=0$. It passes through the point $(1,2,3)$.

$$\therefore \quad 1+4+9-9+\lambda(2+6+12-5)=0, \quad \text{or,} \quad \lambda=-\frac{1}{3}.$$

Thus the equation is $3(x^2+y^2+z^2)-2x-3y-4z-22=0$.]

(h) Find the direction cosines of the line which is equally inclined to the axes.
[**Hints.** See Ex.2 of page 161.]

2. (a) If the general equation of the second degree represents two straight lines, prove that the square of the distance of their point of intersection from the origin is

$$\frac{c(a+b)-f^2-g^2}{ab-h^2}.$$

[**Hints.** See **Corollary 3** of page 24.]

(b) Find the equation of the circle whose diameter is the common chord of the circles $x^2+y^2+2x+3y+1=0$ and $x^2+y^2+4x+3y+2=0$.
[**Hints.** The equation of the common chord is $2x+1=0$.
Let the equation of the circle be $x^2+y^2+2x+3y+1+\lambda(2x+1)=0$.
The centre is $\{-(1+\lambda), -\frac{3}{2}\}$.
It lies on $2x+1=0$. $\therefore -2(1+\lambda)+1=0$, or, $\lambda=-\frac{1}{2}$.
$\therefore$ the equation of the circle is $2(x^2+y^2)+2x+6y+1=0$.]

3. (a) Reduce the equation $7x^2-2xy+7y^2+22x-10y+7=0$ to the normal form and state the nature of the conic.
[**Hints.** Here, $\Delta=-576, D=48$. The equation represents an ellipse. Changing the origin to the centre the equation reduces to $7x^2-2xy+7y^2=12$.
Rotating the axes through an angle θ, where $\cot 2\theta = \frac{7-7}{-2}=0$, or, $\theta=\frac{\pi}{4}$, the equation transforms to $\frac{x^2}{2}+\frac{y^2}{3/2}=1$. It is the normal form.]

(b) Find the polar equation of the chord of the conic $\frac{l}{r}=1-e\cos\theta$ joining the two points on the conic whose vectorial angles are $\alpha+\beta$ and $\alpha-\beta$.
[**Hints.** See the section 8.8 of page 126.]

4. (a) A variable plane passes through the fixed point (a,b,c) and meets the axes at A,B,C. Prove that the locus of the point of intersection of the three planes passing through A,B,C and parallel to the coordinate planes is given by $\frac{a}{x}+\frac{b}{y}+\frac{c}{z}=1$.
[**Hints.** See Ex.9 of page 178.]

(b) Find the equation of the plane through the line $x+y-2z+4=0=3x-y+2z+1$ and parallel to the line $\frac{x+2}{2}=\frac{y-2}{2}=\frac{z-1}{-1}$.
[**Hints.** Let the plane be

$$x+y-2z+4+\lambda(3x-y+2z+1)=0,$$
$$\text{or,} \quad (1+3\lambda)x+(1-\lambda)y-2(1-\lambda)z+4+\lambda=0.$$

If it is parallel to the given line, then

$$2(1+3\lambda)+2(1-\lambda)+1.2(1-\lambda)=0, \quad \text{or,} \quad \lambda=-3.$$

$\therefore$ the equation of the plane is $8x-4y+8z-1=0$.]

5. (a) Find the equation of the cone whose vertex is $(1,2,3)$ and base $ax^2+by^2=1, z=0$.
[**Hints.** Let one generator be

$$\frac{x-1}{l}=\frac{y-2}{m}=\frac{z-3}{n} \tag{1}$$

If this generator meets the base at $(\alpha, \beta, 0)$, then

$$\frac{\alpha-1}{l}=\frac{\beta-2}{m}=\frac{-3}{n}, \quad \text{or,} \quad \alpha=1-3\frac{l}{n}, \beta=2-3\frac{m}{n}.$$

We have $a\alpha^2+b\beta^2=1$.

$$\therefore \quad a\left(1-3\frac{l}{n}\right)^2+b\left(2-3\frac{m}{n}\right)^2=1.$$

Eliminating l, m, n by (1),

$$a\left(1-3\frac{x-1}{z-3}\right)^2+b\left(2-3\frac{y-2}{z-3}\right)^2=1,$$

$$\text{or,} \quad a(z-3x)^2+b(2z-3y)^2=(z-3)^2.$$

It is the equation of the cone.]

(b) Find the equation of the sphere of which the circle given by $x+y+z+3=0$ and $x^2+y^2+z^2=9$ is a great circle.

[Hints. Let the equation of the sphere be $x^2+y^2+z^2-9+\lambda(x+y+z+3)=0$. The centre is

$$\left(-\frac{\lambda}{2}, \frac{-\lambda}{2}, \frac{-\lambda}{2}\right).$$

Here the centre lies on the plane $x+y+z+3=0$.

$$\therefore \quad -\frac{3\lambda}{2}+3=0, \quad \text{or,} \quad \lambda=2.$$

Thus the required equation is $x^2+y^2+z^2+2(x+y+z)-3=0$.]

Group C

11. Answer *any four*:

(a) Show by vector method that angle in a semicircle is a right angle.

[Hints. See Ex.8 of page 302.]

(b) If $\bar{\alpha}, \bar{\beta}, \bar{\gamma}$ be unit vectors such that $\bar{\alpha}+\bar{\beta}+\bar{\gamma}=\bar{0}$, show that

$$\bar{\alpha}\cdot\bar{\beta}+\bar{\beta}\cdot\bar{\gamma}+\bar{\gamma}\cdot\bar{\alpha}=-\tfrac{3}{2}.$$

[Hints.

$$\begin{aligned}
&(\bar{\alpha}+\bar{\beta}+\bar{\gamma})^2=0,\\
\text{or,}\quad &\bar{\alpha}^2+\bar{\beta}^2+\bar{\gamma}^2+2(\bar{\alpha}\cdot\bar{\beta}+\bar{\beta}\cdot\bar{\gamma}+\bar{\gamma}\cdot\bar{\alpha})=0,\\
\text{or,}\quad &1+1+1+2(\bar{\alpha}\cdot\bar{\beta}+\bar{\beta}\cdot\bar{\gamma}+\bar{\gamma}\cdot\bar{\alpha})=0,\\
\text{or,}\quad &\bar{\alpha}\cdot\bar{\beta}+\bar{\beta}\cdot\bar{\gamma}+\bar{\gamma}\cdot\bar{\alpha}=-\tfrac{3}{2}.]
\end{aligned}$$

(c) Show that $[\bar{\alpha}\times\bar{\beta}, \bar{\beta}\times\bar{\gamma}, \bar{\gamma}\times\bar{\alpha}]=[\bar{\alpha}\bar{\beta}\bar{\gamma}]^2$.

[Hints. Let $\bar{\alpha}\times\bar{\beta}=\bar{p}, \bar{\beta}\times\bar{\gamma}=\bar{q}, \bar{\gamma}\times\bar{\alpha}=\bar{r}$. Now,

$$\begin{aligned}
[\bar{\alpha}\times\bar{\beta}, \bar{\beta}\times\bar{\gamma}, \bar{\gamma}\times\bar{\alpha}] &= [\bar{p}\bar{q}\bar{r}]=\bar{p}\cdot(\bar{q}\times\bar{r})=\bar{p}\cdot\{\bar{q}\times(\gamma\times\alpha)\}\\
&= \bar{p}\cdot\{(\bar{q}\cdot\bar{\alpha})\gamma-(\bar{q}\cdot\bar{\gamma})\bar{\alpha}\}\\
&= \bar{p}\cdot[\{(\bar{\beta}\times\bar{\gamma})\cdot\bar{\alpha}\}\bar{\gamma}-\{(\bar{\beta}\times\bar{\gamma})\cdot\bar{\gamma}\}\bar{\alpha}]\\
&= \bar{p}\cdot[\bar{\alpha}\bar{\beta}\bar{\gamma}]\bar{\gamma}=[\bar{\alpha}\bar{\beta}\bar{\gamma}]\{(\bar{\alpha}\times\bar{\beta})\cdot\bar{\gamma}\}=[\bar{\alpha}\bar{\beta}\bar{\gamma}]^2.]
\end{aligned}$$

(d) If three vectors are coplanar, show that their triple product is zero.

[Hints. See the property (iv) of page 298.]

(e) A particle is acted on by constant forces $4\bar{i}+\bar{j}-3\bar{k}$ and $3\bar{i}+\bar{j}-\bar{k}$ and is displaced from the point $3\bar{i}+2\bar{j}+\bar{k}$ to the point $4\bar{i}+\bar{j}+5\bar{k}$. Find the work done by the forces on the particle.
[**Hints.** Resultant of the forces $=(4\bar{i}+\bar{j}-3\bar{k})+(3\bar{i}+\bar{j}-\bar{k})=7\bar{i}+2\bar{j}-4\bar{k}$. The displacement vector $=(4\bar{i}+\bar{j}+5\bar{k})-(3\bar{i}+2\bar{j}+\bar{k})=\bar{i}-\bar{j}+4\bar{k}$.
Work done $=|(7\bar{i}+2\bar{j}-4\bar{k}).(\bar{i}-\bar{j}+4\bar{k})|=|7-2-16|=11$ units.]

(f) If $|\bar{a}-\bar{b}|=|\bar{a}+\bar{b}|$, show that $\bar{a}$ and $\bar{b}$ are perpendicular.
[**Hints.** See Q.11 (b) of 2000.]

2002

Group A

[Answer **Q. No. 1** and **any two** from the rest.]

1. Answer *any five*:

(a) Find the points on the conic $\frac{14}{r}=3-8\cos\theta$ whose radius vector is 2.
[**Hints.**

$$\frac{14}{2}=3-8\cos\theta,\quad \text{or,}\quad 7=3-8\cos\theta,$$
$$\text{or,}\quad \cos\theta=-\frac{1}{2},\quad \text{or,}\quad \theta=\frac{2\pi}{3},\frac{4\pi}{3}.$$

Thus the points are $(2,\frac{2\pi}{3})$ and $(2,\frac{4\pi}{3})$.]

(b) Find the equation of the line $y=2x$ when the axes are rotated through an angle $\frac{\pi}{4}$.
[**Hints.** The equation transforms to

$$x'\sin\frac{\pi}{4}+y'\cos\frac{\pi}{4}=2\left(x'\cos\frac{\pi}{4}-y'\sin\frac{\pi}{4}\right),\quad \text{or,}\quad 3y'=x'.]$$

(c) Define polar of a point with respect to a conic.
[**Hints.** See page 66.]

(d) Find the angle between the lines whose direction ratios are given by $(-1,2,1)$ and $(3,-1,5)$.
[**Hints.** If θ is the angle between the lines, then

$$\cos\theta=\frac{-1.3+2.(-1)+1.5}{\sqrt{1+4+1}\sqrt{9+1+25}}=0.\quad \therefore\quad \theta=\frac{\pi}{2}.]$$

(e) What is the nature of the conic represented by the equation $r=\frac{12}{2-\cos\theta}$.
[**Hints.** See Q.1.(a) of 2000.]

(f) Show that the points $P(2,5,-4), Q(1,4,-3)$ and $R(4,7,-6)$ are collinear.
[**Hints.** The point Q divides PR in the ratio $-1:3$. Hence the points are collinear.]

(g) Find k so that the lines $\frac{x-3}{-3}=\frac{y-2}{2k}=\frac{z-3}{2}$ and $\frac{x-1}{3k}=\frac{y-5}{1}=\frac{z-6}{-5}$ may be perpendicular to each other.
[**Hints.** Lines will be perpendicular if $-9k+2k-10$, or, $k=-\frac{10}{7}$.]

(h) Find the equation of the sphere through the circle $x^2+y^2+z^2=9, x+y-2z=4$ and the origin.
[**Hints.** Let the equation of the sphere be $x^2+y^2+z^2-9+\lambda(x+y-2z-4)=0$. If it passes through the origin $\lambda=-\frac{9}{4}$. Thus the equation is $4(x^2+y^2+z^2)-9x-9y+18z=0$.]

2. (a) Reduce the equation to the normal form and determine the type of the conic represented by $4x^2-4xy+y^2-12x+10y-19=0$.
[**Hints.** Normal form $Y^2=-\frac{8\sqrt{5}}{25}X$, parabola.]

(b) If by rotation of rectangular axes the equation $17x^2 + 18xy - 7y^2 = 1$ reduces to $ax^2 + by^2 = 1$, find a, b.
[**Ans.** $a = 5 + \frac{369}{\sqrt{613}}, b = 5 - \frac{369}{\sqrt{613}}$.]

3. (a) Show that the eccentricity of the ellipse in which the normal at one end of the latus rectum passes through an end of the minor axis is given by the equation $e^4 + e^2 - 1 = 0$.
[**Hints.** See Q.13 of page 75.]

(b) If l and l' are the lengths at the two segments at any focal chord of a parabola, prove that $\frac{1}{l} + \frac{1}{l'} = \frac{1}{a}$.
[**Hints.** See Ex.5 of page 132.]

(c) Prove that the equation of the circle described on the chord $x\cos\alpha + y\sin\alpha = p$ of the circle $x^2 + y^2 = a^2$ as diameter is

$$x^2 + y^2 - a^2 - 2p(x\cos\alpha + y\sin\alpha - p) = 0.$$

[**Hints.** Let the equation be $x^2 + y^2 - a^2 + \lambda(x\cos\alpha + y\sin\alpha - p) = 0$. The centre is

$$\left(-\frac{\lambda\cos\alpha}{2}, -\frac{\lambda\sin\alpha}{2}\right).$$

It lies on $x\cos\alpha + y\sin\alpha = p$. $\therefore$ $\frac{-\lambda}{2}(\cos^2\alpha + \sin^2\alpha) = p$, or, $\lambda = -2p$.
Thus the equation is $x^2 + y^2 - a^2 - 2p(x\cos\alpha + y\sin\alpha - p) = 0$.]

4. (a) Find the equation of the plane which passes through the point $(2,1,4)$ and is perpendicular to each of the planes $x + y + 2z - 4 = 0$ and $2x - 3y + z + 5 = 0$.
[**Hints.** Let the plane be $ax + by + cz + d = 0$. It is perpendicular to the given planes.

$$\therefore \quad a + b + 2c = 0, 2a - 3b + c = 0.$$

By cross multiplication $\frac{a}{7} = \frac{b}{3} = \frac{c}{-5} = k$ (say). The equation is $7x + 3y - 5z + \frac{d}{k} = 0$.
Since it passes through the point $(2,1,4)$, $\frac{d}{k} = 3$.
$\therefore$ the equation is $7x + 3y - 5z + 3 = 0$.]

(b) Find the length of shortest distance between the lines $\frac{x-3}{2} = \frac{y+15}{-7} = \frac{z-9}{5}$ and $\frac{x+1}{2} = \frac{y-1}{1} = \frac{z-9}{-3}$.
[**Hints.** Let the ends of s.d. be

$$(2r_1 + 3, -7r_1 - 15, 5r_1 + 9) \quad \text{and} \quad (2r_1 - 1, r_2 + 1, -3r_2 + 9).$$

The d.rs. are $2r_1 - 2r_2 + 4, -7r_1 - r_2 - 16, 5r_1 + 3r_2$. Since s.d. is perpendicular to both the lines,

$$\begin{aligned} &2(2r_1 - 2r_2 + 4) - 7(-7r_1 - r_2 - 16) + 5(5r_1 + 3r_2) = 0,\\ \text{or,}\quad &13r_1 + 3r_2 + 20 = 0\\ \text{and}\quad &2(2r_1 - 2r_2 + 4) + (-7r_1 - r_2 - 16) - 3(5r_1 + 3r_2) = 0,\\ \text{or,}\quad &9r_1 + 7r_2 + 4 = 0. \end{aligned}$$

From these two $r_1 = -2, r_2 = 2$. Ends of s.d. are $(-1,-1,-1)$ and $(3,3,3)$.
$\therefore$ length of s.d. $= \sqrt{4^2 + 4^2 + 4^2} = 4\sqrt{3}$ units.]

(c) Prove that the general equation of a second degree cone passing through the coordinate axes is $fyz + gzx + hxy = 0$.
[**Hints.** See different approach of page 232.]

5. (a) Find the equation of the cylinder whose generator are parallel to the line $2x = y = 3z$ and which passes through the circle $y = 0, x^2 + z^2 = 8$.
[**Hints.** See Q.5.(b) of 2000.]

(b) Prove that the line $x = pz + q, y = p'z + q'$ intersects the conic

$$z = 0,\ ax^2 + by^2 = 1 \text{ if } aq^2 + bq'^2 = 1.$$

[**Hints.** For $z = 0, x = q, y = q'$. If the line intersects the conic, then $(q, q', 0)$ lies on the conic.

$$\therefore \quad aq^2 + bq'^2 = 1.]$$

(c) If a sphere touches the planes $2x + 3y - 6z + 14 = 0$ and $2x + 3y - 6z + 42 = 0$, and if its centre lies on the line $2x + z = 0, y = 0$, find the equation of the sphere.
[**Hints.** See Ex.19 of page 260.]

Group C

11. Answer *any four*:

(a) Prove by vector method that if two medians of a triangle are equal, the triangle is isosceles.
[**Hints.** See Ex.10 of page 303.]

(b) If $|\bar{a} - \bar{b}| = |\bar{a} + \bar{b}|$, show that $\bar{a}$ and $\bar{b}$ are perpendicular to each other.
[**Hints.** See Q.11.(b) of 2000.]

(c) Show that $[\bar{a} + \bar{b}, \bar{b} + \bar{c}, \bar{c} + \bar{a}] = 2[\bar{a}\bar{b}\bar{c}]$.
[**Hints.** See Ex.18 of page 306.]

(d) If $\bar{\alpha} \times \bar{\beta} + \bar{\beta} \times \bar{\gamma} + \bar{\gamma} \times \bar{\alpha} = \bar{0}$, then show that $\bar{\alpha}, \bar{\beta}, \bar{\gamma}$ are coplanar.
[**Hints.** See Q.17.(a) of page 307.]

(e) If $\bar{\alpha}, \bar{\beta}, \bar{\gamma}, \bar{\delta}$ are four vectors such that $\bar{\alpha} \times \bar{\beta} = \bar{\gamma} \times \bar{\delta}$ and $\bar{\alpha} \times \bar{\gamma} = \bar{\beta} \times \bar{\delta}$, show that $\bar{\alpha} - \bar{\delta}$ and $\bar{\beta} - \bar{\gamma}$ are collinear.
[**Hints.**

$$\begin{aligned} & \bar{\alpha} \times \bar{\beta} + \bar{\beta} \times \bar{\delta} = \bar{\gamma} \times \bar{\delta} + \bar{\alpha} \times \bar{\gamma}, \\ \text{or,}\quad & \bar{\alpha} \times \bar{\beta} - \bar{\delta} \times \bar{\beta} = \bar{\alpha} \times \bar{\gamma} - \bar{\delta} \times \bar{\gamma}, \\ \text{or,}\quad & (\bar{\alpha} - \bar{\delta}) \times \bar{\beta} = (\bar{\alpha} - \bar{\delta}) \times \bar{\gamma}, \\ \text{or,}\quad & (\bar{\alpha} - \bar{\delta}) \times (\bar{\beta} - \bar{\gamma}) = \bar{0}. \end{aligned}$$

$\therefore$ $\bar{\alpha} - \bar{\delta}$ is parallel to $\bar{\beta} - \bar{\gamma}$. Hence $\bar{\alpha} - \bar{\delta}$ and $\bar{\beta} - \bar{\gamma}$ are collinear.]

(f) A particle moves from the point $\bar{i} + 2\bar{j} + 3\bar{k}$ to the point $-4\bar{i} + 6\bar{j} + 8\bar{k}$ on a plane subjected to the force $3\bar{i} + 4\bar{j} + 5\bar{k}$. Find the work done by the force.
[**Hints.** Displacement vector $= (-4\bar{i} + 6\bar{j} + 8\bar{k}) - (\bar{i} + 2\bar{j} + 3\bar{k}) = -5\bar{i} + 4\bar{j} + 5\bar{k}$.
Work done $= (3\bar{i} + 4\bar{j} + 5\bar{k}).(-5\bar{i} + 4\bar{j} + 5\bar{k}) = -15 + 16 + 25 = 26$ units.]

(b) Find the value of 'a' for which the plane $x+y+z=a\sqrt{3}$ touches the sphere $x^2+y^2+z^2-2x-2y-2z-6=0$.

[**Hints.** The centre and radius of the sphere are $(1,1,1)$ and 3.

If the plane touches the sphere, then $\frac{1+1+1-a\sqrt{3}}{\sqrt{3}}=\pm 3$, or, $a=\sqrt{3}\pm 1$.]

(c) Show that the lines $x+7y-z-16=0=x-y+z-4$ and $x+11y-2z=0=x-5y+2z-4$ are parallel.

[**Hints.** Show that the d.rs. of both lines are $3,-1,-4$.]

VECTOR

2004

[Third Paper]

Group A

[Answer **Q. No. 1** and **any one** from the rest.]

1. Answer *any two* questions.

(a) If $\bar{\alpha}=(2,3,4)$ and $\bar{\beta}=(1,-3,2)$, find $\bar{\alpha}+\bar{\beta}$ and $\bar{\beta}-\bar{\alpha}$.

[**Hints.** $\bar{\alpha}+\bar{\beta}=(3,0,6), \bar{\beta}-\bar{\alpha}=(-1,-6,-2)$.]

(b) Determine the value of λ for which the vector $\bar{a}=\lambda\hat{\imath}-4\hat{\jmath}+3\hat{k}$ and $\bar{b}=3\hat{\imath}+\lambda\hat{\jmath}-2\hat{k}$ are perpendicular to each other.

[**Hints.** $\bar{a}\cdot\bar{b}=0$, or, $3\lambda-4\lambda-6=0$, or, $\lambda=-6$.]

(c) Find the scalar area of the triangle with two sides given by the vectors $2\hat{\imath}+\hat{\jmath}+3\hat{k}$ and $3\hat{\imath}-\hat{\jmath}+2\hat{k}$.

[**Hints.** Area $=\frac{1}{2}\left|\left(2\hat{\imath}+\hat{\jmath}+3\hat{k}\right)\times\left(3\hat{\imath}-\hat{\jmath}+2\hat{k}\right)\right|=\frac{5\sqrt{3}}{2}$.]

(d) If two vectors $\bar{a}$ and $\bar{b}$ are such that $|\bar{a}+\bar{b}|=|\bar{a}-\bar{b}|$, find the angle between them.

[**Hints.** $(\bar{a}+\bar{b})\cdot(\bar{a}+\bar{b})=(\bar{a}-\bar{b})\cdot(\bar{a}-\bar{b})$, or, $4\bar{a}\cdot\bar{b}=0$, or, $\bar{a}$ is perpendicular to $\bar{b}$.]

2. (a) Prove by vector method $c=a\cos B+b\cos A$ in a triangle ABC.

[**Hints.** See Ex. 6 of page 302.]

(b) Determine the values of λ and μ for which the vectors $-3\hat{\imath}+4\hat{\jmath}+\lambda\hat{k}$ and $\mu\hat{\imath}+8\hat{\jmath}+6\hat{k}$ are collinear.

[**Hints.** $-\frac{3}{\mu}=\frac{4}{8}=\frac{\lambda}{6}$. $\therefore\ \lambda=3, \mu=-6$.]

(c) Find the moment of the force $\vec{F}=(4,2,1)$ passing through the point $A(5,2,4)$ about the point $B(3,-1,3)$.

[**Hints.** Here $\bar{r}=(5\bar{i}+2\bar{j}+4\bar{k})-(3\bar{i}-\bar{j}+3\bar{k})=2\bar{i}+3\bar{j}+\bar{k}$

Moment $=(2\bar{i}+3\bar{j}+\bar{k})\times(4\bar{i}+2\bar{j}+\bar{k})=\bar{i}+2\bar{j}-8\bar{k}$.]

3. (a) Find the vector equation of the plane passing through the points $(2,0,0),(0,3,0)$ and $(0,0,4)$.

[**Hints.** Let $\bar{a}=(2,0,0), \bar{b}=(0,3,0), \bar{c}=(0,0,4)$. Let $\bar{r}$ be the position vector of a point on the plane. Then the equation of the plane is $(\bar{r}-\bar{a})\cdot\{(\bar{a}-\bar{b})\times(\bar{a}-\bar{c})\}=0$.]

(b) Prove by vector method that the medians of a triangle are concurrent.

[**Hints.** See Ex. 3 of page 286.]

(c) Prove that $\bar{a}\times(\bar{b}\times\bar{c})+\bar{b}\times(\bar{c}\times\bar{a})+\bar{c}\times(\bar{a}\times\bar{b})=\bar{0}$. Find also the condition for which $(\bar{a}\times\bar{b})\times\bar{c}=\bar{a}\times(\bar{b}\times\bar{c})$ will hold.

[**Hints.** $\bar{a}\times(\bar{b}\times\bar{c})+\bar{b}\times(\bar{c}\times\bar{a})+\bar{c}\times(\bar{a}\times\bar{b})$

$$=(\bar{a}\cdot\bar{c})\bar{b}-(\bar{a}\cdot\bar{b})\bar{c}+(\bar{b}\cdot\bar{a})\bar{c}-(\bar{b}\cdot\bar{c})\bar{a}+(\bar{c}\cdot\bar{b})\bar{a}-(\bar{c}\cdot\bar{a})\bar{b}=\bar{0}.$$

For the second part, see Note 2 of page 299.]

2005

[First Paper]

Group C

[Answer **Q. No. 11** and *any two* from the rest.]

11. (a) Answer *any two* questions.

(i) Find the value of K so that the equation $x^2+y^2+2x+K=0$ may represent a pair of straight lines.

[**Hints.** Here $1\cdot1\cdot K-1\cdot1=0$, $K=1$.]

(ii) Find the transformed form of the equation $x^2-y^2=25$ when the axes are turned through an angle of 45° keeping the origin fixed.

[**Hints.** The equation transforms to $(x\cos45°-y\sin45°)^2-(x\sin45°+y\cos45°)^2=25$,

$$\text{or, } x^2(\cos^2 45°-\sin^2 45°)-y^2(\cos^2 45°-\sin^2 45°-4xy\cos 45°\sin 45°=25,$$

$$\text{or, } -2xy=25, \quad \text{or, } \quad 2xy+25=0.]$$

(iii) Find the direction cosines of the line that makes equal angles with the co-ordinate axes.

[**Hints.** See Ex. 2 of page 161.]

(iv) Find the point where the line $\frac{x-1}{2}=\frac{y-2}{1}=\frac{z}{3}$ meets the plane $3x+7y+z=1$.

[**Hints.** Any point on the line is $(2r+1, r+2, 3r)$. If this point lies on the plane, then $3(2r+1)+7(r+2)+3r=1$, or, $r=-1$. Therefore, the point is $(-1,1,-3)$.]

(b) Answer *any one* question.

(i) Determine the nature of the conic $\frac{3}{r}=2+4\cos\theta$ and also find the length of its latus rectum.

[**Hints.** $\frac{3}{r}=2+4\cos\theta$, or, $\frac{3/2}{r}=1+2\cos\theta$. Eccentricity of the conic $=2>1$. Therefore, the conic is a hyperbola. Length of latus rectum $=2\cdot\frac{3}{2}=3$ units.]

(ii) Find the equation of the plane passing through the point $(4,-3,5)$ and containing the y-axis.

[**Hints.** Let the equation of the plane be $ax+cz=0$. If it passes through the point $(4,-3,5)$, then $4a+5c=0$, or, $c=-\frac{4a}{5}$. Thus the equation is $5x-4z=0$.]

12. (a) Reduce the equation $3x^2-5xy+6y^2+11x-17y+13=0$ to the canonical form and determine the nature of the conic.

[**Hints.** See Ex. 5 of page 51.]

(b) Find the condition that the straight line $\frac{l}{r}=A\cos\theta+B\sin\theta$ touch the conic $\frac{l}{r}=1+e\cos\theta$.

[**Hints.** See Ex. 21 of page 139.]

(c) Find the equation to the pair of straight lines joining the origin to the intersection of the line $y = mx + c$ and the circle $x^2 + y^2 = a^2$.

[**Hints.** $x^2 + y^2 = a^2$ is made homogeneous by $y = mx + c$.
The equation is $x^2 + y^2 = a^2 \left(\frac{y-mx}{c}\right)^2$, or, $c^2(x^2 + y^2) = a^2(y - mx)^2$.]

13. (a) Prove that the product of the perpendiculars from the point (x_1, y_1) on the straight lines $ax^2 + 2hxy + by^2 = 0$ is $\dfrac{ax_1^2 + 2hx_1y_1 + by_1^2}{\sqrt{(a-b)^2 + 4h^2}}$.

[**Hints.** See Ex. 7 of page 28.]

(b) Prove that the locus of a point from which perpendicular tangents can be drawn to the circle $x^2 + y^2 = a^2$ is the circle $x^2 + y^2 = 2a^2$.

[**Hints.** See Art. 4.5(c) of page 63.]

(c) If the polar of a point (α, β) with respect to the parabola $y^2 = 4ax$ touches the parabola $x^2 = 4by$, show that the locus of (α, β) is $xy + 2ab = 0$.

[**Hints.** See Ex. 16(c) of page 83.]

14. (a) Find the shortest distance between the lines $\frac{x-3}{2} = \frac{y+15}{-7} = \frac{z-9}{5}$ and $\frac{x+1}{2} = \frac{y-1}{1} = \frac{z-9}{-3}$ and find also the equation of the line of shortest distance.

[**Ans.** $4\sqrt{3}, x = y = z$.]

(b) Obtain the equation of the sphere having the circle $x^2 + y^2 + z^2 + 2x - 4y + 2z + 5 = 0$, $x - 2y + 3z + 1 = 0$ as a great circle.

[**Hints.** See Ex. 8 of page 219.]

(c) If a plane meets the axes in A, B, C and the centroid of $\triangle ABC$ is (α, β, γ), show that the equation of the line is $\frac{x}{\alpha} + \frac{y}{\beta} + \frac{z}{\gamma} = 3$.

[**Hints.** See Ex. 17 of page 181.]

15. (a) Show that the lines $x + 7y - z - 16 = 0 = x - y + z - 4$ and $x + 11y - 2z = 0 = x - 5y + 2z - 4$ are parallel.

[**Hints.** Let l_1, m_1, n_1 be the d.rs. of the first line, then $l_1 + 7m_1 - n_1 = 0$ and $l_1 - m_1 + n_1 = 0$. From these two $\dfrac{l_1}{3} = \dfrac{m_1}{-1} = \dfrac{n_1}{-4}$.
Let l_2, m_2, n_2 be the d.rs. of the second line. Then $l_2 + 11m_2 - 2n_2 = 0$ and $l_2 - 5m_2 + 2n_2 = 0$. From these two $\dfrac{l_2}{3} = \dfrac{m_2}{-1} = \dfrac{n_2}{-4}$. Hence the lines are parallel.]

(b) Find the equation of the two tangent planes to the sphere $x^2 + y^2 + z^2 - 2y - 6z + 5 = 0$ which are parallel to the plane $2x + 2y - z = 0$.

[**Hints.** Let the tangent plane be $2x + 2y - z + k = 0$. The centre of the sphere is $(0, 1, 3)$ and radius is of $\sqrt{5}$ units. The distance of the above plane from $(0, 1, 3)$ is $\sqrt{5}$ units.

$$\therefore \quad \frac{2 - 3 + k}{\sqrt{2^2 + 2^2 + 1}} = \pm\sqrt{5}, k - 1 = \pm 3\sqrt{5}, \quad \text{or,} \quad k = 1 \pm 3\sqrt{5}.$$

Hence the planes are $2x + 2y - z + 1 \pm 3\sqrt{5} = 0$.]

(c) Find the equation of the plane through the straight line $\frac{x-1}{2} = \frac{y+2}{-3} = \frac{z}{5}$ and perpendicular to the plane $x - y + z + 2 = 0$.

[**Hints.** The equation of line can be written as $3x + 2y + 1 = 0 = 5y + 3z + 10$.
Let the plane be $3x + 2y + 1 + \lambda(5y + 3z + 10) = 0$, or, $3x + (2 + 5\lambda)y + 3\lambda z + 10\lambda + 1 = 0$. It is perpendicular to the given plane.
$\therefore\ 3 - 2 - 5\lambda + 3\lambda = 0, \quad \lambda = \frac{1}{2}$.
$\therefore$ the required plane is $2x + 3y + z + 4 = 0$.]

VECTOR

Group D

[Answer **Q. No. 16** and *any one* from the rest.]

16. Answer *any one* question.

(a) Find the vector equation of the straight line passing through the points $(2, 3, 4)$ and $(-1, 3, -2)$.

[**Hints.** Let $\bar{a} = (2, 3, 4)$ and $\bar{b} = (-1, 3, -2)$.
The equation of the line is $\bar{r} = \bar{a} + t(\bar{b} - \bar{a})$.]

(b) Find the work done by the force $4\bar{i} + 5\bar{j} + 3\bar{k}$ which displaces a particle from the point $(-1, -1, 3)$ to the point $(1, 1, 2)$.

[**Hints.** Displacement vector $= 2\bar{i} + 2\bar{j} - \bar{k}$.
$\therefore$ work done $= (4\bar{i} + 5\bar{j} + 3\bar{k}) \cdot (2\bar{i} + 2\bar{j} - \bar{k}) = 15$ units.]

17. (a) In any triangle ABC, with usual notations, prove by vector method that $c^2 = a^2 + b^2 - 2ab\cos C$.

[**Hints.** See Ex. 5 of page 301.]

(b) Show that the three vectors $(4\bar{i} + 2\bar{j} + \bar{k}), (2\bar{i} - \bar{j} + 3\bar{k})$ and $(8\bar{i} + 7\bar{k})$ are coplanar.

[**Hints.** $(2\bar{i} - \bar{j} + 3\bar{k}) \times (8\bar{i} + 7\bar{k}) = -7\bar{i} + 10\bar{j} + 8\bar{k}$
$(4\bar{i} + 2\bar{j} + \bar{k}) \cdot (-7\bar{i} + 10\bar{j} + 8\bar{k}) = -28 + 20 + 8 = 0$
$\therefore$ the vectors are coplanar.]

(c) Find the unit vector perpendicular to both $\bar{i}$ and $\bar{j} + \bar{k}$.

[**Hints.** $\bar{i} \times (\bar{j} + \bar{k}) = -\bar{j} + \bar{k}$. It is perpendicular to $\bar{i}$ and $\bar{j} + \bar{k}$.
The required unit vector $= \frac{-\bar{j}+\bar{k}}{\sqrt{2}}$.]

18. (a) If $\bar{\alpha}, \bar{\beta}, \bar{\gamma}$ are three vectors such that $\bar{\alpha} + \bar{\beta} + \bar{\gamma} = \bar{0}$ and $|\bar{\alpha}| = 2, |\bar{\beta}| = 4, |\bar{\gamma}| = 6$, then show that $\bar{\alpha} \cdot \bar{\beta} + \bar{\beta} \cdot \bar{\gamma} + \bar{\gamma} \cdot \bar{\alpha} = -28$.

[**Hints.** $(\bar{\alpha} + \bar{\beta} + \bar{\gamma})^2$, or, $\bar{\alpha}^2 + \bar{\beta}^2 + \bar{\gamma}^2 + 2(\bar{\alpha} \cdot \bar{\beta} + \bar{\beta} \cdot \bar{\gamma} + \bar{\gamma} \cdot \bar{\alpha}) = 0$,
or, $4 + 16 + 36 + 2(\bar{\alpha} \cdot \bar{\beta} + \bar{\beta} \cdot \bar{\gamma} + \bar{\gamma} \cdot \bar{\alpha}) = 0$
or, $\bar{\alpha} \cdot \bar{\beta} + \bar{\beta} \cdot \bar{\gamma} + \bar{\gamma} \cdot \bar{\alpha} = -28$.]

(b) For any three vectors $\bar{\alpha}, \bar{\beta}, \bar{\gamma}$, prove that $\bar{\alpha} \times (\bar{\beta} \times \bar{\gamma}) = (\bar{\alpha} \cdot \bar{\gamma})\bar{\beta} - (\bar{\alpha} \cdot \bar{\beta})\bar{\gamma}$.

[**Hints.** See the Art. 2.4(b) of page 298.]

(c) Determine the values of λ and μ for which the vectors $\mu\bar{i} + 8\bar{j} + 6\bar{k}$ and $-3\bar{i} + 4\bar{j} + \lambda\bar{k}$ are collinear.

[**Hints.** See the question 2(b) of 2004.]

Burdwan University

2003

[First Paper]

Part I

1. Answer the following:

(ix) Find the co-ordinates of the centre of $x^2 - 4xy - 2y^2 + 10x + 4y = 0$.

[**Hints.** Let $f(x, y) = x^2 - 4xy - 2y^2 + 10x + 4y$.

$$\frac{\partial f}{\partial x} = 2(x - 2y + 5) \quad \text{and} \quad \frac{\partial f}{\partial y} = -4(x + y - 1).$$

If (α, β) is the centre, then $\alpha - 2\beta + 5 = 0$ and $\alpha + \beta - 1 = 0$.

From these equation $\alpha = -1, \beta = 2$.]

(x) Find the equation of a straight line at a distance 5 units from origin in polar co-ordinates.

[**Hints.** See section 8.4 of page 120.]

(xi) Find the distance of $(5, 3, 4)$ from x-axis and z-axis.

[**Hints.** Distance from the x-axis $= \sqrt{3^2 + 4^2} = 5$.

Distance from the z-axis $= \sqrt{5^2 + 3^2} = \sqrt{34}$.]

(xii) Find the co-ordinates of a foot of the perpendicular from the origin to the plane $x + 2y + 4z = 14$

[**Hints.** Equations of the line passing through the origin and perpendicular to the plane are $\frac{x}{1} = \frac{y}{2} = \frac{z}{4}$ $(= r$, say).

Any point on the line is $(r, 2r, 4r)$. If this point lies on the plane, then $r + 4r + 16r = 14$, or, $r = \frac{2}{3}$.

$\therefore$ the co-ordinates of the foot are $\left(\frac{2}{3}, \frac{4}{3}, \frac{8}{3}\right)$.]

(xiii) What is the equation of the sphere having segment joining the points $(1, 2, 1)$ and $(3, 4, 5)$ as diameter.

[**Hints.** The equation of the sphere is $(x - 1)(x - 3) + (y - 2)(y - 4) + (z - 1)(z - 5) = 0$, or, $x^2 + y^2 + z^2 - 4x - 6y - 6z + 16 = 0$.]

Part II

Group B

5. Answer (a) or (b).

(a) (i) Find the co-ordinates of the point where the origin is to be shifted to eliminate the first degree terms of the equation $3x^2 - 8xy + 3y^2 - 2x + 2y - 2 = 0$.

[**Hints.** Let the origin be shifted to the point (h, k) without rotating the axes. The equation transforms to $3(x'+h)^2 - 8(x'+h)(y'+k) + 3(y'+k)^2 - 2(x'+h) + 2(y'+k) - 2 = 0$, or, $3x'^2 - 8x'y' + 3y'^2 + 2(3h - 4k - 1)x' - 2(4h - 3k - 1)y' + 3h^2 - 8hk + 3k^2 - 2h + 2k - 2 = 0$.

To eliminate the first degree terms, $3h - 4k - 1 = 0, 4h - 3k - 1 = 0$. From these (h, k) is $\left(\frac{1}{7}, -\frac{1}{7}\right)$.]

(ii) Determine the nature of the conic section and length of its latus rectum, whose polar equation is $r = \frac{5}{3 - 4\cos\theta}$.

[**Hints.** $r = \frac{5}{3 - 4\cos\theta}$, or, $\frac{5/3}{r} = 1 - \frac{4}{3}\cos\theta$.

$\frac{4}{3}$ is the eccentricity of the conic. It is greater than 1. Therefore, the conic is a hyperbola.

Length of latus rectum $= \frac{10}{3}$ units.]

(b) (i) Find the polar equation of the circle having centre at $(2, \frac{\pi}{2})$ and passing through the origin.

[**Hints.** The equation is $r = 4\cos(\theta - \frac{\pi}{2})$, or, $r = 4\sin\theta$.]

(ii) Suitably reducing the equation to standard form, show that the equation $3x^2 + 2xy + 3y^2 - 16x + 20 = 0$ represents an ellipse.

[**Hints.** See Ex. 2 of page 42.]

6. Answer *any three* of the following questions:

(a) (i) Find the equation of the plane through the line $2x - y = 0 = 3z - y$ and perpendicular to the plane $4x + 5y - 3z = 0$.

[**Hints.** Let the plane be $2x - y + \lambda(3z - y) = 0$, or, $2x - (\lambda + 1)y + \lambda 3z = 0$. It is perpendicular to $4x + 5y - 3z = 0$. $\therefore$ $4 \cdot 2 - 5(\lambda + 1) + 9\lambda = 0$, $\lambda = -\frac{3}{4}$.
$\therefore$ the plane is $8x - y - 9z = 0$.]

(ii) A plane cuts the three axes at A, B, C and the co-ordinates of the centroid of triangle ABC is $(2, 1, 3)$. Find the equation of the plane.

[**Hints.** See Ex. 3 of page 168. The plane is $\frac{x}{2} + \frac{y}{1} + \frac{z}{3} = 3$.]

(b) (i) Find the equation of the projection of the line $\frac{x-1}{2} = \frac{y+1}{-1} = \frac{z-3}{4}$ on the plane $x + 2y + z = 6$.

[**Hints.** Let the line meet the plane at $(2r + 1, -r - 1, 4r + 3)$. Then $2r + 1 - 2r - 2 + 4r + 3 = 6$, or, $r = 1$. $\therefore$ the point is $(3, -2, 7)$.

The equations of the line perpendicular to the plane and passing through the point $(1, -1, 3)$ are $\frac{x-1}{1} = \frac{y+1}{2} = \frac{z-3}{1}$. Let this line meet the plane at $(r' + 1, 2r' - 1, r' + 3)$. Then $r' + 1 + 2(2r' - 1) + r' + 3 = 6$, or, $r' = \frac{2}{3}$. The point is $(\frac{5}{3}, \frac{1}{3}, \frac{11}{3})$. The equation of the projection is the equation of the line passing through the points $(3, -2, 7)$ and $(\frac{5}{3}, \frac{1}{3}, \frac{11}{3})$.]

(ii) Find the distance of the point $(-2, 3, -4)$ from the line $\frac{x+2}{3} = \frac{2y+3}{4} = \frac{3z+4}{5}$ measured parallel to the plane $4x + 12y - 3z + 1 = 0$.

[**Hints.** The equation of the plane parallel to the given plane and passing through the given point is $4x + 12y - 3z = 40$.

Let the line meet this plane at $(3r - 2, \frac{4r-3}{2}, \frac{5r-4}{3})$. Then $4\cdot(3r-2) + 12\frac{4r-3}{2} - 3\cdot\frac{5r-4}{3} = 40$, or, $r = 2$. The point is $(4, \frac{5}{2}, 2)$.

Required distance = distance between the points $(2, 3, -4)$ and $(4, \frac{5}{2}, 2)$

$$= \sqrt{\left\{(4+2)^2 + \left(\tfrac{5}{2} - 3\right)^2 + (2+4)^2\right\}}$$

$$= \sqrt{\frac{289}{4}} \text{ units.]}$$

(c) (i) Prove that $\frac{x-1}{2} = \frac{y-2}{3} = \frac{z-3}{4}$ and $4x - 3y + 1 = 0 = 5x - 3x + 2$ are coplanar.

[**Hints.** Any plane through the second line is $4x - 3y + 1 + \lambda(5x - 3z + 2) = 0$, or, $(4 + 5\lambda)x - 3y - 3\lambda z + 1 + 2\lambda = 0$. If it is parallel to the first line, then $(4 + 5\lambda) \cdot 2 - 3 \cdot 3 - 3\lambda \cdot 4 = 0$, or, $\lambda = -\frac{1}{2}$. The plane is $x - 2y + z = 0$. The point $(1, 2, 3)$ lies on this plane. Therefore, the first line lies on this plane. Hence the lines are coplanar.]

(ii) Find the s.d. beween the lines $\frac{x-1}{2} = \frac{y-2}{3} = \frac{z-3}{4}$ and $\frac{x-3}{3} = \frac{y-3}{4} = \frac{z-4}{5}$.

[**Hints.** The d.cs. of the line joining the points $(1, 2, 3)$ and $(3, 3, 4)$ are $\frac{2}{\sqrt{6}}, \frac{1}{\sqrt{6}}, \frac{1}{\sqrt{6}}$. If l, m, n are d.cs. of s.d., then $2l + 3m + 4n = 0$ and $3l + 4m + 5n = 0$.

From these $\frac{l}{1} = \frac{m}{-2} = \frac{n}{1} = \frac{1}{\sqrt{6}}$.

If the points $(1, 2, 3)$ and $(3, 3, 4)$ are A and B and θ is the angle between AB and s.d., then s.d. $= AB\cos\theta = \frac{1}{\sqrt{6}}$ units.]

(d) (i) Find the radius of the circle $x^2+y^2+z^2+x+y+z-4=0, x+y+z=0$.

[**Hints.** Centre of the sphere is $(-\frac{1}{2},-\frac{1}{2},-\frac{1}{2})$ and radius of it

$$=\sqrt{\frac{1}{4}+\frac{1}{4}+\frac{1}{4}+4}=\sqrt{\frac{19}{4}}.$$

Distance of the plane from the centre $=\left|\dfrac{-\frac{1}{2}-\frac{1}{2}-\frac{1}{2}}{\sqrt{3}}\right|=\dfrac{\sqrt{3}}{2}$.

Radius of the circle $=\sqrt{\frac{19}{4}-\frac{3}{4}}=2$.]

(ii) Find the equation of the tangent plane to the sphere $x^2+y^2+z^2+4x-2y+6z-32=0$ at the point $(1,2,3)$.

[**Hints.** $(1,2,3)$ lies on the sphere. Hence the equation of the tangent plane is

$x\cdot 1+y\cdot 2+z\cdot 3+2(x+1)-(y+2)+3(z+3)-32=0$, or, $3x+y+6z-23=0$.]

(e) (i) Find the equation of the cone whose vertex is at $(-1,2,-3)$ and generators having d.cs. l,m,n satisfying the relation $2l^2-7m^2+9n^2=0$.

[**Hints.** One generator is

$$\frac{x+1}{l}=\frac{y-2}{m}=\frac{z+3}{n}. \qquad (1)$$

Eliminating l,m,n satisfying the relation $2l^2-7m^2+9n^2=0$ by (1), the equation of the cone is $2(x+1)^2-7(y-2)^2+9(z+3)^2=0$.]

(ii) Find the equation of the normal to the surface $\frac{x^2}{a^2}+\frac{y^2}{b^2}+\frac{z^2}{c^2}=1$ at the point

$$\left(\frac{a}{\sqrt{3}},\frac{b}{\sqrt{3}},\frac{c}{\sqrt{3}}\right).$$

[**Hints.** The equation of the tangent plane at the given point is $\frac{x}{a\sqrt{3}}+\frac{y}{b\sqrt{3}}+\frac{z}{c\sqrt{3}}=1$. The normal is the line perpendicular to the tangent plane and passing through the point of contact.

Thus the normal is

$$\frac{x-\frac{a}{\sqrt{3}}}{\frac{1}{a\sqrt{3}}}=\frac{y-\frac{b}{\sqrt{3}}}{\frac{1}{b\sqrt{3}}}=\frac{z-\frac{c}{\sqrt{3}}}{\frac{1}{c\sqrt{3}}}$$

or, $a(\sqrt{3}kx-a)=b(\sqrt{3}y-b)=c(\sqrt{3}z-c)$.]

7. Answer either (a) or (b).

(a) (i) Show that the vectors given by $3\hat{\imath}-7\hat{\jmath}-4\hat{k}, 3\hat{\imath}-2\hat{\jmath}+\hat{k}, \hat{\imath}+\hat{\jmath}+2\hat{k}$ are coplanar. Also find the direction of the normal to the plane containing the three vectors.

[**Hints.** Let $\bar{a}=3\hat{\imath}-7\hat{\jmath}-4\hat{k}, \bar{b}=3\hat{\imath}-2\hat{\jmath}+\hat{k}, \bar{c}=\hat{\imath}+\hat{\jmath}+2\hat{k}$.

$\because\ \bar{a}\cdot(\bar{b}\times\bar{c})=\begin{vmatrix}3&-7&-4\\3&-2&1\\1&1&2\end{vmatrix}=0$, the vectors are coplanar.

Let l,m,n be the direction of the normal.

Then $3l-7m-4n=0, 3l-2m+n=0, l+m+2n=0$. From the first two, $l:m:n$ $1:1:-1$.

Again, $1+1-2=0$. $\quad\therefore$ the direction of the normal is $\frac{1}{\sqrt{3}},\frac{1}{\sqrt{3}},\frac{-1}{\sqrt{3}}$.]

(ii) Show that $[\bar{a}+\bar{b},\bar{b}+\bar{c},\bar{c}+\bar{a}]=2[\bar{a}\bar{b}\bar{c}]$.

[**Hints.** See Ex. 18 of page 306.]

(b) (i) Find the equation of the plane passing through origin and parallel to the vectors $\bar{i}+2\bar{j}+3\bar{k}$ and $2\bar{i}-\bar{j}-\bar{k}$.

[**Hints.** $(\bar{i}+2\bar{j}+3\bar{k})\times(2\bar{i}-\bar{j}-\bar{k})$ is perpendicular to the plane.

$$\left(\bar{i}+2\bar{j}+3\bar{k}\right)\times\left(2\bar{i}-\bar{j}-\bar{k}\right)=\bar{i}+7\bar{j}-5\bar{k}.$$

If $\bar{r}$ is the position vector of any point on the plane, then the equation of the plane is $\bar{r}\cdot(\bar{i}+7\bar{j}-5\bar{k})=0$.]

(ii) Find the torque about the point $\bar{i}+2\bar{j}-\bar{k}$ of a force represented by $3\bar{i}+\bar{k}$ and acting through the point $2\bar{i}-\bar{j}+3\bar{k}$.

[**Hints.** Let the position vectors of the points A and B be $\bar{i}+2\bar{j}-\bar{k}$ and $2\bar{i}-\bar{j}+3\bar{k}$.
Torque $=\overrightarrow{AB}\times(3\bar{i}+\bar{k})=(\bar{i}-3\bar{j}+4\bar{k})\times(3\bar{i}+\bar{k})=-3\bar{i}+11\bar{j}+9\bar{k}$.]

2004

Paper I

1. Answer the following:

(viii) Find the area of the triangle whose vertices are $(-2,4),(2,-6)$ and $(5,4)$.

[**Hints.** Area $=\frac{1}{2}\{-2(-6-4)+2(4-4)+5(4+6)\}=35$ sq. units.]

(ix) Find the centre of the conic given by the equation $3x^2-8xy+7y^2-4x+2y-7=0$.

[**Hints.** To find the centre $6x-8y-4=0$ and $-8x+14y+2=0$. From these equations $x=2, y=1$. $\therefore$ $(2,1)$ is the centre.]

(x) Find the polar co-ordinates of a point whose cartesian co-ordinates are $(\sqrt{3},1)$.

[**Hints.** Let $r,\theta)$ be the polar co-ordinates.
Then $\sqrt{3}=r\cos\theta, 1=r\sin\theta$, $\therefore$ $r=\sqrt{3+1}=2, \theta=\frac{\pi}{6}$.]

(xi) Find the centre and radius of the sphere $x^2+y^2+z^2-2x+4y-6z+5=0$.

[**Hints.** Centre is $(1,-2,3)$. Radius $=\sqrt{1+4+9-5}=3$ units.]

(xii) Find the magnitude of the vector $\bar{a}\times\bar{b}$, where $\bar{a}=6\bar{i}+2\bar{j}+3\bar{k}, \bar{b}=3\bar{i}-6\bar{j}-2\bar{k}$.

[**Hints.** $\bar{a}\times\bar{b}=\begin{vmatrix}\bar{i} & \bar{j} & \bar{k}\\ 6 & 2 & 3\\ 3 & -6 & -2\end{vmatrix}=14\bar{i}+21\bar{j}-42\bar{k}$. $\therefore$ $|\bar{a}\times\bar{b}|=\sqrt{14^2+21^2+42^2}=49$.]

(xiii) Find the equation of the plane through the point $(3,-2,6)$ and containing the x-axis.

[**Hints.** Let the equation of the plane be $by+cz=0$. Since $(3,-2,6)$ lies on the plane $-2b+6c=0$, or, $b=3c$. $\therefore$ $3y+z=0$ is the equation of the plane.]

(xiv) Find the work done by a force $5\bar{i}+10\bar{j}+15\bar{k}$, when it produces a displacement from the point $3\bar{i}-\bar{j}-6\bar{k}$ to $\bar{i}+3\bar{k}$.

[**Hints.** If $\bar{F}$ is the force and $\bar{d}$ is the displacement, then work done $=\bar{F}\cdot\bar{d}$.
$\therefore$ workd done $=(5\bar{i}+10\bar{j}+15\bar{k})\cdot(-2\bar{i}+\bar{j}+9\bar{k})=-10+10+135=135$ units.]

(xv) Express $(2,3,4)$ as a linear combination of $(1,1,1)$ and $(-1,0,1)$.

[**Hints.** Let $(2,3,4)=a(1,1,1)+b(-1,0,1)$.
Then $a-b=2, a=3, a+b=4$, $\therefore$ $b=1$.
Thus $(2,3,4)=3(1,1,1)+(-1,0,1)$.]

(xvi) Show that $\begin{bmatrix} \cos\theta & -\sin\theta \\ \sin\theta & \cos\theta \end{bmatrix}$ is orthogonal.

[**Hints.** Since $\begin{bmatrix} \cos\theta & \sin\theta \\ -\sin\theta & \cos\theta \end{bmatrix}\begin{bmatrix} \cos\theta & -\sin\theta \\ \sin\theta & \cos\theta \end{bmatrix} = \begin{bmatrix} 1 & 0 \\ 0 & 1 \end{bmatrix}$, the given matrix is orthogonal.]

Group B

5. Answer either (a) or (b).

(a) (i) Find the equation of the curve $9x^2 + 4y^2 + 18x - 16y = 11$ referred to parallel axes through the point $(-1, 2)$.

[**Hints.** The transformed equation is $9(x'-1)^2 + 4(y'+2)^2 + 18(x'-1) - 16(y'+2) = 11$, or, $9x'^2 + 4y'^2 = 36$.]

(ii) Find the centre of the circle $r = 4\cos\theta + 3\sin\theta$ in polar co-ordinates. Find also its radius.

[**Hints.** See Ex. 3 of page 123.]

(b) (i) Find the nature of the conic $\frac{2}{r} = 1 - \cos\theta$ and find the length of latus rectum and eccentricity.

[**Hints.** It is a parabola. Length of latus rectum = 4 units and eccentricity = 1.]

(ii) Reduce the equation $x^2 - 5xy + y^2 + 8x - 20y + 15 = 0$ to its standard form and show that it represents a hyperbola.

[**Hints.** Here $\Delta = \frac{21}{4}, D = -\frac{21}{4}$. The equation represents a hyperbola. Changing the origin to the centre the equation reduces to $x^2 - 5xy + y^2 - 1 = 0$.

Rotating the axes through an acute angle θ where $\cot 2\theta = \frac{0}{-5} = 0$, $\therefore\ \theta = \frac{\pi}{4}$, the equation transforms to $(x'\cos\theta - y'\sin\theta)^2 - 5(x'\cos\theta - y'\sin\theta)(x'\sin\theta + y'\cos\theta) + (x'\sin\theta + y'\cos\theta)^2 = 1$, or, $7y'^2 - 3x'^2 = 2$. It is the standard form.]

6. Answer *any three* questions.

(a) (i) Show that the equation of the plane passing through the point $(1, 2, 3)$ and making equal non-zero intercepts on the axes of co-ordinates is $x + y + z = 6$.

[**Hints.** Let the equation of the plane be $x + y + z = a$. Since it passes through the point $(1, 2, 3)$, $a = 6$.]

(ii) A variable plane passes through the point (f, g, h) and meets the axes in A, B, C. If the planes through A, B, C and parallel to the co-ordinate planes meet at P, then show that the locus of P is $\frac{f}{x} + \frac{g}{y} + \frac{h}{z} = 1$.

[**Hints.** See Ex. 9 of page 178.]

(b) (i) Write the equation of the straight lines in symmetrical form $x + 5y - z - 7 = 0$, $2x - 5y + 3z + 1 = 0$.

[**Hints.** Let l, m, n be the d.rs. of the line. Then $l + 5m - n = 0, 2l - 5m + 3n = 0$. By cross multiplication $\frac{l}{2} = \frac{m}{-1} = \frac{n}{-3}$.

Let the line meet the planes $(\alpha, \beta, 0)$. Then $\alpha + 5\beta - 7 = 0, 2\alpha - 5\beta + 1 = 0$. From these $\alpha = 2, \beta = 1$.

$\therefore$ the equations of the line in symmetric form are $\frac{x-2}{2} = \frac{y-1}{-1} = \frac{z}{-3}$.]

(ii) Find the image of the point $(1, 3, 4)$ in the plane $2x - y + z + 3 = 0$.

[**Hints.** The equations of the line through $(1, 3, 4)$ and perpendicular to the plane are $\frac{x-1}{2} = \frac{y-3}{-1} = \frac{z-4}{1}$. Let this line meet the plane at $(2r + 1, -r + 3, r + 4)$.

Then $2(2r+1)-(-r+3)+(r+4)+3=0$, or, $r=-1$. $\therefore$ the point is $(-1,4,3)$.
If (α,β,γ) is the image point, then the midpoint of the join of $(1,3,4)$ and (α,β,γ) is $(-1,4,3)$.
Then $\frac{1+\alpha}{2}=-1$, or, $\alpha=-3$, $\frac{3+\beta}{2}=4, \beta=5$ and $\frac{4+\gamma}{2}=3$, $\gamma=2$.
Thus the image is $(-3,5,2)$.]

(c) (i) Find the s.d. between the straight lines $\frac{x-2}{1}=\frac{y-1}{-2}=\frac{z-6}{1}$ and $\frac{x+3}{7}=\frac{y+3}{-6}=\frac{z+3}{1}$.
[**Hints.** $A(2,1,6)$ and $B(-3,-3,-3)$ are points on the lines. The d.cs. of AB are $\frac{5}{\sqrt{122}},\frac{4}{\sqrt{122}},\frac{9}{\sqrt{122}}$. If l,m,n be the d.cs. of s.d., then $l-2m+n=0, 7l-6m+n=0$.
From these $\frac{l}{2}=\frac{m}{3}=\frac{n}{4}=\frac{1}{\sqrt{29}}$.

$$\begin{aligned}\text{s.d.} &= AB\cdot\cos\theta \quad (\theta \text{ is the angle between } AB \text{ and s.d.})\\ &= 2\sqrt{29} \text{ units.}]\end{aligned}$$

(ii) Find the equation of the straight line perpendicular to both the lines $\frac{x-1}{1}=\frac{y-1}{2}=\frac{z+2}{3}$ and $\frac{x+2}{2}=\frac{y-5}{-1}=\frac{z+3}{2}$ and passing their intersections.
[**Hints.** Let the lines meet at (x_1,y_1,z_1). Then $\frac{x_1-1}{1}=\frac{y_1-1}{2}=\frac{z_1+2}{3}$ and $\frac{x_1+2}{2}=\frac{y_1-5}{-1}=\frac{z_1+3}{2}$. From these relations $x_1=2, y_1=3, z_1=1$.
Let l,m,n be the d.rs. of the perpendicular line. Then $l+2m+2n=0$ and $2l-m+2n=0$.
From these $\frac{l}{7}=\frac{m}{4}=\frac{n}{-5}$.
Thus the equations of the required line are $\frac{x-2}{7}=\frac{y-3}{4}=\frac{z-1}{-5}$.]

(d) (i) The plane $x+2y+3z=6$ meets the co-ordinate axes in A,B,C. Find the equation of the sphere $OABC$.
[**Hints.** Co-ordinates of A,B,C are $(6,0,0),(0,3,0),(0,0,2)$. Let the equation of the sphere be $x^2+y^2+z^2+2ux+2vy+2wz=0$. Since it passes through A,B,C; $2u=-6$, $2v=-3, 2w=-2$.
Thus the equation is $x^2+y^2+z^2-6x-3y-2z=0$.]

(ii) Show that the plane $2x-2y+z+12=0$ touches the sphere $x^2+y^2+z^2-2x-4y+2z-3=0$ and find the point of contact.
[**Hints.** Let the plane touch the sphere at (x_1,y_1,z_1). The equation of the tangent plane at this point is $x(x_1-1)+y(y_1-2)+z(z_1+1)-x_1-2y_1+z_1-3=0$. It is identical with the given plane.

$$\therefore \quad \frac{x_1-1}{2}=\frac{y_1-2}{-2}=\frac{z_1+1}{1}=\frac{-x_1-2y_1+z_1-3}{12}.$$

From these equations, $x_1=-1, y_1=4, z_1=-2$.
Since the unique solution is obtained, the plane touches the sphere at $(-1,4,-2)$.]

(e) (i) Find the equation of the normal to the surface $2x^2+3y^2=4z$ at $(2,2,5)$.
[**Hints.** The point is on the surface. The equation of the tangent plane is $2x+3y-z=5$. Hence the equation of the normal at $(2,2,5)$ is $\frac{x-2}{2}=\frac{y-2}{3}=\frac{z-5}{-1}$.]

(ii) Find the equation of cylinder whose generators are parallel to the line $\frac{x}{1}=\frac{y}{2}=\frac{z}{3}$ and whose guiding curve is the ellipse $x^2+2y^2=1, z=0$.
[**Hints.** Let (x_1,y_1,z_1) be a point on the cylinder. The equation of the generator through this point is $\frac{x-x_1}{1}=\frac{y-y_1}{2}=\frac{z-z_1}{3}$.
Let this generator meet the guiding curve at $(\alpha,\beta,0)$. Then $\frac{\alpha-x_1}{1}=\frac{\beta-y_1}{2}=\frac{-z_1}{3}$.
From this $\alpha=x_1-\frac{z_1}{3}, \beta=y_1-\frac{2z_1}{3}$.
We have $\alpha^2+2\beta^2=1$, or, $\left(x_1-\frac{z_1}{3}\right)^2+2\left(y_1-\frac{2z_1}{3}\right)^2=1$.
The locus of (x_1,y_1,z_1) is the equation of the cylinder.
Thus the equation is $\left(x-\frac{z}{3}\right)^2+2\left(y-\frac{2z}{3}\right)^2=1$, or, $3x^2+6y^2+z^2-8yz-2zx=3$.]

7. Answer either (a) or (b).

(a) (i) Find a vector of magnitude 18 which is perpendicular to both the vectors

$$4\bar{i} - \bar{j} + 3\bar{k} \text{ and } -2\bar{i} + \bar{j} - 2\bar{k}.$$

[**Hints.** $(4\bar{i} - \bar{j} + 3\bar{k}) \times (-2\bar{i} + \bar{j} - 2\bar{k}) = \begin{vmatrix} \bar{i} & \bar{j} & \bar{k} \\ 4 & -1 & 3 \\ -2 & 1 & -2 \end{vmatrix} = -\bar{i} + 2\bar{j} + 2\bar{k}.$

It is perpendicular to both the vectors.

$\therefore$ the required vector is $18\dfrac{-\bar{i} + 2\bar{j} + 2\bar{k}}{\sqrt{(-1)^2 + 2^2 + 2^2}} = 6(-\bar{i} + 2\bar{j} + 2\bar{k}).$]

(ii) Prove by vector method that the internal bisectors of the angles of a triangle are concurrent.

[**Hints.** See Theorem 3 of page 311.]

(b) (i) Find the equation of the plane through the point $2\bar{i} + 3\bar{j} - \bar{k}$ and perpendicular to the vector $3\bar{i} + 4\bar{j} + 7\bar{k}$.

[**Hints.** Let $\bar{r}$ be the position vector of a point on the plane and

$$\bar{a} = 2\bar{i} + 3\bar{j} - \bar{k} \text{ and } \bar{b} = 3\bar{i} + 4\bar{j} + 7\bar{k}.$$

$\bar{r} - \bar{a}$ is a line vector on the plane.

$\therefore$ $(\bar{r} - \bar{a}) \cdot \bar{b} = 0$. It is the equation of the plane.

If $\bar{r} = \bar{i}x + \bar{j}y + \bar{k}z$, then the equation of the plane is

$$\{\bar{i}(x-2) + \bar{j}(y-3) + \bar{k}(z+1)\} \cdot (3\bar{i} + 4\bar{j} + 7\bar{k}) = 0,$$

$$\text{or, } 3(x-2) + 4(y-3) + 7(z+1) = 0,$$

$$\text{or, } 3x + 4y + 7z = 11]$$

(ii) A particle, acted on by constant forces $4\bar{i} + \bar{j} - 3\bar{k}$ and $3\bar{i} + \bar{j} - \bar{k}$, is displaced from the point $\bar{i} + 2\bar{j} + 3\bar{k}$ to the point $5\bar{i} + 4\bar{j} + \bar{k}$. Find the work done by the resultant force.

[**Hints.** See Ex. 7 of page 320.]

2005

[Paper I]

1. Answer the following:

(ix) Transform into cartesian equation: $r^2 \cos 2\theta = a^2$.

[**Hints.** $r^2 \cos 2\theta = a^2$, or, $r^2(\cos^2\theta - \sin^2\theta) = a^2$, or, $x^2 - y^2 = a^2$.]

(x) If $(at_1^2, 2at_1)$ and $(at_2^2, 2at_2)$ be the extremities of any focal chord of the conic $y^2 = 4ax$, prove that $t_1t_2 = -1$.

[**Hints.** $(at_1^2, 2at_1), (a, 0), (at_2^2, 2at_2)$ are collinear.

$$\therefore\ at_1^2(0 - 2at_2) + a(2at_2 - 2at_1) + at_2^2(2at_1 - 0) = 0,$$

or, $t_1t_2(t_2 - t_1) + (t_2 - t_1) = 0$, or, $t_1t_2 + 1 = 0$, or, $t_1t_2 = -1$. (Here $t_1 \neq t_2$).]

(xi) Transform the equation $x^2 + y^2 + 5x - 6y + 3 = 0$ referred to parallel axes through the point $(2, 3)$.

[**Hints.** The equation transforms to $(x' + 2)^2 + (y' + 3)^2 + 5(x' + 2) - 6(y' + 3) = 0$, or, $x'^2 + y'^2 + 9x' + 8 = 0$.]

(xii) Find the symmetrical form of the equation of the straight line $x = 0$ and $z = 0$.

[**Hints.** $\frac{x}{0} = \frac{y}{1} = \frac{z}{0}$.]

(xv) Find a unit vector perpendicular to the plane of vectors $2\bar{i}-6\bar{j}-3\bar{k}$ and $4\bar{i}+3\bar{j}-\bar{k}$.

[**Hints.** $(2\bar{i} - 6\bar{j} - 3\bar{k}) \times (4\bar{i} + 3\bar{j} - \bar{k}) = 15\bar{i} + 10\bar{j} + 30\bar{k}$.

Required unit vector $= \dfrac{5(3\bar{i} + 2\bar{j} + 6\bar{k})}{5\sqrt{3^2 + 2^2 + 6^2}} = \dfrac{3\bar{i} + 2\bar{j} + 6\bar{k}}{7}$.]

(xvi) Are the following vectors $2\bar{i} - 5\bar{j} + \bar{k}, \bar{i} + 4\bar{j}, -3\bar{i} + \bar{j} - \bar{k}$ coplanar? Support the answer.

[**Hints.** $(\bar{i} + 4\bar{j}) \times (-3\bar{i} + \bar{j} - \bar{k}) = -4\bar{i} + \bar{j} + 13\bar{k}$.

$(2\bar{i} - 5\bar{j} + \bar{k}) \cdot (-4\bar{i} + \bar{j} + 13\bar{k}) = -8 - 5 + 13 = 0$.

Since the scalar triple product is zero, the vectors are coplanar.]

Group B

5. Answer either (a) or (b).

(a) (i) If by rotation of rectangular axes about the origin, the expression ax^2+by^2 changes to $a'x^2 + b'y^2$, then prove that $a + b = a' + b'$.

[**Hints.** See Section 1.7 (iv) of page 9.]

(ii) Transform $y^2 = 4a(a - x)$ into polar co-ordinates in the form $r = f(\theta)$.

[**Hints.** Putting $x = r\cos\theta, y = r\sin\theta$, the equation transforms to

$$r^2 \sin^2\theta = 4a(a - r\cos\theta)$$

$$\text{or, } r^2\sin^2\theta + 4ar\cos\theta - 4a^2 = 0$$

$$\therefore\ r = \frac{-4a\cos\theta \pm 4a}{2\sin^2\theta}.$$

Since $r = a$ when $\theta = 0, r = 2a\frac{1-\cos\theta}{1-\cos^2\theta} = \frac{2a}{1+\cos\theta}$.]

(b) (i) Find the nature of the conic $\frac{2}{r} = 3 - 3\cos\theta$.

[**Hints.** $\frac{2}{r} = 3 - 3\cos\theta$, or, $\frac{2/3}{r} = 1 - \cos\theta$. It is a parabola with the focus as pole, axis as the initial line. Length of latus rectum $= \frac{4}{3}$.]

(ii) Transform the following equation into standard form:

$$4x^2 + 24xy - 3y^2 = 312.$$

[**Hints.** Rotating the axes through an acute angle θ, where $\tan 2\theta = \frac{24}{7}$, the equation transforms to

$$4(x'\cos\theta - y'\sin\theta)^2 + 24(x'\cos\theta - y'\sin\theta)(x'\sin\theta + y'\cos\theta)$$
$$-3(x'\sin\theta + y'\cos\theta)^2 = 312$$

$$\text{or, } (4\cos^2\theta + 24\cos\theta\sin\theta - 3\sin^2\theta)x'^2 + (4\sin^2\theta - 24\sin\theta\cos\theta - 3\cos^2\theta)y'^2 = 312$$

$$\text{or, } \left\{2(1 + \cos 2\theta) + 12\sin 2\theta - \frac{3}{2}(1 - \cos 2\theta)\right\} x'^2$$
$$+\left\{2(1 - \cos 2\theta) - 12\sin 2\theta - \frac{3}{2}(1 + \cos 2\theta\right\} y'^2 = \frac{3}{2}$$

$$\text{or, } \left\{2\left(1 + \frac{7}{25}\right) + 12\cdot\frac{24}{25} - \frac{3}{2}\left(1 - \frac{7}{25}\right)\right\} x'^2$$
$$+\left\{2\left(1 - \frac{7}{25}\right) - 12\frac{24}{25} - \frac{3}{2}\left(1 + \frac{7}{25}\right)\right\} y'^2 = 312$$

or, $13x'^2 - 12y'^2 = 312$.]

6. Answer *any three* questions from the following:

(a) (i) Show that the angle between the following planes is 90°:

$$x - y + z = 1,\ 3x + 2y - z + 2 = 0.$$

[**Hints.** D.rs. of the lines perpendicular to the planes are $1, -1, 1$ and $3, 2, -1$. Since $1 \cdot 3 - 1 \cdot 2 + 1 \cdot (-1) = 0$, the angle between the planes is 90°.]

(ii) A variable plane is always at a constant distance p from the origin and meets the axes in A, B, C. Prove that the locus of the point of intersection of three planes through A, B, C, parallel to co-ordinate planes is given by $\frac{1}{x^2} + \frac{1}{y^2} + \frac{1}{z^2} = \frac{1}{p^2}$.

[**Hints.** See Ex. 18 of page 181.]

(b) (i) Show that the equation $2x^2 - 6y^2 - 12z^2 + 18yz + 2zx + xy = 0$ represents a pair of planes.

[**Hints.** Since $\begin{vmatrix} 2 & \frac{1}{2} & 1 \\ \frac{1}{2} & -6 & 9 \\ 1 & 9 & -12 \end{vmatrix} = 0$, the equation represents a pair of planes.]

(ii) Find the image of the point $(1, -2, 3)$ in the plane $2x - 3y + 2z + 3 = 0$.

[**Hints.** Equation of the line passing through the point $(1, -2, 3)$ and perpendicular to the plane are $\frac{x-1}{2} = \frac{y+2}{-3} = \frac{z-3}{2}$. Let this line meet the plane at the point $(2r + 1, -3r - 2, 2r + 3)$. Then $2(2r + 1) - 3(-3r - 2) + 2(2r + 3) + 3 = 0$, or, $r = -1$. The point is $(-1, 1, 1)$. If α, β, γ is the image point, then $(-1, 1, 1)$ is mid-point of the line-segment joining the points $(1, -2, 3)$ and (α, β, γ).

$$\frac{\alpha + 1}{2} = -1, \text{ or, } \alpha = -3,\ \frac{\beta - 2}{2} = 1, \text{ or, } \beta = 4;\ \frac{\gamma + 3}{2} = 1, \text{ or, } \gamma = -1.$$

Thus the image point is $(-3, 4, -1)$.]

(c) (i) Find the s.d. between the lines

$$\frac{x-3}{3} = \frac{y-8}{-1} = \frac{z-3}{1} \text{ and } \frac{x+3}{-3} = \frac{y+7}{2} = \frac{z-6}{4}.$$

[**Hints.** See Ex. 7 of page 192.]

(ii) Find the equation of the plane parallel to the straight line $\frac{x+2}{2} = \frac{y-2}{2} = \frac{z-1}{-1}$ and passing through the straight line $x + y - 2z + 4 = 0 = 3x - y + 2z + 1$.

[**Hints.** Let the plane be

$$x + y - 2z + 4 + \lambda(3x - y + 2z + 1) = 0$$
$$\text{or, } (1 + 3\lambda)x + (1 - \lambda)y + (2\lambda - 2)z + 4 + \lambda = 0.$$

If it is parallel to the given line, then $2(1 + 3\lambda) + 2(1 - \lambda) - (2\lambda - 2) = 0$, or, $\lambda = -3$.

$\therefore$ the required plane is $8x - 4y + 8z - 1 = 0$.]

(d) (i) Find the equation of a sphere having centre $(1, 2, 3)$ and which passes through the point $(3, 4, -5)$.

[**Hints.** Let the sphere be $(x - 1)^2 + (y - 2)^2 + (z - 3)^2 = r^2$. If it passes through the point $(3, 4, -5)$, $(3 - 1)^2 + (4 - 2)^2 + (-5 - 3)^2 = r^2$, or, $72 = r^2$. Thus the equation is $x^2 + y^2 + z^2 - 2x - 4y - 6z - 58 = 0$.]

(ii) Find the equation of the normal to the hyperboloid $3x^2 - 5y^2 + 7y^2 = 30$ at $(2, 3, 3)$.

[**Hints.** The point $(2, 3, 3)$ is on the surface. The equation of the tangent plane at this point is $6x - 15y + 21z = 30$, or, $2x - 5y + 7z = 10$.

$\therefore$ equation of the normal is $\frac{x-2}{2} = \frac{y-3}{-5} = \frac{z-3}{7}$.]

(e) (i) Show that the ellipsoid $x^2 + 2y^2 + 3z^2 = 9$ touches the plane $2x + 2y + 3z = 9$.

[**Hints.** $x^2 + 2y^2 + 3z^2 = 9$, or, $\frac{x^2}{9} + \frac{y^2}{9/2} + \frac{z^2}{3} = 1$.

Since $2^2 \times 9 + 2^2 \cdot \frac{9}{2} + 3^2 \cdot 3 = 36 + 18 + 27 = 81 = 9^2$, the plane $2x + 2y + 3z = 9$ touches the ellipsoid.]

(ii) Find the equation of the cone whose vertex is $(1, 2, 3)$ and guiding curve is the circle $x^2 + y^2 + z^2 = 4, x + y + z = 1$.

[**Hints.** Let $\frac{x-1}{l} = \frac{y-2}{m} = \frac{z-3}{n}$ be a generator. If this generator meets the guiding curve at $(lr + 1, mr + 2, nr + 3)$, then

$$(lr + 1) + (mr + 2) + (nr + 3) = 1,$$

$$\text{or, } (l + m + n)r = -5 \qquad (1)$$

$$\text{and } (lr + 1)^2 + (mr + 2)^2 (nr + 3)^2 = 4,$$

$$\text{or, } (l^2 + m^2 + n^2)r^2 + 2(l + 2m + 3n)r + 10 = 0.] \qquad (2)$$

By (1) and (2), $25(l^2 + m^2 + n^2) - 10(l + 2m + 3n)(l + m + n) + 10(l + m + n)^2 = 0$. Eliminating l, m, n by the equation of the generator the equation of the cone is

$$5\{(x-1)^2 + (y-2)^2 + (z-3)^2\} - 2(x + 2y + 3z - 14)$$
$$(x + y + z - 6) + 2(x + y + z - 6)^2 = 0.]$$

7. Answer either (a) or (b).

(a) (i) For what value of x the following vectors $3\bar{a} - 7\bar{b} - 4\bar{c}, 3\bar{a} - 2\bar{b} + \bar{c}$ and $\bar{a} + \bar{b} + x\bar{c}$ are coplanar? Here $\bar{a}, \bar{b}, \bar{c}$ are three non-coplanar vectors.

[**Hints.** Here $\begin{vmatrix} 3 & -7 & -4 \\ 3 & -2 & 1 \\ 1 & 1 & x \end{vmatrix} = 0$, or, $x = 2$.]

(ii) If $\bar{a} = \bar{i} + \bar{j} + \bar{k}, \bar{c} = \bar{j} - \bar{k}$ and $\bar{b}$ satisfies $\bar{a} \times \bar{b} = \bar{c}$ and $\bar{a} \cdot \bar{b} = 3$, show that $\bar{b} = \frac{5}{3}\bar{i} + \frac{2}{3}\bar{j} + \frac{2}{3}\bar{k}$.

[**Hints.** Let $\bar{b} = \bar{i}x + \bar{j}y + \bar{k}z$. From $\bar{a} \times \bar{b} = \bar{c}, z - y = 0, x - z = 1, y - x = -1$. From $\bar{a} \cdot \bar{b} = 3$, $x + y + z = 3$, or, $x + x - 1 + x - 1 = 3$, $\therefore x = \frac{5}{3}$. Then $y = \frac{2}{3}, z = \frac{2}{3}$.]

(b) (i) Prove by vector method that the medians of a triangle are concurrent.

[**Hints.** See Ex. 3 of page 286.]

(ii) A particle, acted on by constant forces $4\bar{i} + 2\bar{j} + 3\bar{k}$ and $\bar{i} + \bar{j} + \bar{k}$, is displaced from the point $\bar{i} + 2\bar{j}$ to $2\bar{i} - \bar{j} + 3\bar{k}$. Find the work done by the forces.

[**Hints.** $\bar{F}$ = resultant of focus $= (4\bar{i} + 2\bar{j} + 3\bar{k}) + (\bar{i} + \bar{j} + \bar{k}) = 5\bar{i} + 3\bar{j} + 4\bar{k}$. Displacement vector $\bar{d} = (2\bar{i} - \bar{j} + 3\bar{k}) - (\bar{i} + 2\bar{j}) = \bar{i} - 3\bar{j} + 3\bar{k}$. Work done $= \bar{F} \cdot \bar{d} = 5 - 9 + 12 = 8$ units.]

North Bengal University

2003

Group A

[Answer **Q. No. 1** and **any two** from the rest.]

1. Answer *any five.*

(a) Show that the equation $x^2 + 6xy + 9y^2 + 10x + 30y + 25 = 0$, represents a pair of coincident straight lines.

[**Hints.** The equation is $(x + 3y + 5)^2 = 0$. Hence the result follows.]

(b) Transform the equation $(x^2 + y^2)^2 = ax^2y$ to polar form.

[**Hints.** Putting $x = r\cos\theta, y = r\sin\theta$, the equation transforms to $r = a\cos^2\theta\sin\theta$.]

(c) Find the position of the line $x - 7 = 0$ w.r.t. the circle $x^2 + y^2 - 4x - 6y - 12 = 0$.

[**Hints.** The centre of the circle is $(2, 3)$ and radius $= 5$. Distance of the line from the point $(2, 3) = 5$.

$\therefore$ the line touches the circle at $(7, 3)$.]

(d) Determine the nature and the length of the latus rectum of the conic whose polar equation is $r = \frac{8}{4-5\cos\theta}$.

[**Hints.** The equation is $\frac{2}{r} = 1 - \frac{5}{4}\cos\theta$. Here $e = \frac{5}{4}$ and the length of latus rectum $= 4$ units. Sine $e > 1$, the equation represents a hyperbola.]

(e) Find the value of h for which the planes $3x - 2y + hz - 1 = 0$ and $x + hy + 5z + 2 = 0$ may be perpendicular.

[**Hints.** If the planes are perpendicular, $3 \cdot 1 - 2 \cdot h + h \cdot 5 = 0$, or, $h = -1$.]

(f) What are the direction cosines of the straight line joining the points $(3, 4, -1)$ and $(1, 7, -7)$?

[**Hints.** D.cs. are $\frac{2}{7}, \frac{-3}{7}, \frac{6}{7}$.]

(g) Find the radical plane of the two spheres whose equations are $x^2 + y^2 + z^2 + 2u_ix + 2v_iy + 2w_iz + d_i = 0$; $i = 1, 2, \ldots$.

[**Hints.** The radical plane is $2(u_1 - u_2)x + 2(v_1 - v_2)y + 2(w_1 - w_2)z + d_1 - d_2 = 0$.]

2. (a) If the equation $ax^2 + 2hxy + by^2 + 2gx + 2fy + c = 0$ represents a pair of parallel lines, prove that $\dfrac{a}{h} = \dfrac{h}{b} = \dfrac{g}{f}$.

[**Hints.** See Section 2.2 (iii) (B) of page 22.]

(b) If the general equation of second degree $ax^2 + 2hxy + by^2 + 2gx + 2fy + c = 0$ represents two intersecting lines, show that the square of the distance from the point of intersection of the straight lines to the origin is $\dfrac{c(a+b) - f^2 - g^2}{ab - h^2}$.

[**Hints.** See Corollary 3 of page 24.]

(c) Define Director Circle for the ellipse $\dfrac{x^2}{a^2} + \dfrac{y^2}{b^2} = 1$ and find its equation.

[**Hints.** See Section 4.5 of page 62.]

3. (a) Reduce the equation $x^2 - 5xy + y^2 + 8x - 20y + 25 = 0$ to the standard form and determine the conic represented by it.

[**Hints.** Here $\Delta = -\frac{189}{4}, D = -\frac{21}{4}$. Since $D < 0$, the equation represents a hyperbola. Changing the origin to the centre, the equation reduces to

$$x^2 - 5xy + y^2 + 9 = 0. \quad (1)$$

Rotating the axes through an acute angle θ where $\cot 2\theta = 0$, i.e. $\theta = \frac{\pi}{4}$, the equation (1) transforms to

$$\left(x' \cos\frac{\pi}{4} - y' \sin\frac{\pi}{4}\right)^2 - 5\left(x' \cos\frac{\pi}{4} - y' \sin\frac{\pi}{4}\right)\left(x' \sin\frac{\pi}{4} + y' \cos\frac{\pi}{4}\right)$$
$$+\left(x' \sin\frac{\pi}{4} + y' \cos\frac{\pi}{4}\right)^2 + 9 = 0,$$

$$\text{or,}\quad x'^2\left(1 - 5\cdot\frac{1}{\sqrt{2}}\cdot\frac{1}{\sqrt{2}}\right) + y'^2\left(1 + 5\cdot\frac{1}{\sqrt{2}}\cdot\frac{1}{\sqrt{2}}\right) + 9 = 0$$

$$\text{or,}\quad \frac{-3}{2}x'^2 + \frac{7}{2}y'^2 + 9 = 0$$

$$\text{or,}\quad \frac{3}{2}x'^2 - \frac{7}{2}y'^2 = 9.]$$

(b) Prove that the locus of the poles of tangents to the auxiliary circle of the ellipse $\frac{x^2}{a^2} + \frac{y^2}{b^2} = 1$ w.r.t. the ellipse itself is the curve $\frac{x^2}{a^4} + \frac{y^2}{b^4} = \frac{1}{a^2}$.

[**Hints.** Let (x_1, y_1) be the pole. The polar of it w.r.t. the ellipse is $\frac{xx_1}{a^2} + \frac{yy_1}{b^2} = 1$. If this polar is the tangent to the auxiliary circle $x^2 + y^2 = a^2$, then

$$\frac{1}{\sqrt{\frac{x_1^2}{a^4} + \frac{y_1^2}{b^4}}} = a, \text{ or, } \frac{x_1^2}{a^4} + \frac{y_1^2}{b^4} = \frac{1}{a^2}.$$

Hence the locus is $\frac{x^2}{a^4} + \frac{y^2}{b^4} = \frac{1}{a^2}$.]

4. (a) A variable plane passes through the point (α, β, γ) and meets the co-ordinate axes at A, B, C. If the planes through A, B, C and parallel to the co-ordinate planes meet at P, show that the locus of P is $\frac{\alpha}{x} + \frac{\beta}{y} + \frac{\gamma}{z} = 1$.

[**Hints.** See Ex. 9 of page 178.]

(b) Show that the lines $\frac{x-1}{3} = \frac{y+2}{1} = \frac{z+3}{1}$ and $\frac{x-5}{7} = \frac{y+8}{-5} = \frac{z-6}{11}$ are coplanar, and also find the equation of the plane containing them.

[**Hints.** Since $\begin{vmatrix} 5-1 & -8+2 & 6+3 \\ 3 & 1 & 2 \\ 7 & -5 & 11 \end{vmatrix} = \begin{vmatrix} 4 & -6 & 9 \\ 3 & 1 & 2 \\ 7 & -5 & 11 \end{vmatrix} = \begin{vmatrix} 4 & -6 & 9 \\ 3 & 1 & 2 \\ 3 & 1 & 2 \end{vmatrix} = 0$, the lines are coplanar.

The equation of the plane is $\begin{vmatrix} x-1 & y+2 & z+3 \\ 3 & 1 & 2 \\ 7 & -5 & 11 \end{vmatrix} = 0$, or, $21x - 19y - 22z - 125 = 0$.]

(c) Find the equation of the plane which contains the line $\frac{x-1}{2} = \frac{y+1}{-1} = \frac{z-3}{4}$ and is perpendicular to the plane $x + 2y + z = 12$.

[**Hints.** See Ex. 5 of page 199.]

5. (a) Determine the equation of the sphere passing through the circle $x^2 + y^2 + z^2 + 6x - 8y - 4z + 4 = 0, x + 2y + 3z = 6$ and through the point $(2, 3, 1)$. Find also the co-ordinates of the centre of the sphere.

[**Hints.** Let the sphere be $x^2 + y^2 + z^2 + 6x - 8y - 4z + 4 + \lambda(x + 2y + 3z - 6) = 0$. If it passes through the point $(2, 3, 1)$, $\lambda = -\frac{1}{2}$. The equation is $2(x^2 + y^2 + z^2) + 5x - 18y - 11z + 14 = 0$.]

(b) Find the equation of the right circular cone whose vertex is the origin, axis is the positive x-axis and semi-vertical angle is α.

[**Hints.** $y^2 + z^2 = x^2 \tan^2 \alpha$.]

(c) Find the equation of the chord of the conic $\frac{l}{r} = 1 + e\cos\theta$ joining the two points on the conic having vectorial angles $(\alpha - \beta)$ and $(\alpha + \beta)$.

[**Hints.** See Section 8.8 of page 126.]

VECTOR

Group C

11. Answer *any four* questions.

(a) Prove by vector method that the medians of a triangle are concurrent.

[**Hints.** See Ex. 3 of page 286.]

(b) If $\bar{\alpha} = (-2, -2, 4), \bar{\beta} = (-2, 4, -2)$ and $\bar{\gamma} = (4, -2, -2)$, calculate the value of $\bar{\alpha} \cdot (\bar{\beta} \times \bar{\gamma})$. Interpret the result geometrically.

[**Hints.** $\beta \times \gamma = -12\bar{i} - 12\bar{j} - 12\bar{k}$; $\bar{\alpha} \cdot (\bar{\beta} \times \bar{\gamma}) = 24 + 24 - 48 = 0$. Vectors are coplanar.]

(c) Show that $[\bar{a} \times \bar{b}, \bar{b} \times \bar{c}, \bar{c} \times \bar{a}] = [\bar{a}\bar{b}\bar{c}]^2$.

[**Hints.**
$$(\bar{b} \times \bar{c}) \times (\bar{c} \times \bar{a}) = \{(\bar{b} \times \bar{c}) \cdot \bar{a}\}\bar{c} - \{(\bar{b} \times \bar{c}) \cdot \bar{c}\}\bar{a} = [\bar{a}\bar{b}\bar{c}]\bar{c}$$
$$(\bar{a} \times \bar{b}) \cdot \{(\bar{b} \times \bar{c}) \times (\bar{c} \times \bar{a})\} = \{(\bar{a} \times \bar{b}) \cdot \bar{c}\}[\bar{a}\bar{b}\bar{c}] = [\bar{a}\bar{b}\bar{c}][\bar{a}\bar{b}\bar{c}] = [\bar{a}\bar{b}\bar{c}]^2.]$$

(d) A particle acted on by constant forces $4\bar{i} + \bar{j} - 3\bar{k}$ and $3\bar{i} + \bar{j} - \bar{k}$ is displaced from the point $\bar{i} + 2\bar{j} + 3\bar{k}$ to the point $5\bar{i} + 4\bar{j} + \bar{k}$. Find the total work done by the forces.

[**Hints.** See Ex. 7 of page 320.]

(e) If $\bar{a}, \bar{b}, \bar{c}$ are three vectors satisfying the conditions $\bar{a} + \bar{b} + \bar{c} = \bar{0}$ and $|\bar{a}| = 3, |\bar{b}| = 4$ and $|\bar{c}| = 5$, show that $\bar{a} \cdot \bar{b} + \bar{b} \cdot \bar{c} + \bar{c} \cdot \bar{a} = -25$.

[**Hints.** $(\bar{a} + \bar{b} + \bar{c})^2 = 0$, or, $\bar{a}^2 + \bar{b}^2 + \bar{c}^2 + 2(\bar{a} \cdot \bar{b} + \bar{b} \cdot \bar{c} + \bar{c} \cdot \bar{a}) = 0$,
or, $9 + 16 + 25 + 2(\bar{a} \cdot \bar{b} + \bar{b} \cdot \bar{c} + \bar{c} \cdot \bar{a}) = 0$,
or, $\bar{a} \cdot \bar{b} + \bar{b} \cdot \bar{c} + \bar{c} \cdot \bar{a} = -25$.]

(f) Find the vector equation of the line passing through the points whose position vectors are $(\bar{i} - 2\bar{j} + \bar{k})$ and $(3\bar{i} - \bar{k})$.

[**Hints.** The equation is $\bar{r} = t(\bar{i} - 2\bar{j} + \bar{k}) + (1 - t)(3\bar{i} - \bar{k})$,
or, $\bar{r} = 3\bar{i} - \bar{k} - 2t(\bar{i} + \bar{j} - \bar{k})$.]

2004

Group A

[Answer **Q. No. 1** and **any two** from the rest.]

1. Answer *any five*.

(a) For what value of k the equation $2x^2 - y^2 + xy - 2x - 5y + k = 0$ will represent a pair of straight lines?

[**Hints.** For the pair of lines, $2 \cdot (-1) \cdot k + 2 \cdot (-\frac{5}{2}) \cdot (-1) \cdot \frac{1}{2} - 2 \cdot (-\frac{5}{2})^2 - (-1)(-1)^2 - k \cdot (\frac{1}{2})^2 = 0$, or, $k = -4$.]

(b) Find the point on the conic $\frac{15}{r} = 3 - 4\cos\theta$, for which r is least.

[**Hints.** $r = \dfrac{15}{3 - 4\cos\theta}$. Least value of $r = \dfrac{15}{3+4} = \dfrac{15}{7}$ when $\theta = \pi$.]

(c) The foot of the perpendicular from the origin to a plane is $(2, 3, -1)$. Find the equation of the plane.

[**Hints.** Let the equation of the plane be $2x + 3y - z = k$. The point $(2, 3, -1)$ lies on the plane. $\therefore$ $2 \cdot 2 + 3 \cdot 3 - (-1) = k$, or, $k = 14$. Thus the equation is $2x + 3y - z = 14$.]

(d) Find the angles that the line $\dfrac{x-1}{1} = \dfrac{y-1}{1} = \dfrac{z-1}{0}$ makes with the axes.

[**Hints.** The d.cs. of the x-axis are $1, 0, 0$.

$\therefore$ the angle made by the line with the x-axis $= \cos^{-1}\dfrac{1 \cdot 1}{\sqrt{2}} = \cos^{-1}\dfrac{1}{\sqrt{2}} = \dfrac{\pi}{4}$.

Similarly other angles are $\frac{\pi}{4}$ and $\frac{\pi}{2}$.]

(e) What does the following equation represent? $x^2 + y^2 = z^2 \tan^2\alpha$.

[**Hints.** It is the equation of a right-circular cone with the z-axis as the axis and vertex is the origin.]

(f) Find the direction cosines of the straight line $x = 2y = 3z$.

[**Hints.** The line is $\dfrac{x}{6} = \dfrac{y}{3} = \dfrac{z}{2}$. D.cs. are $\dfrac{6}{7}, \dfrac{3}{7}, \dfrac{2}{7}$.]

(g) Find the equation of a plane passing through the point $(1, 2, -3)$ and normal to the line joining the points $(-1, 3, 4)$ and $(5, 2, -1)$.

[**Hints.** See Ex. 1 of page 176.]

2. (a) Find the equation of bisectors bisecting the angle between the lines represented by $ax^2 + 2hxy + by^2 = 0$.

[**Hints.** See Section 2.1 (iii) of page 18.]

(b) If the normal at $(at_1^2, 2at_1)$ meets the parabola $y^2 = 4ax$ again at $(at_2^2, 2at_2)$, show that $t_2 = -t_1 - \frac{2}{t_1}$.

[**Hints.** The equation of the normal at $(at_1^2, 2at_1)$ is

$$y + t_1 x = 2at_1 + at_1^3 \tag{1}$$

To find the points of intersection between (1) and $y^2 = 4ax$,

$$y^2 = 4a\frac{2at_1 + at_1^3 - y}{t_1}$$

$$\text{or,}\quad y^2 t_1 - 8at_1 - 4a^2 t_1^3 + 4ay = 0$$

$$\text{or,}\quad (y - 2at_1)\{t_1(y + 2at_1) + 4a\} = 0$$

$$y = 2at \text{ and } y = 2a\left(-t_1 - \frac{2}{t_1}\right) \quad \therefore\ t_2 = -t - \frac{2}{t_1}.]$$

(c) Find the condition that the chord of the circle $x^2 + y^2 = a^2$ intercepted on the line $y = mx + c$ $(c \neq 0)$ may subtend a right angle at the centre of the circle.

[**Hints.** The equation of the lines joining the origin to the points of intersection between $x^2 + y^2 = a^2$ and $y = mx + c$ is $x^2 + y^2 = a^2\left(\frac{y - mx}{c}\right)^2$. If these lines are at right angles, then $\left(1 - \dfrac{a^2 m^2}{c^2}\right) + \left(1 - \dfrac{a^2}{c^2}\right) = 0$, or, $a^2(1 + m^2) = 2c^2$.]

3. (a) Find the polar equation of the chord of the conic $\frac{l}{r} = 1 - e\cos\theta$ joining the two points on the conic, whose vectorial angles are $\alpha - \beta$ and $\alpha + \beta$.

[**Hints.** See Section 8.8 of page 126.]

(b) Reduce the equation to the canonical form and determine the type of the conic represented by $4x^2 - 4xy + y^2 - 12x + 10y - 19 = 0$.

[**Hints.** The equation can be written as $(2x-y)^2 = 12x - 10y + 19$.
Now $(2x - y + \lambda)^2 = (12 + 4\lambda)x - (10 + 2\lambda)y + \lambda^2 + 19$.
λ is chosen in such a way that the lines $2x - y + \lambda = 0$ and $(12 + 4\lambda)x - (10 + 2\lambda)y + \lambda^2 + 19 = 0$ are at right angles. For this $2(12 + 4\lambda) + (10 + 2\lambda) = 0$, or, $\lambda = -\frac{17}{5}$.

$$\therefore \quad \left(2x - y - \frac{17}{5}\right)^2 = -\frac{8}{5}x - \frac{16}{5}y + 764$$

$$\text{or,} \quad (10x - 5y - 17)^2 = -4(10x + 20y - 191)$$

$$\text{or,} \quad \left(\frac{10x - 5y - 17}{\sqrt{10^2 + 5^2}}\right)^2 \times 125 = -4 \times \sqrt{10^2 + 20^2} \times \frac{10x + 2y - 191}{\sqrt{10^2 + 20^2}}$$

$$\text{or,} \quad Y^2 = -16X.$$

It is the canonical form. It is a parabola.]

4. (a) Show that the acute angle between the lines whose direction cosines are given by the relation $l + m + n = 0$ and $l^2 + m^2 - n^2 = 0$ is $\frac{\pi}{3}$.
[**Hints.** See Ex. 5 of page 162.]

(b) If $2c$ is the s.d. between the lines $\frac{x}{l} - \frac{z}{n} = 1, y = 0$ and $\frac{y}{m} + \frac{z}{n} = 1, x = 0$, then show that $\frac{1}{l^2} + \frac{1}{m^2} + \frac{1}{n^2} = \frac{1}{c^2}$.
[**Hints.** See Ex. 9 of page 201.]

(c) Find the equation of the plane which contains the line $x + 2y + 3z - 4 = 0$, $2x + y - z + 5 = 0$ and perpendicular to the plane $5x + 3y + 6z + 8 = 0$.
[**Hints.** See Q. 11 (c) of C.U. 2004.]

5. (a) Find the equation of the cone whose vertex is at the point $(1, 2, 3)$ and guiding curve is the ellipse $\frac{x^2}{9} + \frac{y^2}{4} = 1, z = 0$.
[**Hints.** Let one generator be

$$\frac{x-1}{l} = \frac{y-2}{m} = \frac{z-3}{n}. \tag{1}$$

If it meets the guiding curve at $(x_1, y_1, 0)$, then $\frac{x_1 - 1}{l} = \frac{y_1 - 2}{m} = \frac{-3}{n}$, $x_1 = 1 - 3\frac{l}{n}$, $y_1 = 2 - 3\frac{m}{n}$, we have $\frac{x_1^2}{9} + \frac{y_1^2}{4} = 1$.

$$\therefore \quad \frac{\left(1 - 3\frac{l}{n}\right)^2}{9} + \left(2 - 3\frac{m}{n}\right)^2 = 1.$$

Eliminating l, m, n by (1), the equation of the cone is obtained.]

(b) Find the equation of the sphere, for which the circle $x^2 + y^2 + z^2 + 2x - 4y + 2z + 5 = 0$, $x - 2y + 3z + 1 = 0$ is a great circle.
[**Hints.** See Ex. 8 of page 219.]

(c) Find the equation of the cylinder whose generators are parallel to the line $2x = y = 3z$ and which passes through the circle $y = 0, x^2 + z^2 = 8$.
[**Hints.** Let (α, β, γ) be a point on the cylinder. The generating line through this point is $\frac{x - \alpha}{3} = \frac{y - \beta}{6} = \frac{z - \gamma}{2}$. It meets the directrix, where

$$\frac{x - \alpha}{3} = \frac{0 - \beta}{6} = \frac{z - \gamma}{2}, \text{ or, } x = \alpha - \frac{\beta}{2}, z = \gamma - \frac{\beta}{3}.$$

We have, $\left(\alpha - \frac{\beta}{2}\right)^2 + \left(\gamma - \frac{\beta}{3}\right)^2 = 8$.

Hence the equation of the cylinder is $\left(x - \frac{y}{2}\right)^2 + \left(z - \frac{y}{3}\right)^2 = 8$.]

VECTOR

Group C

11. Answer *any four* questions.

(a) Show by vector method that the angle in a semi-circle is at right angle.

[**Hints.** See Ex. 8 of page 302.]

(b) Show by vector method that $\sin(\alpha+\beta)=\sin\alpha\cos\beta+\cos\alpha\sin\beta$.

[**Hints.** Solve by the help of Ex. 12 of page 304.]

(c) Show that $[\bar{\alpha}+\bar{\beta},\bar{\beta}+\bar{\gamma},\bar{\gamma}+\bar{\alpha}]=2[\bar{\alpha}\bar{\beta}\bar{\gamma}]$.

[**Hints.** See Ex. 18 of page 306.]

(d) If $\bar{\alpha}+\bar{\beta}+\bar{\gamma}=\bar{0}$, prove that $\bar{\alpha}\times\bar{\beta}=\bar{\beta}\times\bar{\gamma}=\bar{\gamma}\times\bar{\alpha}$.

[**Hints.** $\bar{\alpha}\times(\bar{\alpha}+\bar{\beta}+\bar{\gamma})=\bar{0}$, or, $\bar{\alpha}\times\bar{\beta}+\bar{\alpha}\times\bar{\gamma}=\bar{0}$, or, $\bar{\alpha}\times\bar{\beta}=\bar{\gamma}\times\bar{\alpha}$
$\bar{\beta}\times(\bar{\alpha}+\bar{\beta}+\bar{\gamma})=\bar{0}$, or, $\bar{\beta}\times\bar{\alpha}+\bar{\beta}\times\bar{\gamma}=\bar{0}$, or, $\bar{\beta}\times\bar{\gamma}=\bar{\alpha}\times\bar{\beta}$.]

(e) Find the vector equation of the straight line passing through the points whose position vectors are $2\bar{i}-\bar{j}+\bar{k}$ and $-3\bar{i}+4\bar{j}-4\bar{k}$.

[**Hints.** The equation is $\bar{r}=t(2\bar{i}-\bar{j}+\bar{k})+(1-t)(-3\bar{i}+4\bar{j}-4\bar{k})$.]

(f) Prove that $\bar{x}\times(\bar{y}\times\bar{z})=(\bar{x}\cdot\bar{z})\bar{y}-(\bar{x}\cdot\bar{y})\bar{z}$.

[**Hints.** See Section 2.4 (b) of page 298.]

2005

[First Paper]

Group C

[Answer **Q. No. 9** and **any five** from **Q. No. 10**.]

9. (a) If $ax+by$ transforms to $a'x+b'y$ under rotation of axes, then show that $a^2+b^2=a'^2+b'^2$.

[**Hints.** See Ex. 5 of page 13.]

(b) Determine the nature of the conic $\frac{8}{r}=4-5\cos\theta$.

[**Hints.** The equation is $\frac{2}{r}=1-\frac{5}{4}\cos\theta$. Here $e=\frac{5}{4}>1$ and length of latus rectum $=4$ units. It is a hyperbola.]

(c) If $\cos\alpha,\cos\beta,\cos\gamma$ are d.cs. of a line, show that $\sin^2\alpha+\sin^2\beta+\sin^2\gamma=2$.

[**Hints.** See Corollary 2 of Section 2.2 of page 158.]

(d) Find the equation of the sphere through the points $(0,0,0),(a,0,0),(0,b,0),(0,0,c)$.

[**Hints.** See Ex. 1 of page 216.]

(e) What do you mean by the s.d. between two skew lines?

[**Hints.** See Section 4.6 of page 190.]

10. (a) Reduce the equation $11x^2-4xy+14y^2-58x-44y+71=0$ to its canonical form.

[**Hints.** See Ex. 2 of page 49.]

(b) Show that the triangle formed by the lines $ax^2+2hxy+by^2=0$ and $lx+my=1$ is right angled if $(a+b)(al^2+2hlm+bm^2)=0$.

[**Hints.** See Ex. 11 of page 30.]

(c) In any conic $\frac{l}{r} = 1 - e\cos\theta$, prove that the sum of the reciprocals of two perpendicular focal chords is constant.

[**Hints.** See Ex. 6 of page 133.]

(d) Find the equation of the sphere passing through the point $(1, 0, -3)$ and through the circle represented by $x^2 + y^2 + z^2 - 4x - 6y + 2z - 16 = 0, 3x + y + 3z - 4 = 0$.

[**Hints.** Let the sphere be $x^2 + y^2 + z^2 - 4x - 6y + 2z - 16 + \lambda(3x + y + 3z - 4) = 0$. Since it passes through the point $(1, 0, -3)$, $\lambda = -\frac{8}{5}$.]

(e) Find the equation of the lines of intersection of the plane $3x + 4y + z = 0$ and the cone $15x^2 - 32y^2 - 7z^2 = 0$.

[**Hints.** Let the plane cut the cone by the line $\frac{x}{l} = \frac{y}{m} = \frac{z}{n}$. Then $3l + 4m + n = 0$, $15l^2 - 32m^2 - 7n^2 = 0$, $n = -(3l + 4m)$. Putting this value of n in the second $15l^2 - 32m^2 - 7(3l + 4m)^2 = 0$, or, $2l^2 + 7lm + 6m^2 = 0$, or, $(l + 2m)(2l + 3m) = 0$. By the help of these lines are $\frac{x}{2} = \frac{y}{-1} = \frac{z}{2}$ and $\frac{x}{3} = \frac{y}{-2} = \frac{z}{1}$.]

(f) From $P(\alpha, \beta, \gamma)$ perpendiculars PA, PB, PC are drawn on the co-ordinate planes. Show that the equation of the plane ABC is $\frac{x}{\alpha} + \frac{y}{\beta} + \frac{z}{\gamma} = 2$.

[**Hints.** See Ex. 20 of page 181.]

(g) What is the locus of the point P if its polar with respect to the hyperbola $\frac{x^2}{a^2} - \frac{y^2}{b^2} = 1$ be equally inclined to the co-ordinate axes.

[**Hints.** Let (x_1, y_1) be the co-ordinates of P. The polar is $\frac{xx_1}{a^2} - \frac{yy_1}{b^2} = 1$. For the given condition $\frac{y_1}{b^2} = \pm\frac{x_1}{a^2}$, or, $y_1 = \pm\frac{b^2}{a^2}x_1$. The locus is $y = \pm\frac{b^2}{a^2}x$.]

VECTOR

Group D

11. Answer *any three* questions.

(a) Prove by the vector method that the angle in a semi-circle is right angle.

[**Hints.** See Ex. 8 of page 302.]

(b) Find the moment of a force $4\bar{i} + 2\bar{j} + \bar{k}$ through the point $5\bar{i} + 2\bar{j} + 4\bar{k}$ about the point $3\bar{i} - \bar{j} + 3\bar{k}$.

[**Hints.** Moment $= (2\bar{i} + 3\bar{j} + \bar{k}) \times (4\bar{i} + 2\bar{j} + \bar{k}) = \bar{i} + 2\bar{j} - 8\bar{k}$.]

(c) If $\bar{\alpha} \times \bar{\beta} + \bar{\beta} \times \bar{\gamma} + \bar{\gamma} \times \bar{\alpha} = \bar{0}$, then show that $\bar{\alpha}, \bar{\beta}, \bar{\gamma}$ are coplanar.

[**Hints.** $\bar{\gamma} \cdot (\bar{\alpha} \times \bar{\beta}) + \bar{\gamma} \cdot (\bar{\beta} \times \bar{\gamma}) + \bar{\gamma} \cdot (\bar{\gamma} \times \bar{\alpha}) = 0$, or, $\gamma \cdot (\alpha \times \beta) = 0$. Hence $\bar{\alpha}, \bar{\beta}, \bar{\gamma}$ are coplanar.]

(d) If $\bar{\alpha} = (2, -10, 2), \bar{\beta} = (3, 1, 2)$ and $\bar{\gamma} = (2, 1, 3)$, then find $\bar{\alpha} \times (\bar{\beta} \times \bar{\gamma})$ and interpret the result geometrically.

[**Hints.** $\bar{\beta} \times \bar{\gamma} = \begin{vmatrix} \bar{i} & \bar{j} & \bar{k} \\ 3 & 1 & 2 \\ 2 & 1 & 3 \end{vmatrix} = \bar{i} - 5\bar{j} + \bar{k}$. $\bar{\alpha} \times (\bar{\beta} \times \bar{\gamma}) = \begin{vmatrix} \bar{i} & \bar{j} & \bar{k} \\ 1 & -5 & 1 \\ 2 & -10 & 2 \end{vmatrix} = 0$.

β is parallel to $\bar{\gamma}$, or, $\bar{\alpha}$ is perpendicular to $\bar{\beta}$ and $\bar{\gamma}$.]

(e) Find the vector equation of the line passing through the points $(1, 1, 1)$ and $(3, 2, -1)$.

[**Hints.** $\bar{r} = t(1, 1, 1) + (1 - t)(3, 2, -1)$.]

